SPRINGER
LABORATORY

H.- J. Gabius   S. Gabius (Eds.)

Lectins and Glycobiology

With 128 Figures

Springer-Verlag
Berlin Heidelberg New York London Paris
Tokyo Hong Kong Barcelona Budapest

Professor Dr. HANS-JOACHIM GABIUS
Dr. SIGRUN GABIUS

Institut für Physiologische Chemie
der Ludwig-Maximilians-Universität
Veterinärstraße 13
D-80539 München
Germany

ISBN 3-540-56211-7 Springer-Verlag Berlin Heidelberg New York
ISBN 0-387-56211-7 Springer-Verlag New York Berlin Heidelberg

Library of Congress Cataloging-in-Publication Data. Lectins and glycobiology / H.-J. Gabius (Eds.). p. cm. –
(Springer laboratory) Includes bibliographical references and index. ISBN 3-540-56211-7 (alk. paper). – ISBN
0-387-56211-7 (alk. paper) 1. Lectins. 2. Glycoconjugates. I. Gabius, H.-J. (Hans-Joachim), 1955– II. Gabius, S.
(Sigrun), 1960–. III. Series. [DNLM: 1. Lectins. 2. Glycoconjugates. 3. Glycosylation. QW 640 L4707 1993]
QP552.L42L423 1993 574.19′245–dc20 DNLM/DLC for Library of Congress 93-24827

© Springer-Verlag Berlin Heidelberg 1993
Printed in Germany

Typesetting: Macmillan India Ltd., Bangalore
31/3145-5 4 3 2 1 0 – Printed on acid-free paper

Preface

The intriguing complexity, precision, and regulation of the wide range of biological processes is determined by intricate mechanisms of molecular recognition. Their nature is under intense scrutiny. In addition to the well-appreciated interaction of proteins either with amino acid or nucleotide sequences, the investigation of their interplay with carbohydrate elements of cellular glycoconjugates currently exerts increasing attraction. In the group of carbohydrate-binding proteins, lectins are distinguished from antibodies or ligand-affecting enzymes, according to the most recent definition. The thorough analysis of their structure and function is considered as a focus to collect a critical mass of information for delineating details of a further array of biochemical processes with pivotal physiological impact. Following an already century-long history of scientific description, reflected by subjectively chosen highlights (see the Brief History of Lectin Research at page VI), the excitement in glycobiological research that prevails today can easily be explained by our growing awareness of the multifarious significance of a sugar-code system of biological information. This present notion unmistakably has an impact on lines of research in diverse disciplines like cell and molecular biology, histochemistry, or clinical sciences. It also prompts inherent practical questions such as how to obtain lectins, or how to employ them as instruments in various assay systems with the best possible results.

Thus, this book is devoted intentionally to cover the techniques in different research fields that deal with lectins. The first parts place emphasis on the purification of lectins, the methods for determination of their specificity, and their application as laboratory instruments to analyze as well as to localize defined carbohydrate ligands biochemically, cytologically, and histochemically. These studies are a prerequisite to test the potential of lectins as drug carriers or as targets for drug delivery. The following parts offer state-of-the-art presentations of experimental systems to elucidate lectin-mediated biosignaling on the level of cells or organisms and to assess lectin-dependent cell adhesion as well as relevant techniques of molecular biology.

The individual chapters are accompanied by our sincere wish that the description will prove to be valuable as technical guideline to gratifying progress, thereby addressing the attractive challenge to meticulously map so far unknown territory in the area of lectinology and glycobiology.

SIGRUN GABIUS
HANS-JOACHIM GABIUS

Table 1. A brief history of lectin research

1888	description of toxic, cell-agglutinating protein in seed extract of Ricinus communis, termed ricin (H. Stillmark)
1891	application of toxic plant agglutinins as model antigens (P. Ehrlich)
1898	introduction of the term "hemagglutinin" for plant proteins that agglutinate erythrocytes (M. Elfstrand)
1902	detection of bacterial agglutinins (R. Kraus)
1907	discovery of non-toxic plant agglutinins (K. Landsteiner, H. Raubitschek)
1913	application of cells for lectin isolation and desorption by acid (R. Kobert)
1919	crystallization of concanavalin A (J. B. Sumner)
1936	definition of a carbohydrate group as ligand for concanavalin A (J. B. Sumner, S. F. Howell)
1941	detection of viral agglutinins (G. K. Hirst)
1947/48	description of blood group specificity of certain agglutinins (W. C. Boyd, K. O. Renkonen)
1954	introduction of the term "lectin" for antibody-like proteins (W. C. Boyd)
1960	detection of elicitation of a mitogenic response of lymphocytes by binding of a lectin to the cell surface (P. C. Nowell)
1963	introduction of affinity chromatography to lectin isolation (I. J. Goldstein)
1972	sequencing and analysis of the three-dimensional structure of the first lectin (G. M. Edelman, K. D. Hardman, C. F. Ainsworth)
1974	purification of the first mammalian lectin from liver (G. Ashwell)
1980	definition as carbohydrate-binding protein of non-immune origin that agglutinates cells (I. J. Goldstein et al.)
1988	definition as carbohydrate-binding protein other than an antibody or an enzyme (S. H. Barondes)

Contents

Lectins as Tools for the Characterization of Glycoconjugates

Determination of Expression of Lectins and Their Ligands

Lectins and Neoglycoconjugates
in Histochemical and Cytochemical Analysis

Lectins and Biosignaling

Lectins and Cell Adhesion

Molecular Genetics of Lectins

List of Contributors

H. ALLEN
Roswell Park Cancer Institute
New York State Department
of Health
Elm and Carlton Streets
Buffalo, NY 14263-0001
USA

H. AUGUSTIN-VOSS
University of Göttingen
Cell Biology Laboratory
Deptpartment of Gynecology
and Obstetrics
Robert-Koch-Str. 40
D-37075 Göttingen
Germany

J. BEUTH
Hygiene Institut der Universität
Goldenfelsstr. 19–21
D-50935 Köln
Germany

K. B. BODMAN
Department of Immunology
Arthur Stanley House
University College and
Middlesex School of Medicine
40–50 Tottenham Street
London W1P 9PG
UK

N. V. BOVIN
Shemyakin Institute of
Bioorganic Chemistry
ul. Miklukho-Maklaya 16/10
117871 GSP-7 Moscow V-437
Russian Federation

C. F. BREWER
Albert Einstein College of Medicine
Jack Pearl Resnick Campus
1300 Morris Park Avenue
Bronx, NY 10461
USA

M. CARON/R. JOUBERT-CARON
Université Paris Nord
Laboratoire de Biochemie et
Technologie des Protéines
74, rue Marcel Cachin
F-93012 Bobigny Cedex
France

E. VAN DAMME
Katholieke Universiteit Leuven
Lab. voor Plantenbiochemie
Willem De Croylaan 42
B-3001 Heverlee
Belgium

A. DANGUY
Université Libré de Bruxelles
Faculté des Sciences
Lab. de Biologie Animale et
Histologie Compareé
808 Route de Lennik CP 620
B-1070 Bruxelles
Belgium

E. VAN DRIESSCHE
Vrije Universiteit Brussel
Instituut voor Moleculaire Biologie
Paardenstraat 65
B-1640 St.-Genesius-Rode
Belgium

G. EGEA
Institut Municipal d'Investigacio
Medica
Passeig Maritim 25–29
ES-08003 Barcelona
Spain
Present address:
Electron Microscopy Unit
Centro de Biologia Molecular
Facultad de Ciencias - UAM
ES-28049 Madrid
Spain

C. EGENHOFER
Medizinische Einrichtungen der
Heinrich-Heine-Universität
Abteilung für Immunbiologie
Moorenstraße 5
D-40225 Düsseldorf
Germany

M. ETZLER
Deptartment of Biochemistry
and Biophysics
University of California
Davis, CA 95616
USA

H.-J. GABIUS
Institut für Physiologische Chemie
der Ludwig-Maximilians-Universität
Veterinärstraße 13
D-80539 München
Germany

D. GAUSS
Max-Planck-Institut für
Experimentelle Medizin
Hermann-Rein-Straße 3
D-37075 Göttingen
Germany

N. GILBOA-GARBER
Dept. of Life Sciences
Bar-Ilan University
52900 Ramat-Gan
Israel

G. HERRLER
Institut für Virologie der
Philipps-Universität
Robert-Koch-Straße 17
D-35037 Marburg
Germany

R.L. HILL
Duke University Medical Center
Deptartment of Biochemistry
Durham, NC 27710
USA

J. HIRABAYASHI
Teikyo University
Deptartment of Biol. Chemistry
Faculty of Pharmaceutical Sciences
Sagamiko
Kanagawa 199-01
Japan

S.-C. HO
Michigan State University
Deptartment of Biochemistry,
Biochemistry Bld.
East Lansing, MI 48824-1319
USA

T. IRIMURA
Division of Chemical Toxicology
and Immunochemistry
Faculty of Pharmaceutical Sciences
The University of Tokyo
7-3-1 Hongo, Bunkyo-ku
Tokyo 113
Japan

S.S. JOSHI
Dept. of Cell Biology & Anatomy
University of Nebraska
Medical Center
600 South 42nd Street
Omaha, NE 68198-6395
USA

H. E. JUNGINGER
Center for Bio-Pharmaceutical
Sciences
Dept. of Pharmaceutical
Technology
Leiden University
Einsteinweg 5, P.O. Box 9502
NL-2300 RA Leiden
The Netherlands

K. KAYSER
Abteilung Pathologie
Thoraxklinik der LVA Baden
Amalienstr. 5
D-69126 Heidelberg
Germany

D. KILPATRICK
Edinburgh and S. E. Scotland Blood
Transfusion Service
Blood Donor Centre Lauriston Bldg.
41 Lauriston Place
Edinburgh EH3 9HB
UK

V. KOLB-BACHOFEN
Medizinische Einrichtungen
der Universität
Abteilung für Immunbiologie
Moorenstraße 5
D-40225 Düsseldorf 1
Germany

E. KÖTTGEN
Freie Universität Berlin
Universitätsklinikum R. Virchow
Spandauer Damm 130
D-14050 Berlin
Germany

S. KOJIMA
Research Institute for Biosciences
Science University of Tokyo
2669 Yamazaki
Noda City
Chiba 278
Japan

Y. C. LEE
Johns Hopkins University
Deptartment of Biology
34th & N. Charles Street
Baltimore, MD 21218-2685
USA

R. LOTAN
University of Texas
Deptartment of Tumor Biology (108)
MD Anderson Cancer Center
1515 Holcombe Boulevard
Houston, TX 77030
USA

P. L. MANN
University of New Mexico
Dept. of Radiopharmacy
College of Pharmacy
Albuquerque, NN 87131-1071
USA

E. ORTEGA-BARRIA
Stanford University
School of Medicine
Department of Pediatrics
Stanford, CA 94305-5119
USA

M. E. A. PEREIRA
New England Medical Center
Hospitals
Division of Geographic Medicine
750 Washington Street, Box 41
Boston, MA 02111
USA

J. M. RHODES
The University of Liverpool
University Dept. of Medicine
P. O. Box 147
Liverpool L69 3BX
UK

S. ROSEMAN
Johns Hopkins University
Deptartment of Biology
34th & N. Charles Street
144 Mudd Hall
Baltimore, MD 21218-2685
USA

P. RUDD
Glycobiology Institute
Department of Biochemistry
University of Oxford
South Parks Road
Oxford OX1 3QU
UK

H. RÜDIGER
Institut für Pharmazie und
Lebensmittelchemie der Universität
Am Hubland
D-97074 Würzburg
Germany

H.C. SIEBERT
Bijvoet Center for
Biomolecular Research
P.O. Box 80075
NL-3508 TB Utrecht
The Netherlands

F. SINOWATZ
Institut für Tieranatomie
der Universität
Veterinärstraße 13
D-80539 München
FRG

W.R. SPRINGER
Department of Veterans Affairs
Medical Center
3350 La Jolla Village Drive
San Diego, CA 92161
USA

N. SUMAR
Department of Surgery
St. George's Hospital Medical School
Cranmer Terrace
London SW17 0RE
UK

H. THALMANN
Karlstraße 27
D-22085 Hamburg
Germany

G. UHLENBRUCK
Institut für Immunbiologie
der Universität zu Köln
Kerpener Straße 15
D-50937 Köln
Germany

F. VIDAL-VANACLOCHA
Deptartment of Cell Biology
and Morphological Sciences
School of Medicine and Dentistry
University of the Basque Country
Leioa
ES-48940 Vizcaya
Spain

A. VILLALOBO
Instituto de Investigaciones
Biomedicas, C.S.I.C.
Arturo Duperier 4
ES-28029 Madrid
Spain

H. WALZEL
Institut für Biochemie der
Universität, Bereich Medizin
Schillingallee 70
D-18057 Rostock
Germany

J. WANG
Michigan State University
Deptartment of Biochemistry
Biochemistry Bld.
East Lansing, MI 48824-1319
USA

K. YAMAOKA
Tokyo Metropolitan Institute
of Medical Science
18-22 Honkomagome 3-chome
Bunkyo-ku
Tokyo 113
Japan

N. YAMAZAKI
National Institute of Materials
and Chemical Research
Functional Molecules Laboratory
1-1 Higashi, Tsukuba-city
Ibaraki 305
Japan

C. YU
Johns Hopkins University
Department of Biology
144 Mudd Hall
34th & N. Charles Street
Baltimore, MD 21218-2685
USA

J. P. ZANETTA
Lab. de Neurobiologie Moleculaire
des Interactions Cellulaires
CNRS-UPR 416
Centre de Neurochimie du CNRS
5, rue Blaise Pascal
F-67000 Strasbourg
France

F.-Y. ZENG/A. TIMOSHENKO
Institut für Pharmaz. Chemie
der Philipps-Universität
Marbacher Weg 6
D-35037 Marburg
Germany

B. ZIPSER
Michigan State University
Deptartment of Physiology
111 Giltner Hall
East Lansing, MI 48824-1101
USA

Abbreviations

AIBN	azodiisobutyronitrile
AP	alkaline phosphatase
ASF	asialofetuin
ATCC	American Type Culture Collection
BCA	bicinchoninic acid
BCIP (= X-gal)	5-bromo-4-chloro-3-indolyl-β-galactopyranoside
BSA	bovine serum albumin
CAPS	3-(cyclohexylamino)-1-propanesulfonic acid
CHAPS	3-([3-cholamidopropyl]dimethylammonio)-1-propanesulfonate
DCC	dicyclohexylcarbodiimide
DMEM	Dulbecco's modified Eagle medium
DMF	dimethylformamide
2,6-DPIP	2,6-dichlorophenolindophenol
DVS	divinyl sulfone
EDAC	1-ethyl-3-(3-dimethylaminopropyl)carbodiimide
EDTA	ethylenediaminetetraacetic acid
EEDQ	N-ethoxycarbonyl-2-ethoxy-1,2-dihydroquinoline
EGTA	ethylene glycol-bis(2-aminoethyl ether)tetraacetic acid
ELISA	enzyme-linked immunoadsorbent assay
GalAH	6-aminohexyl-β-D-galactopyranoside
GP-NH$_2$	glycopeptide
HA	hemagglutinating activity
Hepes	4-(2-hydroxyethyl)-1-piperazineethanesulfonic acid
HPLC	high-performance liquid chromatography
HRP	horseradish peroxidase
IBA	iodosobenzoic acid
IP	inositol-1-monophosphate
IPIG	isopropylthio-β-galactopyranoside
ME	2-mercaptoethanol
MST	mean survival time
MTT	3-(4,5-dimethylthiazol-2-yl)-3,5-diphenyltetrazolium bromide
NBT	nitroblue tetrazolium
NEPHGE	non-equilibrium pH gradient electrophoresis
PA	phosphatidic acid
PC	phosphatidylcholine
PE	phosphatidylethanolamine
PI	phosphatidylinositol
PMSF	phenylmethylsulfonyl fluoride
PTU	pseudothiourea
PVDF	poly(vinylidene difluoride)
RBC	red blood cell

RCF	relative centrifugal force
SDS	sodium dodecyl sulfate
TCA	trichloroacetic acid
TEA	triethylamine
TEMED	N,N,N',N'-tetramethylethylenediamine
TFA	trifluoroacetic acid
TLC	thin layer chromatography
Tris	tris(hydroxymethyl)-aminomethane

Introduction

1 Lectins: Insights into the State of Knowledge by Literature Searches

DIETER H. GAUSS

Computers connected online to literature databases need only a few seconds to state the existence of about 20 000 documents (publications, patents, reports, theses, etc.) dealing with lectins or at least mentioning lectins (chemistry sector: *Chemical Abstracts* about 12 200; biology sector: *Biological Abstracts* about 17 000; medicine sector: *medline* about 15 500; they will be largely identical). The numbers are increasing annually, e.g., *Chemical Abstracts* and *medline* 1991 by 1023 and 601 documents, 1990 by 853 and 578 documents, and 1980 by 395 and 411 documents. Thus a steadily increasing importance of lectin research and application can be anticipated. Altogether, the documents mentioning a lectin in any respect may count for about 0.1% of all documents in the chemistry-biology-medicine field. Those documents dealing exclusively with lectins (or at least mainly with lectins), as characterized by exhibiting "lectin" in the title, are a subgroup of the documents labeled with the keyword lectin, and are present in considerably less numbers, namely about 6000 in CA, about 8000 in BA, and about 5000 in *medline*. In summary, there is considerable knowledge in the lectin field coupled with a notable lack of knowledge; this is reflected in the increasing interest in these molecules.

In the fields of chemistry or biochemistry, besides detecting so far unknown lectins, in particular the carbohydrate binding capacities of the lectins are subject of research and of corresponding documents (about 1200 cases). Furthermore, the amino acid sequences of the lectins, and, in the case of those lectins which bind distinct oligosaccharides, the sequences of the oligosaccharides are of interest (about 300 documents). A sample search to find citations on the three dimensional structure of lectins, as established by X-ray crystallography, is given below:

```
       (FILE 'HOME' ENTERED AT 09:00:47 ON 28 FEB 92)
       FILE 'CA' ENTERED AT 09:00:56 ON 28 FEB 92
L1       11908 S LECTIN OR LECTINS
L2       11387 S X(W)RAY(W)ANALYSIS
L3        2672 S X(W)RAY(W)STRUCTURE
L4       14041 S L2 OR L3
L5           5 S L1 AND L4
=> d 15 1-5 an au ti so

L5    ANSWER 1 OF 5   COPYRIGHT 1992 ACS
AN    CA116(5):36568u
AU    Bourne, Yves; Rouge, Pierre; Cambillau, Christian
TI    X-ray structure of a biantennary octasaccharide-lectin complex
      refined at 2.3-.ANG. resolution
SO    J. Biol. Chem., 277(1), 197-203

L5    ANSWER 2 OF 5   COPYRIGHT 1992 ACS
AN    CA115(25):275114d
AU    Rackovsky, S.
TI    Quantitative organization of the known protein x-ray structures.  I.
      Methods and short-length-scale results
SO    Proteins:  Struct., Funct., Genet., 7(4), 378-402

L5    ANSWER 3 OF 5   COPYRIGHT 1992 ACS
AN    CA115(5):44525x
AU    Zaluzec, Eugene J.; Zaluzec, Marianne M.; Olsen, Kenneth W.;
      Pavkovic, Stephen F.
TI    Crystallization and preliminary x-ray analysis of peanut
      agglutinin-N6-benzylaminopurine complex
SO    J. Mol. Biol., 219(2), 151-3
```

```
L5    ANSWER 4 OF 5   COPYRIGHT 1992 ACS
AN    CA113(23):207074z
AU    Bourne, Yves;  Rouge, Pierre;  Cambillau, Christian
TI    X-ray structure of a (.alpha.-Man(1-3).beta.-Man(1-4)GlcNAc)-lectin
      complex at 2.1-.ANG. resolution.  The role of water in sugar-lectin
      interaction
SO    J. Biol. Chem., 265(30), 18161-5

L5    ANSWER 5 OF 5   COPYRIGHT 1992 ACS
AN    CA100(7):49859j
AU    Wright, Christine Schubert;  Gavilanes, Francisco;  Peterson, Darrell
      L.
TI    Primary structure of wheat germ agglutinin isolectin 2.  Peptide
      order deduced from x-ray structure
SO    Biochemistry, 23(2), 280-7
```

Lectins have been found in the plant and animal kingdoms and in microorganisms (including prokaryotes). e.g., both *Biological Abstracts* and *medline* report 2500 documents regarding mice, about 100 documents regarding snails, about 20 documents regarding *Drosophila*, about 100 documents regarding yeasts, and about 200 documents regarding *Escherichia coli*. Evolution is dealt with or mentioned in about 100 documents, cell communication or signaling phenomena in up to 300 documents. Single lectins of importance are abrin, mentioned in more than 200 documents, ricin, mentioned in up to 1700 documents, and, last but not least, concanavalin, mentioned as a research tool in more than 20 000 documents; these three lectins are an essential subject of research or essential reagent in about 5000 documents.

In the field of medicine, participation of lectins in infection processes may be a relevant question to a database; lectins and infection are mentioned together in more than 500 documents; however, there are only 27 detailed studies of the role of lectins in infectious processes. In contrast, the oncology area is the subject of more than 1000 documents. In a typical search, *medline* was asked for documents dealing with lectins and brain. The result showed 156 publications fulfilling these conditions, and there are 46 publications exhibiting both "brain" and "lectin" within the title of the document. A printout of authors, titles and sources of a few of these documents may illustrate their relevance:

```
s s=6;f=au;ti;so
6.00/000001 DIMDI: -MEDLINE /COPYRIGHT NLM
AU: Joubert R;  Caron M;  Avellana-Adalid V;  Mornet D;  Bladier D
TI: Human brain lectin: a soluble lectin that binds actin.
SO: J Neurochem, 58 (1) 200-3  /1992 Jan/ IMD=9203
6.00/000002
AU: Sasaki A;  Nakanishi Y;  Nakazato Y;  Yamaguchi H
TI: Application of lectin and B-lymphocyte-specific monoclonal antibodies for
    the demonstration of human microglia in formalin-fixed, paraffin-embedded
    brain tissue.
SO: Virchows Arch A Pathol Anat Histopathol, 419 (4) 291-9  /1991/ IMD=9202
MORE
6.00/000003 DIMDI: -MEDLINE /COPYRIGHT NLM
AU: Lutsik BD;  Stekhnovich IV;  Lutsik AD
TI: Lektinogistokhimiia mikroglii golovnogo mozga cheloveka pri bakterial'nykh
    meningoentsefalitakh.
    (Use of lectins in histochemical examination of microglia of the human
    brain in bacterial meningoencephalitis)
SO: Zh Nevropatol Psikhiatr, 91 (2) 41-4  /1991/ IMD=9110
6.00/000004
AU: Debbage P;  Marguth F;  Gabius HJ
TI: Glycohistochemical visualisation of mannose-binding lectins in the
    microvasculatures of brain and brain tumours.
SO: Acta Histochem Suppl, 40 113-6  /1990/ IMD=9108
MORE
6.00/000005 DIMDI: -MEDLINE /COPYRIGHT NLM
AU: Takamatsu H;  Tani Y;  Akiyama M;  Nakata Y;  Segawa T
TI: Characterization of the carbohydrate chain on the substance P receptor in
    the rat brain cortex: effect of lectins on [3H]substance P binding.
SO: J Neurochem, 56 (4) 1452-4  /1991 Apr/ IMD=9106
6.00/000006
AU: Kaur C;  Ling EA;  Wong WC
TI: Lectin labelling of amoeboid microglial cells in the brain of postnatal
    rats.
```

```
SO: J Anat, 173 151-60  /1990 Dec/ IMD=9106
MORE
6.00/000007 DIMDI: -MEDLINE /COPYRIGHT NLM
AU: Avellana-Adalid V;  Joubert R;  Bladier D;  Caron M
TI: Biotinylated derivative of a human brain lectin: synthesis and use in
    affinoblotting for endogenous ligand studies.
SO: Anal Biochem, 190 (1) 26-31  /1990 Oct/ IMD=9105
6.00/000008
AU: Eloumami H;  Bladier D;  Caruelle D;  Courty J;  Joubert R;  Caron M
TI: Soluble heparin-binding lectins from human brain: purification, specificity
    and relationship to an heparin-binding growth factor.
SO: Int J Biochem, 22 (5) 539-44  /1990/ IMD=9009
```

Obviously, lectin research so far has not notably contributed to psychology and psychiatry, as indicated by the lack of corresponding hits in that same search, as traced through its research protocol (the second column therein displays the number of documents found):

```
 1.00     15188  CT=LECTINS
 2.00    127234  CT=BRAIN
 3.00       156  FIND 1 AND 2
 4.00      5179  FIND LECTIN$/TI
 5.00     80417  FIND BRAIN/TI
 6.00        46  FIND 4 AND 5
 7.00     12029  CT=PSYCHIATRY
 8.00      2828  CT=PSYCHOANALYSIS
 9.00      4491  CT=PSYCHOLOGY
10.00         0  FIND 4 AND 7
11.00         0  FIND 4 AND 8
12.00         0  FIND 4 AND 9
13.00         0  FIND 1 AND 7
14.00         0  FIND 1 AND 8
15.00         0  FIND 1 AND 9
```

The nomenclature term "lectin" and online research were established about 25 years ago. However, knowledge of lectins is far older. Searches for lectins in general in earlier literature have to be made under the entries agglutinins, hemagglutinins, phytohemagglutinins, and precipitins.

These introductory remarks have reviewed some facts on size, storage, and successful retrieval of available information on lectins. However, a really good knowledge retrieval requires a lectin facts database rather than a literature database. There is no such database, and, concomitantly, the inherently interdisciplinary lectin field so far lacks well-established detailed nomenclature terms that are well accepted by the scientists of different lectin disciplines, and with which the facts could be classified exactly. As the spectrum of the individual chapters of this book testify, the field is still unfolding, and is still too new to meet these requirements.

Chemical Synthesis of Lectin Ligands

2 Synthetic Ligands for Lectins

Reiko T. Lee and Yuan Chuan Lee

Synthetic carbohydrate ligands are an indispensable research tool for lectinology, especially in the realm of purification and specificity characterization of lectins. Naturally occurring glycoproteins usually contain complex and heterogeneous mixtures of oligosaccharide chains, and this fact makes it extremely difficult to obtain definitive conclusions from experiments using natural glycoproteins or glycopeptides and oligosaccharides derived from them. In contrast, neoglycoproteins can be easily prepared in large amounts, each neoglycoprotein carries a known sugar structure, and often a wide range of the degree of sugar substitution on proteins can be accomplished.

Most useful neoglycoproteins are those that contain multiple copies of mono- or oligo-saccharide determinant for a certain lectin, since affinity of ligand often increases geometrically with increasing sugar valency. For certain lectins, affinity increase can also be attained with small, synthetic ligands that carry two or three sugar determinants. These small cluster ligands are useful for the determination of binding stoichiometry, ligand targeting, etc.

In this chapter, we describe preparations of three types of neoglycoprotein as well as some small cluster ligands.

2.1 General Methods

Some methods repeatedly used in the preparations of carbohydrate ligands are described here.

Gel filtration columns used are Sephadex G-25 (5×210 cm) and sephadex G-15 (2.5×135 cm), both equilibrated and eluted with 0.1 M acetic acid. Water-insoluble compounds are fractionated on a column of Sephadex LH-20 (5×190 cm), equilibrated and eluted with 95% ethanol. All Sephadex beads are obtained from Sigma. Fraction size is 20 ml for columns of 5-cm diameter and 6 ml for the column of 2.5-cm diameter. After each use, the columns are thoroughly washed until all the input components are eluted off so that they can be reused. Monitoring of carbohydrate material in the effluent is by the phenol-sulfuric acid method (McKelvy and Lee 1969), in which a sample (0.5 ml) contained in a 13×100-mm test tube is mixed with 0.3 ml of 5% phenol, and then 2.8 ml of concentrated sulfuric acid is rapidly added to it with a repipet (Labindustries). The color formed is read at 480 nm after tubes are cooled to room temperature. Thin layer chromatography (TLC) is done on silica gel G-60 (F-254) layer on aluminum plate (EM Science). After chromatography, plates are air dried, UV absorbing components are detected under UV lamp, and then plates are sprayed with 15% sulfuric acid in 50% ethanol and heated on a hot plate ($\sim 140°$). Evaporation of solvents is carried out with a rotary evaporator under suction by water aspirator.

De-O-acetylation is done by treating compounds in dry methanol containing 10 mM sodium methoxide for several hours to overnight. Depending on the subsequent step, the reaction mixture is either acidified with a few drops of 60% acetic acid and evaporated, or alternatively, it can be neutralized with Dowex 50 × 8 (H⁺ form) suspended in a small volume of water, the resin is filtered off and the filtrate is evaporated. After extraction of water-immiscible organic solution with aqueous solutions, the organic solution is routinely dried by addition of solid anhydrous sodium sulfate (enough to clarify the solution). After a few minutes, the salt is filtered off and the filtrate is evaporated. Crystalline products are harvested by filtration through a Büchner funnel and washed with a small amount of solvent (usually chilled) used for crystallization. Solid or syrupy products are routinely dried in a vacuum desiccator containing fresh NaOH pellets, which is evacuated with an aspirator. When a volatile base (e.g., triethylamine) is present, a beaker containing concentrated sulfuric acid (ca. 1-cm depth) is also placed in the desiccator.

2.2 Preparation of Neoglycoproteins

Many methods are available for the preparation of neoglycoproteins (for review, see Stowell and Lee 1980). The three methods described here (2.1, 2.2, and 2.3) utilize amino groups of proteins for linking. The advantages of utilizing amino groups over other groups (e.g., phenolic and sulfhydryl groups) are their relative abundance and accessibility, so that a series of neoglycoproteins with a wide range of sugar substitution level can be prepared. The linking process in all three methods preserves the positive charge of the amino group. Thus, no apparent denaturation (e.g., precipitation) is observed even with large number of amino groups modified (Lee and Lee 1980).

Sections 2.1 and 2.2 describe methods suitable for coupling a large number of easily available mono- and di-saccharides, while the third method (2.3) is suitable for efficient coupling of precious glycopeptides. Bovine serum albumin (BSA), which is inexpensive and nonglycosylated, is used as example for preparation of all three types of neoglycoproteins. However, other proteins, including glycoproteins, can be used and have been used by us successfully.

2.2.1 Imidate Coupling (Lee et al. 1976)

Reaction scheme for the preparation of this type of BSA derivative starting from galactose (Gal) is shown in Fig. 1.

Materials   Galactose, BSA: from Sigma Chem. Co.
Acetic anhydride, HBr in glacial acetic acid, thiourea, 2-chloroacetonitrile, and sodium methoxide: from Aldrich Chem. Co.

Procedure   *Step 1.* 2,3,4,6-Tetra-O-acetyl-1-bromo-1-deoxy-α-D-galactopyranose [1-Br-Gal (OAc)$_4$] (Kartha and Jennings 1990): A cold solution of HBr ($\sim$ 30%) in glacial acetic acid (30 ml) is added to a 500-ml round-bottomed flask containing galactose

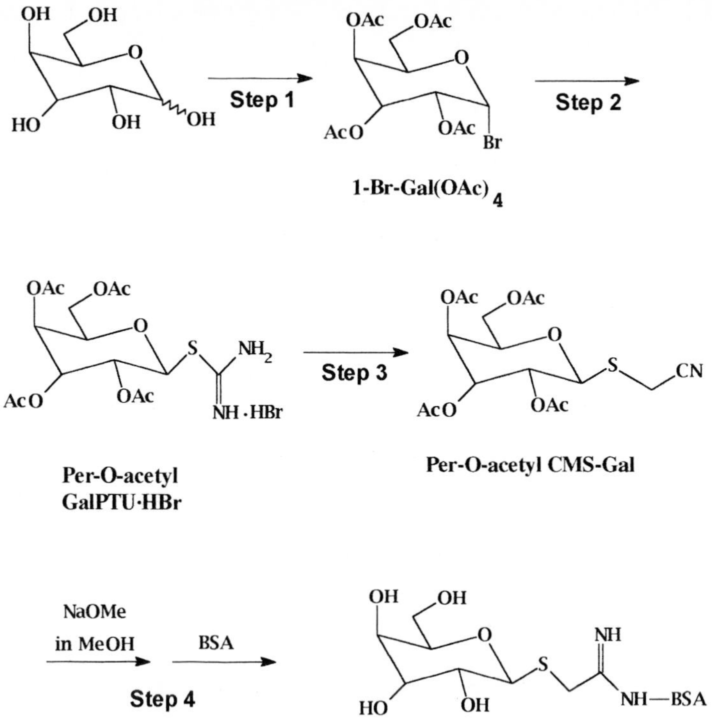

Fig. 1. Reaction scheme for the preparation of neoglycoproteins by the imidate coupling method

(18 g, 100 mmol) and acetic anhydride (100 ml). The mixture is stirred at room temperature until galactose completely dissolves (~ 1 h). One hundred and twenty ml of HBr in glacial acetic acid is added and the mixture is stirred for a total of 4 h. The amber-colored solution is evaporated to a thick syrup, mixed well with 80 ml of toluene and evaporated. The coevaporation with toluene is repeated two more times to yield syrupy 1-Br-Gal(OAc)$_4$.

Step 2. 2-S-(2,3,4,6-Tetra-O-acetyl-β-D-galactopyranosyl)-2-thiopseudourea hydrobromide [per-O-acetyl Gal-PTU·HBr]: The syrupy 1-Br-Gal(OAc)$_4$ is dissolved in 30 ml of dry acetone and transferred into a 100-ml round-bottomed flask with a stirring bar. Thiourea (6.7 gms, 88 mmol) is added and the mixture is stirred and heated to reflux. Refluxing is continued for 15–30 min at which point the PTU derivative crystallizes out. Heating is stopped immediately, and after cooling to room temperature, crystals are harvested. Dried crystals can be stored at room temperature.

Step 3. Per-O-acetylated cyanomethyl 1-thio-β-D-galactopyranoside [CMS-Gal(OAc)$_4$]: The PTU derivative (9.74 g, 20 mmol) and 2-chloroacetonitrile (5 ml, 79 mmol) are added to 40 ml of 1:1 (v/v) water-acetone mixture. When they are nearly in solution, potassium carbonate (3.2 g, 23.2 mmol) and sodium bisulfite (4.0 g, 40.4 mmol) are added and the mixture is stirred at room temperature. The mixture is then poured into ice-water (160 ml) and stirred for 2 h. The precipitate is filtered, washed with cold water, air-dried, and crystallized from hot methanol (80 ml) to yield 5.8 g (14.4 mmol, 72%) of CMS-Gal(OAc)$_4$.

Step 4. Coupling of CMS-Gal(OAc)$_4$ to BSA via imidate intermediate: To a solution of CMS-Gal(OAc)$_4$ (400 mg, 1 mmol) in 10 ml of dry methanol is added 250 μl of 4 M NaOMe in dry methanol. The mixture is left at room temperature for 24 h and then evaporated without heat being applied. A solution of BSA (660 mg, 10 μmol) in 30 ml of 0.25 M sodium borate buffer, pH 8.5 is immediately added to the syrupy residue, and the solution is left at room temperature for 24 h. The solution is dialyzed against several changes of water and the dialyzed BSA solution is freeze-dried to yield BSA which bears approximately 20 Gal residues per molecule. To obtain Gal-BSA containing approximately 10, 30, and 40 Gal residues per molecule, 40, 400, and 1500 molar excess of the CMS derivative over BSA should be used.

Helpful Hints
1. For evaporation of reaction mixture in the 1-Br-Gal(OAc)$_4$ preparation, the solvent trap should be chilled with ice–water bath, and the evaporator should be rinsed thoroughly immediately after use. 1-Br-Gal(OAc)$_4$ can also be prepared by treating 39 g (100 mmol) of per-O-acetylated galactose in 30 ml of glacial acetic acid with 30% HBr in glacial acetic acid (80 ml) for 3 h, and following the evaporation procedure as described above.
2. If per-O-acetyl Gal-PTU · HBr does not crystallize out from the reaction mixture, the mixture should be concentrated to 1/2 volume and kept in the cold to obtain crystals. The crystals should be washed and dried thoroughly to remove any free HBr remaining. This will prolong the shelf life of the PTU derivative.
3. Sodium metabisulfite (Na$_2$S$_2$O$_5$) can be substituted for sodium bisulfite. Use the same weight amount.
4. Conversion of cyanomethyl aglycon to methyl imidate is usually carried out with CMS-sugar derivative at 0.1 M in dry methanol containing 0.1 M NaOMe. However, because of the lower solubility of the CMS derivatives of N-acetyl-hexosamines, the corresponding conversion should be carried out with concentrations of both the CMS sugar and NaOMe at 0.05 M for 48 h. Yield of the imidate derivative is ∼ 50% under these conditions.
5. A thiosugar linkage can be cleaved under a mild condition in the presence of mercuric ion (Krantz and Lee 1976). Thus, the degree of sugar substitution can easily be determined even for neoglycoproteins prepared from natural glycoproteins.

2.2.2 Reductive Alkylation Using a Galactopyranoside with an ω-Aldehydo Aglycon (Lee and Lee 1979)

Reaction scheme for this coupling is shown in Fig. 2.

Materials
2-Chloroacetic anhydride, 2-aminoacetaldehyde dimethyl acetal, triethylamine (TEA): from Aldrich
Sodium cyanoborohydride: from Sigma
Per-O-acetyl Gal-PTU · HBr: see Sect. 2.2.1, Step 3

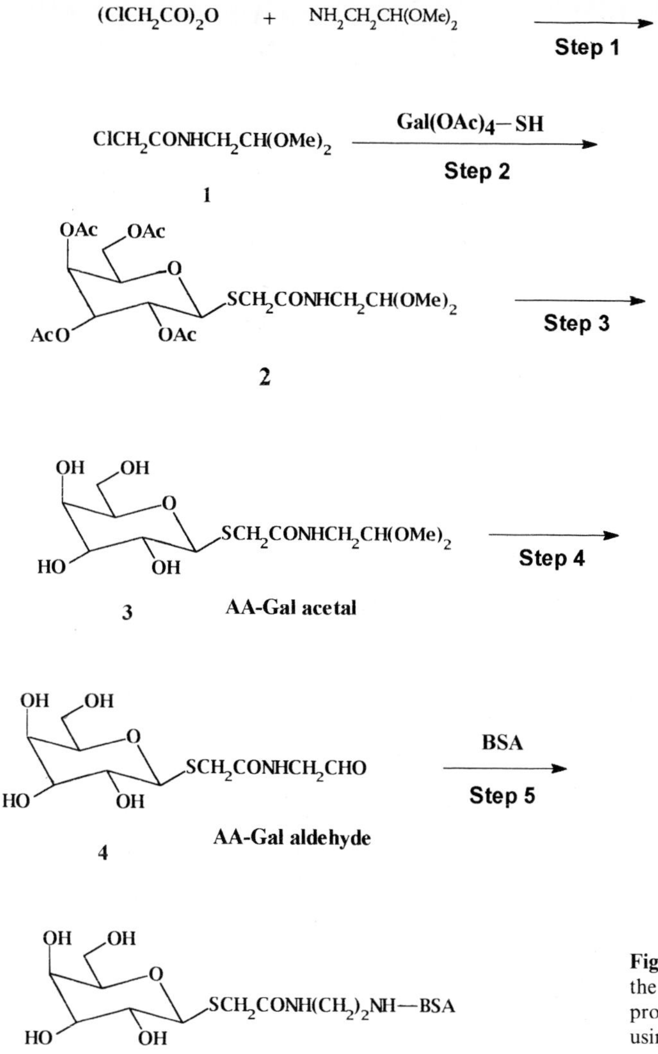

Fig. 2. Reaction scheme for the preparation of neoglycoproteins by reductive alkylation using ω-aldehydoglycoside

Procedure

Step 1. N-(Chloroacetyl) aminoacetaldehyde dimethyl acetal: 2-Chloroacetic anhydride (7.52 g, 44 mmol) is slowly added to an ice-cooled solution of aminoacetaldehyde dimethyl acetal (4.32 ml, 40 mmol) in dry methanol (200 ml) while being stirred. After 30 min, the mixture is brought to neutral pH by addition of triethylamine and evaporated. The residue is dissolved in cold chloroform (100 ml) and the solution is washed successively with 30 ml each of cold water and cold, saturated sodium hydrogen carbonate, dried and evaporated.

Step 2. Per-O-acetyl AA-Gal acetal: The residue obtained above is dissolved in 140 ml of 1:1 (v/v) acetone-water mixture and Gal(OAc)$_4$-PTU·HBr (9.75 g, 20 mmol), potassium carbonate (3.32 g, 24 mmol) and sodium bisulfite (2.08 g,

20 mmol) are added in that order. After overnight at room temperature, the mixture is partitioned between chloroform (160 ml) and water (80 ml). The aqueous layer is extracted once with chloroform (60 ml) and the combined chloroform solution is washed with 1 M NaCl (2 × 50 ml), dried and evaporated.

Step 3. AA-Gal acetal: The syrupy product in 80 ml of methanol is de-O-acetylated with NaOMe, acidified with acetic acid and evaporated. The residue is dissolved in 40 ml of water and fractionated in two batches on the Sephadex G-25 column. Carbohydrate material in the effluent is evaporated. After drying overnight, the residue is crystallized from hot isopropanol to yield 4.03 g (11.8 mmol) of AA-Gal acetal. Additional 0.62 g crystals can be obtained by adding ether to the mother liquor.

Step 4. AA-Gal aldehyde: A solution of AA-Gal acetal (1 g, 2.9 mmol) in 50 mM HCl (2 ml) is heated for 20 min in a boiling water bath, cooled, made neutral with 2 M NaOH (50 µl), and stored overnight in the cold. Crystals are filtered, washed with cold water and 95% ethanol and dried, yielding 0.4 g (1.35 mmol) of AA-Gal aldehyde. An additional amount of this product can be recovered in pure form from the mother liquor by gel filtration through the Sephadex G-15 column.

Step 5. Coupling of AA-Gal aldehyde to BSA (Lee and Lee 1980): AA-Gal aldehyde (29.5 mg, 0.1 mmol) is dissolved into a solution of BSA (66 mg, 1 µmol) in 6 ml of 0.2 M sodium phosphate buffer, pH 7. Sodium cyanoborohydride (119 mg, 1.7 mmol) is dissolved into this solution and the mixture is kept at 37 °C for 24 h. The solution is dialyzed thoroughly against water in the cold, and freeze-dried. The level of Gal incorporation is ∼ 20 mol/mol of BSA. To obtain BSA derivatives containing approximately 10, 30, 40, and 50 mol of Gal per mol of BSA, 18-, 180-, 500-, and 800-fold molar excess, respectively, of the reagent (AA-sugar aldehyde) over BSA should be used. Sodium cyanoborohydride should be added at least in fivefold molar excess of AA-sugar aldehyde.

Helpful Hints

1. AA-Gal aldehyde should be stored in the freezer. AA-Gal acetal, which can be stored at room temperature, is better for long-term storage. Coupling to protein is carried out with a freshly generated aldehyde.
2. Of 59 amino groups present in a molecule of BSA, up to 45 residues can be modified by the imidate method (2.2.1). Further increase of the reagent in the reaction mixture does not increase the incorporation. With the reductive alkylation method (2.2.2), however, the initially formed secondary amino-derivative (see Fig. 2) can accept a second molecule of AA-sugar aldehyde to form a tertiary amine derivative, when a large excess of the reagent in present. Thus, BSA derivatives containing more than 50 Gal residues can be prepared (Lee and Lee 1980).
3. Pyridine-borane complex, which is less toxic, can be substituted for sodium cyanoborohydride in the reductive alkylation, see Section 2.2.2, Step 5.

2.2.3 Efficient Coupling of Glycopeptides Using a Hetero-Bifunctional Reagent
(Lee et al., 1989a)

A universally applicable method for efficient attachment of a glycopeptide to proteins is described. Since a highly purified glycopeptide is a precious commodity,

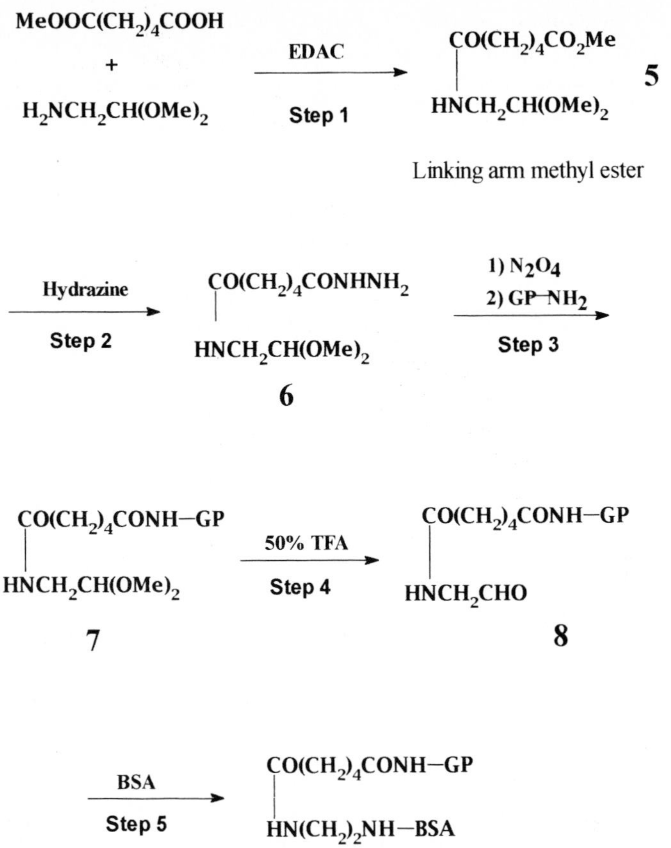

$$MeOOC(CH_2)_4COOH$$

$$+$$

$$H_2NCH_2CH(OMe)_2$$

→ EDAC, Step 1 →

$$\underset{|}{CO(CH_2)_4CO_2Me}$$
$$HNCH_2CH(OMe)_2$$ **5**

Linking arm methyl ester

→ Hydrazine, Step 2 →

$$\underset{|}{CO(CH_2)_4CONHNH_2}$$
$$HNCH_2CH(OMe)_2$$

6

→ 1) N_2O_4 2) GP—NH_2, Step 3 →

$$\underset{|}{CO(CH_2)_4CONH-GP}$$
$$HNCH_2CH(OMe)_2$$

7

→ 50% TFA, Step 4 →

$$\underset{|}{CO(CH_2)_4CONH-GP}$$
$$HNCH_2CHO$$

8

→ BSA, Step 5 →

$$\underset{|}{CO(CH_2)_4CONH-GP}$$
$$HN(CH_2)_2NH-BSA$$

Fig. 3. Reaction scheme for conjugating glycopeptides to proteins using a heterobifunctional reagent

efficiency of the glycopeptide utilization is the most important factor here. The reaction scheme is shown in Fig. 3.

Materials

Adipic acid methyl monoester, 2-aminoacetaldehyde dimethyl acetal, 1-(3-dimethylaminopropyl)-3-ethylcarbodiimide hydrochloride (EDAC), hydrazine, trifluoroacetic acid (TFA), nitrogen dioxide, pyridine borane, TEA: from Aldrich.

Procedure

Step 1. Linking arm methyl ester, **5**: A mixture containing aminoacetaldehyde dimethyl acetal (2.18 ml, 20 mmol), adipic acid methyl monoester (2.96 ml, 20 mmol) and EDAC (4.98 g, 20 mmol) in dichloromethane (75 ml) is stirred until a clear solution is obtained. After 24 h at room temperature, the reaction mixture is extracted with cold solutions (50 ml each) of 0.75 M sulfuric acid, 1 M NaCl, saturated sodium hydrogen carbonate and 1 M NaCl, in that order, dried, and evaporated to yield syrupy product (70% yield), which is used in the next step without further purification.

Step 2. Protected heterobifunctional linking arm: The above residue (14 mmol) is dissolved in 11 ml of dry methanol and to this is added hydrazine (0.6 ml, 19 mmol). After overnight at room temperature, the mixture is evaporated to a solid residue, which is dissolved in 12 ml of water and extracted three times with an equal volume of dichloromethane. The aqueous solution is evaporated to dryness and further dried in a desiccator to give crystalline product in 71% yield. This product can be stored at room temperature.

Step 3. Attaching a glycopeptide (GP–NH$_2$) to the linking arm: The hydrazide obtained above (16 mg, 65 µmol) in 0.5 ml of dimethyl formamide (DMF) is chilled to $-40\,°C$ (dry ice-acetone) and 0.46 M nitrogen dioxide in dichloromethane (0.18 ml, 83 µmol) is added. After 15 min at $-15\,°C$, this solution is added to a solution of GP–NH$_2$ (10 µmol) in 0.5 ml of 0.2 M sodium borate buffer, pH 8.5. Triethylamine (8 µl, 57 µmol) is added and the mixture is kept overnight in the cold. After extraction with equal volumes of cholroform (two times), the aqueous solution is purified on a column (1.5 × 70 cm) of Sephadex G-25 in a pyridine acetate buffer, pH 4.5 (10 mM in acetate). Carbohydrate-containing fractions are combined and freeze-dried to yield GP-linking arm conjugate in 88% yield.

Step 4. Conversion of acetal to aldehyde: The GP-linking arm conjugate (10 mg) is dissolved in 5 ml of 50% TFA. After 2 h at room temperature, the mixture is either freeze-dried or rotary evaporated with 20 volumes of absolute ethanol without heating.

Step 5. Conjugation to BSA: To a solution of BSA (10 mg, 0.15 µmol) and a portion of the aldehyde-containing conjugate (1.5 µmol) in 0.2 M sodium phosphate buffer, pH 7, (0.7 ml) is added 20 µl of 0.3 M methanolic solution of pyridine-borane (75 µl of pyridine-borane in 2.5 ml of methanol), and the mixture is left at room temperature for 24–40 h. The solution is dialyzed thoroughly against water in the cold and freeze-dried to yield a GP–BSA conjugate having ~ 8.3 molecules of GP per molecule of BSA (coupling efficiency = 83%).

Helpful Hints
1. Solution of nitrogen dioxide in dichloromethane should be tightly sealed and kept in freezer and handling of this reagent should be done in a well-ventilated hood, since nitrogen dioxide is quite toxic. Alternatively, coupling of the hydrazide to GP can also be done with t-butyl nitrite (Lemieux et al. 1975).
2. If an acid-labile group (e.g., sialyl group) is present on the glycopeptide, freeze-drying method is recommended after 50% TFA treatment.
3. For preparation of a larger amount of neoglycoproteins, a proper amount of pyridine-borane should be added as is, rather than as a methanolic solution.

2.3 Preparation of Small Cluster Ligands

2.3.1 Cluster Ligands Based on 2-Amino-2-(Hydroxymethyl)-1,3-Propanediol (Tris) (Lee 1978)

The reaction sequence for this preparation is shown in Fig. 4.

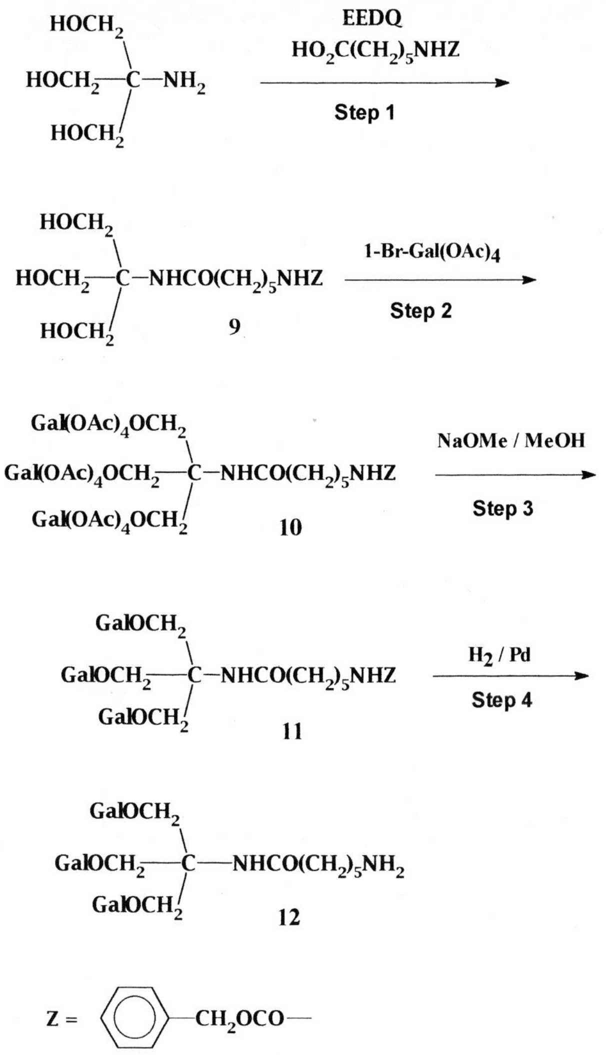

HOCH$_2$

HOCH$_2$—C—NH$_2$ → $\xrightarrow[\text{Step 1}]{\begin{array}{c}\text{EEDQ}\\ \text{HO}_2\text{C(CH}_2)_5\text{NHZ}\end{array}}$

HOCH$_2$

HOCH$_2$

HOCH$_2$—C—NHCO(CH$_2$)$_5$NHZ $\xrightarrow[\text{Step 2}]{\text{1-Br-Gal(OAc)}_4}$

HOCH$_2$ **9**

Gal(OAc)$_4$OCH$_2$

Gal(OAc)$_4$OCH$_2$—C—NHCO(CH$_2$)$_5$NHZ $\xrightarrow[\text{Step 3}]{\text{NaOMe / MeOH}}$

Gal(OAc)$_4$OCH$_2$ **10**

GalOCH$_2$

GalOCH$_2$—C—NHCO(CH$_2$)$_5$NHZ $\xrightarrow[\text{Step 4}]{\text{H}_2\text{ / Pd}}$

GalOCH$_2$ **11**

GalOCH$_2$

GalOCH$_2$—C—NHCO(CH$_2$)$_5$NH$_2$

GalOCH$_2$ **12**

Z = ⟨phenyl⟩—CH$_2$OCO—

Fig. 4. Reaction scheme for the preparation of the Tris-type trivalent glycoside

Tris and sodium borohydride: from Sigma **Materials**
2-Ethoxy-1-ethoxycarbonyl-1,2-dihydroquinoline (EEDQ) and palladium on carbon: from Aldrich
N-Cbz-aminohexanoic acid: from Bachem California
1-Br-Gal(OAc)$_4$: 2.2.1, Step 1

Step 1. Formation of **9**: A mixture of Tris (6.06 g, 50 mmol), cbz-aminohexanoic acid **Procedure**
(14.6 g, 55 mmol), and EEDQ (14.8 g, 60 mmol) in absolute ethanol (500 ml) is boiled
under reflux for 5 h, allowed to cool to room temperature, and evaporated to a

syrup. Addition of ether (250 ml) yields the product as crystals, which is re-crystallized from ethyl acetate to yield pure **9** in 64% yield (11.7 g, 31.8 mmol).

Step 2. Formation of **10**: A mixture of **9** (3.68 g, 10 mmol), 1-Br-Gal(OAc)$_4$ (12.3 g, 30 mmol) and mercuric cyanide (7.57 g, 30 mmol) in 200 ml of dry nitromethane-toluene mixture (1:1, v/v) is stirred at room temperature for 24 h, at which time 10 mmol each of 1-Br-Gal(OAc)$_4$ and mercuric cyanide are added and the mixture is stirred for 2 more days. The mixture is evaporated, and the resulting syrup is dissolved in chloroform (150 ml), washed four times with 1 M NaCl (cold), dried, and evaporated. The syrup obtained is dissolved in 46 ml of 95% ethanol, and purified in three batches on the Sephadex LH-20 column. The trigalactosylated product (**10**), which is contained in the carbohydrate peak emerging first, is evaporated to a syrup (9.0 g, 66.3% yield).

Step 3. De-O-acetylation: A solution of per-O-acetyl trisglycoside **10** (9.0 g, 6.6 mmol) is deacetylated in the usual fashion in 60 ml of dry methanol and neutralized with Dowex 50 resin. Repeated coevaporation of the syrupy residue with toluene produces hygroscopic crystals, **11** (5.1 g, 5.9 mmol).

Step 4. Hydrogenolysis: Hydrogenolysis of **11** (3.0 g, 3.5 mmol) in 60% acetic acid (50 ml) in the presence of 10% palladium on carbon (0.35 g) is performed in a Brown hydrogenator (Brown and Brown 1966). After 6–7 h reaction time, the catalyst is filtered off, and the filtrate is concentrated to ~10 ml and purified on the Sephadex G-25 column. The fractions containing the major carbohydrate peak are combined and evaporated to yield amorphous product, **12** (2.2 g, 3.05 mmol).

Helpful Hints

1. If both the bis- and tris-glycosylated derivatives are desired, the second addition of 1-Br-Gal(OAc)$_4$ in Step 2 is omitted. These derivatives can be separated on the Sephadex LH-20 column, and can be identified by TLC using a 3:2:1 (v/v) mixture of ethyl acetate-acetic acid-water. The Tris-glycoside elutes first from the column and has the lowest R_f, followed by the bis-glycoside with a higher R_f. The mono-glycosylated product, which elutes next, is contaminated with other byproducts.
2. Generation of amino group in Step 4 is for the purpose of attaching another group, such as a fluorescent or a radioactive probe. Therefore, this step can be omitted, if the amino function is not needed.
3. If hydrogenolysis is to be avoided, one can start the synthesis with 6-trifluoroacet-amidohexanoic acid instead of cbz-aminohexanoic acid. The former compound is prepared from 6-aminohexanoic acid using trifluoroacetic anhydride. The amino group is regenerated, in this case, by TEA treatment, as described in Section 2.3.2.

2.3.2 Cluster Glycosides Based on Aspartic Acid (Lee et al. 1984)

This type of cluster glycosides utilizes two carboxylic acid groups of Asp for attachment of an amino-terminated glycoside, such as 6-aminohexyl galactopyrano-side. Frequently, a glycine residue is attached to each carboxylic acid of Asp for further elongation of arms. We observed that this type of divalent Gal glycoside has much stronger affinity toward mammalian hepatic lectins than the di- and tri-valent

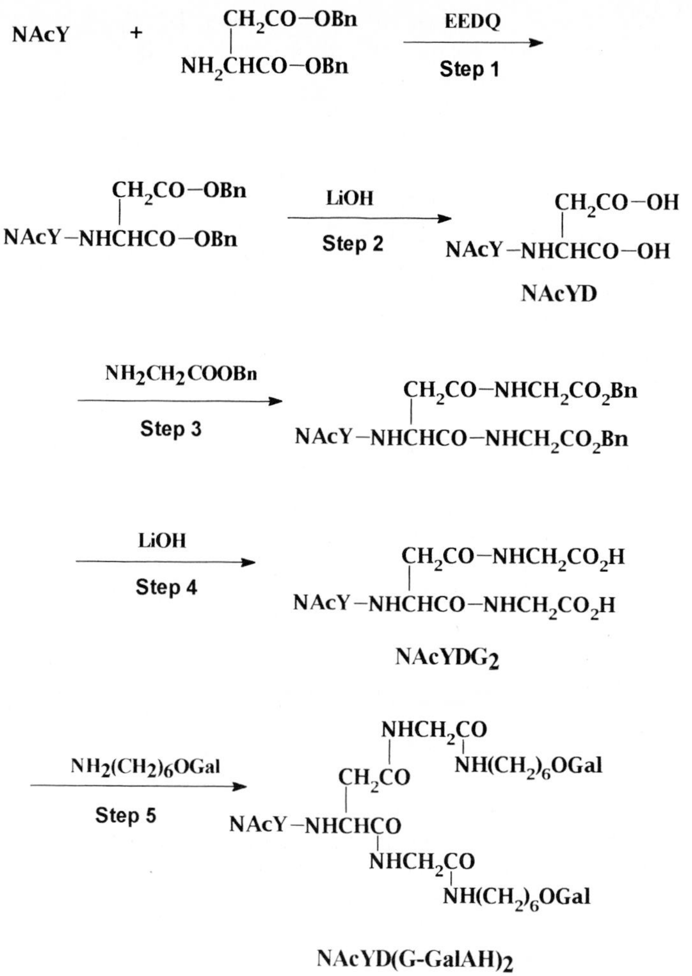

NAcY   +   CH$_2$CO$-$OBn   **EEDQ**
            | $\longrightarrow$
           NH$_2$CHCO$-$OBn   **Step 1**

CH$_2$CO$-$OBn   **LiOH**   CH$_2$CO$-$OH
| $\longrightarrow$ |
NAcY$-$NHCHCO$-$OBn   **Step 2**   NAcY$-$NHCHCO$-$OH

                                              NAcYD

NH$_2$CH$_2$COOBn   CH$_2$CO$-$NHCH$_2$CO$_2$Bn
$\longrightarrow$ |
Step 3   NAcY$-$NHCHCO$-$NHCH$_2$CO$_2$Bn

LiOH   CH$_2$CO$-$NHCH$_2$CO$_2$H
$\longrightarrow$ |
Step 4   NAcY$-$NHCHCO$-$NHCH$_2$CO$_2$H

                                      NAcYDG$_2$

                                    NHCH$_2$CO
                                    |
NH$_2$(CH$_2$)$_6$OGal   CH$_2$CO   NH(CH$_2$)$_6$OGal
$\longrightarrow$ |
Step 5   NAcY$-$NHCHCO
                              |
                         NHCH$_2$CO
                         |
                         NH(CH$_2$)$_6$OGal

NAcYD(G-GalAH)$_2$

Fig. 5. Reaction scheme for the preparation of the Asp-type divalent glycoside

Tris-type galactosides. In fact, the three-dimensional separation of Gal residues in the Asp-type galactosides is more similar to that of natural, Gal-terminated bian-tennary oligosaccharides than the Tris-type galactosides. The reaction scheme for the preparation is shown in Fig. 5.

Materials

Aspartic acid dibenzyl ester p-toluenesulfonate, glycine benzyl ester p-toluenesulfon-ate: from Research Organics

N-Acetyl tyrosine: from Sigma

EEDQ, EDAC, dicyclohexyl carbodiimide (DCC), 1-hydroxybenzotriazole mono-hydrate (1-OH-Bt), 6-aminohexanol, ethyl trifluoroacetate, TEA, N-methyl-morpholine: from Aldrich

1-Br-Gal(OAc)$_4$: 2.2.1, Step 1

Procedure   Preparation of 6-aminohexyl β-D-galactopyranoside (Weigel et al. 1979): 6-Amino-hexanol (30 g, 256 mmol) is added to 36 ml of ethyl trifluoroacetate and the mixture is stirred at room temperature for 5 h. The mixture is poured into 500 ml of water while being stirred vigorously, and then stirred gently overnight in the cold. Crystalline product [6-(trifluoroacetamido)hexanol] is filtered and washed with cold water. The filtrate is evaporated to ~ 100 ml, and the crystals formed are harvested. A mixture containing 20 mmol each of 1-Br-Gal(OAc)$_4$ (8.22 g), 6-(trifluoroacetamido)hexanol (4.26 g) and mercuric cyanide (5.05 g) in a 1:1 (v/v) mixture of dry toluene-nitromethane (100 ml) is stirred for 24 h at room temperature. The mixture is processed as described in 2.3.1, Step 2. After purification on the column of Sephadex LH-20, the glycoside, 6-(trifluoroacetamido)hexyl 2,3,4,6-tetra-O-acetyl-β-D-galactopyranoside, is crystallized from 95% ethanol. The glycoside is deacetylated and neutralized by the resin. The evaporated product is dissolved in 45 ml of water, to which are added 5 ml each of 95% ethanol and TEA. The mixture is left at room temperature overnight, and then evaporated. The residue is dissolved in water, and fractionated on the Sephadex G-15 column. Carbohydrate-containing fractions are combined, evaporated to a syrup and dried. The residue is dissolved in 20 ml of 50% ethanol and to this solution is added Dowex 1×8, OH$^-$ form, (~ 20 mEq). The supernatant solution should be basic (pH > 9) after this treatment. 6-Aminohexyl-β-D-galactopyranoside can be obtained in crystalline form from 95% ethanol.

Step 1. Preparation of NAcYD dibenzyl ester: EEDQ (3.21 g, 13 mmol) is added to a solution containing N-acetyl tyrosine (2.23 g, 10 mmol), aspartic acid dibenzyl ester p-toluenesulfonate (6.31 g, 13 mmol) and TEA (2 ml, 14 mmol) in absolute ethanol (150 ml), and the mixture is stirred for 10 h at room temperature. After evaporation, the residue is partitioned in chloroform and water. The organic layer is washed with cold solutions of 0.75 M sulfuric acid, saturated sodium hydro-gencarbonate, and 1 M NaCl, dried and evaporated. Crystallization from hot 95% ethanol gives NAcYD dibenzyl ester in 47% yield (2.42 g).

Step 2. Saponification of the benzyl ester: To a cold suspension of NAcYD dibenzyl ester (2.42 g, 4.67 mmol) in methanol (70 ml) is added a solution of LiOH $\cdot$ H$_2$O (905 mg, 21.6 mmol) in water (20 ml), and the mixture is stirred for 6 h at room temperature. The mixture is brought to $\sim$ pH 2 by addition of Dowex 50, the resin is filtered off, and the filtrate is evaporated.

Step 3. Attachment of glycine benzyl ester: To a solution of the above residue in DMF (50 ml) are added glycine benzyl ester p-toluenesulfonate (3.94 g, 11.7 mmol), TEA (1.63 ml, 11.7 mmol), anhydrous 1-OH-Bt (1.57 g, 11.7 mmol) and EDAC (2.68 g, 14 mmol) at -10 °C. The mixture is stirred for 10 h at room temperature, evaporated and fractionated on the Sephadex LH-20 column. The major UV-positive fractions are combined and evaporated. The product (NAcYDG$_2$ dibenzyl ester) is obtained from water-acetone as crystals (2.87 g, 4.53 mmol, 97%).

Step 4. Saponification: NAcYDG$_2$ dibenzyl ester (1.2 g, 1.8 mmol) in 20 ml of methanol is saponified as above by addition of LiOH $\cdot$ H$_2$O (506 mg, 12 mmol) in 10 ml of water. The residue obtained from the Dowex 50 treatment is crystallized from water-acetone to yield NAcYDG$_2$ (520 mg, 1.15 mmol, 64%).

Step 5. Coupling of GalAH to NAcYDG$_2$: NAcYDG$_2$ (525 mg, 1.16 mmol) and anhydrous 1-OH-Bt (322 mg, 2.4 mmol) are dissolved in dry DMF (5 ml) and cooled in ice. DCC (530 mg, 2.57 mmol) is added, and the mixture is stirred in an ice–NaCl mixture ($\sim -10\,°C$) for 40 min. A solution of GalAH (718 mg, 2.57 mmol) in DMF (2–3 ml) and N-methylmorpholine (265 μl, 2.4 mmol) are added to this mixture, stirred in ice-water bath for 30 min, then at room temperature overnight. To this mixture is added ~ 100 ml of toluene to precipitate out the product. After overnight at room temperature, the liquid is decanted off and the precipitate is washed with ether, and air-dried. The precipitate is dissolved in 50% ethanol (10 ml) and any insoluble material is filtered off. The filtrate is evaporated and the residue is fractionated on the Sephadex G-15 column. The fractions containing the major carbohydrate component are combined and evaporated to yield NAcYD(G-GalAH)$_2$ in 70% yield.

Helpful Hints

1. Commercially obtained 1-OH-Bt monohydrate is dried by dissolving it in 100% ethanol and coevaporating it with toluene.
2. Three different methods have been tried in our laboratory for joining of the carboxylic acid of an amino acid to the amino group of sugar-AH. These are EEDQ method, a mixed anhydride method using methyl chloroformate (Lee et al. 1984), and the 1-OH-Bt assisted carbodiimide coupling method described here (Section 2.3.2, Step 5). The highest yield is obtained when the 1-OH-Bt method is used.
3. A glycine is attached to each carboxylic acid group of Asp for the purpose of elongating the arm. This is because for hepatic lectins, the divalent ligand with longer arm [NAcYD(G-sugar-AH)$_2$] has higher affinity than the shorter version [NAcYD(sugar-AH)$_2$] (Lee et al. 1989b). One can omit two steps (Steps 3 and 4) in the reaction scheme by reacting NAcYD (obtained from Step 2) with sugar-AH under the conditions used in Step 5 to yield the shorter divalent ligand [NAcYD(sugar-AH)$_2$].

2.4 Suppliers

Aldrich Chemical Co., Inc., 1001 W. St. Paul Ave., Milwaukee, WI 53233 USA
Bachem California, 3132 Kashiwa Street, Torrance, CA 90505 USA
EM Science, 480 Democrat Rd., Gibbstown, NJ 08027 USA
Labindustries, Inc., 620 Hearst Ave., Berkeley, CA 94710 USA
Research Organics, Inc., 4353 E. 49th Street, Cleveland, OH 44125 USA
Sigma Chemical Company, P.O. Box 14508, St. Louis, MO 63178 USA

References

Brown CA, Brown HC (1966) Catalytic hydrogenation. II. A new, convenient technique for laboratory hydrogenations. A simple, automatic device for atmospheric pressure hydrogenations. J Org Chem 31: 3989–3995
Johnson TB, Coward JK (1987) Synthesis of oligophosphopeptides and related ATP γ-peptides esters as probes for c-AMP-dependent protein kinase. J Org Chem 52: 1771–1779

Kartha KPR, Jennings HJ (1990) A simplified, one-pot preparation of acetobromosugars from reducing sugars. J Carbohydr Chem 9: 777–781

Krantz MJ, Lee YC (1976) Quantitative hydrolysis of thioglycosides. Anal Biochem 71: 318–321

Lee RT, Lee YC (1979) Preparation of 1-thioglycosides having ω-aldehydo aglycons useful for attachment to proteins by reductive amination. Carbohydr Res 77: 149–156

Lee RT, Lee YC (1980) Preparation and some biochemical properties of neoglycoproteins produced by reductive amination of thioglycosides containing an ω-aldehydo aglycon. Biochemistry 19: 156–163

Lee RT, Lin P, Lee YC (1984) New synthetic cluster ligands for galactose/N-acetylgalactosamine-specific lectin of mammalian liver. Biochemistry 23: 4255–4261

Lee RT, Wong TC, Lee R, Yue L, Lee YC (1989a) Efficient coupling of glycopeptides to proteins with a heterobifunctional reagent. Biochemistry 28: 1856–1861

Lee RT, Rice KG, Rao NBN, Ichikawa Y, Barthel T, Piskarev V, Lee YC (1989b) Binding characteristics of N-acetylglucosamine-specific lectin of the isolated chicken hepatocytes: Similarities to mammalian hepatic galactose-N-acetylgalactosamine-specific lectin. Biochemistry 28: 8351–8358

Lee YC (1978) Synthesis of some cluster glycosides suitable for attachment to proteins or solid matrices. Carbohydr Res 67: 509–514

Lee YC, Stowell CP, Kranz MJ (1976) 2-Imino-2-methoxyethyl 1-thioglycosides: new reagents for attaching sugars to proteins. Biochemistry 15: 3956–3963

Lemieux RU, Bundle DR, Baker DA (1975) The properties of a "synthetic" antigen related to the human blood-group Lewis a. J Am Chem Soc 97: 4076–4083

McKelvy JF, Lee YC (1969) Microheterogeneity of the carbohydrate group of Aspergillus oryzae α-amylase. Arch. Biochem. Biophys. 132: 99–110

Stowell CP, Lee YC (1980) Neoglycoproteins: The preparation and application of synthetic glycoproteins. Adv Carbohydr Chem Biochem 37: 225–281

Weigel PH, Naoi M, Roseman S, Lee YC (1979) Preparation of 6-aminohexyl D-aldopyranosides. Carbohydr Res 70: 83–91

3 Sugar-Polyacrylamide Conjugates as Probes for Cell Lectins

N.V. Bovin

Glycoconjugates on the cell surface or in soluble form take part in many vitally important processes of recognition. Synthetic analogs of glycoconjugates – neoglycoproteins, neoglycolipids, and pseudopolysaccharides are successfully used for studying and modeling these processes. An advantage of synthetic glycoconjugates is the possibility of creating molecules with the programmed properties, e.g., solubility, molecular weight, ligand density, etc. Polyacrylamide-based glycoconjugates have a number of essential differences from traditionally used BSA-based neoglycoproteins. These differences are the following: chemical and immunological inertness (low nonspecific interaction with proteins) and also the flexibility of the polymer which is a random coil, allowing the carbohydrate ligands to re-arrange themselves and to interact with the carbohydrate-binding protein in an optimal way.

Pseudopolysaccharides proved to be convenient reagents when there is a need to study lectin sensitivity to epitope density of carbohydrate ligands: the ligand content in neoglycoconjugate is set by its initial ratio and a smooth change in ligand density is possible. A simultaneous or subsequent introduction of biotin label as 6-aminohexylbiotin makes it possible to provide sensitive probes for the solid-phase study of the fine specificity of purified lectins and to reveal carbohydrate-binding molecules on the cell surface.

3.1 Synthesis of Pseudopolysaccharides by Condensation of 3-Aminopropylglycosides with Activated Polymer

To synthesize polyacrylamide pseudopolysaccharides, the strategy of introduction of an olefinic group (allyl or acryloyl) in a carbohydrate was earlier employed followed by copolymerization with acrylamide (Bovin et al. 1993 and the papers cited therein). Our experience in performing double and triple copolymerization and literature data analysis showed the essential limits of this approach, the most important of them being low reproducibility of copolymerization (regarding yield, ratio of ligands inserted into polymer and molecular weight of copolymers), and technical difficulties when performing the reaction with quantities of about 0.1–0.5 mg. The approach detailed below is based on condensation of poly(4-nitrophenylacrylate) (i.e., fully activated polyacrylic acid) with aminoalkyl glycosides.

3.1.1 Chemicals

Azodiisobutyronitrile (AIBN), 2-ethanolamine, triethylamine from Merck; 4-Nitrophenylacrylate (monomer) and (N-biotinyl)hexamethylenediamine trifluoroacetate (BiotNH$_2$) synthesized according to Bovin et al. (1993), the first crystallized from

Chemicals

heptane twice; Benzene for polymerization reaction was purified by vigorous shaking with concentrated sulfuric acid for 2 h followed by water, sodium bicarbonate, water washing, calcium chloride drying, and distillation on sodium.

3.1.2 Preparation of Initial Activated Polymer, Poly (4-Nitrophenylacrylate)

Preparation   AIBN corresponding to 3% (W/W) of the monomer was added to 1 M solution of 4-nitrophenylacrylate in dry benzene under a stream of nitrogen and kept for 50 h at 70 °C. Benzene solution was decanted and the viscous brown residue remaining on the flask walls was dissolved in DMF or DMSO to obtain 1–2% solution. The polymer was re-precipitated with five volumes of methanol. Precipitation was repeated and the clean white residue was washed with methanol and dried. The polymer did not contain residual monomer, short oligomers, or AIBN according to TLC (EtOH/Et$_2$O, 2:1).

3.1.3 Condensations of Activated Polymer

Condensations   3.1.3.1 Synthesis of Polyacrylamide-Based Pseudopolysaccharides

Attachment of Trisaccharide A as an Example: PAA with 10% Inclusion of Oligosaccharide. A solution of 0.54 mg (1.0 μmol) of 3-aminopropyl glycoside of trisaccharide A, GalNAcα1-3(Fucα1-2) Gal (Syntesome, Germany), in 0.5 ml DMF and 10 μl triethylamine was added to a solution of 1.93 mg (10 μg-Eq.) poly(4-nitrophenylacrylate) in 0.5 ml DMF; the resulting solution was kept for 16 h at 25 °C, and a 30-fold molar excess of 2-ethanolamine was added and left for 16 h. The solution was applied to a Sephadex LH-20 column and the conjugate was eluted with a mixture of acetonitrile/water 1:1 with RI control of fractions. The yield of liophilyzed conjugate was more than 90%.

3.1.3.2 Synthesis of Polyacrylic Acid-Based Pseudopolysaccharides
Preparation of Conjugates Series of 5-, 10-, 20-, and 30% (mol)
N-Acetylneuraminic Acid Content

3.5 μmol sialic acid derivative, Neu5Aeα2–OCH$_2$C$_6$H$_4$NHCOCH$_2$NH$_2$ (p) (Byramova et al. 1991), in 0.2 ml of DMF were mixed with 0.07 ml of 10% solution of poly(4-nitrophenylacrylate) (35 μg-Eq.) in DMF, 0.02 ml of triethylamine were than added and the mixture was kept at 20 °C for 48 h. The resulting conjugate was further modified by addition of 2 ml 0.1 N aqueous NaOH. The solution was applied to the column with Sephadex LH-20 (1 × 25 cm) and the conjugate was eluted with the mixture of acetonitrile-water 1:1 with RI control of fractions. Polymer with 10% Neu5Ac molar content was thus obtained. To obtain 5-, 20-, and 30%-conjugates, 3.5 μmol of sialic acid derivative in DMF were mixed with 0.14, 0.035, and 0.023 ml of 10% poly(4-nitrophenylacrylate) in DMF, respectively, the unreacted nitrophenyl groups were hydrolyzed by NaOH and final conjugates were purified as above.

3.1.3.3 Synthesis of Biotinylated Probes

TF (Thomsen-Friedenreich) Disaccharide as an Example: The solution of 5 mg (11 μmol) Galβ1–3GalNAcα1–OCH$_2$CH$_2$CH$_2$NH$_2$ (Bovin et al. 1985) in 0.6 ml of DMF was added to 11 mg (55 μg-Eq.) poly(4-nitrophenylacrylate) in 0.4 ml DMF, then 20 μl triethylamine and 1.3 mg (2.8 μmol) BiotNH$_2$ in 0.1 ml of DMF were added: the mixture was kept for 24 h at 40 °C, the absence of carbohydrate and BiotNH$_2$ was shown by TLC (CHCl$_3$/EtOH/H$_2$O 3:8:2); 0.05 ml ethanolamine was added, the mixture kept for 24 h at room temperature, the conjugate was purified as described for pseudopolysaccharides, the yield 90%.

Synthesis

3.1.4 Quality Control

The relation of initial polymer to poly(4-nitrophenylacrylate) follows from NMR, IR, and UV spectra data: its purity can easily be controlled by determination of the number of 4-nitrophenolic groups in a fixed weight: the weight is dissolved in DMF, hydrolyzed by excess of alkali and the quantity of nitrophenol formed, which should be about 95–101%, is determined by the standard spectrophotometrical method using a calibration curve. The molecular weight of poly(4-nitrophenylacrylate) depends on a number of factors: reagents purity (particularly benzene and AIBN), their ratio and reaction temperature, conditions used permit obtaining the polymer with a polymerisation degree of about 200–400, which corresponds to the molecular weights of typical pseudopolysaccharides and probes of 20–40 kDa interval. It is convenient to synthesize, fractionate by re-sedimentation, and characterize several grams of polymer, which can be stored well at 4 °C, and can be thus used as a standard reagent for many years.

Quality Control

The molecular weight of the polymers was determined by: (1) HPLC on a column HEMA 1000 Bio (Tessek, CSFR) in water with the flow rate 1.0 ml/min and RI-detection, (2) preparative liquid chromatography on TSK HW 50S column (1.5 × 43 cm) in buffer 0.01 M ammonium acetate-0.2 M sodium chloride (pH 6.1), (3) ultrafiltration on PM 30, XM 100 and XM 300 membranes (Amicon) in water with polymer concentration 1 mg/ml and RI detection. The above approaches gave quite corresponding results: polymers nonsubstituted with carbohydrate ligands had molecular weight between 15–20 kDa and 26–32 kDa for typical probes.

3.1.5 Troubleshooting

As the process of removal of excess nitrophenylic groups (about 95% may remain after the binding of specific ligands) is accompanied by significant change in the hydrophilic-hydrophobic properties of the polymer, some quantity of nitrophenylic groups (percent) can remain within the polymeric coil and not be replaced by ethanolamine. This takes place rather seldom and depends on the nature of the attached ligand; it is expressed visually by yellowing of the substance during storage (nitrophenol is released). A final product guaranteed free of remaining nitrophenylic groups is obtained if the procedure of ethanolamine treatment is repeated after the first column purification.

Troubleshooting

3.2 Solid-phase Study of Influenza Virus Hemagglutinin Specificity

The development of competitive inhibitors of microbial attachment to the host cells may provide a novel approach to chemoprophylaxis and the treatment of infection diseases. It may be envisaged that in order to block a multi-point cooperative microbial reception, the inhibitors must also be capable of poly- or oligovalent binding to the pathogen. In an effort to produce potent synthetic inhibitors of influenza virus and *Mycoplasma pneumoniae*, attachment polyvalent sialic acid containing pseudopolysaccharide was prepared by coupling spacered aminobenzyl sialoside to a polyacrylate carrier. The viruses (more than 30 strains were studied) were adsorbed into the wells of plastic micro-plate coated with fetuin, and the binding of sialic acid containing conjugates by solid phase-attached virus was quantified by their competition with the standard fetuin-horseradish peroxidase conjugate. The activity of Neu5Ac-containing pseudopolysaccharides in the test system show distinct maximum at 10 mol% Neu5Ac, the polymeric sialoside of optimal Neu5Ac content was three to four orders of magnitude more active than monovalent ligand, and its inhibitory activity was comparable to that of the best natural inhibitor, equine α_2-macroglobulin.

3.2.1 Chemicals and Viruses

Chemicals and Viruses Sialyllactose, tween-80, horse-radish peroxidase (HRP) from Serva, bovine fetuin from Fluka, fetuin-HPR conjugate was synthesized by the periodate method (Boorsma and Streefkerk 1979). All other reagents were of analytical grade. Buffers: PBS is 0.14 M sodium chloride + 0.01 M sodium phosphate (pH 7.3). PBST is PBS containing 0.01% tween-80. Seed stocks of influenza virus were obtained from the Virus Collection of the Ivanovsky Institute of Virology, Moscow; viruses were grown in 9-day-old embryonated chicken eggs and purified by sucrose density gradient centrifugation (Virology, a practical approach 1985).

3.2.2 Fetuin-Binding Inhibition Assay

Assay The assay was carried out as follows. The microtiter plate wells (Linbro) were coated with 100 µl of fetuin solution (10 µg/ml) in PBS at 4 °C overnight. The plate was washed with PBST and water, and air-dried. The virus (0.2 µg) in 100 µl of PBST was allowed to bind to the immobilized fetuin for 2 h at 4 °C, then the plate was washed with PBST. The PBST-dissolved inhibitors (including polyacrylic acid-based pseudopolysaccharide (Sect. 3. 1.3.2.) in various concentrations or PBS alone (50 µl) were added for 10 min followed by 50 µl of fetuin-HRP conjugate (1 µg/ml fetuin). After incubation for 2 h at 4 °C, the plate was washed, filled with o-phenylenediamine, and further operations were carried out as usual.

3.2.3 Results

Results In order to study the influence of the density of polymer-bound sialic acid residues on the influenza virus-neutralizing properties of the polymer, a number of Neu5Ac

Table 1. Inhibition of influenza virus strains by synthetic polyacrylic derivatives of N-acetylneuraminic acid

Sialoside	Conc. for 50% inhibition, μmol	
	Strain BK/79	Strain TX/77
Sialyllactose	2100	1500
Neu5Acα2–OCH$_2$C$_6$H$_4$NHCOCH$_2$NHAc	450	300
Polyacrylic acid, 5% Neu5Ac	30	10
Polyacrylic acid, 10% Neu5Ac	0.2	0.2
Polyacrylic acid, 20% Neu5Ac	0.6	1.7
Polyacrylic acid, 30% Neu5Ac	6	4
Polyacrylamide, 10% Neu5Ac	140	5
Fetuin (natural glycoprotein)	7	14

derivatives were synthesized (Table 1). The activity of substances was expressed as the concentration of sialic acid residues required to inhibit fetuin-HRP binding to the virus by 50%. The coupling of the monovalent Neu5Ac derivative with the polymeric carrier markedly increased its ability to block viral receptor-binding activity, the polymer of 10% molar substitution being three orders of magnitude more active than the monomer. Ten percent molar Neu5Ac polymer had an optimal ligand density for interaction with the multivalent viral hemagglutinin; additional increase of the Neu5Ac residue content in the polymers lowered the inhibitory activity substantially. The influence of the macromolecular carrier on the inhibitory activity is demonstrated: polyacrylic acid-based polymer (see Sect. 3.1.3.2) is 20–1000-fold (dependently from virus strain) more active than polyacrylamide-based one (synthesized according to Sect 3.1.3.1) of the same Neu5Ac content.

3.2.4 Quality Control and Troubleshooting

See Sects. 3.1.4 and 3.1.5.

3.3 Revealing of Surface Lectins (Mice Cell Lines P-338 and EL-4)

Migration of lymphocytes and white blood cells in inflammatory lesions, neutrophil-platelet adhesion, and a number of other processes of adhesion and recognition in which hemopoietic and lymphoid cells take part are mediated by the interaction of the surface lectins of one cell with the surface glycoconjugates of another. An increased expression level of endogenous lectins was observed for the transformed cells, and a significant role in metastases formation is presumed. The carbohydrate parts of the cell surface glycoconjugates are usually complex oligosaccharides, so it can be expected that their partners in complementary intercellular recognition have oligosaccharidic specificity. The use of specific oligosaccharidic probes (see Sect. 3.1.3.3) to reveal their presence as well as low molecular weight saccharides for inhibition permits the study of fine carbohydrate specificity of endogenous lectins.

Table 2. Inhibition of binding cell lines P-338 and EL-4 with TF probe (0.1 mg/ml) by low molecular sugars (1 mM)

Inhibitor		Degree of inhibition
Galβ1-3GalNAcα-sp[a]	+	Complete inhibition for all cells
Galβ1-3GalNAcα-OCH$_2$C$_6$H$_5$	+	
Galβ1-3GalNAc	+ $-$	Partial inhibition
Galβ1-OCH$_3$	$-$ +	
Galβ1-3GalNAcβ-sp	$-$	No inhibition
GalNAcα-sp	$-$	

[a]sp = OCH$_2$CH$_2$CH$_2$NHCOCF$_3$.

3.3.1 Chemicals

Chemicals Ten percent formalin, hydrogen peroxide, 3,3'-diaminobenzidine, and β-methyl galactopyranoside from Sigma, streptavidin-horse radish peroxidase conjugate from Amersham. The synthesis of saccharides used is described in Bovin et al. (1985).

3.3.2 Inhibition of Probe Binding

Inhibition Cells (mice lymphoid cell lines P-338 and EL-4, approximately 4×10^6 cells) were dried, fixed in 10% formalin vapors for 3 min, and incubated with probe solution (0.1 mg/ml in PBS) for 1 h in a humid chamber at room temperature. After three washes with PBS, incubation with streptavidin-peroxidase was performed under the same conditions. Visualization of the bound probes was achieved by 3,3'-diaminobenzidine/H$_2$O$_2$ reaction. In inhibition experiments, low molecular weight inhibitors were added to the solution of probes before incubation with cells in ten fold molar concentration as compared with the probe.

3.3.3 Results

Results Lack of interaction of cells studied with β-Glc probe shows that non specific sorption was low. At the same time, the high oligosaccharide specificity is revealed by disaccharide probe TF (carbohydrate ligand being Galβ1-3GalNAcα). It may be seen from Table 2 that its binding is inhibited only by α-disaccharide and not β-disaccharide or monosaccharides (Galβ1-OCH$_3$, GalNAcα-sp).

References

Boorsma PM, Streefkerk JG (1979) J Immunol Meth 30, 245–255

Bovin NV, Korchagina EY, Zemlyanukhina TV, Byramova NE, Galanina OE, Zemlyakov AE, Ivanov AE, Zubov VP, Mochalova LV (1993) Synthesis of polymeric neoglycoconjugates based on N-substituted polyacrylamides. Glycoconjugate J 10: 142–151

Bovin NV, Zemlyanukhina TV, Khorlin AY (1985) Synthesis of neoglycoproteins having haptens (Galβ1-3GalNAcα1-) Ser, Galβ1-3GalNAcα1, Galβ1-3GalNAcβl, (GalNAcα1-) Ser. Bioorgan Khim 11: 1256–1264

Byramova NE, Mochalova LV, Belyanchikov IM, Matrosovich MN, Bovin NV (1991) Synthesis of sialic acid pseudopolysaccharides by coupling of spacer-connected Nau5Ac with activated polymer. J. Carbohydr Chem 10: 691–700

Mahy BW (1985) Virology, a practical approach. Ed., IRL Press, Oxford

Purification and Characterization of Lectins

4 Isolation of Plant Lectins

H. Rüdiger

Lectins are carbohydrate-binding proteins of nonimmune origin which occur throughout the biosphere (Goldstein and Poretz 1986). Their most obvious property, hemagglutination, has been known for more than a century (Franz 1988), whereas the art of purifying lectins is much younger. The great breakthrough came with the introduction of affinity chromatography, first with homopolymer polysaccharides which are suitable for only certain lectins (Agrawal and Goldstein 1965,

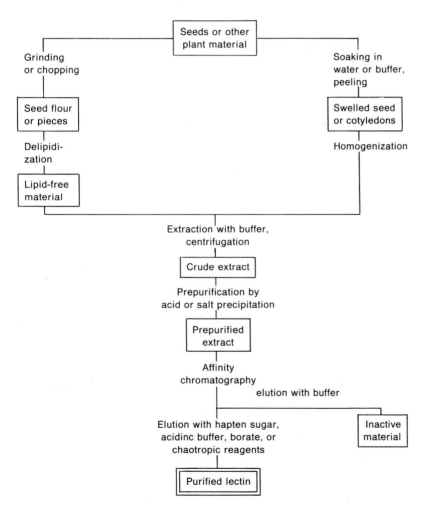

Fig. 1. General scheme for the isolation of lectins

1967), later with tailor-made affinity adsorbents prepared by immobilizing mono-saccharides, oligosaccharides, glycopeptides, or glycoproteins to hydrophilic and water-insoluble matrices (Lis and Sharon 1981; Rüdiger 1988). The present chapter deals with the isolation of lectins primarily from plants. The procedures are, however, applicable to lectins from other sources as well.

4.1 Isolation of Lectins

4.1.1 Plant Material

Plant materials used for preparation of lectins are mostly seeds. They may either be bought from seed dealers, collected in the open field, or grown in a private garden or field. Whenever possible, commercial sources are to be preferred because collecting or growing seeds needs much effort. Moreover, crops from wild-growing plants usually are very poor, the seeds are small and often contaminated. After having been collected, intact seeds have to be separated from damaged ones, especially if they contain insect larvae which may give rise to unidentifiable and unreproducible protein bands. Be sure that you have collected the right species. In case of doubt, ask a botanist.

If seeds of common vegetable plants are used, these may readily be bought from local food suppliers. Care should be taken to buy enough at once so as not to be forced to change the cultivar during an experiment if the original one is no longer in stock. The lectin content, the isolectin composition, the quality of accompanying proteins, and hence the optimal procedure for lectin purification may vary from one cultivar to another. Moreover, one cannot be sure that the species name given on the label is correct. For example, the name "bean" does not say anything except that the seeds have the shape and size similar to that of the common garden bean, *Phaseolus vulgaris*. The name "bean" is used for totally different seeds which may not even belong to the same family: *Canavalia* (Jack beans), *Glycine* (soya or soybeans), *Phaseolus* (garden or French beans), *Vigna* (mung beans) or *Vicia* (broad beans) are Leguminosae, whereas *Ricinus* (castor bean) belongs to the Euphorbiaceae. Some-times vegetable seeds are sold under fancy "bean" names without the supplier or manufacturer being able to give information as to the actual species. We detected mung beans sold under the name of soybeans. In such cases it is safer not to buy the seeds in food stores, but from reliable seed suppliers even if seeds intended for gardening are more expensive than food seeds.

If vegetative tissue is used, a sufficient stock of the desired property should be bought at once. Vegetative material may differ extremely in quality depending on season or year of harvesting.

The usual garden or vegetable seeds are available from most local seed suppliers. More unusual seeds an other plant material are offered in Germany by Bornträger (Offstein, Germany). Vegetable seeds not cultivated in Europe are available from Atlee Burpee (Warminster Pa., USA). Shrub and tree seeds can be bought from Eichenberg (Miltenberg, Germany) or Schumacher (Sandwich, Mass., USA). To our knowledge, some seeds are only available from suppliers of biochemicals, such as, e.g., *Canavalia ensiformis* seeds (Sigma Chemie, Deisenhofen, Germany) or *Griffonia* (formerly also *Bandeiraea*) *simplicifolia* (Calbiochem, Bad Soden, Germany). In contrast to other seeds, the *Griffonia* seeds are not viable.

4.1.2 Pretreatment of the Plant Material

In seeds, lectins usually occur in the cotyledons or in the endosperm, but scarcely ever in the embryo and never in the seed coat. Therefore it is recommended to remove the coat and, if possible, also the embryo. Many seed coats contain tannins, dyes, and other detrimental substances which, if extracted, may diminish the yield of lectin by unspecific precipitation, may lead to brownish colored products, or may adsorb to column materials and spoil them irreversibly. The removal of the seed coat is difficult with small seeds from wild-growing plants and with very hard seeds. In leguminous seeds, the embryo should also be removed because it is a rich source of many kinds of proteins, but contains practically no lectin.

If the seeds or cotyledons are not too hard and brittle enough, they can be ground in a regular coffee mill with rotating blades, otherwise special mills with a breaker must be used. Grinding of dry seeds ensures that life processes come to an end immediately, and that proteolytic processes which may accompany swelling and germinating are reduced. For the isolation of most lectins, however, this is not very important because they are comparatively stable toward endogenous proteases, in contrast to the labile storage proteins.

If the seeds are rich in fat and other lipids, the total or crushed seeds or the meal has to be extracted with an organic solvent (formerly dichoromethane, now mostly halogen-free solvents such as petroleum ether or diethylether are used). Separation of the solid material from the solvent should be done by filtration through pre-pleated filter paper rather than by centrifugation because of explosion risk. The extracted material is dried overnight under a hood at room temperature. In order to facilitate homogenization, seeds can also be swelled beforehand if the lectin resists this procedure unchanged. Swelling is usually done in unbuffered water or the extraction buffer. Some seeds have been shown to lose or to actively excrete proteinaceous material on swelling, among others lectin. Swelled seeds are soft enough to be homogenized in a household mixer or a similar laboratory equipment such as, e.g., an Ultraturrax. Some seeds, especially from wild-growing plants, do not take up water unless they are damaged artificially with a knife or a file.

Pretreatment

4.1.3 Extraction and Prepurification of Lectins

Extracts are often prepared in unbuffered saline. This should be avoided because the pH of the extract thus obtained is unpredictable. Mostly, it is slightly acidic, which may not be optimal for solubilizing the lectin. Although the lectin may be fairly soluble, this does not necessarily apply for the majority of seed proteins (storage proteins). If they remain undissolved, they may enclose some of the lectin, thus decreasing its yield. Slightly alkaline buffer (e.g., pH 8) of not too low ionic strength has been approved. In any case, the pH of the crude extract should be controlled and adjusted if necessary in order to guarantee complete extraction. Some plant materials, in particular of vegetative origin, contain phenols which are slowly oxidized and change the color of the extract to dark brown. Several methods have been proposed to avoid phenol oxidation, in part by adding reductants (ascorbic acid, metabisulfite, dithionite) or polyphenol oxidase inhibitors (diethyldithiocarbamate, thiourea) to the extraction buffer (van Driessche et al. 1983) or to adsorb phenols by extracting in the presence of insoluble polyvinyl-pyrrolidone (Polyclar, Loomis and Bataille 1966). If the lectin activity is to be assayed by hemagglutination, the ionic strength

Extraction
Prepurification

should be high enough to prevent hemolysis, e.g., 0.05 M Tris/HCl pH 8, supplemented with 0.1 M NaCl. In order to preserve the extracts from microbial growth, addition of NaN_3 (0.02%) or other preservatives is necessary. Because of the high reactivity of azide, addition of NaN_3 should be repeated after a while if the crude extract has to stand for longer periods of time. Some lectins need bivalent metal ions for full activity. As a precaution, many researchers therefore add these ions (Ca^{2+}, Mg^{2+} and/or Mn^{2+} as chlorides) in millimolar concentrations to the extraction buffers after pH adjustment. Mn^{2+} is a somewhat delicate ion because, even at slightly alkaline pH, it is oxidized immediately and irreversibly in air-saturated buffers to MnO_2, which forms brown colloidal solutions and cannot be resolubilized again. Therefore the pH should never be adjusted towards the alkaline side if Mn^{2+} is present. Phosphate and pyrophosphate are quite inexpensive. Their disadvantage, however, is that they are able to precipitate or scavenge metal ions which may be necessary for lectin activity. Buffers prepared from these salts should therefore only be used if pilot experiments have shown that they do not diminish this activity.

The crude extract is cleared by centrifugation at a moderate speed if small or medium-sized amounts are processed. For large-scale preparations, filtering, preferably in a filter press, is the method of choice. If the seed material is very expensive, the pellet or filter cake may be re-extracted in order to raise the yield.

In many cases it has proved good to remove most of the storage proteins by precipitation. Storage proteins generally form the bulk of seed proteins. They are often proteolytically quite unstable. Since their fragments may be less soluble than the native proteins, they may give rise to a continuous slight precipitate of residual storage proteins during all the following purification steps. This not only leads to impure lectin preparations but may also clog affinity and other columns. Usually, storage proteins can be removed by adjusting the clarified crude extract to pH 5 by slowly adding aqueous acetic acid under constant stirring, spinning down the precipitation, which consists of the storage proteins, to readjust the supernatant to pH 8 and to dialyze it. After this, the solution is ready for affinity chromatography.

A further method to remove proteinaceous and other impurities and to concentrate the lectin is to fractionate the prepurified extract by ammonium sulfate precipitation. This procedure, however, takes more time because it requires at least two centrifugation and one dialysis steps.

Example: Extraction of pea seeds

Materials   Tris, HCl, $CaCl_2$, $MgCl_2$, HOAc, NaOH, NaN_3, glucose, dialysis bags, Sephadex G-50 or G-100

0.05 M Tris/HCl, pH 8.0, 1 mM $CaCl_2$, 1 mM $MgCl_2$, 0.02% NaN_3, the same buffer with 0.25 M glucose, 1 M HOAc, 1 M NaOH

Household mixer and/or high-speed homogenizer (Ultraturrax, IKA, Staufen), cold room, pH meter, medium speed centrifuge, column (glass or plastic), magnetic stirrer

Procedure   Ten g of dry pea (*Pisum sativum*) seeds is swelled in water at 4 °C overnight, the seed coat and the embryo removed, and the remaining cotyledons homogenized in 100 ml of 0.05 M Tris/HCl, pH 8.0, 1 mM $CaCl_2$, 1 mM $MgCl_2$, and 0.02% NaN_3 first with a household mixer for 30 s, then with an Ultraturrax for an additional 30 s, and stirred for 2 h at 4 °C. The suspension is filtered through nylon tissue and centrifuged (20 min at 23 500 g). Storage proteins are precipitated by slowly adding 1 M acetic acid to the stirred solution until pH 5.0 is reached. After an additional hour of stirring at 4 °C, the suspension is centrifuged (20 min at 23 500 g) again and the supernatant readjusted to pH 8.0 by addition of 1 M NaOH. This solution can be directly applied to the affinity column. If importance is attached to a lectin preparation that is absolutely free from traces of storage proteins (e.g., for the production of anti-lectin antibodies that should not cross-react with storage proteins), and if the capacity of the affinity column is to be utilized completely, it is recommendable to dialyze the prepurified extract exhaustively against the extraction buffer, which will remove all low molecular weight substances which might interfere with lectin binding, and then to spin down the slight precipitate forming from residual storage protein prior to affinity chromatography.

Lectin activity is usually assayed by agglutination of erythrocytes. This method will be described and discussed under the heading "Characterization" below.

4.1.4 Purification by Affinity Chromatography

Standard procedures for chromatographic protein purification are covered in numerous monographs (e.g., Scopes 1982; Williams and Wilson 1984; Wilson and Goulding 1986; Sternbach 1991) and will not be outlined here.

Of these procedures, gel filtration can only process small amounts and needs long periods of time if not performed in the HPLC mode. Ion exchange resins have higher capacities and can be run much faster.

Nevertheless, it is not recommended to start with ion exchange because it may resolve isolectins before other contaminating proteins have been removed. This will complicate the purification procedure because after the ion exchange separation, every impure but lectin-containing fraction will have to be subjected to affinity chromatography separately.

For the same reason, chromatofocusing and hydrophobic chromatography should not be chosen as the first chromatographic steps.

The method of choice for purifying lectins is affinity chromatography on adsorbents bearing immobilized carbohydrates. The method is treated in several books dealing with protein purificaion in general but also in special monographs, e.g., Dean et al. (1985).

4.1.4.1 Insoluble Polysaccharides

Insoluble polysaccharides are suitable affinity adsorbents for many lectins. They are generally inexpensive and need not even be chemically homogeneous. They have, however, to be absolutely free from material that might leak out during the course of a chromatography. Luck et al. (1992) used extensively washed plant cell ghosts for purifying galactose-binding lectins. GlcNAc-binding lectins have often been purified by the use of chitin. This material needs extensive prepurification. An insufficiently prewashed chitin may release agglutinating substances which may be mistaken for a lectin (Whitmore 1992). Even if exhaustively washed, e.g., by the procedure of van Driessche et al. (1993), chitin may bind nonlectinic GlcNAc-binding proteins as chitinases which frequently occur in plant material. Their biological effects may then be erroneously ascribed to the lectin (Schlumbaum et al. 1986).

If the polysaccharides to be used are soluble (guar gum, arabic gum, locust gum, dextran), they can easily be transformed into hydrophilic gels by cross-linking agents, e.g., epichlorohydrin (Lönngren et al. 1976) or divinylsulfone (Young and Leon 1978).

The naturally occurring insoluble and the cross-linked polysaccharides are not obtained in spherical form. This restricts their use to runs in short columns or to batch procedures. Since, however, in affinity chromatography long columns are usually unnecessary, this disadvantage is tolerable.

Example: Preparation of cross-linked polysaccharides (Young and Leon 1978)

Materials Guar gum (from *Cyanopsis tetragonolobus* locust gum (from *Ceratonia siliqua*) or dextran (from *Leuconostoc mesenteroides*), NaOH, divinylsulfone, NaN_3

0.2 M NaOH

Magnetic stirrer, Ultraturrax homogenizer, broad spatula

Procedure Ten g of the polysaccharide is slowly added to 200 ml (dextran: 70 ml) of 0.2 N NaOH under stirring with a magnetic bar. It is important to take care that the powder is evenly distributed and does not clog. After complete swelling has taken place, 0.75 ml divinylsulfone is slowly added dropwise while the solution is vigorously stirred, preferably by hand with a broad spatula. Stirring is continued until the solution becomes thick. The gel is kept at room temperature for 90 min, then it is added to 1.5 l demineralized water and carefully chopped with an Ultraturrax (IKA) homogenizer at low speed in order not to generate too much fines. Gel particles are allowed to sediment by gravity, the supernatant being removed by suction and replaced by water. Washing is continued until the pH of the supernatant has fallen to 8–9. After equilibrating the gel with 0.05 M Tris/HCl pH 8.0, it is ready for use. Gels should be stored at 4 °C in a buffer containing a preservative, e.g., 0.02% NaN_3, which has to be supplemented from time to time.

The leguminous mannose/glucose-binding lectins from *Canavalia ensiformis*, *Pisum sativum*, *Lens culinaris*, and *Vicia faba* can be purified using cross-linked dextran, cross-linked guar and locust gums are suitable for the galactose-binding lectins from *Arachis hypogaea*, *Griffonia simplicifolia*, *Ricinus communis*, *Sophora japonica*, and *Wisteria floribunda*.

4.1.4.2 Immobilized Glycoconjugates

Another method to prepare affinity adsorbents is to immobilize mono- or oligosaccharides to hydrophilic gelatinous matrices. This can be done by several methods which are listed by Porath (1974), Lis and Sharon (1981), and Dean et al. (1985). A very simple though time-consuming method is the reductive amination of di- and other oligosaccharides (maltose, lactose, melibiose) to gels bearing primary amino groups (Gary 1974; Baues and Gray 1977). This leads to adsorbents in which the initially reducing sugar residue serves as a spacer, whereas the nonreducing one forms the affinity site. A procedure that is not too difficult to follow starts from allylglycosides prepared by the Fischer method. These are then copolymerized with acrylamide and N,N'-methylene-bis-acrylamide at predetermined rates (Hořejší and Kocourek 1973, 1974). This allows a strict control of the amount of carbohydrate incorporated and therefore has been used not only for preparative but also for analytical purposes (Hořejší et al. 1977, 1979).

There are many papers dealing with immobilization to gels of tailor-made glycosides bearing a spacer of suitable length and reactivity. Most of these methods, however, require some experience in preparative organic chemistry and will not be discussed here. Some of these affinity adsorbents or the defined mono- or oligosacc-harides are available from several companies, e.g., from Biocarb Chemicals, Chembiomed, Dextra Laboratories, Medac, Pierce, Seikagaku, or Sigma.

Since the affinity of lectins towards oligosaccharides usually is much higher than towards mono- or disaccharides, naturally occurring glycoproteins or glycoprotein mixtures (e.g., thyreoglobulin, fetuin, glycophorin, immunoglobulins, ovomucoid, submaxillar and gastric mucins, red cell stroma) or glycopeptides derived therefrom have been used by many authors (Lis and Sharon 1981 and literature cited therein). In our experience, it is not necessary to first prepare glycopeptides because also the undegraded glycoproteins are suitable as affinity adsorbents.

Commercially available glycoproteins useful for affinity chromatography are hog gastric mucin, thyreoglobulin, fetuin, and hen egg ovomucoid (trypsin inhibitor). We have observed that ovomucoid freshly prepared from egg white is superior to commercial preparations with regard to binding capacity, trypsin inhibitor activity, and price. For its isolation, the method of Lineweaver and Murray (1947) is followed:

Example: Purification of hen egg ovomucoid

Egg white, trichloroacetic acid, acetone, diethyl ether, prepleated filter paper   **Materials**

0.5 M aqueous trichloroacetic acid/acetone (1 + 2)

Büchner funnel, water pump
Avoid centrifugation because of explosion risk.

Egg white from several eggs is separated from the yolk, collected and adjusted to   **Procedure** pH 3.5 under constant stirring by addition of an equal volume of 0.5 M aqueous trichloroacetic acid/acetone (1 + 2). After stirring for an additional 30 min, the thick creamy suspension is filtered through a prepleated filter paper in the cold. The precipitate is discarded. If further material precipitates from the filtrate on standing, it should be refiltered. A sample should remain clear on heating (5 min/80 °C).

Ovomucoid is precipitated from the clear solution by slow addition of the 2–2.5-fold volume of acetone. After sedimenting, the supernatant is removed by siphoning or suction even if it is not completely clear. The sediment is washed with acetone several times. If sedimenting is very slow, the precipitate may be filtered through a Büchner funnel. The washed precipitate is suspended in as little water as possible (5–10% of the original volume of the filtrate) at pH 4.5, the protein again precipitated by addition of acetone, collected by filtration, redissolved in water and dialyzed in order to remove residual trichloracetic acid. Finally, the protein is again precipitated by acetone, the precipitate washed first with acetone, then with diethyl ether, and dried at the open air under a hood at room temperature. Three hundred mg of ovomucoid per 100 ml egg white are obtained.

If the protein is not used immediately, it should be dried over $CaCl_2$ in a vacuum desiccator and stored in a tightly closed bottle at -20 °C.

Pretreatment of Glycoproteins: Most plant lectins not only do not recognize an acid   **Pretreatment** but also are prevented by terminal sialic acid from binding to subterminal sugar residues. This applies especially for mucins. In order to improve the capacity of the affinity gels, it is recommended to remove the sialic acid before coupling. This can be done enzymatically, but the most simple and inexpensive way is to subject the glycoprotein to a gentle acid treatment. One has to make a compromise: too gentle treatment removes only little sialic acid, too harsh treatment leads to hydrolysis also of other glycosidic bonds.

Example: Partial desialylation of a glycoprotein (Svennerholm 1958)

Materials   Hog gastric mucin, H_2SO_4, NaOH

0.15 N H_2SO_4, 2 M NaOH

Ultraturrax homogenizer, magnetic stirrer, heatable water bath, pH meter, centrifuge

Procedure   One g of commercial mucin is evenly distributed by the use of an Ultraturrax homogenizer in the three to five fold amount of 0.15 N H_2SO_4. The suspension is heated (80 °C) under constant stirring for 90 min, starting from the time when the solution has warmed up. Evaporated water should be continuously replaced. After this time, the solution has turned brown and viscous. It is cooled to 4 °C, brought to pH 8.9 by the addition of 2 N NaOH and centrifuged (15 min 40 000 g). The loss in sialic acid is about 30–35%, determined according to Jourdian et al. (1971). The supernatant is used for immobilization as described below.

Preparation of General Affinity Adsorbent by Immobilization of Glycoproteins (Freier et al. 1985):

Preparation   *Example:* Activation of Sepharose 4B and immobilization of a glycoprotein according to Kohn and Wilchek (1982)

Materials   Sepharose 4B, acetone, ice-salt mixture, cyanogen bromide, triethanolamine, HCl, $NaHCO_3$, triethylamine, hog gastric mucin, ovomucoid, NaH_2PO_4, NaOH, Tris

0.05 M triethanolamine/HCl buffer pH 8.9 containing 0.5 M NaCl, 0.25 M $NaHCO_3$ pH 8.9, 10% cyanogen bromide in 60% acetone (w/v), 0.05 M sodium phosphate buffer pH 8.5, 0.05 M Tris/HCl buffer pH 8.0,

Hood, motor with a propeller, Ultraturrax homogenizer, medium speed centrifuge, 1 mm quartz cuvets, spectrophotometer with facility to record derivative spectra

Activation   *Activation:* Under a well-drawing hood, 10 ml of commercial Sepharose 4B suspension is sucked dry in a Büchner funnel which leads to 7 g of moist gel. The gel is washed successively with water, 30% acetone, and 60% acetone, each solution chilled to 0 °C, and resuspended in 60% acetone at 0 °C. The suspension is chilled down to − 15 °C in a vessel with ice/salt mixture and stirred with a propeller mounted to a motor. Do not use a magnetic bar because it will disrupt the spherical agarose particles. Now a solution of 120 mg cyanogen bromide in 1.2 ml 60% acetone at − 15 °C is added and the stirring speed raised. Within 1 min, 180 mg triethylamine in 1.2 ml 60% acetone (− 15 °C) are added dropwise, the suspension filtered through a chilled Büchner funnel and washed with 60% acetone (− 15 °C), 30% acetone (− 10 °C), water (0 °C), and coupling buffer (0.05 triethanolamine, 0.5 M NaCl pH 8.9, or 0.25 M $NaHCO_3$ pH 8.9, 4 °C) as quickly as possible.

Washings should follow without allowing the gel to run dry. The activated gel has to be used immediately.

Coupling: One g hog gastric mucin (preferably desialylated as described above) is evenly distributed in the three- to five fold amount of 0.05 M sodium phosphate buffer pH 8.5 by the use of an Ultraturrax homogenizer and stirred for an additional 2 h at 4 °C. If after this time a thick paste has formed, more buffer may be added. The turbid solution is centrifuged (40 000 g, 15 min), the pH of the supernatant brought to 8.9, and the activated Sepharose 4B added. The suspension is stirred for 2 days at 4 °C, washed with a suitable buffer, e.g., 0.05 M Tris/HCl pH 8.0 which at the same time blocks all residual activated groups of the gel. By recording the derivative spectra of the gels in 1 mm cuvets, the concentration of the immobilized protein can be determined (Schurz and Rüdiger 1982). In essentially the same manner, ovomucoid may be coupled to activated Sepharose.

Coupling

If a matrix less expensive than Sepharose is desired and a slow column flow can be tolerated, gels prepared by divinyl sulfone cross-linking (see above) can be activated with cyanogen bromide and coupled to glycoproteins in essentially the same manner as described for Sepharose 4B.

Affinity Purification:

Example 1. Isolation of the lectin from *Pisum sativum* (pea)

Purification

Sephadex G-100 (all Sephadex types from G-50 to G-200 are suitable), Tris, HCl, CaCl$_2$, MgCl$_2$, MnCl$_2$, NaN$_3$, glucose

Materials

0.05 M Tris/HCl buffer pH 8.0 with 1 mM CaCl$_2$, 1 mM MgCl$_2$, 1 mM MnCl$_2$, 0.02% NaN$_3$; the same buffer with 0.25 M glucose

Column 2.8 × 15 cm, fraction collector

A column is filled with the affinity gel (Sephadex G-100) and equilibrated with the pH 8.0 buffer. To the column, the prepurified extract is applied at 60 ml/h. The column is washed with the buffer at the same speed until the A$_{280}$ has fallen to < 0.05. Then the buffer is exchanged for the glucose-containing one which desorbs the lectin from the column. The fractions containing the lectin are combined on the basis of A$_{280}$, concentrated by ultrafiltration if necessary and dialyzed against Tris buffer pH 8.0. The lectin is stable for several weeks in this buffer if stored at 4 °C. If it has to be stored for longer periods of time, the solution is first dialyzed against water, frozen, lyophilized, the powder packed into a tightly closed bottle and kept at − 20 °C.

Procedure

From 10 g of seeds, about 14 mg of lectin will be obtained.

Example 2. Isolation of the lectin from *Phaseolus lunatus limensis* (Lima beans)

Sepharose 4B containing immobilized desialylated hog gastric mucin (6 mg/ml gel), Tris, HCl, CaCl$_2$, MgCl$_2$, MnCl$_2$, NaN$_3$, N-acetylgalactosamine

Materials

0.05 M Tris/HCl buffer pH 8.0 with 1 mM CaCl$_2$, 1 mM MgCl$_2$, 1 mM MnCl$_2$, 0.02% NaN$_3$; the same buffer with 0.1 M N-acetylgalactosamine

Column 1.0×10 cm, fraction collector.

Procedure Ten g of lima beans is soaked in water at 4 °C, shelled, homogenized, extracted with buffer, the crude extract subjected to acid precipitation, the supernatant readjusted to pH 8.0, and dialyzed as described above for peas.

To the column filled with the affinity gel and equilibrated with the pH 8.0 buffer, the prepurified extract is applied and the column washed with the same buffer at 10 ml/h until the A_{280} has fallen to < 0.05. The eluate should be free from agglutinating activity against blood group A. Now the lectin is desorbed with the GalNAc-containing buffer. The lectin containing fractions are combined on the basis of A_{280}, concentrated by ultrafiltration if necessary and dialyzed against the Tris buffer.

The yield of lectin from 10 g of seeds is about 16 mg.

Purification protocols should be documented by a table as follows:

Purification step	Protein amount	Lectin amount	Yield (%)	Specific activity	Purif. factor
Crude extract (10 g seeds) pH 5	4400	6400	100	1.5	1
Supernatant pH 5	576	5120	80	9	6
Precipitate	3630	340	5	0.1	0.07
Affinity chromatgraphy	16	4980	78	311	207

Avoid incorporating data which are more or less accidental. The only necessary information are protein and lectin amounts, the latter being calculated from the titer and the volume. For the sake of simplicity, it will often suffice to calculate an approximate protein amount from the A_{280} and the volume. Of course, there are other UV-absorbing substances in many seeds but this does not change the balance too much. Moreover, since determination of lectin activity is far from being precise, it is useless to invest much effort in protein determination at this stage.

By the use of desialylated hog gastric mucin or ovomucoid immobilized to Sepharose (about 6 mg glycoprotein/ml gel), lectins from the following plants could be purified (Freier et al. 1985; Fleischmann et al. 1985; Fleishmann and Rüdiger 1986):

Lectins Leguminosae. *Arachis hypogaea, Bauhinia purpurea, Canavalia ensiformis, Caragana arborescens, Cytisus scoparius, Dolichos biflorus, Glycine max, Griffonia simplicifolia, Laburnum alpinum* and *vulgare, Lens culinaris, Lotus tetragonolobus, Phaseolus lunatus limensis, Phaseolus vulgaris, Pisum sativum, Robinia pseudoacacia, Sophora japonica, Ulex europaeus, Vicia cracca, Vicia faba, Wisteria floribunda* and *sinensis*
Celestraceae. *Euonymus europaeus*
Moraceae. *Maclura pomifera*
Phytolaccaceae. *Phytolacca americana* (roots)
Euphorbiaceae. *Ricinus communis*
Gramineae. *Triticum vulgare*

Capacities of the affinity gels range from 0.2 to 7 mg lectin/ml gel, depending on the lectin and gel types. For most lectins, immobilized desialylated mucin has higher capacities than immobilized ovomucoid. In some cases, however, the reverse is true: the *Bauhinia purpurea*, *Griffonia simplicifolia*, *Phaseolus vulgaris*, *Ricinus communis*, and *Robinia pseudoacacia* lectins are bound better by ovomucoid than by mucin.

Lectins of differing specificities which occur together in the same seeds can be separated on the same affinity column in one run by using different desorbent successively. This applies to the lectins from *Vicia cracca*, *Griffonia simplicifolia*, *Phaseolus vulgaris*, *Robinia pseudoacacia*, and *Ulex europaeus*.

For some lectins which bind easily to glucose or galactose homopolymers such as those from *Arachis hypogaea*, *Canavalia ensiformis*, *Lens culinaris*, *Pisum sativum*, *Vicia faba*, *Vicia cracca* (lectin II), and *Ricinus communis*, Sephadex, Sepharose, or divinylsulfone cross-linked dextran or galactomannans (see above) are better affinity adsorbents than the immobilized glycoproteins.

4.2 Characterization of Lectins

4.2.1 Determination of Lectin Activity

4.2.1.1 Agglutination

The most simple, though not most reliable, way to demonstrate the presence of lectins and to determine their amounts is hemagglutination. Usually, human red cells are used because outdated blood conserves are available from blood banks. Though human blood mostly can be regarded as a harmless liquid, it should be handled with utmost care as a potentially infectious material. Thus, it may never be pipeted by mouth, and direct contact with the skin, in particular if it contains small lesions, has to be absolutely avoided. In spite of elaborate controls in blood banks, an absolute safety can never be guaranteed. Otherwise, animal blood (rabbits, sheep, cattle etc.) can be used. The sensitivity of red cells from different species or even of different human blood groups towards hemagglutination by a particular lectin may differ considerably. Often, rabbit red cells are more sensitive than human cells, but this is not necessarily so in every case.

Whole blood conserves are quite stable. At $4\,^{\circ}C$, the original plastic bags can be kept for many weeks without loss of activity. In contrast, diluted red cell suspensions ready for use have to be prepared freshly every day because hemolysis occurs within a few hours even under isotonic conditions. They may, however be stabilized by treatment with glutardialdehyde as described by Turner and Liener (1975).

Red cells from old blood conserves tend to yield higher titers than fresh ones, presumably because hidden surface receptors become unmasked by proteolytic processes. Finally, however, when the blood turns into a highly viscous fluid and the red cells do not sediment in a regular manner, the sample should be discarded and replaced.

Some lectins do not react with native red cells unless they are enzymatically processed. Thus, the lectin from *Arachis hypogaea* seeds does not agglutinate native but neuraminidase-treated erythrocytes. Improvement of the agglutinating capacity after protease (trypsin or papain) treatment of the cells has been observed in many cases.

Hemagglutination can be performed in test tubes or on microscopic slides. It is, however, most convenient to use plastic microtiter plates which are available from many manufacturers.

Agglutination in microtiter plates is simple to perform but not very precise. If higher accuracies are desired it can be followed spectrophotometrically (Lis and Sharon 1972; Turner and Liener 1975), but this requires more material and expenditure.

Lectins agglutinate not only erythrocytes but also other cells. In relatively few cases, also yeast or bacterial cells have been used to determine lectin concentrations. Mostly, however, the aim of such studies was to analyze microbial strains for their surface carbohydrates rather than to detect lectins.

Example. Hemagglutination

Materials   Human or animal blood, Tris, $CaCl_2$, $MgCl_2$, $MnCl_2$, HCl, NaCl

0.05 M Tris/HCl buffer pH 8.0 containing 0.1 M NaCl, 1 mM $CaCl_2$, 1 mM $MgCl_2$, 1 mM $MnCl_2$ and 0.02% NaN_3, 2–4% erythrocyte suspension, prepared by washing on a centrifuge 0.4-0.8 ml whole blood several (at least three) times with about 5 ml of the Tris buffered saline until the supernatant remains colorless, and filling up to 10 ml with the buffer. This suspension has to be prepared freshly every day.

Low speed centrifuge, microtiter plates ($130 \times 85 \times 15$ mm) containing 8×12 wells with V-shaped bottoms, microliter pipets or multipets, dropping pipets.

Procedure   The lectin to be tested should be dissolved in an isotonic buffer at a pH not too far from 8, e.g., in the buffer used for diluting and washing the erythrocytes. 25 µl of the lectin solution is pipeted in triplicates into the wells of a microtiter plate and 25 µl of the red cell suspension added. As controls, the red cell suspension is also added to buffer alone. The result can be read after 30 min. If the plates cannot be evaluated at this time, they should be protected from evaporation by covering with a plastic sheet or tape. If agglutination has occurred, the red cells aggregate with the lectin and sediment in the form of a uniform layer which covers the whole bottom of the well including the slopes. On inspection from above, the whole cross-section of the well is uniformly colored in light red. Care has to be taken not to mistake hemolysis for agglutination. The difference between these phenomena can be easily seen if the wells are looked at from the side. In hemagglutination, the red layer covers only the bottom whereas the supernatant is colorless. If hemolysis has occurred, the whole well is filled with a red and transparent solution.

If agglutination has failed to take place, the cells sediment individually. Since they are smooth enough to move freely, they roll down the slopes of the bottom and concentrate at the apex of the well. Looked at from above, a red dot is observed in the center of the well. In case of weak agglutination, the dot is surrounded by some agglutinated cells which cover only a part of the bottom.

Hemagglutination not only provides a yes-or-no answer, but also an at least half-quantitative estimate of the lectin amount. To this end, from the original lectin solution a series of decreasing concentrations is prepared and the agglutination capacity tested. First, all wells of row except the first one are filled with 25 µl of buffer, then two 25 µl aliquots of the lectin solution are pipeted into the first and second wells. Starting from the second well, 25 µl of the lectin dilution is removed and mixed with the buffer of the third well, then from the third into the fourth well, etc. With multichannel pipets several dilution series can be prepared in parallel. Finally, each well is provided with 25 µl of the red cell suspension and inspected after 30 min or later. Usually, it is sufficient to drop rather than pipet the red cell suspension into the wells if a suitable dropping pipet with a calibrated mouth is used.

The reciprocal of the concentration in the last well of the row in which agglutination can be recognized is the titer of the original solution. It represents a relative concentration measure. For example, if the dilution schedule above has been followed, and the fourth well is the last one in which the cells agglutinate, the titer of the original solution is 8 because in this well it has been diluted to $1/2^3 = 1/8$.

Dilution in wells is not very precise, since inaccuracies in the transfer propagate in geometrical progression. Therefore it is recommendable not to dilute more than four steps in a plate. This applies in particular if solutions are rich in protein and therefore tend to foam. If higher dilutions are required, predilution should be performed outside the plate in wider steps (e.g., 1/10) and in test tubes. This enables one to use volumes which can be handled and mixed with more accuracy, e.g., in the milliliter range. The prediluted samples are then assayed separately in the plate as above.

Agglutination assays should be run at least in triplicates. Even then, only accuracies of at the most ± 1 dilution step can be expected.

4.2.1.2 Other Methods

Agglutination of red cells, although simple, is not very reliable. Therefore many attempts have been made to replace this method by others. Instead of cells, carbohydrate covered latex particles have been used (Pongor and Riedl 1983; Kaul et al. 1991) which are more stable than native red cells and respond only to carbohydrate-binding proteins but not tannins and other agglutinins. The latex material, however, is quite expensive, and derivatization with glycoconjugates has to be performed beforehand.

A sensitive assay has been developed for some galactose-binding lectins. In this, the decrease in turbidity is determined which occurs if bacterial cell walls are hydrolyzed by lysozyme. If an enzyme preparation is used to which galactose residues had been coupled chemically, the enzymatic breakdown of the cell walls can be inhibited by the lectin from *Ricinus communis* and by other galactose-binding lectins (Ghosh et al. 1979).

If anti-lectin antibodies are available, the amount of a lectin can be determined with high sensitivity by RIA (Felsted et al. 1982; Ramakrishnan et al. 1982) or ELISA (Kolberg et al. 1983; Díaz et al. 1984; Nilsson et al. 1988) methods. In order to exclude possible false-positive results with lectin-crossreactive materials, it is advantageous to perform a sandwich ELISA with two antibodies directed against different epitopes of the lectin. One of these could be a conventionally prepared polyclonal, the other a monoclonal antibody. Care has to be taken not to use antibodies that are directed against the carbohydrate moieties of the lectin because such structures may occur in many otherwise totally unrelated proteins even from other species (Lauriére et al. 1989).

4.2.2 Carbohydrate Specificity

4.2.2.1 Agglutination

Lectins are defined as carbohydrate binding proteins. Hemagglutination by lectins is brought about by interaction of the lectin with glycoconjugates of the red cell surface. If the lectin-binding carbohydrate is present freely dissolved and at a sufficiently high concentration, it will compete with the red cell glycoconjugates for the lectin and agglutination will not take place. Many other reagents such as, e.g., tannins, some lipids, and bivalent ions, also may cause red cells to aggregate. Therefore, in order to confirm the presence of a lectin, in particular in crude plant extracts, it is important to find out whether some mono- or oligosaccharide is able to inhibit the agglutination. This, however, may be difficult if the lectin in question does not recognize simple sugars but only more complex structures.

The carbohydrate binding specifity can be tested in the simplest way by inhibition of agglutination. To this end, a dilution series of the sugar to be tested is pipeted or prepared in the wells of a microtiter plate in the same way as described above for the dilution of the lectin. Each well then receives a lectin solution of a titer between 2 and 4, which is strong enough to give a distinct agglutination. After a short incubation time (15–30 min), the erythrocyte suspension is added as described before. Inhibiting concentrations of lectins by monosaccharides are mostly in the millimolar range.

The specificity can also be determined by direct agglutination if a broad panel of carbohydrates immobilized to latex particles is available (Pongor and Riedl 1983; Kaul et al. 1991).

4.2.2.2 Other Methods

For more precise measurements, inhibition of lectin-mediated polysaccharide or glycoprotein precipitation by carbohydrates can be used (e.g., So and Goldstein 1967, 1968; Kaku et al. 1991). Since the degree of precipitation is followed by turbidimetry or nephelometry, it allows narrower steps to be tested than agglutination does.

Binding constants have been determined by affinity electrophoresis (Hořejší et al. 1977, 1979), binding constants and stoichiometry of binding by equilibrium dialysis, usually by the use of a radioactive carbohydrate (Scatchard 1949; Wilson and Goulding 1986). Other procedures are outlined in this book.

References

Agrawal BBL, Goldstein IJ (1965) Specific binding of concanavalin A to cross-linked dextran gels. Biochem J 96, 23C

Agrawal BBL, Goldstein IJ (1967) Protein carbohydrate interaction. VI. Isolation of concanavalin A by specific adsorption on cross-linked dextran gels. Biochim Biophys Acta 147: 262–271

Baues RJ, Gray GR (1977) Lectin purification on affinity columns containing reductively aminated disaccharides. J Biol Chem 252: 57–60

Causse H, Lemoin A, Rougé P (1986) Quantification of soybean seed lectin in soybean tissues during the life-cycle of the plant, by an enzyme linked immunosorbent assay (ELISA). In: Bøg-Hansen TC, van Driessche E (eds) Lectins—biology, biochemistry, clinical biochemistry Vol 5, p667–675

Dean PDG, Johnson WS, Middle FA (1985) Affinity chromatography. A practical approach. IRL Press, Oxford

Díaz CL, Lems-van Kan P, van der Schaal IAM, Kijne JW (1984) Determination of pea (*Pisum sativum*) root lectin using an enzyme-linked immunoassay. Planta 161: 302–307

Felsted RL, Pokrywka G, Chen C, Egorin MJ, Bachur NR (1982) Radioimmunoassay and immunochemistry of *Phaseolus vulgaris* phytohemagglutinin: verification of isolectin subunits. Arch Biochem Biophys 215: 89–99

Fleischmann G, Rüdiger H (1986) Isolation, resolution and partial characterization of two *Robinia pseudo-acacia* seed lectins. Biol Chem Hoppe-Seyler 367: 27–32

Fleischmann G, Mauder I, Illert W, Rüdiger H (1985) A one-step procedure for isolation and resolution of the *Phaseolus vulgaris* isolectins by affinity chromatography. Biol Chem Hoppe-Seyler 366: 1029–1032

Franz H (1988) The ricin story. Adv Lectin Res Vol 1: 10–25

Freier T, Fleischmann G, Rüdiger H (1985) Affinity chromatography on immobilized hog gastric mucin and ovomucoid. A general method for isolation of lectins Biol Chem Hoppe-Seyler 366: 1023–1028

Ghosh M, Bachhawat BK, Surolia A (1979) A rapid and sensitive assay for detection of nanogram quantities of castor bean (*Ricinus communis*) lectins. Biochem J 183: 185–188

Goldstein IJ, Poretz RD (1986) Isolation, physicochemical characterization, and carbohydrate-binding specificity of lectins. In: Liener IE, Sharon N, Goldstein IJ (eds) The lectins, properties, functions, and applications in biology and medicine Academic Press, New York, p33–247

Gray GR (1974) The direct coupling of oligosaccharides to proteins and derivatized gels. Arch Biochem Biophys 163: 426–428

Hořejší V, Kocourek J (1973) Studies on phytohemagglutinins. XII. O-Glycosyl polyacrylamide gels for affinity chromatography of phytohemagglutinins. Biochim Biophys Acta 297: 346–351

Hořejší V, Kocourek J (1974) O-Glycosyl polyacrylamide gels: application to phytohemagglutinins. Methods in Enzymol 34: 361–367

Hořejší V, Tichá M, Kocourek J (1977) Studies on lectins. XXXI. Determination of dissociation constants of lectin-sugar complexes by means of affinity electrophoresis. Biochim Biophys Acta 499: 290–300

Hořejší V, Tichá M, Kocourek J (1979) Affinity electophoresis. TIBS 4, N6–N7

Jourdian GW, Dean L, Roseman S (1971) The sialic acids. XI. A periodate resorcinol method for the quantitative estimation of free sialic acids and their glycosides. J Biol Chem 246: 430–435

Kaku H, Goldstein IJ, Oscarson S (1991) Interactions of five D-mannose-specific lectins with a series of synthetic branched trisaccharides. Carbohydr Res 213: 109–116

Kaul R, Read J, Mattiasson B (1991) Screening for plant lectins by latex agglutination tests. Phytochemistry 30: 4005–4009

Kohn J, Wilchek M (1982) A new approach (cyano transfer) for cyanogen bromide activation of Sepharose at neutral pH, which yields activated resins, free of interfering nitrogen derivatives Biochem Biophys Res Commun 107: 878–884

Kolberg J, Michaelsen TE, Sletten K (1983) Properties of a lectin purified from the seeds of *Cicer arietinum*. Hoppe-Seyler's Z physiol Chem 364: 655–664

Lauriére M, Lauriére C, Chrispeels MJ, Johnson KD, Sturm A (1989) Characterization of a xylose-specific antiserum that reacts with the complex asparagine-linked glycans of extracellular and vacuolar glycoproteins. Plant Physiol 90: 1182–1188

Lineweaver H, Murray CW (1947) Identification of the trypsin inhibitor of egg white. J Biol Chem 171: 565–581

Lis H, Sharon N (1972) Soy bean (*Glycine max*) agglutinin. Methods in Enzymol 28: 360–368

Lis H, Sharon N (1981) Affinity chromatography for the purification of lectins. J Chromatogr 215: 361–372

Lönngren J, Goldstein IJ, Bywater R (1976) Cross-linked guaran: a versatile immunosorbent for D-galactopyranosyl binding lectins. FEBS Letters 68: 31–34

Loomis WD, Bataille J (1966) Plant phenolic compounds and the isolation of plant enzymes. Phytochemistry 5, 423–438

Luck K, Ehwald R, Ziska P, Koppitz H (1992) Affinity sorbent for galactose-specific lectins consisting of deproteinized plant cell wall ghost. Plant Sci 82: 29–35

Nilsson B, Larsson AC, Påhlsson K, Öhgren Y, Back M, Nilsson K (1988) An amplification technology for improving sensitivity when measuring components in biological samples. J Immunol Methods 108: 237–244

Pongor S, Riedl Z (1983) A latex agglutination test for lectin binding. Anal Biochem 129: 51–56

Porath J (1974) General methods and coupling procedures. Affinity chromatography. Methods in Enzymol 34: 13–30

Ramakrishnan R, Eagle MR, Houston LL (1982) RIA of ricin A- and B-chain applied to samples of ricin A-chain prepared by chromatofocussing and by DEAE Bio-Gel A chromatography. Biochim Biophys Acta 719: 341–348

Rüdiger H (1988) Preparation of plant lectins. In: Franz H (ed) Adv Lectin Res Vol 1, p26–72

Scatchard G (1949) The attraction of proteins for small molecules and ions Ann N Y Acad Sci 51: 660–672

Schlumbaum A, Mauch F, Vögeli U, Boller T (1986) Plant chitinases are potent inhibitors of fungal growth. Nature 324: 365–367

Schurz H, Rüdiger H (1982) A spectrophotometric determination of protein immobilized to affinity gels. Anal Biochem 123: 174–177

Scopes R (1982) Protein purification. Principles and practice. Springer, New York Heidelberg Berlin

So LL, Goldstein IJ (1967) Protein-carbohydrate interaction. IV. Application of the quantitative precipitin method to polysaccharide-concanavalin A interaction. J Biol Chem 242: 1617–1622

So LL, Goldstein IJ (1968) Protein-carbohydrate interaction. XIII. The interaction of concanavalin A with α-mannans from a variety of microorganisms. J Biol Chem 243: 2003–2007

Sternbach H (1991) Chromatographische Methoden in der Biochemie. Thieme, Stuttgart New York

Svennerholm L (1958) Quantitative estimation of sialic acids. III. An anion exchange resin method. Acta Chem Scand 12: 547–554

Turner RH, Liener IE (1975) The use of glutaraldehyde-treated erythrocytes for assaying the agglutinating activity. Anal Biochem 68: 651–653

van Driessche E, Beeckmans S, Dejaegere R, Kanarek L (1983) A critical study on the purification of potato lectin (*Solanum tuberosum* L.) In: Bøg-Hansen TC, Spengler GA (eds) Lectins–biology, biochemistry, clinical biochemistry Vol 3, p629–638

Vang O, Larsen KP, Bøg-Hansen TC (1986) A new quantitative and highly specific assay for lectin binding activity. In: Bøg-Hansen TC, van Driessche E (eds) Lectins—biology, biochemistry, clinical biochemistry Vol 5, p638–644

Whitmore FW (1992) A hemagglutinating substance in chitin. BioTechniques 12: 202–208

Williams BL, Wilson K (1984) Methoden der Biochemie. Thieme, Stuttgart New York

Wilson K, Goulding KH (1986) A biologist's guide to principles and techniques of practical biochemistry. Edward Arnold, London

Young NM, Leon MA (1978) Preparation of affinity chromatography media from soluble polysaccharides by cross linkage with divinyl sulfone. Carbohydr Res 66: 299–302

5 A General Procedure for the Purification of Fimbrial Lectins from *Escherichia coli*

E. Van Driessche, R. Sanchez, S. Beeckmans, F. De Cupere, G. Charlier, P. Pohl, P. Lintermans, and L. Kanarek

The presence of lectins on the surface of *E. coli* has been known since Duguid et al. (1955) described in the mid-1950's the hemagglutinating properties of some *E. coli* strains (reviewed by Duguid and Old 1980). Especially during the past years, much effort has been devoted to the purification and characterization of surface lectins from several bacterial species (Sharon 1987; Jann and Jann 1990; Bertels et al. 1991; Gilboa-Garber and Avichezer 1993). Without doubt, the *E. coli* fimbrial lectins are among the most thoroughly investigated. This is not surprising, since although *E. coli* are advantageous commensals of the large intestine of mammals and birds, some strains are able to colonize other niches as well, and are the causative agents of both intestinal and extra-intestinal diseases in man and his domestic animals. It is now generally recognized that the attachment of bacteria to the host's mucosae is an initial step in pathogenesis and that attachment or adhesion is mediated by bacterial surface lectins which can be of either fimbrial or nonfimbrial nature. Consequently, inhibition of bacterial adhesion is the way of choice to prevent infection (Beachy 1981).

Until now, the immunological approach has proved to be the most successful and needs the availability of bulk amounts of purified fimbrial lectins.

The purification procedure described here allows the isolation of fimbriae from enterotoxigenic *E. coli* strains that cause diarrhea in calves as well as of fimbriae from an *E. coli* strain that is responsible for edema disease in weaned piglets. Preliminary results have shown that this procedure is also useful for the isolation of fimbriae from *Klebsiella* and *Serratia* species isolated from humans.

5.1 Materials and Methods

5.1.1 Storage and Origin of *E. coli* Strains

E. coli strains 25KH09, 111KH86, B41MC, and B41 were from the collection of the Nationaal Instituut voor Diergeneeskundig Onderzoek (N.I.D.O) and previously isolated from the feces of calves suffering from diarrhea. *E. coli* strain K514 is a reference strain expressing Type-1 (F1) fimbriae, and is kept in collection at the N.I.D.O. The fimbriated mutant 107/86 of strain O139:K12:H1 associated with edema disease in weaned piglets was kindly provided by Professor Bertschinger (Switzerland). Serotypes of all strains are given in Table 1.

All strains used are kept at $-80\,^{\circ}\mathrm{C}$ in sterilized skimmed milk until use, except for the edema strain which is kept on Dorset egg medium at $4\,^{\circ}\mathrm{C}$; Dorset medium is commercially available from Difco.

Table 1. Percentage of ammonium sulfate used for the purification of different types of *E. coli* fimbriae

E. coli strain	Serotype	Type of fimbriae expressed	Ammonium sulfate (% saturation)	
K514	O?:K?:F1	Type-1 (F1)	20	
25KH09	O101:K$^+$:F17	F17	20	
111KH86	O101:K?:F(Att111)	F111	20	
B41MC	O101:K$^-$:F41	F41	20	
B41	O101:K$^-$:F41, K99	K99, F41	20	→(F41)
			20–40	→(K99)
107/86	O139:K12:H1	F107	20	

5.1.2 Growth of Bacteria

1. *E. coli* strains 25KH09, 111KH86, and K514, expressing respectively F17, F111, and Type-1 (F1) fimbriae, were grown overnight at 37°C in liquid LB medium until the culture reached the early stationary phase.
   LB medium is composed of 5 g/l yeast extract (Difco), 10 g/l bactotryptone (Difco) and 10 g/l NaCl (Merck). The pH of the medium is adjusted to 7.3 using 0.1 M NaOH.
2. *E. coli* B41MC and B41, expressing respectively F41- and F41 + K99-fimbriae, were grown overnight in a minimal medium with the following composition: 4.5 g/l KH_2PO_4, 5.7 g/l K_2HPO_4, 2 g/l NH_4Cl, 0.2 g/l $MgSO_4.7H_2O$, 5 mg/l $FeSO_4.7H_2O$, 5 mg/l citric acid, 11 mg/l $CaCl_2$, and 1.98 g/l glucose (all Merck reagents) adjusted to pH 7.3 with 0.1 M NaOH.
3. *E. coli* 107/86 expressing F107 fimbriae were grown overnight at 37°C in tryptic soy broth (Oxoid, U.K.).

Before being inoculated, the growth media are autoclaved for 20 min. When glucose is to be used as a component of the growth medium, this sugar should be dissolved in water and autoclaved separately; then the appropriate amount can be added to the other autoclaved components to make up the final medium.

5.2 Protocols

5.2.1 Electron Microscopy of *E. coli*: Negative Staining

Electron Microscopy Since the expression of fimbrial adhesins critically depends on the growth conditions such as composition of the medium, pH, and temperature, before starting the purification procedure outlined below, the bacteria should be examined by electron microscopy. Indeed, hemagglutination assays or adhesion tests will only provide information on the presence of surface adhesins which can be of fimbrial and/or nonfimbrial nature. Negative staining is an easy technique that allows the demonstration of fimbriated cells.

The procedure used by us is performed as follows:

1. *E. coli* are collected from the growth medium by centrifugation (10 000 g, 20 min, 4°C) and the pelleted bacteria are washed three times with phosphate buffered saline.

2. Bacteria are adsorbed onto copper grids coated with pioloform and strengthened with carbon. Adsorption is allowed to proceed for 10 min by placing the grid on a drop of the concentrated bacterial suspension.
3. The grids are washed twice on a drop of saline (a few seconds each).
4. Then the grids are transferred on a drop of uranylacetate (2%, w/v, in distilled water) for 5 to 9 s. The exact time needed for optimal negative staining should be determined experimentally. If the grids are left too long in contact with uranyl-acetate, a positive instead of a negative staining will be obtained with a concomitant loss of resolution on electron microscopical examination.

   After each incubation step, the excess liquid is removed by gently touching the grids with a sheet of filter paper. The specimens are now ready for examination in the transmission electron microscope. A final magnification of 100 000 usually allows clear observation of fimbriae.

The observation of the bacteria by electron microscopy also offers the opportunity to assess the presence of flagella which might interfere with the purification procedure described in this chapter. Removal of these organelles from partially purified fimbriae will be described below. Moreover, besides host-specific fimbriae, Type-1 (F1) fimbriae can easily be identified on negatively stained E. coli. Unlike host-specific fimbriae, which are very thin and flexible appendices protruding from the bacterial surface, Type-1 (F1) fimbriae are easily recognized because of their rigid and rather thick appearance.

5.2.2 Identification of Fimbriae

Electron microscopy provides information only on the fimbriation and/or flagell-ation of the bacteria, and consequently identification of the fimbriae present should be performed. This is often achieved by immunogold labeling techniques using specific antisera directed against known fimbriae. However, this technique is rather cumbersome and usually we prefer to use the highly sensitive agglutination assay. In this assay 25 µl of the bacterial suspension is mixed on a glass microscope slide with 25 µl specific antiserum. Agglutination of the bacteria in big clumps will occur nearly instantaneously and can be observed with the naked eye. When very small aggreg-ates are formed, a light microscope at low magnification should be used to observe the result.

5.2.3 Purification of Fimbrial Lectins

The purification scheme described below has proved to be useful for the isolation of several types of fimbriae of different strains of E. coli and Klebsiella and Serratia species, and is essentially based on precipitation of solubilized fimbriae by low concentrations of ammonium sulfate.

   The purification procedure can be divided in three different steps, i.e.:

– solubilization of fimbriae,
– ammonium sulfate precipitation, and
– storage of purified fimbriae.

Solubilization   Solubilization of fimbriae can be achieved in different ways, i.e.:

a) by incubation of *E. coli* at temperatures around 60 °C for 30 min in buffer (Morris et al. 1977);
b) by ultrasonication (De Graaf et al. 1980);
c) by incubation in chaotropic agents such as 3 M KSCN at 60 °C (Altmann et al. 1982);
d) by mixing the bacterial suspension for 2×40 s at 4 °C in a Waring blendor (this procedure) or other blending devices such as Ultra Turrax, Virtis homogenizer, etc. (Korhonen et al. 1980)

A critical comparison of these procedures for the purification of F17 fimbriae from *E. coli* revealed that procedure (d) is the method of choice, since it is a very mild one in which fimbriae are released by "shearing" off from the bacterial surface and eliminates the risks of protein denaturation by high temperatures or by dissociation when chaotropic solvents are used. Moreover, fimbriae released by blending are only slightly contaminated by other proteins, and consequently easy to purify.

Obviously sonication results in a complete desintegration of the bacteria yielding a complex mixture as starting material for the subsequent purification procedure.

Solubilization of fimbriae by blending is performed as follows [according to method (d) above]:

1. before any further manipulation, the bacterial cultures are cooled to 4 °C and stored at this temperature for at least 30 min; this treatment renders the bacteria resistant to lysis in the blending conditions used;
2. bacteria are collected from the growth medium by centrifugation (10 000 g, 20 min, 4 °C) and washed twice with 50 mM phosphate buffered saline;
3. the pelleted bacteria are then resuspended in 50 mM ice-cold potassium phosphate buffer pH 7.3 to obtain a dispersion value of about 30 at 660 nm;
4. this suspension is mixed in a pre-cooled Waring blendor (maximal speed) for 2×40 s at 4 °C.

Ammonium Sulfate Precipitation

Precipitation   1. After solubilization of the fimbriae by blending bacteria are removed by centrifugation (10 000 g, 20 min, 4 °C) and the supernatant is filtered through a Durapore membrane of 0.22 μ pore size;
2. to the clear filtrate, solid ammonium sulfate is added to obtain 20% saturation at 4 °C;
3. precipitation is allowed to proceed overnight at 4 °C;
4. precipitated proteins are collected by centrifugation (48 000 g, 20 min, 4 °C) and the pellet is washed three times with 20% saturated ammonium sulfate in 50 mM potassium phosphate buffer pH 7.3.

The procedure described above is of general use when only one type of fimbriae is expressed. From strains that express two types of fimbriae such as, for example, *E. coli* strain B41, which produces both K99 and F41 fimbriae, purification to homogeneity can still be achieved by differential ammonium sulfate precipitation. F41 fimbriae can first be precipitated at 20% saturation of ammonium sulfate and removed by centrifugation. Afterwards, K99 fimbriae can be precipitated by adding the appropriate amount of ammonium sulfate to the supernatant to obtain 40% saturation (see Table 1).

Fig. 1. General procedure for the purification of *E. coli* fimbriae

The general procedure described above for the purification of bacterial fimbriae is summarized in Fig. 1. It should be noted that the amount of ammonium sulfate at which the particular type of fimbriae precipitates should be determined experimentally for each type.

Storage of Purified Fimbriae

Purified fimbriae can be stored at 4 °C either as an ammonium sulfate precipitate, or in 50 mM potassium phosphate buffer or phosphate buffered saline at pH 7.3. Sodium azide (0.2%, w/v) should be added in order to prevent all bacterial and/or fungal growth.

Storage

5.2.4 SDS Electrophoresis of Purified Fimbriae

Electrophoresis   The homogeneity of the purified fimbriae preparation can easily be checked by SDS-electrophoresis in 10 or 12% polyacrylamide gels according to Laemmli (1970). Except for Type-1 (F1) fimbriae, boiling of the fimbriae for 3 min in SDS sample solution will result in the complete dissociation into subunits. Although most fimbriae consist of "major" and "minor" subunits (De Graaf et al. 1990), only the major subunits will be revealed by SDS electrophoresis, even after the gels are stained with the sensitive silver-staining procedure. Indeed, the "minor" subunits or adhesins are present in too low quantities to allow their detection.

Since, besides host specific fimbriae, many *E. coli* strains also express "common" or Type-1 (F1) fimbriae which also precipitate at low (20% saturation) concentrations of ammonium sulphate, it is important to test the purified fimbriae for Type-1 (F1) fimbriae contamination.

This can be done in two ways, i.e.:

a) by specific anti-Type-1 (F1) antibodies in an agglutination assay with the bacteria;
b) by SDS-electrophoresis of the purified fimbriae. When procedure (b) is used, which is advisable even if procedure (a) gives a negative result, it should be kept in mind that the major subunits of Type-1 (F1) fimbriae are not resolved upon SDS electrophoresis unless the proteins are treated in a special way before being loaded on the gels, i.e., they should be dissociated in guanidine-HCl (Eshdat et al. 1981) before being boiled in SDS sample solution as follows:

1. Precipitated fimbriae are dialyzed against bidistilled water and lyophilized.
2. The lyophilized powder is dissolved in saturated (ca. 8.6 M) guanidine-HCl and incubated at 37 °C for 2 h.
3. The dissociated proteins are extensively dialyzed overnight against 5 mM Tris-HCl buffer pH 8.0 containing 0.1 M EDTA. Dialysis should be performed in Spectrapore membranes (molecular weight cut off 3500).
4. After dialysis, an equal volume of SDS sample buffer is added to the specimen, which is then boiled for 3 min and loaded onto the SDS-polyacrylamide gel.
5. The major subunit of Type-1 (F1) fimbriae can be visualized after staining the gel with Coomassie brilliant blue.

    For some reason not understood until now, sometimes reaggregation of Type-1 (F1) fimbriae occurs after dissociation with guanidine-HCl. Aggregates formed are so large that they fail to enter the polyacrylamide gel, and generally most of them are hardly detectable upon staining the gel. To overcome this problem, proteins separated by SDS electrophoresis are electroblotted onto nitrocellulose membranes and aggregated Type-1 (F1) major subunits are visualized with rabbit anti-Type-1 (F1) antibodies and Protein-A peroxidase as follows.

6. The protein sample is treated with guanidine-HCl as described above, submitted to SDS-electrophoresis according to Laemmli, and electroblotted on nitrocellulose paper (100 V, 75 min; Mini-Protean-II apparatus, Biorad).
7. The blot is incubated for 1 h in 3% (w/v) bovine serum albumin in phosphate buffered saline pH 7.5 in order to saturate free protein-binding sites on the paper.
8. The paper is washed twice (10 min each) in phosphate buffered saline containing 0.05% (v/v) Tween-20 (further called Tween-PBS).

9. The nitrocellulose membrane is then incubated overnight with rabbit anti-Type-1 (F1) fimbriae serum properly diluted (at least 100 times) in Tween-PBS containing 3% (w/v) bovine serum albumin.
10. After five washes with Tween-PBS, the membrane is incubated for 1 h in a Protein-A peroxidase (Sigma Chem. Comp.) solution containing 1 µg/ml of the conjugate in phosphate buffered saline.
11. The membrane is washed five times (10 min each) with Tween-PBS, and twice with phosphate buffered saline; the color is finally developed by incubating the membrane in the color-development solution (85 ml phosphate buffered saline, 60 µl cold 30% H_2O_2, 15 ml methanol and 50 mg 4-chloro-1-naphtol (Biorad)).
12. Once the bands are clearly visible, the enzymatic reaction is stopped by washing the membrane in bidistilled water.
13. The membrane is then dried between two sheets of filter paper and stored in the dark.

5.2.5 Conclusion and Some Remarks

By using the procedure described in this paper, it is possible to prepare large quantities of highly purified fimbrial lectins from *E. coli*. These fimbriae have been shown in our laboratories to be excellent antigens to be used in immunization schemes for the production of antibodies in rabbits, hens, or in other animals.

In SDS polyacrylamide gel electrophoresis the F17, F111, F41, F107, K99, and Type-1 (F1) fimbriae prepared as described here migrate as a single band, corresponding to the major subunit of the fimbriae. Moreover, N-terminal sequence analysis of these preparations revealed only one single sequence without any trace of contaminating polypeptides.

When compared to other purification procedures in which gel filtration or sucrose gradient centrifugation steps are used to achieve homogeneity, the fimbriae prepared as described here are of equal quality. Without any doubt, the attractivity of our procedure lies in its simplicity and the possibility for scaling-up, even at the industrial level, if necessary. It should be emphasized, however, that in cases where different types of fimbriae are expressed on a particular *E. coli* strain, the procedure may fail to yield pure fimbriae of either type present. However, when the different fimbriae types precipitate quite differently, such as, for instance, in the case of F41 and K99, differential ammonium sulfate precipitation can still be used. Alternatively, separation starting from precipitated material can be achieved by ion exchange or hydrophobic chromatography. Especially the latter technique might prove to be useful, since different types of fimbriae often differ in their hydrophobic properties.

When the *E. coli* strain under study also expresses flagella, these structures will be released as well by blending the bacteria and they will co-precipitate with the fimbriae. However, thanks to the difference in stability between fimbriae and flagella in denaturing agents such as urea, flagella can be dissociated while fimbriae remain intact (Korhonen et al. 1980). The fimbriae will still precipitate at low saturation of ammonium sulfate while the flagella subunits will remain in the supernatant.

As a result of the strong tendancy of solubilized fimbriae to aggregate, gel filtration can be a technique to be considered only if small amounts of protein have to be prepared. Indeed, it is common knowledge that scaling-up of this technique is cumbersome and should be avoided whenever possible.

Acknowledgements. The Belgian "Instituut tot Aanmoediging van het Wetenschappelijk Onderzoek in Nijverheid en Landbouw", and the "Nationaal Fonds voor Wetenschappelijk Onderzoek" are kindly acknowledged for their financial support. S.B. is Research Associate of the latter foundation.

References

Altmann K, Pyliotis NA, Mukkur TK (1982) A new method for the extraction and purification of K 99 pili from enterotoxigenic *Escherichia coli* and their characterization. Biochem J 201: 505–513

Beachy EH (1981) Bacterial adherence: adhesin-receptor interactions mediating the attachment of bacteria to mucosal surfaces. J Infect Diseases 143: 325–345

Bertels A, De Greve H, Lintermans P (1991) Function and genetics of fimbrial and nonfimbrial lectins from *Escherichia coli*. In: Kilpatrick DC, Van Driessche E, Bøg-Hanseń TC (eds) Lectin reviews, vol 1. Sigma Chem. Comp., St. Louis MO, USA, pp. 53–67

De Graaf FK, Klemm P, Gaastra W (1980) Purification, characterization, and partial covalent structure of *Escherichia coli* adhesive antigen K99. Infec Immun 33: 877–883

Duguid JP, Old DC (1980) Adhesive properties of Enterobacteriaceae. In: Beachy EH (ed) Bacterial adherence, Chapman and Hall. London, Methuen, New York, pp. 184–217

Duguid JP, Smith IW, Dempster G, Edmunds TN (1955) Non flagellar filamentous appendages ("fimbriae") and hemagglutinating activity in *Bacterium coli*. J Pathol Bacteriol 70: 335–348

Eshdat Y, Silverblatt FJ, Sharon N (1981) Dissociation and reassembly of *Escherichia coli* Type-1 pili. J Bacteriol 148: 308–314

Gilboa-Garber N, Avichezer D (1993) Effects of *Pseudomonas aeruginosa* PA-I and PA-II lectins on tumor cells. In: Gabius HJ, Gabius S (eds) Lectins and Glycobiology, Springer-Verlag, pp. 380–395

Jann K, Jann B (eds) (1990) Current topics in microbiology and immunology, vol 151. Springer Verlag, Berlin-Heidelberg.

Korhonen T, Nurmiaho EL, Ranta H, Svanborg Edén C (1980) New method for isolation of immunologically pure pili from *Escherichia coli*. Infec Immun 27: 569–575

Laemmli UK (1970) Cleavage of structural proteins during the assembly of the head of bacteriophage T4. Nature 227: 680–685

Morris JA, Stevens AE, Sojka WJ (1977) Preliminary characterization of cell-free K99 antigen isolation from *Escherichia coli* B41. J Gen Microbiol 99: 353–357

Sharon N (1987) Bacterial lectins, cell-cell recognition and infectious disease. FEBS Letters 217: 145–157

6 Isolation, Detection, and Ligand Binding Specificity of a Lectin Uniquely Found in Rat Kupffer Cells

Jeffery D. Greene and Robert L. Hill

The rat Kupffer cell receptor was originally isolated as a result of studies examining the clearance to liver of fucose-containing glycoproteins from the blood of rats and mice (Prieels et al. 1978; Furbish et al. 1980; Lehrman et al. 1986b). In order to isolate the receptor or receptors involved in this clearance process, Triton X-100 extracts of rat liver were chromatographed on an affinity adsorbent of the neoglycoprotein L-fucose-β-BSA conjugated to agarose. Three proteins were specifically retained on this column: one protein of $M_r = 32\,000$ corresponding to the mannose/N-acetylglucosamine lectin (Mizuno et al. 1981) thought to be involved in antibody-independent defensive reactions against pathogens (Ikeda et al. 1987; Ezekowitz et al. 1988); a second protein of $M_r = 162\,000 - 180\,000$ corresponding to the mannose receptor (Haltiwanger and Hill 1986) involved in the receptor-mediated endocytosis of mannose-containing glycoconjugates (Haltiwanger et al. 1986; Taylor et al. 1990) and phagocytosis of yeast cells (Ezekowitz et al. 1990); and the third protein, made up of two polypeptides of $M_r = 88\,000$ and $77\,000$ originally named the fucose-binding protein to distinguish it from other hepatic lectins (Lehrman and Hill 1986). The lower molecular weight species is now known to be a proteolytic degradation product of the higher molecular weight species (Hoyle and Hill 1988) and this protein has been renamed the Kupffer cell receptor since immunofluorescence staining of liver thin sections and immunoblots of other cell and tissue types indicated that it is expressed uniquely in Kupffer cells, the resident liver macrophages (Haltiwanger et al. 1986).

The ligand binding properties of detergent-solubilized Kupffer cell receptor were examined in a binding assay with $[^{125}\text{I}]$ L-fucose-β-BSA as ligand. The receptor requires calcium for binding and has a pH optimum between 7.6 and 8.6. The best neoglycoprotein ligands for the receptor are β-fucosides and galactosides and the inhibition of binding by monosaccharides is in the following order: N-acetyl-D-galactosamine = D-fucose > L-arabinose > D-galactose > D-mannose > L-fucose (Lehrman et al. 1986a). Both the cDNA (Hoyle and Hill 1988) and the gene sequences (Hoyle and Hill 1991) of the receptor have been determined and the derived amino acid sequence of the receptor shares homology with the Type II integral membrane protein subgroup of the C-type lectin family (Hoyle and Hill 1988; Bezouska et al. 1991).

The ligand binding specificity and functional properties of the receptor in cells were examined by transfection of COS-1 cells with the cDNA encoding the receptor. Receptor-mediated adhesion of these cells to glycolipids immobilized in microtiter wells was strongest with Gb_4Cer (GalNAcβ1-3Galα1-4 Galβ1-4Glcβ1-Cer), Gb_5Cer (GalNAcα1-3GalNAcβ1-3Galα1-4 Galβ1-4Glcβ1-Cer), and LacCer (Galβ1-4Glcβ1-Cer), whereas gangliosides were not good ligands (Tiemeyer et al. 1992). Receptor expressed in these cells failed to mediate endocytosis and degradation of ligand (J. Greene and R. Hill unpubl.) and is thus thought to have a different function than the other hepatic C-type lectins. Other possible functions consistent with the properties of the receptor are cell adhesion or signal transduction but participation of the receptor in these processes has not been shown.

The methods described below detail the procedures for purification of the Kupffer cell receptor from rat liver by affinity chromatography and the assay for the binding activity of the detergent-solubilized receptor. A method for the detection of the receptor in rat liver thin sections by immunofluorescent staining is also included.

6.1 Equipment and Reagents

Equipment
4 °C cold room or cold box
Blender (Waring, 4 l capacity)
Medium speed centrifuge (Sorvall RC3 with HG-4L rotor or equivalent)
High speed centrifuge (Sorvall RC5 with SS34 and GSA rotors or equivalent)
Glass columns (Bio Rad Econo Columns, 1, 2.5 × 20 cm; 1, 1.5 × 20 cm; 1, 0.7 × 10 cm)
Fraction collector
Apparatus for polyacrylamide gel electrophoresis
Vacuum manifold (Amicon)
Polystyrene tubes, 75 × 12 mm (Sarstedt)
Whatman GF/C filter disks
Gamma counter
Cryostat
Microscope slides
Microscope with epifluorescence capability

Reagents
Rat livers, 300–400 g (Pel-Freeze)
Rat for sacrificing to obtain fresh liver
L-fucose-β-BSA, 50 mol L-fucose/mol BSA[a]
L-fucose-β-BSA-Agarose, 35 ml of 2.0 mg/ml and 4 ml of 10.0 mg/ml[a]
Peroxide-free Triton X-100, 20% stock solution[a]
Polyethylene glycol 6000
Aprotinin (Sigma)
EDTA
Tris
Freund's adjuvent (complete and incomplete)
Bovine serum albumin
Human fibrinogen (Kabi)
Enzymobeads (Bio Rad)
OCT tissue imbedding compound (Miles)
Protein A-Sepharose (Sigma)
Anti-rabbit IgG (whole molecule) FITC conjugate F(ab')$_2$ fragment of goat antibody (Sigma)
Glycine
Sodium acetate
Sodium azide
Sodium bicarbonate
Sodium borate

[a]Detailed procedures for the preparation of these reagents can be found in Lehrman and Hill (1983).

Sodium chloride
Calcium chloride
Sodium [^{125}I] (350–600 µCi/ml, Amersham)
Sodium phosphate
N-acetyl-D-glucosamine, D-galactose, other monosaccharides of interest

6.2 Purification

This section describes the procedure for the purification of the Kupffer cell receptor from rat liver.

Prepare the following buffers:

2 l homogenization buffer (HB):   50KIU/ml Aprotinin   **Buffers**
                                   0.25 mM EDTA
                                   0.02% sodium azide
                                   Adjust to pH 7.7 with sodium bicarbonate

2 l binding buffer (BB):   0.02 M Tris-Cl, pH 7.8
                                   0.2 M sodium chloride
                                   0.01 M calcium chloride
                                   0.5% Triton X-100
                                   0.02% sodium azide

1 l Elution buffer (EB):   0.02 M Tris-Cl, pH 7.8
                                   0.2 M sodium chloride
                                   0.001M EDTA
                                   0.5% Triton X-100
                                   0.02% sodium azide

4 l Elution buffer without Triton X-100 (EB no Triton)

1 l Regeneration buffer 1 (Regen 1): 0.1 M sodium acetate, pH 4.5
                                      0.5 M sodium chloride
                                      0.5% Triton X-100
                                      0.02% sodium azide

1 l Regeneration buffer 2 (Regen 2): 0.1 M sodium borate, pH 8.5
                                      0.5 M sodium chloride
                                      0.5% Triton X-100
                                      0.02% sodium azide

20 ml 1.0 M calcium chloride

50 ml 1.0 M Tris base

Figure 1 summarizes the steps in the purification. Unless otherwise stated, all procedures are performed at 4 °C and the pH of all buffer solutions is the pH of the solution at 4 °C.

Step 1. Crude Homogenate: Rat livers (300 to 400 g) are homogenized on high speed in 2 l of HB in a blender. Peroxide-free Triton X-100 (200 ml of 20% stock solution) is added and the homogenate is adjusted to pH 7.8 with 1 M Tris base. The mixture

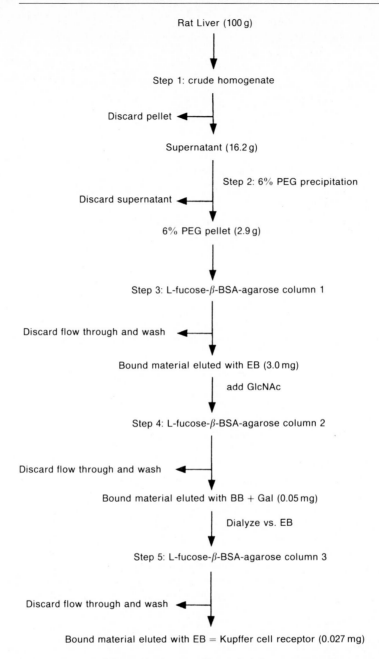

Rat Liver (100 g)

Step 1: crude homogenate

Discard pellet ◄

Supernatant (16.2 g)

Step 2: 6% PEG precipitation

Discard supernatant ◄

6% PEG pellet (2.9 g)

Step 3: L-fucose-β-BSA-agarose column 1

Discard flow through and wash ◄

Bound material eluted with EB (3.0 mg)

add GlcNAc

Step 4: L-fucose-β-BSA-agarose column 2

Discard flow through and wash ◄

Bound material eluted with BB + Gal (0.05 mg)

Dialyze vs. EB

Step 5: L-fucose-β-BSA-agarose column 3

Discard flow through and wash ◄

Bound material eluted with EB = Kupffer cell receptor (0.027 mg)

Fig. 1. Summary of the purification steps for the isolation of the Kupffer cell receptor from rat liver

is stirred for 1 h and then filtered through four layers of cheesecloth into 500 ml centrifuge bottles. After centrifuging at 7100 g for 20 min, the resulting supernatant is treated as described in step 2.

Step 2. 6% Polyethylene Glycol Precipitation: Solid PEG-6000 is added to the supernatant from step 1 to give a 6% solution (the following equation can be used to calculate the amount of PEG to add with a correction for the volume of the PEG: g PEG/volume of the supernatant $= 0.638$). The PEG is added very slowly with vigorous stirring over the course of 10 min. The mixture is stirred vigorously for an additional 15 min and the precipitation is allowed to continue without stirring for 30 min. After centrifuging at $7100\,g$ for 90 min the clear, red-brown supernatant is discarded and the tan-colored precipitate is resuspended by vigorous stirring for 30 min in 400 ml of BB containing an additional 0.02 M calcium chloride. The mixture is centrifuged in 250-ml centrifuge bottles at $16\,300\,g$ for 20 min and the supernatant is used in step 3.

Step 3. Affinity Chromatography on L-fucose-β-BSA-Agarose Column 1: The supernatant from step 2 is loaded on L-fucose-β-BSA-Agarose column 1 (35 ml, 2.0 mg L-fucose-β-BSA/ml) pre-equilibrated in BB and chromatographed at a flow rate of 1.5 ml/min. The top of the column can be stirred as necessary to unclog the gummy precipitate which collects over time. After the sample has been loaded, the column is washed with 350 ml of BB and the adsorbed lectins are eluted with 100 ml of EB at the same flow rate.

Step 4. Affinity Chromatography on L-fucose-β-BSA-Agarose Column 2: Solid N-acetyl-D-glucosamine (1.11 g) and 1.0 M calcium chloride (1 ml) are added to the 100 ml eluate from column 1 to give final concentrations of 0.05 M and 0.01 M respectively. This solution is applied to L-fucose-β-BSA-Agarose column 2 (3 ml, 10.0 mg L-fucose-β-BSA/ml) pre-equilibrated in BB containing 0.05 M N-acetyl-D-glucosamine and chromatographed at a flow rate of 0.2 ml/min. After the sample has been loaded, the column is washed with 15 ml of BB containing 0.05 M N-acetyl-D-glucosamine followed by 15 ml of BB and then eluted with 15 ml of BB containing 0.04 M D-galactose. Bovine serum albumin (1.5 mg) is added to the eluate and the mixture is dialyzed against 4 l of EB (no Triton) overnight.

Step 5. Affinity Chromatography on L-fucose-β-BSA-Agarose Column 3: The volume of the dialyzed material from step 4 is measured and the solution is adjusted to 0.01 M calcium chloride by the addition of the appropriate amount of 1.0 M calcium chloride. This solution is applied to L-fucose-β-BSA-Agarose column 3 (1.0 ml, 10.0 mg L-fucose-β-BSA/ml) pre-equilibrated in BB and chromatographed at a flow rate of 0.2 ml/min. After the sample has been loaded, the column is washed with 10 ml of BB and eluted with EB. Fifteen 0.5 ml fractions are collected and the Kupffer cell receptor is generally found in fractions 4–9, and can be detected by SDS-PAGE followed by silver staining or by using the assay described in Section 6.3. The purified receptor is unstable when stored at 4 °C but is stable for at least 2 months when stored at $-20\,°C$ in EB at a protein concentration of 0.02 mg/ml or greater. Losses on thawing and refreezing vary from 0 to 25%, and thus solutions of pure receptor are not normally frozen and thawed more than three times.

Additional Notes: The affinity columns can be used several times if cared for properly. Immediately after each use the columns are eluted with 10 volumes of Regen 1 followed by 10 volumes of Regen 2 and 10 volumes of EB. The columns can be stored in EB until ready for use. The yield of receptor is decreased significantly if partially pure fractions are left at 4 °C and thus it is best if the purification procedure is done in as short a time as possible.

6.3 Assay for Receptor Activity and the Determination of Ligand Binding Specificity

This section describes the assay used to measure receptor activity and the ligand binding specificity of the receptor.

Prepare the following solutions:

Solutions   50 ml 2 × Assay Stock Solution: 0.04 M Tris-Cl, pH 7.8
0.4 M sodium chloride
2 mg/ml BSA
1% Triton X-100
0.1% sodium azide

20 ml 1.0 M calcium chloride

50 ml 5 mg/ml human fibrinogen in 0.02 M Tris-Cl, pH 7.8
0.2 M sodium chloride
0.02% sodium azide

$[^{125}I]$ L-fucose-β-BSA (20–40 µ Ci/µg prepared using Enzymobeads as described by the manufacturer)

Final assay mix (prepared fresh before each series of assays):
Per sample to be assayed:
0.5 ml 2 × assay stock solution
0.02 ml 1.0 M calcium chloride
10 ng of $[^{125}I]$ L-fucose-β-BSA
distilled water such that volume of sample to be assayed + volume of water = 0.48 ml

100 ml disk soaking solution: 0.02 M Tris-Cl, pH 7.8
0.5 mg/ml BSA
0.01 M calcium chloride
0.2 M sodium chloride
0.02% sodium azide

1 l rinse solution: 0.01 M sodium glycinate, pH 10
0.2 M sodium chloride
0.01 M calcium chloride
0.02% sodium azide

Column fractions (usually 25 µl) or pure receptor are incubated with final assay mix in a total volume of 1.0 ml for 30 min at 25 °C in polystyrene tubes. Human fibrinogen (65 µl/tube) is added and the mixture incubated for an additional 10 min. The mixture is then filtered by vacuum filtration on Whatman GF/C filter discs that have been pre-soaked for 30 min in disk soaking solution. The assay tubes and filters are washed twice with 5 ml per wash of rinse solution and then counted in a gamma counter.

The binding specificity of the Kupffer cell receptor can be determined by measuring the inhibition of $[^{125}I]$ L-fucose-β-BSA binding by various monosaccharides or neoglycoproteins at different concentrations of inhibitor and then calculating the amount of inhibitor required to give 50% inhibition. Representative data obtained by this method are shown in Table 1.

Table 1. The inhibition of binding of [^{125}I] L-fucose-β-BSA to Kupffer cell receptor by different monosaccharides and glycosides

Inhibitor	Concentration for 50% Inhibition (mM)
N-acetyl-D-galactosamine	3.0
D-fucose	3.0
β-methyl-D-galactoside	3.5
L-arabinose	4.0
D-galactose	5.0
D-mannose	6.0
L-fucose	12
α-methyl-D-galactoside	12
D-talose	15
L-galactose	18
D-galactosamine	19
D-glucose	26
D-glucosamine	28
D-mannose-6-phosphate	54
N-acetyl-D-glucosamine	300

6.4 Detection of the Kupffer Cell Receptor in Rat Liver Thin Sections

This section describes the preparation of anti-Kupffer cell receptor antibodies and their use in immunofluorescent staining of rat liver thin sections.

Prepare the following buffers (Additional reagents required for affinity purification of IgG and the preparation of F(ab')$_2$ fragments are found in the references to these procedures.):

20 ml 0.1 M sodium phosphate, pH 7.5, 0.15 M sodium chloride   **Buffers**
50% sodium hydroxide
100 ml phosphate buffered saline (PBS): 0.01 M sodium phosphate, pH 7.4
                                        0.15 M sodium chloride
                                        0.02% sodium azide
10 ml phosphate buffered saline containing 1% BSA

Preparation of Anti-Kupffer Cell Receptor Antibody. Kupffer cell receptor purified as described in Section 6.2 can be used to immunize rabbits for the purpose of obtaining polyclonal antibodies to the receptor. The receptor is prepared for injection by precipitating the protein from EB in 80% ethanol at 0 °C for 30 min. The precipitate is collected by centrifugation at 16 000 g and redissolved in 1 ml of 0.1 M sodium phosphate, pH 7.5 containing 0.15 M sodium chloride. This protein solution is emulsified with an equal volume of Freund's complete adjuvant. An initial sub-cutaneous injection of 100 µg of receptor emulsified in complete adjuvant followed by boost injections of the same amount of protein in incomplete adjuvant at 8 and 14 days after the initial injection were sufficient to yield a high titer antiserum when blood was drawn at 21 days after the initial injection. The IgG fraction of the rabbit

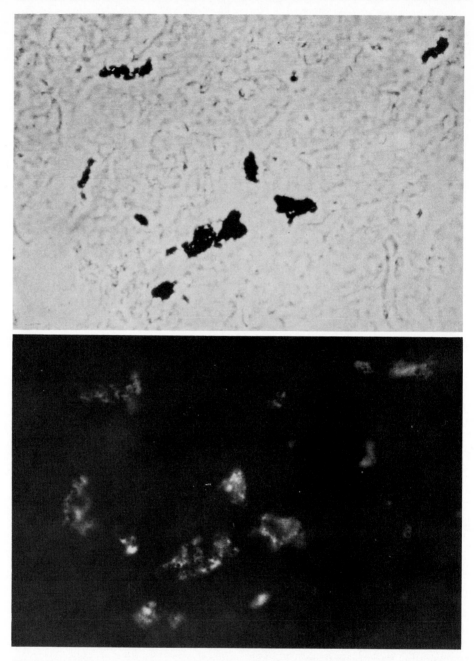

Fig. 2. Immunofluorescent staining of rat liver thin section from rat injected with colloidal iron particles. The sections were examined by fluorescence (*bottom*) or phase contrast (*top*) microscopy of the same field to reveal the labeled antibody and colloidal iron particles, respectively. (Haltiwanger et al. 1986)

anti-receptor antiserum is obtained by affinity chromatography on Protein A-Sepharose (Goding 1976). After elution from the Protein A column, IgG solutions are adjusted to pH 7.8 with 50% sodium hydroxide and 1.0 M Tris and stored at $-20\,°C$ until needed. Whole IgG is not suitable for thin section staining due to the presence of Fc receptors on Kupffer cells and thus $F(ab')_2$ fragments of anti-receptor IgG prepared by pepsin digestion (Weir 1978) are used in the staining procedure.

Immunofluorescent Staining of Rat Liver Thin Sections. Freshly isolated rat liver is immersed in OCT compound and quickly frozen on solid CO_2. Frozen sections (8 µm) are cut on a cryostat and fixed onto glass microscope slides with absolute ethanol at $-20\,°C$ for 20 min. The sections are then washed three times for 10 min per wash in PBS. Each section is then covered with one drop of PBS plus 1% BSA containing 50 µg/ml of anti-receptor $F(ab')_2$ and incubated at 25 °C for 2 h. The sections are washed with PBS as before and then incubated with one drop of PBS plus 1% BSA containing 50 µg/ml goat anti-rabbit (whole molecule) IgG FITC conjugate $F(ab')_2$ fragments at 25 °C for 2 h. The sections are then washed as before, mounted and viewed in a microscope with epifluorescence capability. Figure 2 shows the results that are obtained by this method. Note that the Kupffer cells in this figure have been labeled by injection of the rat with colloidal iron (Wincek et al. 1975) prior to harvesting the liver. It is not necessary to do this every time, but is a good check to show that the antibody only reacts with Kupffer cells.

References

Bezouska K, Crichlow GV, Rose JM, Taylor ME, Drickamer K (1991) Evolutionary conservation of intron position in a subfamily of genes encoding carbohydrate-recognition domains. J Biol Chem 266: 11604–11609

Ezekowitz RAB, Day LE, Herman GA (1988) A human mannose-binding protein is an acute-phase reactant that shares sequence homology with other vertebrate lectins. J Exp Med 167: 1034–1046

Ezekowitz RAB, Sastry K, Bailly P, Warner A (1990) Molecular characterization of the human macrophage mannose receptor: demonstration of multiple carbohydrate recognition domains and phagocytosis of yeasts in COS-1 cells. J Exp Med 172: 1785–1794

Furbish FS, Krett NL, Barranger JA, Brady RO (1980) Fucose plays a role in the clearance and uptake of glucocerebroside by rat liver cells. Biochem Biophys Res Commun 95: 1768–1774

Goding JW (1976) Conjugation of antibodies with fluorochrome: modifications to the standard methods. J Immunol Methods 13: 215–226

Haltiwanger RS, Hill RL (1986) The ligand binding specificity and tissue localization of a rat alveolar macrophage lectin. J Biol Chem 261: 15696–15702

Haltiwanger RS, Lehrman MA, Eckhardt AE, Hill RL (1986) The distribution and localization of the fucose-binding lectin in rat tissues and the identification of a high affinity form of the mannose/N-acetylglucosamine-binding lectin in rat liver. J Biol Chem 261: 7433–7439

Hoyle GW, Hill RL (1988) Molecular cloning and sequencing of a cDNA for a carbohydrate-binding receptor unique to rat Kupffer cells. J Biol Chem 263: 7487–7492

Hoyle GW, Hill RL (1991) Structure of the gene for a carbohydrate-binding protein unique to rat Kupffer cells. J Biol Chem 266: 1850–1857

Ikeda K, Sannoh N, Kawasaki N, Kawasaki T, Yamashina I (1987) Serum lectin with known structure activates complement through the classical pathway. J Biol Chem 262: 7451–7454

Lehrman MA, Hill RL (1983) Purification of rat liver fucose-binding protein. In: Fleischer S, Fleischer B (eds) Methods in enzymology, Vol 98 Academic Press, New York, pp 309–320

Lehrman MA, Hill RL (1986) The binding of fucose-containing glycoproteins by hepatic lectins. J Biol Chem 261: 7419–7425

Lehrman MA, Haltiwanger RS, Hill RL (1986a) The binding of fucose-containing glycoproteins by hepatic lectins. J Biol Chem 261: 7426–7432

Lehrman MA, Pizzo SV, Imber MJ, Hill RL (1986b) The binding of fucose containing glycoproteins by hepatic lectins. J Biol Chem 261: 7412–7418

Mizuno Y, Kozutsumi Y, Kawasaki T, Yamashina I (1981) Isolation and characterization of a mannan-binding protein from rat liver. J Biol Chem 256: 4247–4252

Prieels JP, Pizzo SV, Glasgow LR, Paulson JC, Hill RL (1978) Hepatic receptor that specifically binds oligosaccharides containing fucosyl α1-3 N-acetylglucosamine linkages. Proc Nat Acad Sci USA 75: 2215–2219

Taylor ME, Conary JT, Lennartz MR, Stahl PD, Drickamer K (1990) Primary structure of the mannose receptor contains multiple motifs resembling carbohydrate-recognition domains. J Biol Chem 265: 12156–12162

Tiemeyer M, Brandley BK, Ishihara M, Swiedler SJ, Greene JD, Hoyle GW, Hill RL (1992) The binding specificity of normal and variant rat Kupffer cell (lectin) receptors expressed in COS cells. J Biol Chem 267: 12252–12257

Weir DM (1978) Handbook of experiment biology, Vol. I, 3rd Ed., pp 6.19–6.21, Blackwell Scientific Publications, Oxford

Wincek TJ, Hupka AL, Sweat FW (1975) Stimulation of adenylate cyclase from isolated hepatocytes and Kupffer cells. J Biol Chem 250: 8863–8873.

7 Galaptin: Isolation, Electrophoretic Analysis, Peptide Preparation for Sequencing, and Carbohydrate-Binding Site Specificity

Howard J. Allen, Ashu Sharma, and Edward C. Kisailus

By 1980, the presence of galactoside-binding lectins in several different vertebrate animals and tissues had been described. The term galaptin was applied to a group of low molecular weight, cation-independent, galactoside-binding lectins that could be extracted from vertebrate tissues in the presence of lactose (Harrison and Chesterton 1980). Drickamer (1988) subsequently classified animal lectins into C-type (calcium-dependent, homologous carbohydrate recognition domains) and S-type. The S-type lectins are nearly equivalent to the galaptin group in that they are soluble, generally cation-independent proteins, and are found in a wide variety of animal tissues and cells. Studies of the physicochemistry of the S-type lectins have revealed a variety of proteins with N-acetyllactosamine-binding specificity, some being co-expressed in the same tissue or cells. This group of proteins has also been termed S-Lac lectins (Leffler et al. 1989).

The major S-type (S-Lac, galaptin) lectin frequently isolated from mammalian tissues is a 30 kDa galactoside-binding dimeric lectin comprised of identical subunits joined by noncovalent bonds (Allen et al. 1987). Occasionally, the monomeric form is also isolated. The carbohydrate-binding activity of this lectin is cation-independent in that it remains active in the presence of EDTA. This lectin requires the presence of thiol agents (mercaptoethanol, dithiothreitol) to retain carbohydrate-binding activity during and after isolation. This requirement may be removed, however, by mild alkylation of the lectin with iodoacetamide. For convenience, we define this specific lectin as galaptin. It is now known that galaptin has been well conserved through evolution; and homologous amino acid sequences may be present in higher molecular weight lactose-binding proteins (Lotan 1992).

We have used spleen as the major source organ for our studies on galaptin. This organ is more convenient to use than some others in that it lacks cartilage, large blood vessels, mucin and adipose tissues, all of which make tissue work-up more difficult when present. Other organs, tissues and cells may be used as a galaptin source, however.

Fresh spleen is frozen at $-20\,^\circ$C until utilized. Galaptin may be successfully isolated, in yields nearly comparable to fresh spleen, from specimens frozen for as long as 7 years.

7.1 Isolation of Galaptin

7.1.1 Preparation of Crude Lectin for Affinity Chromatography

7.1.1.1 By Extraction Into Lactose-Containing Buffer

Frozen spleen is placed on a cutting board and is cut into small fragments (upper limit ∼5 g) with a standard metal worker's hacksaw. The fragments are placed in the homogenizing vessel and ice cold extraction buffer is added (4 ml/g tissue). The sample is made 0.25 mM PMSF by addition from a stock 125 mM PMSF solution in isopropanol (i.e., 1 g tissue + 4 ml extraction buffer + 2 µl PMSF stock solution). The sample is immediately homogenized with the vessel immersed in ice. To avoid excess heat generation, cycle the homogenizer 30 s on followed by 60 s off.

For small sample workup, homogenization may be carried out with a Brinkmann Polytron homogenizer using a beaker or flask as the homogenizing vessel. For larger scale workup, an Omnimixer with a sealed (screw cap) stainless steel homogenizing vessel is satisfactory. If working in a cold room, the use of a kitchen style blender (Waring, Oster) also suffices. The homogenate is stirred in the cold for 2–24 h to allow displacement of galaptin from endogenous receptors and inhibitors.

The homogenate is centrifuged at 12 000–18 000 g, 30 min, 4 °C. The supernate is carefully decanted through glass wool loosely packed in the stem of a funnel. This helps to remove fine particles and aggregated lipid. The filtrate is then dialyzed against lectin buffer to remove lactose present in the extraction buffer. A precipitate may form during dialysis and this is removed by centrifugation as above.

The retentate-supernate is assayed qualitatively for the presence of lactose-inhibitable HA. For this assay, place one drop of sample + one drop of rabbit RBC on a microscope slide and mix. A strong HA reaction (granular clumping of RBC) is apparent to the naked eye in several seconds. Weaker reactions are confirmed with the aid of a microscope. Control reactions are buffer + RBC, and sample + 0.1 M lactose + RBC.

HA negative samples are usually due to insufficient dialysis to remove lactose. The HA positive sample is ready for affinity chromatography.

Buffers   Extraction buffer. 100 mM NaCl/50 mM PO_4/10 mM ME/2 mM EDTA/100 mM lactose/0.5% Triton X-100/0.05% NaN_3, pH 7.3.
Lectin buffer. 100 mM NaCl/50 mM PO_4/10 mM ME/2 mM EDTA/0.05% NaN_3, pH 7.3.

The route taken to arrive at the appropriate PO_4 concentration and pH is not critical. We use the tetrasodium salt of EDTA since this goes into solution very readily.

Notes   – PMSF is a toxic serine-dependent protease and acetylcholinesterase inhibitor and is to be handled with caution. The stock solution (125 mM PMSF in isopropanol) can be stored at − 20 °C between uses. The PMSF that crystalizes out will redissolve at RT.
  – Tissues and organs from humans are to be considered potentially infectious and universal precautions are to be taken when working with these materials. This includes the wearing of latex gloves, lab coat, and face mask. All surfaces, implements, and containers in actual or presumed contact are to be disinfected after working with human-derived materials. Bear in mind that the use of some homogenizers creates aerosols, as do some centrifugation procedures.
  – We use Triton X-100 in our extraction buffers to solubilize galaptin, in addition to lactose to displace galaptin from endogenous insoluble receptors. The use of Triton X-100 for extraction of galaptin is frequently not necessary. However, if one wants to extract other β-galactoside-binding lectins for which the solubility properties have not been well defined, the use of Triton X-100 is recommended. Under the conditions described here, considerable Triton X-100 remains in the dialysis bag after dialysis of the extract to remove lactose.

– For HA assay of galaptin, rabbit RBC are the cells of choice. For qualitative assay, the RBC may be fresh. Washing and storage of the RBC in Alsever's solution is satisfactory and the RBC are usable for 2–4 weeks when stored cold. For HA assay, a 1–2% RBC suspension is prepared. For semiquantitative assay and for convenience, trypsinized, and glutaraldeyde-fixed RBC are preferred. These are prepared as described by Nowak et al. (1976). Trypsinization of the RBC increases the sensitivity of the HA assay five-ten fold. Such cells may be stored in 100 mM NaCl/50 mM PO_4/1% BSA, pH 7.4, at 4 °C for up to 3 years. When HA data are to be compared, RBC of similar age should be used for the assays.

PMSF, α- or β-lactose, Triton X-100, grade 1 glutaraldehyde (25%), BSA (crystal-lized 1× and lyophilized), tetrasodium EDTA (99.5%)-Sigma.   **Materials**

7.1.1.2 By Preparation of Acetone Powder

Extraction of large weights of source material as described in Section 7.1.1.1 generates cumbersome volumes of sample for centrifugation, dialysis and affinity chromatography. To minimize these volumes, source material is extracted with acetone to remove water and lipid and to reduce the weight of material subsequently used for buffer extraction.

Reagent grade acetone is made 10 mM ME (0.7 ml ME/1 acetone). The acetone is chilled to near 0 °C by immersion in ice. Alternatively, the acetone may be placed in a cold room overnight if the cold room is properly constructed for housing flammable volatiles. Frozen spleen is cut into fragments (upper limit ~ 2.5 g) as described in Section 7.1.1.1. The fragments are homogenized in 10 volumes of ice-cold acetone/10 mM ME (1 g sample/10 ml acetone) using the OmniMixer as described in Section 7.1.1.1. The homogenate is rapidly filtered under vacuum through Whatman filter paper (No. 1, 2, or 5 is satisfactory) on a Buchner funnel of at least 15 cm diameter. The residue is filtered until nearly dry. This filtration is done in a fume hood at room temperature. About 300–500 ml of homogenate is filtered through a single filter paper. The filter paper with the acetone-insoluble residue is immediately placed over NaOH in a vacuum dessicator under continuous evacuation. Additional filter papers with residue may be stacked up like pancakes in the dessicator.

If the acetone powder is not going to be immediately extracted, it remains under vacuum in the dessicator overnight in the cold. The powder is then scraped into screw-top containers and stored at − 20 °C for later use. Such acetone powders are usable for at least up to 5 years from preparation.

The acetone-insoluble residues (about 1/3 the weight of starting frozen material) may be taken to immediate buffer extraction as described in Section 7.1.1.1 (1 g acetone powder/4 ml extraction buffer). When scraping the acetone powder from the filter paper, the powder will contain contaminating paper shreds. This is of no consequence or concern. The extract is then processed as given in Section 7.1.1.1 to give a lactose-free HA-positive extract for affinity chromatography.

ME-Sigma; acetone-VWR Scientific; Whatman filter paper-Fisher Scientific.   **Materials**

7.1.1.3 By Low Salt Buffer Extraction and DEAE Sephacel Chromatography

Acetone extraction of source material and dialysis of extracts is avoided by adsorbing crude galaptin to DEAE Sephacel to concentrate the galaptin and to wash away lactose present in the extract.

Frozen spleen cut into fragments, or acetone powder, is the starting material for homogenization in low salt extraction buffer. Homogenization and centrifugation are carried out as given in Section 7.1.1.1. Dialysis is not carried out. The supernate is diluted 1:1 with extract dilution buffer. The diluted supernate is then applied directly to DEAE Sephacel in the cold. The loading ratio is the equivalent of 1 g of wet spleen/1 ml of DEAE Sephacel. We typically load the equivalent of 100–125 g of wet spleen onto 125 ml of DEAE Sephacel packed into a 5×10 cm column. The flow rate for loading is 300 ml/h achieved with the aid of a peristaltic pump. The DEAE Sephacel, $C1^-$ form, is pre-equilibrated with crude galaptin DEAE wash buffer. The nonadsorbed effluent is collected in bulk, not in fractions.

An aliquot of this effluent may be desalted on a small Sephadex G-25 or BioGel P-2 column and assayed for lactose-inhibitable HA to confirm that galaptin adsorbed to the DEAE Sephacel. If HA-negative, this effluent is discarded.

After loading, the DEAE Sephacel is washed with five to ten bed volumes of crude galaptin DEAE wash buffer at 300 ml/h to remove lactose and some contaminating protein. The wash is collected in bulk and is discarded. Galaptin is eluted from the DEAE Sephacel with crude galaptin DEAE elution buffer at 50 ml/h. This eluate is collected in 10 ml fractions.

The fractions are assayed for HA activity as given in Section 7.1.1.1. The active fractions elute somewhat broadly and usually occur over the range of fraction 10–18. Higher flow rates with crude DEAE elution buffer increases the range over which galaptin elutes from the column. The HA-active fractions are pooled for affinity chromatography.

Buffer   Low salt extraction buffer: 10 mM Tris/10 mM ME/100 mM lactose/0.5% Triton X-100/0.01% thimerosal, pH 7.3

Extract dilution buffer: 10 mM Tris/10 mM ME/0.1% Triton X-100/0.01% thimerosal, pH 7.3

Crude galaptin DEAE wash buffer: 20 mM NaCl/10 mM Tris/10 mM ME/0.1% Triton X-100/0.01% thimerosal, pH 7.3

Crude galaptin DEAE elution buffer: 2.0 M NaCl/10 mM Tris/10 mM ME/0.1% Triton X-100/0.01% thimerosal, pH 7.3

The above buffers are designed to eliminate anions, such as EDTA and N_3^-, that may interfere with DEAE Sephacel chromatography.

Notes   Galaptin will adsorb to DEAE Sephacel under low salt conditions. Affinity purified galaptin can be eluted with 200 mM NaCl. However, crude galaptin, when extracted and loaded onto DEAE Sephacel as described in Section 7.1.1.3, is slowly eluted with 200 mM NaCl over a very broad elution volume. The galaptin can be eluted in a much smaller volume with 2.0 M NaCl.

Materials   DEAE Sephacel, Sephadex G-25-Pharmacia LKB; BioGel P2, 5×10 cm Econo-Columns-Bio-Rad. Other sized Econo-Columns are available. Flex-Columns from Kontes are also convenient.

7.1.2 Affinity Chromatography

Affinity chromatography of galaptin may be carried out on asialofetuin-Sepharose, lactose-Sepharose, or lactose-Sepharose CL. We prefer the lactose gels because they are easy to prepare, have a high binding capacity, and are stable. We have had no success with acid-treated Sepharose 6B for the isolation of galaptin.

Affinity chromatography is carried out in a cold room. The affinity gel is assumed to have a binding capacity of 1 mg galaptin/ml gel, although in our hands, this is an underestimate. The appropriate volume of gel is selected for the estimated galaptin content of the sample to be loaded. This estimate is derived from prior experience or from literature values. The gel may be placed in any of a variety of columns including Pasteur pipets. When practical, we maintain a loading flow rate of no greater than 30 ml/h to allow sufficient time for galaptin to interact with the gel ligand.

The conditions of affinity chromatography are designed to eliminate nonspecific adsorption and elution of proteins, to remove cation-dependent lectins, and to allow the lactose eluate containing galaptin and other minor β-galactoside-binding lectins to be directly applied to DEAE Sephacel.

The samples prepared as described in Sections 7.1.1.1 and 7.1.1.2 are ready for loading on the affinity gel. Samples prepared as described in Section 7.1.1.3 are made 2 mM EDTA by addition of dry Na_4EDTA and the pH adjusted to 7.3 if necessary. Samples may be loaded by gravity feed or with the aid of a peristaltic pump. The nonadsorbed fraction is collected in bulk. Periodically during loading, a drop of effluent from the affinity column is collected on a microscope slide and tested for the lack of lactose-inhibitable hemagglutinating activity. This assay is to monitor the efficiency of the column during loading. In our experience, other hemagglutinating activities from mammalian tissues are not present at this point. This nonadsorbed fraction is discarded.

After loading the sample, the affinity gel is washed with five to ten bed volumes of a phosphate/Triton X-100 wash buffer or with a Tris/Triton X-100 wash buffer. Either buffer is satisfactory. The affinity gel is then washed with five bed volumes of DEAE equilibration buffer. These washes are collected in bulk and discarded.

The galaptin plus some contaminating lectins are eluted with 0.1 M lactose in DEAE equilibration buffer. For affinity columns with a large dead volume above the gel, the lactose elution is easily carried out manually. We find it convenient to collect one fraction equivalent to $1.5 \times$ the affinity gel volume for small columns. For larger gel columns (> 10 ml), we discard the first amount of lactose eluate equivalent to 75% of the bed volume. We collect the next 75% of bed volume in one fraction. Individual fractions of any size may be collected, however. Protein in the lactose-eluted fraction is assayed with one of the commercially available Coomassie blue dye binding reagents using albumin as a standard. In this assay, the absorption coefficient for galaptin is $A_{1\,cm,\,595\,nm} = 700$ for a 1% solution (final chromophore volume = 1 ml).

Buffers

Phosphate/Triton X-100 wash buffer: 500 mM NaCl/50 mM PO_4/10 mM ME/2 mM EDTA/0.05% NaN_3/0.5% Triton X-100, pH 7.3.

Tris/Triton X-100 wash buffer: 500 mM NaCl/10 mM Tris/10 mM ME/2 mM EDTA/0.05% NaN_3/0.5% Triton X-100, pH 7.3.

DEAE equilibration buffer: 20 mM NaCl/10 mM Tris/10 mM ME/0.01% thimerosal, pH 7.3.

Notes

– Asialofetuin is prepared by hydrolysis of fetuin in 0.1 N H_2SO_4 for 70 min at 80 °C followed by dialysis and lyophilization. Asialofetuin is coupled to CNBr-activated Sepharose 2B or 4B using previously described procedures (Allen and Johnson 1980). In this procedure, CNBr is predissolved in acetonitrile, aliquoted into sealed serum vials, and stored at $-20\,°C$ until used. For use, a vial is warmed to RT to dissolve the CNBr and the desired amount is removed by syringe. The remainder is re-frozen. The activation of Sepharose is carried out in $2.5\,M\,K_3PO_4$, pH 12, (drained Sepharose mixed with an equal volume of $5.0\,M\,K_3PO_4$) and the monitoring and frequent adjustment of pH is not necessary. A Celstir flask is convenient for stirring the Sepharose suspensions. The pH adjustment of the $5.0\,M$ phosphate buffer is done by monitoring the pH of ten fold dilutions of the stock. When the coupling procedure is carried out at 10 mg asialofetuin/ml Sepharose, coupling efficiency is typically 95%. Fresh asialofetuin-Sepharose is stable when stored cold in PBS, losing protein from the gel at the rate of 1% /month. However, this affinity gel frequently loses galaptin-binding activity after two to five runs with crude tissue extracts.

– Lactose-Sepharose is prepared as previously described for L-fucose (Allen and Johnson 1977) via divinylsulfone (also termed vinyl sulfone) activation of Sepharose 6B, 4B, or CL-6B. The divinyl sulfone should be clear or light yellow. The stock reagent is aliquoted into sealed serum vials and stored as for CNBr/acetonitrile. The coupling procedure calls for 20% lactose. This concentration yields a lactose suspension that should be used. The fresh affinity gel is very stable when stored cold in PBS and can be used after at least 2 years. This gel can also be re-used repeatedly without loss of function.

– A variety of nonspecific adsorption and elution phenomena can occur during affinity chromatography. This can lead to considerable artifact when identifying trace polypeptides present in extracts as lectins. Such artifact is readily demonstrable when working with metabolically radiolabeled extracts where radioassay is used to identify polypeptdies. To minimize this problem, we utilize Triton X-100 and high NaCl concentration in our affinity column wash buffers.

Materials   Fetuin (lyophilized), CNBr-Sigma; Sepharose 2B, 4B, Cl-6B-Pharmacia LKB; Coomassie blue dye binding protein assay reagent-Bio-Rad or Pierce; acetonitrile-Pierce; vinyl sulfone-Aldrich; Wheaton Celstir flask-VWR Scientific.

7.1.3 DEAE Sephacel Chromatography of Lactose-Eluted Lectin

The lactose eluate from affinity columns typically contain galaptin as the major lectin. Other β-galactoside-binding lectins may account for 5–20% of the lactose-eluted protein. These minor lectins do not adsorb to DEAE Sephacel under low salt conditions (20–25 mM NaCl), whereas galaptin does. Affinity purified galaptin may be recovered from DEAE Sephacel by elution with 200 mM NaCl. This is a 10x less concentration of NaCl required for the elution of crude galaptin from DEAE Sephacel described in Section 7.1.1.3.

We assume the DEAE gel to have a binding capacity of 10 mg/ml but this is an underestimate. We carry out the chromatography manually using disposable polypropylene columns selected to have a large reservoir volume (3–10 ml) above the DEAE gel.

The appropriate amount of DEAE Sephacel (less than 0.5 ml is inconvenient to work with) is placed in the column and is washed with several volumes of DEAE equilibration buffer given in Section 7.1.2.1. We find it convenient to set up several columns and load the equivalent of 1 mg of galaptin/column. The lactose eluate containing the β-galactoside-binding lectins is applied to the gel by filling the reservoir above the gel and allowing it to drain through. The flow rate at this point is relatively unimportant. The flow-through containing unabsorbed protein is collected in one fraction and is assayed for protein to estimate the amount of galaptin on the column. The DEAE gel is washed with five to ten bed volumes of DEAE equilibration buffer. The DEAE gel containing galaptin may now be equilibrated with DEAE equilibration buffer/0.1 M lactose/50% glycerol (Sect. 7.1.2.1) and stored at − 20 °C until use. It is convenient to reach 50% glycerol by going through buffers of 10, 20, 30, 40 and 50% glycerol; otherwise the flow rate is extremely slow and the gel tends to float to the top of the buffer. Galaptin is stable under these conditions for several months.

To recover galaptin from the DEAE Sephacel, wash out the glycerol and lactose with DEAE equilibration buffer (Sect. 7.1.2.1). The initial flow rate will be very slow. Galaptin may then be eluted in one fraction with purified galaptin DEAE elution buffer. With 0.5 ml of DEAE gel, a 1-ml elution fraction is convenient. The hemagglutinating activity is semiquantitated with a two-fold serial dilution assay using trypsinized, glutaraldehyde-fixed erythrocytes. The concentration of galaptin giving detectable hemagglutination is usually in the range of 0.1–1.0 µg/ml.

Buffer Purified galaptin DEAE elution buffer: 200 mM NaCl/10 mM Tris/10 mM ME/0.05% NaN$_3$, pH 7.3.

Materials Disposable polypropylene columns (Bio-Spin, Poly-Prep, Econo-pac)-Bio-Rad; glycerol (certified)-Fisher Scientific.

7.1.4 Mild Alkylation of Galaptin to Remove the Thiol Requirement for Retention of Carbohydrate Binding Activity.

Galaptin may be mildly alkylated with iodoacetamide to maintain carbohydrate binding activity in the absence of thiol agents. This alkylation may be done with galaptin adsorbed to DEAE Sephacel. If the galaptin is going to be reacted with radioactive iodoacetamide, the DEAE Sephacel should be pre-treated with unlabeled iodoacetamide.

The DEAE Sephacel containing adsorbed galaptin, in a foil-wrapped column to afford protection from light, is washed with three to five bed volumes of galaptin-DEAE alkylation buffer at room temperature. Then 1.2x the DEAE gel bed volume of freshly prepared 0.1 M iodoacetamide in galaptin-DEAE alkylation buffer is applied to the column and allowed to drain through the gel. The column is stopped and an additional 0.3x the bed volume of buffer is applied. After 30 min, the DEAE Sephacel is washed with galaptin-DEAE alkylation buffer to remove alkylation by-products. The DEAE Sephacel containing the alkylated galaptin may be equilibrated with galaptin-DEAE alkylation buffer/50% glycerol for storage at − 20 °C as given in Section 7.1.3. Galaptin is stable under these conditions for several months. The alkylated galaptin is eluted from the DEAE Sephacel as given in Section 7.1.3.

Buffer Galaptin-DEAE alkylation buffer: 20 mM NaCl/10 mM Tris/0.1 M lactose, pH 7.5.

Notes − Titration (semiquantitative) of the HA activity of galaptin is carried out as previously described (Allen et al. 1977). Galaptin is serially diluted twofold in microtiter plates (U bottom or V bottom) followed by the addition of an equal volume of RBC. The plate is sealed with Mylar film. For mixing, strike the plate sharply on the lab bench or place on a rotator for several minutes. Read the plates 2–24 h after preparation. For comparison of data, the time should be consistent for all experiments. Alternatively, HA assays may be carried out in small round bottom test tubes. A tight RBC button or pellet is a negative reaction. A positive reaction (which varies in degree) is noted by a dispersed, granular-appearing pellet. Very high galaptin concentrations may give a false negative reaction.

- In our experience, galaptin behaves similarly on DEAE Sephacel in the presence of 20 or 25 mM NaCl.
- Dithioerythritol or DTT (1 mM) may replace 10 mM ME as a thiol agent for galaptin. These compounds are less odorous than ME but are more expensive.
- Purified galaptin, in the absence of a protein carrier, has a pronounced tendency to adsorb to glass or plastic that can lead to a significant loss of lectin. This can be prevented by the addition of BSA to galaptin solutions.

Materials   Iodoacetamide, dithioerythritol, DTT-Sigma; polystyrene microtiter plates for HA assay, Mylar film-Dynatech Laboratories.

7.2 Electrophoretic Analysis

7.2.1 Introduction

SDS-PAGE is used to assess purity of isolated galaptin and to prepare samples for amino acid sequence analysis. Galaptin must be reduced and alkylated prior to analysis to ensure the absence of aggregates. Satisfactory sequence information can be obtained by using "electro-phoretic purity" reagents from several different sources (BioRad, BM Biochemicals, etc) and standard SDS-PAGE protocols. Addition of 1 mM thioglycollate in the upper buffer (cathode) chamber of the electrophoretic apparatus and pre-electrophoresis for 15 min before the application of the sample greatly helps in the yields of the free N-terminus signal. Thioglycollate helps in scavenging the free radicals remaining in the gel which otherwise could modify the reactive amino acids or the amino termini of the peptide during electrophoresis.

7.2.2 Reductive Alkylation

Dissolve galaptin at a concentration of 1 mg/ml in iodoacetamide alkylation buffer in a 1.5 ml microfuge tube. Make the solution 30 mM DTT by addition of 1.0 M DTT. Mix the contents of the tube by vortexing, flush with N_2 to displace O_2, and quickly close the cap to seal the tube. Incubate at 37 °C for 2–3 h. Following incubation, add $0.1 \times$ volume of freshly prepared 1 M iodoacetamide in iodoacetamide alkylation buffer to give a final concentration of 100 mM iodoacetamide. Alternatively, solid iodoacetamide may be added to the reduced sample. Vortex, flush with N_2, and incubate at 37 °C for 45 min in the dark.

Galaptin may also be alkylated with iodoacetic acid as follows. Galaptin is dissolved in iodoacetic acid alkylation buffer at 1–2 mg/ml. ME, 8 μl/mg galaptin, is added and the sample is incubated under N_2 at 37 °C for 4 h. Then 4 mg of iodoacetic acid/mg galaptin is added. The sample is incubated in the dark under N_2 at 37 °C for 30 min.

The reduced and alkylated samples may be equilibrated with 10 mM ammonium acetate by (a) dialysis, (b) buffer exchange on a BioGel P-2 column, or (c) with the aid of a Centricon 10 centrifugal filtration device. The samples are then lyophilyzed and stored in the dark, dessicated at − 20 °C until used. If to be used immediately, the samples are equilibrated with SDS-PAGE sample buffer (Sect. 7.2.3.1) instead of with ammonium acetate.

Iodoacetamide alkylation buffer: **Buffers**
   10 M deionized urea/50 mM Tris/1.0 mM EDTA, pH 8.2.
Iodoacetic acid alkylation buffer:
   10 M deionized urea/550 mM Tris/37 mM EDTA, pH 8.5.
Highest grade reagents are used for reductive alkylation.

It is essential that samples, both purified and crude galaptin, be rigorously re-   **Notes**
ductively alkylated prior to SDS-PAGE. Otherwise, multiple bands may be present upon staining for protein or in Western blots probed with anti-galaptin serum. Boiling of samples in sample buffer for at least 5 min immediately before loading the gels is also important. The 1.0 M DTT stock solution may be prepared in deionized water, flushed with N_2 and stored in aliquots at − 20 °C. Deionized urea is prepared by making a 6 M urea solution in deionized water and passing this through a mixed bed ion exchange column. The effluent is collected, flushed with N_2, and lyophilized. The deionized urea is stored dessicated or in an air-tight container at − 20 °C.

Iodoacetamide, Iodoacetic acid-Sigma; urea (certified)-Fisher Scientific; Centricon   **Materials**
centrifugal filtration devices – Amicon; mixed bed (deionizing) resin AG 11A8-Bio-Rad.

7.2.3 SDS-PAGE

1. Wash glass plates (14 × 10 cm) with detergent, rinse with water, and air dry.   **Protocol**
   Assemble plates by using a 1.5 mm spacer in a Hoeffer or similar gel aparatus according to manufacturer's guidlines.
2. Prepare 12% acrylamide resolving gel by mixing 24 ml of 30% stock with 15 ml of 4x resolving buffer stock and 21 ml of water. Add 400 µl of 10% APS and degas under vacuum. Add 37 µl or TEMED and mix gently. This 60 ml solution is good for two gels. Pour between the glass plates to the desired height, gently tap to remove air bubbles, and overlay with 1x resolving gel buffer using a fine Pasteur pipet. Save about 100 µl of the above resolving gel solution in a pipet tip to check for polymerization. A different concentration gel can be prepared accordingly by adjusting the volumes of 30% stock of acrylamide and water.
3. In about 1/2 h, when the polymerization is complete, evident from the 100 µl solution saved above and also from a clear interface between the gel and overlay solution, decant off the overlay solution by inverting and remove the traces of liquid by whisking with filter paper. Prepare stacking gel by mixing 2 ml of 30% acrylamide stock, 3 ml of 4x stacking buffer stock, 7 ml of water. Add 100 µl of 10% APS and degas. Add 20 µl of TEMED and mix gently. This 12 ml solution is sufficient for two gels. Pour between the glass plates and insert the comb, making sure that no air bubbles are trapped beneath. Save 100 µl as above to check for

polymerization. Allow at least 45 min for complete polymerization of this low concentration gel.

4. Remove the combs, rinse well with water to remove unpolymerized material and place the gel in the reservoir tank. Add 1x running buffer in the lower reservoir and 1x running buffer with 1 mM thioglycollate in the upper buffer tank (cathode). Remove any air bubbles from the bottom of the gel and circulate cold water at 10 °C. Prerun the gel at 100 V for 15 min.

5. In the meantime, boil samples in 1x sample buffer for 5 min. At the end of the pre-electrophoresis, load samples and electrophorese at 25 mA per gel until the dye reaches the bottom of the gel. Normally, up to 500 µg of protein per band can be loaded. Remove gel from the glass plate and nick a corner to mark the orientation. Stain gel as below or proceed to electroblotting.

6. Stain gel in gel staining solution for 1 h at room temperature and destain in the destaining solution until the background is clear. To enhance the destaining process rock the gel and change destaining solution often.

7.2.4 Electroblotting

A variety of buffer systems have been used to transfer proteins from the acrylamide gel to the PVDF membranes. The following protocol describes the CAPS buffer using tank type transfer apparatus. The CAPS buffer has distinctive advantage over other buffer systems for sequencing because it does not contain any glycine which otherwise causes "ghost" peaks in sequencing runs. It is also important to use two PVDF membranes (one of them acting as a backup) in blotting and each one of them should be stained to determine the behavior of the protein under study.

Protocol 1. Immediately after electrophoresis the gel is soaked in CAPS transfer buffer for 30 min at room temperature. Mark the orientation of the gel by clipping a corner with a razor.

2. Two sheets of PVDF membrane are cut to match the size of the gel and labeled with pencil. Wet the dry PVDF membrane first with 100% methanol for 10 s and then soak in transfer buffer for 5 min. Make sure that the membrane is completely wetted and it should not be allowed to dry. Cut six sheets of Whatman #3 paper to match the size of the gel and soak in transfer buffer for 5 min.

3. Assemble the electroblotting cassette in a large tray of transfer buffer at 4 °C in the following order:

   Porous foam pad
   Three filter paper sheets
   The gel
   Two PVDF membranes
   Three filter paper sheets
   Second porous foam pad

   It is important to remove any air bubbles trapped between gel and membranes; close the cassette and load into the electroblotting tank so that the PVDF membranes are closest to the anode (+). Fill the tank with transfer buffer and circulate cooling water through the tank. Alternatively, the blotting can be performed at 4 °C with the transfer buffer stirred with the magnetic stir bar.

4. Electroblotting can be carried out at 1.5 amps for 45 min. Make sure that heat produced as a result is controlled by circulating cold water.

5. After transfer, wash the blots with deionized water for 5 min; stain for 5 min with 0.1% Coomassie blue R250 in 50% methanol (v/v) and destain in 50% methanol/10% acetic acid until the bands are clearly visible. After washing with deionized water the blots can be dried under vacuum. The band of interest is cut with the help of a clean razor blade, dried in vacuo and stored at $-20\,^{\circ}$C until utilized.

7.3 Sequencing

Protein sequencing is carried out in a gas phase or liquid-pulse sequencer with an on-line analyzer for phenylthiohydantoin derivatized amino acids. The protein band blotted onto PVDF is placed on top of the trifluoroacetic acid activated glass fiber filter coated with Polybrene. For the exact sequencing procedures, consult the instrument manuals.

7.3.1 In Situ V8 Protease Digestion

Protein bands of interest visualized after staining can be digested in situ using *Staphylococcus aureus* V8 protease.

Protocol

1. Cut out the band of interest with a clean razor blade from a preparative gel after destaining. Several similar bands can be pooled together if the yields of protein per band are not good. Total protein of approximately 100–200 µg is enough.
2. Soak the acrylamide piece in distilled water for 15–20 min to reduce the methanol and acetic acid contents. Recover the acrylamide by aspirating off the liquid and transfer the gel piece to a fresh 1.5 ml Eppendorf tube.
3. Add an equal volume of V8 protease buffer containing 20% glycerol (v/v)/50 mM Tris, pH 6.8/0.2% bromophenol blue/0.1% SDS/V8 protease enzyme at 10 µg/ml (0.05 U/Ml). Grind the gel pieces using an Eppendorf tube micropestle; this procedure greatly improves protein digestion.
4. Load the above slurry in a large well (l0× 29 × 1.5 mm) of a 15% gel, pre-electrophoresed in the presence of 1 mM thioglycolate. Load standard marker proteins in a side lane. Circulate water at 37 °C through the electrophoresis chamber.
5. Allow the acrylamide/protein/enzyme mixture to incubate in the well for 15 min and then run the bromophenol blue to the stacking/resolving gel interface at low voltage (50–75 V). When it reaches the interface, stop the current and let it incubate for 30–45 min to allow enzyme digestion. The amount of digestion could be varied by varying the time of incubation at the interface. In this way, the relative amount of various peptides could be changed to obtain internal sequence data from different parts of the protein.
6. Resume electrophoresis, typically at 200 V, until the dye front reaches the bottom of one gel. Be aware that some small peptides may run faster than bromophenol blue.

Notes

In our experience, galaptin is very resistant to digestion by proteases (see also Abbott and Feizi 1991). The digestion achieved with V8 protease is limited, although sufficient to obtain peptides for sequencing. We could not generate peptides with trypsin digestion or with arg C digestion.

7.3.2 In Situ CNBr Clevage

Cyanogen bromide cleavage at internal methione residues can be used to obtain internal amino acid sequences of the N-terminal blocked proteins.

Protocol
1. For chemical cleavage in Eppendorf tubes, cut the blotted protein band into small pieces (2 × 4 mm) and suspend them into 100–200 µl of 70% formic acid (enough to cover the pieces). Prepare a CNBr solution (70 mg cyanogen bromide per ml of 70% formic acid), add an appropriate amount into the tube (1 ml/mg protein) and incubate in the dark for 16–18 h at 37 °C.
2. Pour the cleavage solution into a fresh tube. Wash the membrane pieces three times with 100–200 µl of distilled water each time and pool the washes with cleavage solution saved above. Dry the pooled fraction in a vacuum centrifuge.
3. Elute the CNBr released polypeptides from the membrane by adding 100–150 µl of elution solution (70% isopropanol/5% trifluoroacetic acid). Vortex and incubate at room temperature for 2 h with occasional agitation. Transfer the elution solution to the dried fraction saved in step 2 and repeat two more times. Dry the eluted polypeptides in a vacuum centrifuge.
4. Electrophorese the CNBr polypeptides on a 15–20% SDS–PAGE gel and blot onto a PVDF membrane as above. Cut out the band of interest after staining/destaining and subject it to sequencing.

7.3.3 In Situ Iodosobenzoate Cleavage

Cleavage at internal tryptophan residues can be performed with an IBA reagent.

Protocol
1. Place the blotted protein band on a Teflon support, protein side up, elevated about 1 cm above the Petri dish bottom.
2. Cover the band with IBA reagent (6 mg iodosobenzoate in 1 ml of 4 M guanidine hydrochloride prepared in 80% acetic acid) to wet the membrane and place excess reagent in the Petri dish. In this way the membrane is prevented from drying. Incubate the dish overnight at room temperature in the dark.
3. Dry the membrane in vacuo and wash it twice with deionized water. Store the dried membrane at − 20 °C until utilized for sequencing.
4. Alternatively, the released polypeptides can be eluted off the membrane and sequenced (follow steps 3 and 4 of CNBr cleavage).

a) SDS-PAGE

Buffers and Reagents
1. N,N,N′,N′-tetramethylethylenediamine (TEMED)
2. 10% ammonium persulfate (APS) : make fresh daily.
3. 30% acrylamide stock: 29.9%/0.8% (w/w) acrylamide/bis acrylamide, filter through Whatman #1 filter paper and store protected from light.
4. 4x Resolving buffer stock: 1.5 M Tris base, 0.4% sodium dodecyl sulfate (SDS), adjust pH to 8.8 with hydrochloric acid.

5. 1x Resolving buffer overlay: dilute 4x resolving gel buffer stock to 1x.
6. 4x Stacking buffer overlay: 0.5 M Tris-HCl, 0.4% (w/v) SDS, adjust pH to 6.7 with sodium hydroxide.
7. 10x Running buffer: 0.25 M Tris base, 1.92 M glycine, 1% SDS.
8. 2x Sample buffer: 0.25 M Tris base, 4% SDS, 20% 2-mercaptoethanol or 0.1 M dithiothreitol (DTT), 40% (v/v) glycerol, 0.05% (w/v) bromophenol blue.
9. SDS-PAGE stain: 30% methanol, 10% acetic acid, 0.1% Coomassie blue, dissolve and filter through Whatman #1 filter paper.
10. SDS-PAGE destain: 20% methanol, 10% acetic acid.

b) Electroblotting

1. Electroblotting buffer: 10 mM CAPS/10% methanol, pH 11.0
2. PVDF Coomassie stain: 50% methanol, 10% acetic acid and 0.1% (w/v) Coomassie brilliant blue-250, dissolve and filter through Whatman #1 filter paper.

TEMED, APS, acrylamide, bis acrylamide, SDS, CAPS, Coomassie brilliant blue R250-Bio-Rad; PVDF (Immobilon P) membranes-Millipore; V8 protease-Calbiochem; iodosobenzoic acid-Sigma. **Materials**

7.4 Galaptin Carbohydrate-Binding Site Specificity

For these studies, it is convenient to use galaptin that has been mildly alkylated with iodoacetamide to eliminate the thiol requirement for retention of carbohydrate-binding activity. For the ELISAs, each investigator must determine and/or confirm optimal conditions for assay and quantitative validity of the results. It is also important to confirm the linearity of plate reader responses as a function of increasing absorbance (concentration) of a test solution as well as to confirm the absorbance range (maximum valid absorbance) of the plate reader with a test solution. Although asialofetuin is a rather poor ligand for galaptin, it is suitable for the assays described here.

7.4.1 ELISA Using Anti-Galaptin Serum

In this method, compounds are evaluated for their ability to inhibit binding of galaptin to plastic-adsorbed asialofetuin. Galaptin bound to plastic-adsorbed asialofetuin is detected with rabbit polyclonal anti-galaptin serum via an anti-rabbit IgG-peroxidase complex. The use of single- and multi-channel repetitive pipetors simplifies the procedures.

Immulon I micro ELISA plates are coated with asialofetuin in 100 mM Na_2CO_3/0.02% NaN_3, pH 9.6, (2.0 µg/100 µl/well) at 37 °C for 3 h. The covered plates can then be stored at 4 °C for several days prior to further processing. After complete aspiration of the asialofetuin solution, the wells are fixed with 2% formaldehyde in PBS (150 µl/well) for 45 min at RT. The wells are washed thrice with buffer A and then blocked with 0.4% BSA in buffer A (200 µl/well) for 30 min at 37 °C. The blocking solution is aspirated or flicked out and the wells are washed thrice with buffer A. Galaptin, 18 µg in 60 µl of buffer A in duplicate is mixed with an equal volume of inhibitor in buffer A and the mix is preincubated for 1 h at 37 °C. **Protocol**

Aliquots, 100 µl, of the mix are then added to duplicate wells of the blocked ELISA plates containing adsorbed asialofetuin. The plates are then incubated for 1 h at 37 °C. After washing the wells three to five times with cold buffer A, the wells are fixed again with 2% formaldehyde to prevent loss of bound galaptin. Rabbit anti-galaptin serum (100 µl/well of a 1 : 5000 dilution in 5% BSA/buffer A) is added to the wells followed by incubation for 1 h at 37 °C. The wells are then washed three to five times with buffer A and incubated with 100 µl goat anti-rabbit IgG-peroxidase conjugate (2.0 µg/ml in 1% BSA/buffer A). Following incubation for 1 h at 37 °C, the wells are washed five times with buffer A. ABTS substrate reagent, 100 µl, is added and the plates are incubated for 15 min at 37 °C. The reaction is stopped by addition of either 50 µl of 20% SDS or 100 µl of 200 mM citric acid. The plates are read at 405 nm on a plate reader.

To assay concentration-dependent inhibition, inhibitors are two-fold serially diluted in microfuge tubes prior to mixing with an equal volume of galaptin solution.

Controls include duplicate wells containing no asialofetuin nor inhibitor (background binding = blank), and duplicate wells containing no inhibitor but an equivalent volume of buffer (100% galaptin binding). When first establishing the assay, other controls include wells lacking the anti-galaptin serum to monitor nonspecific binding of the IgG-peroxidase complex, and wells lacking galaptin to monitor nonspecific binding of anti-galaptin serum.

To compare the relative potency of various inhibitors, graphs are constructed showing percent inhibition on the ordinate (relative to 100% galaptin binding) versus inhibitor concentration (after being mixed with galaptin) on the abscissa. It is usually convenient to plot the abscissa on a log scale. The concentration of inhibitor giving 50% inhibition of binding (I_{50}) is interpolated from the graphs; the lower the I_{50} value, the higher the inhibitor potency.

For assays done on different days, a complete lactose inhibition curve is obtained each day and all I_{50} values for each day are normalized relative to the corresponding lactose I_{50} value; i.e., lactose I_{50} is set to 1.0. This compensates for day to day variations in the ELISA.

When analyzing multivalent ligands such as glycoconjugates as inhibitors, the concentration of the nonreducing termini of oligosaccharides should be used to compare relative inhibitory potencies.

Buffer and Reagents   Buffer A: 120 mM NaCl/10 mM PO_4/0.01% thimerosal/0.05% Tween 20, pH 7.3. ABTS substrate reagent: 1.0 mM 2,2'-azinobis (3-ethylbenzothiazoline-6-sulfonate), diammonium salt/50 mM Na citrate, pH 4.5. Just before use, make the ABTS 2 mM H_2O_2.

7.4.2 Microplate Assay Using Galaptin-Peroxidase Conjugates

The use of a galaptin-enzyme conjugate for microplate assay is convenient in that it eliminates some of the steps that are required in the antibody-linked reactions. In this method, alkylated galaptin is conjugated to horseradish peroxidase with glutaraldehyde. The I_{50} for lactose in this assay is similar to that in the preceding ELISA.

Protocol   Galaptin (1 mg) is mixed with 3 mg peroxidase in 1.0 ml of 200 mM NaCl/8 mM PO_4/100 mM lactose, pH 6.8. Glutaraldehyde (120 µl of a 1% solution) is added and the conjugation proceeds overnight (15–17 h) at 4 °C. The conjugate is diluted nine-fold with 10 mM Tris, pH 7.3, and incubated 4–6 h to block residual glutaraldehyde end groups. The conjugate mix is then applied to 1 ml DEAE Sephacel and washed with 8 mM PO_4, pH 7.3, to remove lactose. The conjugate is eluted with 200 mM NaCl/8 mM PO_4, pH 7.3, and assayed for HA. Conjugate containing active galaptin, and perhaps some free galaptin, is obtained by affinity chromatography on lactose-Sepharose. The purified conjugate is made 1% BSA/50% glycerol and is stored at – 20 °C. Aliquots are removed as needed for use. Prior to use, lactose is removed by microdialysis or by DEAE Sephacel chromatography, and HA titer and peroxidase activity is determined.

A modification of this conjugation procedure that gives somewhat higher yields is as follows. After the overnight incubation with glutaraldehyde, 1.0 M Tris, pH 7.3 (100 µl), is added and the sample is incubated for 4–6 h at 4 °C. BSA (10 mg) is added and the sample is dialyzed against 500 mM NaCl/10 mM Tris, pH 7.3. After dialysis, conjugate containing active galaptin, and perhaps some free galaptin, is obtained by affinity chromatography on lactose-Sepharose. The lactose eluate is made 0.1% BSA and concentrated by centrifugal filtration with a Centricon 30 device. The concentrate is made 50% glycerol and stored at – 20 °C.

Assay of the inhibition of galaptin binding by ligands is carried out with microplates coated with asialofetuin and blocked as given in Section 7.3.1. Galaptin-peroxidase conjugate (24 ng/120 µl) in Tween buffer is mixed with an equal volume (120 µl) of ligand of varying concentrations (usually a two-fold serial dilution series). After incubation for 60 min at 4 °C, 100 µl of conjugate-ligand mix is added to duplicate wells and the plates are incubated for 60 min at 4 °C. The plates are then washed 5 × with cold Tween buffer. Bound peroxidase is assayed with ABTS as described. I_{50} values for the various ligands are determined as in Section 7.3.1.

Buffer

Tween buffer: 120 mM NaCl/10 mM PO_4/0.05% Tween 20/0.01% thimerosal, pH 7.3.

Notes

– Ligands to be tested for inhibitory potency may be dissolved in water for use. We have found that the outer row of wells in microtiter plates tend to give variable results, therefore, we leave these unused. This variability may be due to manufacture of the plates and/or the characteristics of the plate reader.
– Cold buffers are used in plate washing. The washing may be done in several ways: (i) using an automatic washer; (ii) use of a hand-held aspirator; (iii) immersing the plate in a large container of buffer and flicking the buffer from the wells.
– Avoid the use of NaN_3 in buffers that may contact the peroxidase-containing reagents, since this is an inhibitor of the enzyme.
– Avoid the use of Triton X-100 in buffers for microplate assays since this tends to remove protein adsorbed to the microplates and introduces variability in the results. Gelatin (2%) may replace BSA as a blocking agent. The use of 2% formaldehyde stabilizes the bonding of asialofetuin to plastic and stabilizes the bonding of galaptin to asialofetuin, giving far more consistent results than otherwise obtained.
– The IgG-peroxidase solution with 1% BSA is stable for several weeks when stored cold.
– The use of citric acid to stop the peroxidase reaction is preferred over 20% SDS. Pipeting of the SDS tends to form bubbles in the wells that interfere with reading of the plates. The ABTS solution, without peroxide added, is stable for several weeks if stored cold and in the dark.

– Electroelution cups are very convenient for the dialysis of small samples (up to about 1 ml). The device is positioned on the lip of a heavy walled beaker with a piece of Tygon tubing between the device legs and wall of the beaker. This applies sufficient friction to hold the device in place. Several cups may be placed around the edge of a beaker. Dialysis buffer is placed in the beaker to rise just above the dialysis membrane on the bottom of the cup. The cups are covered with a glass or plastic plate to prevent evaporation of sample.

Materials   Immulon I micro ELISA plates-Dynatech Laboratories; electroelution cups-Isco; formaldehyde-Fisher Scientific; gelatin (EIA grade)-Bio-Rad; thimerosal, Tween 20, horseradish peroxidase-Sigma; ABTS-Aldrich; goat anti-rabbit IgG-peroxidase conjugate, pre-prepared two-component ABTS reagent-Kirkegaard & Perry.

References

Abbott WM, Feizi T (1991) Soluble 14-kDa β-galactoside-specific bovine lectin. J. Biol Chem 266: 5552–5557

Allen HJ, Johnson EAZ (1977) A simple procedure for the isolation of L-fucose-binding lectins from *Ulex europeus* and *Lotus tetragonolobus*. Carbohydr Res 58: 253–265

Allen HJ, Johnson EAZ (1980) Chromatographic behaviour of pea lectin receptors from the 6C3HED murine ascites tumor. Biochim Biophys Acta 600: 320–331

Allen HJ, Johnson EAZ, Matta KL (1977) A comparison of the binding site specificities of the lectins from *Ulex europeus* and *Lotus tetragonolobus*. Immunol Commun 6: 585–601

Allen HJ, Cywinski M, Palmberg R, DiCioccio RD (1987) Comparative analysis of galactoside-binding lectins isolated from mammalian species. Arch Biochem Biophys 256: 523–533.

Drickamer K (1988) Two distinct classes of carbohydrate-recognition domains in animal lectins. J Biol Chem 263: 9557–9560

Harrison FL, Chesterton CJ (1980) Factors mediating cell-cell recognition and adhesion. Galaptins, a recently discovered class of bridging molecules. FEBS Lett 122: 157–165

Leffler H, Masiarz FR, Barondes SH (1989) Soluble lactose-binding vertebrate lectins: A growing family. Biochemistry 28: 9222–9229

Lotan R, (1992) β-Galactoside-binding vertebrate lectins: synthesis, molecular biology, function. In: Allen, HJ, Kisailus, EC (eds) Glycoconjugates: composition, structure, and function. Marcel Dekker, Inc., New York, pp 635–671

Nowak TP, Haywood PL, Barondes SH (1976) Developmentally regulated lectin in embryonic chick muscle and a myogenic cell line. Biochem Biophys Res Comm 68: 650–657

8 Determination of Carbohydrate Specificity in Solid-Phase Assays

F.-Y. Zeng and H.-J. Gabius

The carbohydrate specificity of lectins is customarily determined by measuring the ability of different monosaccharides, oligosaccharides, or glycopeptides (or glycoproteins) to inhibit lectin-mediated hemagglutination or glycoconjugate precipitation. These methods require relatively large quantities of glycoconjugates which can be difficult to obtain in sufficient quantities for inhibition studies. Suitable ligands for agglutination may only be present on the surface of certain types of cells or after chemical conjugation of adequately tailored structures to cell surfaces. Moreover, monovalent carbohydrate-binding proteins cannot agglutinate cells or precipitate polysaccharides. Thus, solid-phase assays to analyze the sugar specificity of sugar-binding proteins present an attractive approach to overcome these obstacles. Binding of a lectin to an immobilized ligand which can then be separated from unbound lectin is one practical approach (see Allen et al. chapter 7 of this Vol.). Moreover, a lectin is immobilized onto a nitrocellulose membrane by spotting or an ELISA plate surface by adsorption, the immobilized lectin then specifically interacts with labeled (e.g., biotinylated) ligand in the absence or presence of inhibitors and, subsequently, labeled ligand is visualized by peroxidase (conjugated with streptavidin) or colloidal gold (coated with streptavidin) (Kohnke-Godt and Gabius 1989; Zeng and Gabius 1991). Alternatively, a suitable ligand is immobilized on the surfaces and labeled lectin is added which will bind to it. Labeling of lectins is performed according to procedures described by Joubert-Caron, this Volume, and by Kayser et al., this Volume. These assays provide a simple and reliable means to determine the carbohydrate specificity of lectins (including monovalent carbohydrate-binding proteins).

8.1 Experimental Part

8.1.1 Spotting-Type Assay

Nitrocellulose (0.2 μm, Scheicher & Schuell, Dassel, FRG)   **Materials**
24 wells tissue culture plate (Greiner, Nürtingen, FRG)
Microtiter syringe
Video densitometer

Buffer A: 20 mM Tris-HCl, pH 7.5, containing 150 mM NaCl, 20 mM $CaCl_2$ (only in cases of Ca^{2+}-requirement) and 0.2% carbohydrate-free bovine serum albumin (BSA)

Buffer B: 10 mM Tris-HCl, pH 7.5, containing 150 mM NaCl and 3% BSA

Biotinylated probe like fetuin: prepared by biotinylation with the N-hydroxy-succinimide ester (optimal concentration to be determined in initial tests)

Solution of streptavidin-peroxidase conjugate: 5-10 µg/ml in buffer A

Substrate solution: freshly mix components before use
A: 2% (w/v) 4-chloro-1-naphthol in methanol (1.5 ml)
B: 50 mM Tris-HCl, pH 7.5 (50 ml)
C: 30% H_2O_2 (30 µl)

Procedure

1. Nitrocellulose membrane is soaked in distilled water for 1 min and taken out to dry in the air, then cut into small squares (1×1 cm) and placed in separate wells of tissue culture plates.
2. One µl protein solution (1 mg/ml in 10 mM Tris-HCl, pH 7.5, e.g., calcyclin) is carefully spotted in the middle of the squares with a 25 µl Hamilton syringe, and allowed to dry for 30 min.
3. The membrane squares are briefly washed with 1 ml of buffer A and the residual binding sites are blocked with 0.3 ml of buffer B at room temperature (RT) for 1 h.
4. The squares are washed with buffer A for 3×5 min, incubated with 0.3 ml of 20 µg/ml biotinylated fetuin in buffer A at RT for 2 h under gentle shaking. For inhibition tests, the squares are preincubated with the inhibitors at different concentrations at RT for 1 h prior to the addition of labeled fetuin to this solution, then coincubated with 20 µg/ml biotinylated fetuin in buffer A at RT for another 2 h.
5. After washing with 1 ml buffer A in each of the three rapid buffer changes and three changes after 10 min to withdraw labeled ligand, the squares are incubated with 0.3 ml of streptavidin-peroxidase solution at RT for 90 min.
6. To visualize the bound probe, 0.5 ml of substrate solution is added to each well after extensive washing with buffer A. The color development is stopped after 10–20 min by replacing the substrate solution with water and further washing with distilled water. Finally, the squares are dried in the air and quantitative evaluation of extent of bound probe is performed by scanning with a video densitometer.

8.1.2 ELBA-type Assay (Enzyme-Linked Lectin Binding Assay)

Materials

96-well immuno-microplate (Greiner, Nürtingen, FRG)

ELISA plate reader (Dynatech, Cologne, FRG)

Coating buffer: 0.1 M Na_2CO_3, pH 9.6

Buffer A: 10 mM phosphate-buffered saline (PBS), pH 7.6, containing 0.1% Tween 20

Buffer B: PBS containing 3% BSA

Buffer C: PBS containing 0.5% BSA and 0.1% Tween 20

Biotinylated probe like galactoside-specific mistletoe lectin-I (ML-I) : prepared by biotinylation with the N-hydroxysuccinimide ester

Solution of avidin-alkaline phosphatase conjugate: 3 µg/ml in buffer C

Substrate solution: 1 mg p-nitrophenyl phosphate (Sigma 104 phosphate substrate) per ml 50 mM Na_2CO_3, pH 9.8, containing 1 mM $MgCl_2$

1. 100 µl of 2 µg/ml asialofetuin in coating buffer is added into each well of a microplate and incubated at 4 °C overnight.   **Procedure**
2. After washing four times with buffer A, the residual protein-binding sites are blocked with buffer B at 37 °C for 1 h.
3. One hundred µl of biotinylated lectin (0–50 µg/ml for determination of K_d and 5 µg/ml for inhibition test) is added and incubated at 37 °C for 2 h. To assess ligand properties of a glycosubstance, the wells are incubated with a serial dilution of the reagent in the presence of 5 µg/ml of labeled lectin at 37 °C for 2 h.
4. The wells are washed four times for 6 min at each step with buffer A and incubated with 100 µl of avidin-alkaline phosphatase solution at 37 °C for 1 h.
5. After washing five times for 6 min with buffer A and further washing once with 50 mM Na_2CO_3, pH 9.8, containing 1 mM $MgCl_2$, 100 µl of substrate solution is added to each well. The reaction is allowed to proceed at room temperature for 10–20 min before stopping the reaction with 50 µl of 1 M NaOH.
6. The absorbance at 410 nm is determined using a microplate reader, equipped with a narrow bandpass filter for 410 nm.

8.2 Results

An optimal lectin concentration for spotting or coating should be determined before carrying out an inhibition test. To obtain the concentration of inhibitor required for

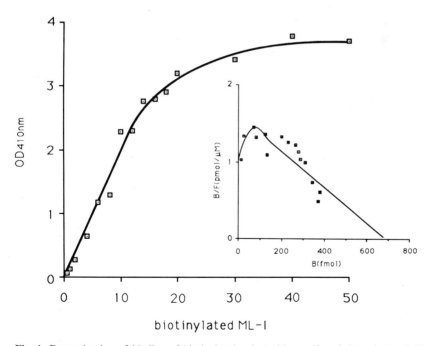

Fig. 1. Determination of binding of biotinylated galactoside-specific mistletoe lectin (0–50 µg/ml) to immobilized asialofetuin and transformation of the binding data to allow calculation of the K_d value

Table 1. Inhibition of binding of biotinylated fetuin to immobilized calcyclin by sugars and glycosubstances in a solid-phase assay. (Zeng and Gabius 1991)

Sugars or glycosubstances	Concentration required for 50% inhibition (mM or mg/ml)
Neu5Ac[a]	> 100
Neu5Gc[a]	3
Sialyllactose[a]	25
Lactose[a]	50
Galacturonic acid[a]	100 (20%)
Glucuronic acid[a]	50 (20%)
Man-6-p[a]	25
Gal-6-p[a]	30
Glc-6-p[a]	40
Fetuin[b]	0.20
Asialofetuin[c]	> 10
BSM[b]	0.10
Sialic acid-BSA[b]	0.15
Glucuronic acid-BSA[b]	0.15
Man-6-P-BSA[b]	0.12
Gal-6-P-BSA[b]	0.12
Heparin[c]	5.0 (20%)
Fucoidan[c]	2.0
Dextran sulfate[c]	1.5

[a] mM, [b] mM (concentration is given in terms of sialic acid or saccharide). [c] mg/ml.

Abbreviations: BSM = bovine submaxillary mucin, BSA = bovine serum albumin, glc-6-p = glucose-6-phophate, gal-6-p = galactose-6-phosphate, man-6-p = mannose-6-phosphate.

50% inhibition, a standard curve for each individual compound is measured. To exemplify the value of the assay, a compilation of the potency of ligands to reduce calcyclin's ability to bind fetuin is given (Table 1). The enzyme-linked lectin binding assay (ELBA) in ELISA-plates results in similar values. It also allows to determine the properties of ligand binding, as shown for a Scatchard analysis of the binding of galactoside-specific mistletoe lectin (ML-I) to immobilized asialofetuin (Fig. 1). A dissociation constant K_d of 410 nM is determined for this interaction.

8.3 Remarks

Nonspecific binding may take place and may vary from protein to protein. It can be reduced by optimizing the incubation and washing conditions, e.g., by addition of low concentration of nonionic detergent (Tween 20 or Triton X-100). When the lectin is immobilized, the results of test series should be validated by performing the assay with labeled lectin in solution. Perturbation of its structure by surface immobilization is thus excluded. However, the effect of the chemical alteration by labeling on the binding activity has to be assessed. Studies with group-specific reagents are helpful to define the optimal procedure for labeling, resulting in adequate label incorporation without reducing ligand binding properties (see chapter 10).

References

Kohnke-Godt B, Gabius HJ (1989) Heparin-binding lectin from human placenta: purification and partial molecular characterization and its relationship to basic fibroblast growth factors. Biochemistry 28: 6531–6538.

Zeng FY, Gabius HJ (1991) Carbohydrate-binding specificity of calcyclin and its expression in human tissues and leukemic cells. Arch Biochem Biophys 289: 137–144

9 Analysis of Isoelectric Variants of Carbohydrate Binding Protein 35

ELIZABETH A. COWLES, RICHARD L. ANDERSON, and JOHN L. WANG

Carbohydrate Binding Protein 35 (CBP35) is a galactose-specific lectin belonging to the L-30 group of the S-type family of animal lectins (Wang et al. 1991; Anderson and Wang 1992). The polypeptide chain ($M_r \sim 35\,000$) consists of two distinct domains: a proline- and glycine-rich domain at the amino-terminal half and a carbohydrate recognition domain at the carboxyl-terminal half. The amino acid sequence information also indicates that CBP35 is identical to (within a given species) or homologous with (between species) proteins isolated and studied under other names: (1) L-34, a tumor cell lectin; (2) human and rat lung lectins, HL-29 and RL-29; (3) IgE-binding protein, εBP; (4) a nonintegrin type laminin-binding protein, LBP; and (5) Mac-2, a cell surface marker of thioglycollate-elicited macrophages.

In mouse 3T3 fibroblasts, the majority of CBP35 is intracellular, being found in both the cytoplasm and the nucleus (Moutsatsos et al. 1986). Analysis of CBP35 in the cytosol and in the nucleoplasm suggests that the lectin is associated with a ribonucleoprotein-complex, as indicated by its position of sedimentation on sucrose and cesium sulfate gradients and by the sensitivity of this position of migration to treatment with micrococcal nuclease (Wang et al. 1992). Using a cell-free assay for the splicing of intervening sequences from pre-mRNA, Patterson et al. (1991) found that saccharide ligands and antibodies that bind to CBP35 inhibited the formation of the spliced mRNA product, whereas control reagents failed to yield the same effect. These results provoke the intriguing possibility that CBP35 may play a role in the processing of mRNA precursors in the nucleus.

The pI of the murine CBP35 polypeptide is 8.7, as determined both by calculation from the deduced amino acid sequence and experimentally by isoelectric focusing of recombinant CBP35 obtained by expression of a cDNA clone in *Escherichia coli* (Cowles et al. 1990). When cell extracts are subjected to two-dimensional gel electrophoresis and immunoblotting, however, two spots are observed, corresponding to pI values of 8.7 and 8.2. The pI 8.2 form represents a posttranslational modification of the pI 8.7 polypeptide by the addition of a single phosphate group, probably O-linked. CBP35 in human cells also occurs in these two isoelectric forms (Hamann et al. 1991). The isoelectric variants exhibit differential expression and localization in cells of both species (see below).

As will be evident from the following discussion, many questions regarding the mechanism, control, localization, and function of CBP35 phosphorylation and dephosphorylation emerge as challenges for future studies. Therefore, it would be useful to describe in detail the experimental protocols by which the polypeptide and its posttranslationally modified derivative can be distinguished and analyzed. In this chapter, we provide a collection of such protocols, with specific comments on the rationale of certain procedures, as well as some pitfalls.

9.1 Proliferation-Dependent Expression and Localization of CBP35

Quiescent cultures of 3T3 cells expressed a low level of CBP35; the polypeptide was almost exclusively in the phosphorylated form (pI 8.2) and was located predominantly in the cytoplasm (Moutsatsos et al. 1987; Cowles et al. 1990). The addition of serum to these cells increased the expression of CBP35, in terms of elevated transcription rate of the gene, increased accumulation of the mRNA, and increased amount of the protein (Mountsatsos et al. 1987; Agrwal et al. 1989). These proliferating cells had an increased level of the phosphorylated polypeptide, both in the cytoplasm and the nucleus. More striking, however, was the increase in the level of the unmodified form (pI 8.7), which was confined to the nucleus.

The expression and intracellular localization of CBP35 has also been investigated in human fibroblasts (Hamann et al. 1991), which, in contrast to 3T3 cells, normally have a finite replicative life span. The question addressed was whether any change in CBP35 expression occurs in human fibroblasts as they acquire a reduced replicative capacity (i.e., become "senescent") by passage of culture through many generations. The levels of CBP35 mRNA and protein, as well as the intracellular distribution of the two isoelectric variants of the polypeptide, in "young" human fibroblasts behaved similarly to those of mouse 3T3 cells. In contrast, "older" human fibroblasts failed to exhibit the correlation between the level of CBP35 and the proliferation state of the culture. The levels of CBP35 mRNA, as well as protein, remained high (no down-regulation) in serum-starved cells, and serum addition resulted in a decrease rather than the expected increase in CBP35 expression. The unphosphorylated form of the CBP35 polypeptide (pI 8.7) was not observed in late passage cultures, whether they are serum-stimulated or not. These results establish that the expression of CBP35 becomes altered as human fibroblasts acquire reduced replicative capacities.

9.2 Cells, Cell Extracts, and Subcellular Fractions

9.2.1 Culture of Mouse 3T3 Fibroblasts

Mouse Fibroblasts

Swiss 3T3 fibroblasts (American Type Culture Collection, CCL92, Rockville, MD) are cultured in Dulbecco's modified Eagle's medium (DMEM) supplemented with 100 U/ml penicillin, 100 µg/ml streptomycin, and 10% (v/v) calf serum (Microbiological Associates, Walkersville, MD). The cells are incubated in Corning plastic tissue culture flasks (Corning Glass Works, Corning, NY), maintained at 37 °C in a humidified atmosphere of 10% CO_2. In general, the volume of the culture medium was 0.1 ml/cm^2 of growth surface. The 3T3 cells are passaged and used for a maximum of 3 months, after which they are discarded and a fresh sample is grown from frozen cultures kept at -80°C.

Routinely, our stock culture of 3T3 cells is passaged every 2 days, before the cells reached a confluent state (5×10^4 cells/cm^2). The growth medium is first removed, and the cells are washed and then incubated in phosphate-buffered saline (PBS) containing 4×10^{-4} M EDTA at 37°C for 3 min. This solution is discarded and a solution of PBS containing 0.25% (w/v) trypsin (Nutritional Biochemicals, Cleveland, OH) and 4×10^{-4} M EDTA is added. After wetting the monolayer of cells, this trypsin solution is immediately decanted. The culture flask is then incubated at 37°C (in a humidified incubator) for 3 min. The cells are dislodged from the growth surface by the addition of DMEM-10% calf serum, diluted, and seeded at the desired density.

Confluent monolayers and sparse cultures of these cells are obtained by seeding at a density of 5×10^4 cells/cm^2 and 0.7×10^4 cells/cm^2, respectively. The cells are allowed to attach overnight in DMEM-10% calf serum, followed by a 24 h period in DMEM-0.2% calf serum. For comparing serum-starved versus serum-stimulated cultures, cells are seeded at an initial density of 1.5×10^4 cells/cm^2, starved by deprivation of serum (DMEM-0.2% calf serum for 48 h), followed by addition of serum to 10%. Cells are collected for analysis 16 h after serum addition.

9.2.2 Culture of Human SL66 Fibroblasts

Human Fibroblasts

Normal human fibroblasts, SL66 (Drinkwater et al. 1982), were gifts from Drs. J.J. McCormick and V.M. Maher (Michigan State University, East Lansing, MI). The cells were received at passage 6. Between passage 6 and passage 25, the regime for subculture was alternating 1:2 split ratio (every 3 days) and 1:3 split ratio (every 4 days). At passage 26–27, the cells entered phase III (Hayflick 1965) and did not survive well under sparse conditions; therefore, all subcultures beyond passage 26 were carried out at a 1:2 split ratio. Finally, the SL66 fibroblasts failed to progress beyond passage 36. For SL66 fibroblasts, passage numbers 11, 17, 27, and 31 correspond to cumulative population doublings of 18, 33, 55, and 62, respectively.

The SL66 cells are cultured in Eagle's minimum essential medium containing 20% (v/v) fetal calf serum (GIBCO BRL, Gaithersburg, MD). For each passage, the growth medium is removed and the cells are washed with PBS containing 4×10^{-4} M EDTA (5 min at 37°C). After discarding the EDTA solution, PBS containing 0.04% (w/v) trypsin and 4×10^{-4} M EDTA is used to wet the monolayer and immediately decanted. The culture flask is incubated at 37°C for 5 min (in a humidified incubator). Addition of the growth medium then dislodges the cells for dilution and/or reseeding. Sparse cultures of these cells are seeded at a density of 1.0×10^4 cells/cm^2. Following overnight attachment, the cultures are synchronized by washing twice in minimum essential medium without serum and then incubated for 48 h in the same medium containing 0.2% fetal calf serum. The human cells are stimulated for 17 h with 20% fetal calf serum.

9.2.3 Preparation of Extracts from Whole Cells

Extracts

The cells are washed in PBS, scraped off the culture dishes, and pelleted by centrifugation (1330 g, 3 min). The cells are then resuspended in 10 mM Tris (pH 7.5) containing 2 mM EDTA, 1 mM phenylmethylsulfonyl fluoride, 1 µg/ml soybean trypsin inhibitor, and 1 U/ml aprotinin. The cells are incubated on ice for 20 min, followed by sonication (four times, 15 s each). Aliquots are then taken for protein determination and for marker enzyme assays, as described below.

9.2.4 Subcellular Fractionation

Fractionation

The cells are washed in PBS, scraped off the culture dishes, and pelleted by centrifugation (1330 g, 3 min). For the isolation of subcellular fractions, $\sim 10^6$ cells

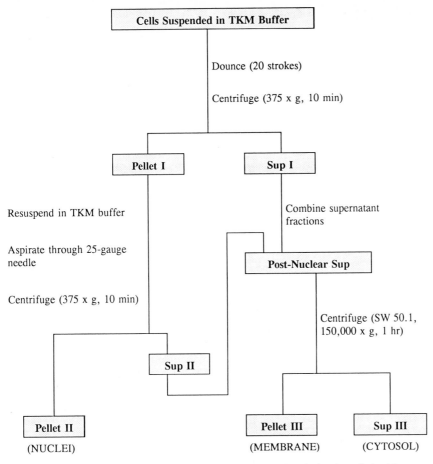

Fig. 1. Flow chart used for the isolation of nuclear and cytoplasmic fractions derived from mouse 3T3 fibroblasts and human SL66 fibroblasts. TKM buffer, 20 mM Tris (pH 7.2), 5 mM KCl, 1 mM MgCl$_2$, 0.1 mM EDTA, 1 µg/ml soybean trypsin inhibitor, 1 U/ml aprotinin, and 1 µg/ml leupeptin *Sup* supernatant fraction from centrifugation step

are resuspended in 0.5 ml of TKM buffer (20 mM Tris, pH 7.2, 5 mM KCl, 1 mM MgCl$_2$, 0.1 mM EDTA, 1 µg/ml soybean trypsin inhibitor, 1 U/ml aprotinin, and 1 µg/ml leupeptin) (see Fig. 1). The cells are lysed in a Dounce homogenizer (20 strokes) and the lysate is centrifuged in a DuPont/Sorvall HB-4 rotor (Wilmington, DE) (375 g, 10 min, 4 °C). The pellet (pellet I in Fig. 1) contains mostly nuclei and some unbroken cells. Therefore, this pellet is resuspended in 400 µl of TKM buffer, aspirated through a 25-gauge needle five times to lyse the cells, and resedimented (375 g, 10 min, 4 °C). The pellet from this step represents the nuclear pellet (pellet II in Fig. 1).

The supernatant fractions from the low speed sedimentations (sup I and sup II in Fig. 1) are combined to yield the postnuclear supernatant fraction. This is centrifuged in a Beckman SW 50.1 rotor (Beckman Instruments, Palo Alto, CA) (150 000 g, 1 h at 4 °C) to yield the membrane fractions of the cell (pellet III in Fig. 1) and the cytosol (sup III in Fig. 1). Aliquots of each of the fractions are taken for protein determination and for marker enzyme assays.

9.2.5 Protein and Marker Enzyme Assays

Enzyme Assays

Protein concentration is determined by the Bradford assay (Bradford 1976). Subcellular fractions are characterized by specific marker assays: (a) enzyme assay for lactate dehydrogenase (EC 1.1.1.27) (Kaplan and Cahn 1962), a cytoplasmic compartment marker; (b) enzyme assay for NADH-diaphorase (EC 1.6.99.1) (Hay 1971), a marker for the endoplasmic reticulum; and (c) DNA binding assay for the dye Hoechst 33258 (Cesarone et al. 1979) in the nuclear compartment. The protocols for these general assays will not be detailed here.

9.3 Recombinant CBP35 and ^{32}P-Labeled CBP35

9.3.1 Isolation of ^{32}P-labeled CBP35 from 3T3 Fibroblasts

Isolation of CBP35

Confluent cultures of 3T3 fibroblasts are washed in phosphate-free DMEM. To each of ten 150-cm^2 flasks is added 10 ml of phosphate-free DMEM containing 50 µCi/ml ^{32}PO$_4$ (DuPont/New England Nuclear, Boston, MA). An additional 10-20 flasks of the same cells are cultured in parallel in DMEM without radioactive phosphate. After 24 h of labeling, the medium is decanted and the cells are washed with 75 mM Tris, 50 mM CaCl$_2$, pH 7.5. The cells are then scraped in a minimal volume of column buffer (10 mM Tris, 10 mM β-mercaptoethanol, 0.2% Triton X-100, pH 7.8). The cells from the radiolabeled flasks are combined with those from unlabeled flasks. They are homogenized with 20 strokes in a Dounce homogenizer, followed by sonication (four times, 15 s each). After centrifugation (3 000 g, 15 min), the supernatant is subjected to affinity chromatography on a column (1.3 × 4 cm) containing asialofetuin-derivatized (Raz et al. 1987) Affi-gel 15 (BioRad, Richmond, CA). The bound material is eluted with 10 mM Tris (pH 7.8), 10 mM β-mercaptoethanol, 0.3 M lactose, pooled, and characterized by sodium dodecyl sulfate polyacrylamide gel electrophoresis (SDS-PAGE), as well as by two-dimensional gel electrophoresis.

9.3.2 Recombinant CBP35 from *Escherichia coli*

Recombinant CBP35

The cDNA for CBP35 is excised form plasmid pWJ31 (Agrwal et al. 1989) using EcoRI digestion, treated with mung bean nuclease to form blunt ends, and ligated to NcoI linkers (d[GCCATGGC]) (Pharmacia, Piscataway, NJ). The resulting cDNA is digested with NcoI and then ligated to the pKK-233-2 expression vector (Amann and Brosius 1985; Pharmacia) at the NcoI site. The correct orientation is determined by restriction enzyme analysis. The resulting expression plasmid is designated prCBP35. The *E. coli* strain JM105 is used for transformation and expression of the recombinant protein. One liter of L-broth (0.5% Difco yeast extract, 1% Difco tryptone, and 1% NaCl) containing 75 µg/ml ampicillin is inoculated with fresh overnight culture of prCBP35 transformed *E. coli*. The cells are grown at 37 °C with aeration for 2 h, followed by addition of isopropyl-β-D-thiogalactoside to a final concentration of 1 mM. The bacteria are then cultured overnight (12–16 h) at 37 °C with rapid agitation.

To harvest the cells, the *E. coli* are pelleted by centrifugation, washed with phosphate-buffered saline, and resuspended in buffer A (75 mM Tris-HCl, pH 7.0, containing 2 mM EDTA, 75 mM NaCl, 10 mM β-mercaptoethanol, 1 µg/ml soybean trypsin inhibitor, 1 U/ml aprotinin, 0.5 µg/ml leupeptin, and 1 mM phenylmethyl sulfonyl fluoride). The cells are disrupted using a French Press at 16 000 p.s.i at 4 °C. The lysate is then incubated with deoxyribonuclease I (0.5 µg/ml) and ribonuclease A (50 µg/ml) on ice for 30 min. The lysate is centrifuged at 90 000 g for 30 min and the supernatant is subjected to precipitation by ammonium sulfate (50% of saturation). The precipitate is dialyzed against buffer A and then subjected to affinity chromatography on asialofetuin-derivatized Affi-gel 15. The bound material is eluted with 0.4 M lactose in buffer A and concentrated in an Amicon filter using a PM10 membrane. This material will be hereafter designated recombinant CBP35 (rCBP35).

9.4 Two-Dimensional Gel Electrophoresis

9.4.1 Sample Preparation

All samples are dissolved in lysis buffer (O'Farrell 1975), which contains 9.5 M urea, **Samples** 2% (w/v) Triton X-100, 2% ampholines (see below), and 5% β-mercaptoethanol. Recrystallized ultrapure urea (Boehringer-Mannheim, Indianapolis, IN), free of cyanate, should be used for two-dimensional gels. The lysis buffer can be stored as frozen aliquots.

For the analysis of cell extracts and subcellular fractions, aliquots containing 100–200 µg of total protein are used. These aliquots are first digested with ribonuclease A (50 µg/ml; Boehringer-Mannheim) and deoxyribonuclease I (50 µg/ml; Boehringer-Mannheim) on ice for 30 min and then diluted into lysis buffer. For the analysis of purified CBP35 preparations, 30–50 ng of the protein are dissolved in lysis buffer.

An alternative method of sample preparation is to dissolve the lyophilized protein in 8 M urea. The sample is then mixed with the acrylamide solution, which is then polymerized into a gel containing the protein(s). This procedure should be used with caution because the heat produced during the polymerization reaction (Dunn 1987) may result in carbamylation of the protein(s), particularly if undeionized urea containing cyanate is used in the preparation of the buffers (Cowles et al. 1990).

9.4.2 Equilibrium Isoelectric Focusing Analysis

In the first method of two-dimensional gel electrophoretic analysis, equilibrium **IEF** isoelectric focusing (IEF), is performed in the first dimension, followed by SDS-PAGE in the second dimension. The details of the method are described by O'Farrell (1975). The ampholines used are supplied as 40% (w/v) solutions (Pharmacia). For our standard IEF gels, the 2% ampholines used in the lysis buffer, in the overlay buffer, and in the acrylamide solution consisted of the following mixture: (a) 0.8% pH 4–6.5 ampholine; (b) 0.8% pH 2–5 ampholine; (c) 0.4% pH 3–10 ampholine. The first dimension IEF gel is electrophoresed at 400 V for 16 h and at 800 V for 1 h. The anode electrode solution is 10 mM H_3PO_4 and the cathode electrode solution is 10 mM NaOH, 5 mM $Ca(OH)_2$. The second dimension SDS-PAGE had 12 cm of separating gel containing 12.5% acrylamide.

In order to save sample (50–100 μg total protein) and time, the two-dimensional IEF analysis was adapted to a mini-gel system (BioRad). In this modification, only 2% pH 3–10 ampholines are used and the IEF gel is electrophoresed for 10 min at 500 V and 3.5 h at 750 V. The anode solution is 25 mM H_3PO_4 and the cathode solution is 50 mM NaOH. The second dimension SDS-PAGE had 5 cm of separating gel containing 12.5% acrylamide.

9.4.3 Nonequilibrium pH Gradient Electrophoretic Analysis

NEPHGE   A second method of two-dimensional gel analysis developed by O'Farrell et al. (1977) uses nonequilibrium pH gradient electrophoresis in the first dimension, followed by SDS-PAGE in the second dimension. In the standard NEPHGE gels, 2% pH 3-10 ampholines are used. The anode solution is 10 mM H_3PO_4 and the cathode solution is 10 mM NaOH, 5 mM $Ca(OH)_2$. The position of migration of CBP35 remains the same, relative to standards with known pI values (e.g., glyceraldehyde 3-phosphate dehydrogenase, pI 8.5), irrespective of whether the NEPHGE gel is electrophoresed for 3, 6, or 16 h. Therefore, although the value obtained from NEPHGE analysis may not actually represent an equilibrium isoelectric point, the pH in the gel to which a polypeptide migrates is, nevertheless, referred to as its pI value. Routinely, our standard NEPHGE gels are electrophoresed for 5 h at 400 V.

To adapt the NEPHGE analysis to the mini-gel system, the 2% ampholine used consisted of a 2 : 1 ratio of pH 3-10 ampholines and pH 7-9 ampholines. The running conditions are as follows: (a) basic electrode solution, 50 mM NaOH; (b) acidic electrode solution, 25 mM H_3PO_4; (c) running times of 1 h at 400 V, 75 min at 650 V and 15 min at 750 V.

9.4.4 Detection of the Position of Protein Migration

Protein Detection   Proteins are detected by Coomassie blue (Diezel et al. 1972) and silver (Wray et al. 1981) staining of the gel. ^{32}P-Labeled samples are detected by autoradiography, using Kodak X-OMat (Eastman Kodak, Rochester, NY) film. Immunoblotting is performed as described previously (Towbin et al. 1979) with the following modifications. Proteins are transferred to Immobilon-P (Millipore) and incubated in Tris (10 mM Tris, pH 7.5, 0.5 M NaCl, 0.05% thimerosal) containing 3% bovine serum albumin for at least 1 h. The blots are incubated in rabbit anti-CBP35 antiserum (1: 150) overnight at 4 °C. The blots are washed in the same Tris buffer containing 0.2% (v/v) Tween-80. Positive bands are revealed with the colored products of the horseradish peroxidase conjugated goat anti-rabbit immunoglobulin (BioRad).

To compare the isoelectric variants on a quantitative basis, a direct positive is made of the immunoblot on LPD4 film (Kodak). Relative intensities of the spots are then determined by scanning densitometric analysis on a Gelman ACD-18 automatic computing densitometer.

9.5 Isoelectric Variants of CBP35

9.5.1 Two-Dimensional Gel Electrophoretic Analysis of CBP35

Electrophoresis   Both IEF and NEPHGE two-dimensional gel analyses were carried out in our studies on CBP35 because certain proteins do not focus at the high pH end of the isoelectric gradient under IEF analysis, whereas others are not well resolved at the low end during NEPHGE analysis. To illustrate the relevance of this point to the present study, seven protein standards were subjected to IEF analysis. Coomassie blue staining of the gel yielded only four identifiable spots consistent with the molecular weights and pI values of the protein standards (Fig. 2). When the same mixture of seven proteins was subjected to NEPHGE analysis, seven identifiable spots corresponding to the molecular weights and pI values of the protein standards were observed (Fig. 2). On the basis of comparing the IEF versus the NEPHGE results, we find that proteins with a basic pI, such as carbonic anhydrase, phosphorylase a, and phosphoglycerate kinase, do not focus on an IEF gel with its limited pH range. Although all seven protein standards were identified in the NEPHGE analysis, the resolution at the low pH end was poor (compare, for example, positions of spots 1,2, and 3). Therefore, both IEF and NEPHGE analyses were carried out in parallel on certain samples reported in the present study.

Extracts of mouse 3T3 fibroblasts were subjected to electrophoresis on IEF and NEPHGE gels, followed by immunoblotting with anti-CBP35. On IEF gels, no immunoreactive spots were observed (Fig. 3A). This result was identical to that obtained with rCBP35 (Fig. 3C). Parallel analysis of the same extracts on NEPHGE gels revealed two spots with pI values of 8.7 and 8.2 (Fig. 3B). The pI 8.7 species corresponds to that observed on NEPHGE analysis of rCBP35 (Fig. 3D). Since the production of rCBP35 by the transformed E. coli most probably did not result in posttranslational modification of the polypeptide chain, the pI 8.7 value is most likely the isoelectric point of the CBP35 polypeptide. This value agrees with the pI of the polypeptide, calculated from its amino acid composition (Cowles *et al.* 1990). The

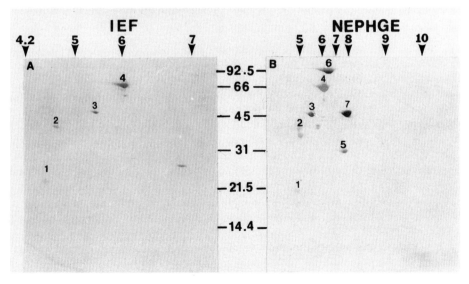

Fig. 2. Two-dimensional electrophoretic analysis of protein standards on IEF and NEPHGE gels. The protein standards were revealed by Coomassie blue staining. The numbers 1-7 correspond to. *1*) soybean trypsin inhibitor (8 μg); *2*) rabbit muscle tropomyosin (10 μg); *3*) rabbit muscle actin (10 μg); *4*) bovine serum albumin (5 μg); *5*) bovine erythrocyte carbonic anhydrase (10 μg); *6*) rabbit muscle phosphorylase a (14 μg); *7*) yeast phosphoglycerate kinase (8 μg). The *numbers at the top* indicate the pH values of the ampholine gradient. The *numbers down the middle* indicate positions of migration of molecular weight markers

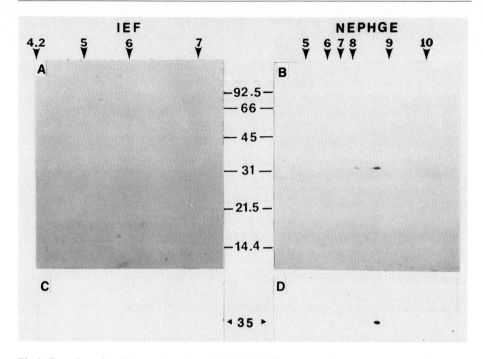

Fig. 3. Two-dimensional electrophoretic analysis of CBP35 in extracts of mouse 3T3 fibroblasts (A and B) and rCBP35 (C and D) on IEF and NEPHGE gels. In **A** and **B**, extracts of 3T3 cells ($\sim$200 µg total protein) were electrophoresed and in **C** and **D**, purified rCBP35 ($\sim$30 ng) was electrophoresed. CBP35 was revealed by immunoblotting with rabbit anti-CBP35. The *numbers at the top* indicate the pH values of the ampholine gradient. The *numbers down the middle* indicate positions of migration of molecular weight markers. The *arrows* indicate positions of migration of authentic CBP35 ($M_r \sim 35\,000$)

pI 8.2 spot, therefore, most probably represents a posttranslationally modified isoelectric variant of the polypeptide.

If the sample of CBP35 purified from 3T3 cells was dissolved in buffer prepared using undeionized urea, mixed with the acrylamide solution which was then polymerized into a gel, IEF analysis yielded two spots corresponding to pI values of 4.7 and 4.5 (Roff and Wang 1983). This result has now been ascribed to artifactual carbamylation of the pI 8.7 and 8.2 isoelectric species due to the cyanate in the deionized urea solution (Cowles et al. 1990). Such artifacts form the basis for urging caution concerning the use of undeionized urea and the use of the polymerization technique for sample loading onto the isoelectric focusing gel.

9.5.2 Posttranslational Modification of CBP35

Modification   Extracts of 3T3 cells, cultured for 24 h in ^{32}P-labeled phosphate were fractionated on affinity columns containing asialofetuin. A peak of ^{32}P radioactivity was observed upon the addition of lactose. This ^{32}P-labeled sample of CBP35 was subjected to NEPHGE analysis, followed by immunoblotting with rabbit anti-CBP35, as well as autoradiography of the nitrocellulose membrane (Fig. 4A). The immunoblot yielded two protein spots, pI 8.7 and pI 8.2; only the pI 8.2 species, however, was radioactive.

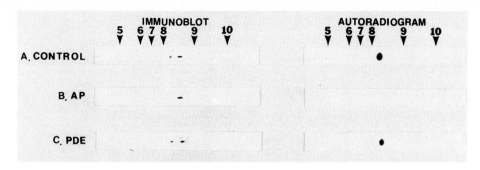

Fig. 4. Two-dimensional NEPHGE gel analysis of purified ^{32}P-labeled CBP35 (5×10^5 cpm). The protein was revealed either by immunoblotting with rabbit anti-CBP35 or by autoradiography of the ^{32}P label. Prior to electrophoresis, the samples were incubated for 2 h at 37 °C in 40 mM Tris, 15 mM MgCl$_2$, 1 mM ZnCl$_2$, pH 8.0 (*A. CONTROL*) or in the same buffer containing 2 U/ml of calf intestine alkaline phosphatase (*B. AP*) or 0.2 U/ml of phosphodiesterase I (*C. PDE*). The *numbers at the top* indicate the pH values of the ampholine gradient

The ^{32}P-labeled material was dialyzed against buffer containing 40 mM Tris (pH 8.0), 15 mM MgCl$_2$, 1 mM ZnCl$_2$, 1 mM phenylmethylsulfonyl fluoride, 1 µg/ml soybean trypsin inhibitor, and 1 U/ml aprotinin. The dialyzed sample was then digested either with 2 U/ml calf intestine alkaline phosphatase (Boehringer-Mannheim) or with 0.2 U/ml phosphodiesterase I (Boehringer-Mannheim) for 2 h at 37 °C. The reactions were stopped by lyophilizing the samples, resuspension in lysis buffer, and analysis by NEPHGE. Alkaline phosphatase treatment reduced the immunoblot to a single spot (pI ~8.7; Fig. 4B); there was no radioactive spot in the autoradiogram. Parallel treatment with phosphodiesterase did not have any effect on the ^{32}P-labeled sample (Fig. 4C). We interpret these results to indicate that the pI 8.2 spot is a posttranslationally modified product of the pI 8.7 polypeptide chain by the addition of a single phosphate. The sensitivity of this ^{32}P-labeling to alkaline phosphatase suggests that it is an O-linked phosphate group. The introduction of one phosphate group to the amino acid composition of CBP35 converts the calculated pI from a value of 8.7 to 8.2 (Cowles et al. 1990).

9.5.3 Expression of Isoelectric Variants of CBP35 and Their Differential Localization

Mouse 3T3 fibroblasts exhibit the phenomenon of density-dependent inhibition of growth (Todaro and Green 1963). In sparse cultures (cell density of $< 5 \times 10^4$ cells/cm^2, for example), the cells undergo DNA synthesis and cell division. Unless specifically synchronized by experimental manipulation, the cells are found distributed in different phases of the cell cycle. In confluent monolayers (cell density 5×10^4 cells/cm^2), there is little or no DNA synthesis and cell division. These cells are reversibly arrested in a quiescent state; they can be reactivated by dilution to low cell density (Todaro and Green 1963) or by stimulation of serum growth factors (Antoniades et al. 1979). Using this paradigm, we had previously shown (Cowles et al. 1990; Moutsatsos et al. 1987; Agrwal et al. 1989) that the expression of CBP35 and its subcellular localization were dependent on the proliferation state of the cells by

Isoelectric Variants

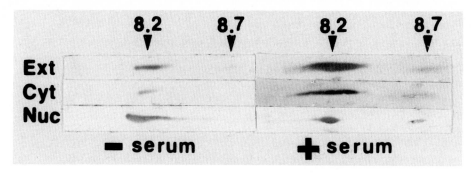

Fig. 5. Comparison by NEPHGE gels of the isoelectric variants of CBP35 in confluent monolayers of 3T3 fibroblasts serum-starved (0.2% calf serum for 48 h) (− serum) and serum-stimulated (10% calf serum for 16 h) (+ serum). *Ext* whole cell extract; *Cyt* cytosol fraction; *Nuc* nuclear pellet. Approximately 100 μg of each sample were electrophoresed and the protein was detected by immunoblotting with rabbit anti-CBP35. The positions of migration of the pI 8.2 and pI 8.7 species are indicated *at the top*

comparing: (a) quiescent, confluent cultures versus proliferating sparse cultures; (b) quiescent, serum-starved sparse cultures versus proliferating serum-stimulated sparse cultures; (c) quiescent, confluent cultures versus confluent cultures stimulated to undergo one more round of DNA synthesis by serum growth factors. We will now use the latter pair of conditions to compare the isoelectric variants in terms of their expression and differential localization.

The extract of confluent cultures without serum yielded a single spot at pI 8.2 upon NEPHGE analysis; the extract of serum-stimulated confluent cultures yielded a spot at pI 8.7, in addition to a prominent pI 8.2 species (Fig. 5). The cytosol fraction from cultures with and without serum both yielded the pI 8.2 spot. The intensity of this spot was much greater in serum-stimulated cells than in quiescent, confluent cells, reflecting the increase in the expression of the CBP35 polypeptide in proliferating cells. Finally, the nuclei fraction of confluent cells stimulated with serum showed both the pI 8.7 and the pI 8.2 spots, whereas parallel cultures without serum yielded only the pI 8.2 species. Thus, addition of serum to density-arrested monolayers of 3T3 cells resulted in an elevation in the expression of CBP35. This elevation is reflected by both an increase in the intensity of the pI 8.2 spot (relative to the corresponding spot in confluent cells without serum), as well as the appearance of the pI 8.7 spot.

Dramatic differences in the expression of the pI 8.7 species, as well as its nuclear localization, have also been demonstrated between sparse versus confluent cultures of 3T3 cells, between serum-starved versus serum-stimulated mouse 3T3 fibroblasts and human SL66 fibroblasts, and finally, between "young" (low passage) versus "old" (high passage) human SL66 cells. Particularly striking was the observation that the unphosphorylated (pI 8.7) form of the protein could not be detected in the "old" cells, whether they are serum-stimulated or not. Thus, these intriguing observations form the basis for new questions regarding the mechanism, regulation, and localization of the enzyme(s) responsible for phosphorylation and dephosphorylation of CBP35 and their physiological role in the life cycle of a cell.

Acknowledgments: The authors thank Mrs. Linda Lang for her help in the preparation of the manuscript. This work was supported by grants GM-38740 and GM-27203 from the National Institutes of Health.

References

Agrwal N, Wang JL, Voss PG (1989) Carbohydrate binding protein 35. Levels of transcription and mRNA accumulation in quiescent and proliferating cells, J Biol Chem 264: 17236–17242

Amann E, Brosius J (1985) 'ATG vectors' for regulated high-level expression of cloned genes in *Escherichia coli*. Gene (Amst) 40: 183–190

Anderson RL, Wang, JL (1992) Carbohydrate binding protein 35. Trends Glycosci Glycotech 4: 43–52

Antoniades HN, Scher CD, Stiles CD (1979) Purification of human platelet-derived growth factor. Proc Natl Acad Sci USA 76: 1809–1813

Bradford MM (1976) A rapid and sensitive method for the quantitation of microgram quantities of protein utilizing the principle of protein dye binding. Anal Biochem 72: 248–254

Cesarone CF, Bolognesi C, Santi L (1979) Improved microfluorometric DNA determination in biological material using 33258 Hoechst. Anal Biochem 100: 188–197

Cowles EA, Agrwal N, Anderson RL, Wang JL (1990) Carbohydrate binding protein 35. Isoelectric points of the polypeptide and a phosphorylated derivative. J Biol Chem 265: 17706–17712

Diezel W, Kopperschlager G, Hoffman E (1972) An improved procedure for protein staining in polyacrylamide gels with a new type of Coomassie brilliant blue. Anal Biochem 48: 617–620

Drinkwater NR, Corner RC, McCormick JJ, Maher VM (1982) An in situ assay for induced diphtheria toxin resistant mutants. Mutat Res 106: 277–287

Dunn A (1987) Two-dimensional polyacrylamide gel electrophoresis. In: Chrambach A, Dunn MJ, Radola BJ (eds) Advances in electrophoresis. VCH Publishers, New York (vol 1) p 441

Hamann KK, Cowles EA, Wang JL, Anderson RL (1991) Expression of carbohydrate binding protein in human fibroblasts: variations in the levels of mRNA, protein, and isoelectric species as a function of replicative competence. Exp Cell Res 196: 82–91

Hay AJ (1971) Studies on the formation of the influenza virus envelope. Virology 60: 398–418

Hayflick L (1965) The limited *in vitro* lifetime of human diploid cell strains. Exp Cell Res 37: 614–636

Kaplan NO, Cahn RD (1962) Lactic dehydrogenases and muscular dystrophy in the chicken. Proc Natl Acad Sci USA 48: 2123–2130

Moutsatsos IK, Davis JM, Wang JL (1986) Endogenous lectins from cultured cells: subcellular localization of carbohydrate-binding protein 35 in 3T3 fibroblasts. J Cell Biol 102: 477–483

Moutsatsos IK, Wade M, Schindler M, Wang JL (1987) Endogenous lectins from cultured cells: nuclear localization of carbohydrate-binding protein 35 in proliferating 3T3 fibroblasts. Proc Natl Acad Sci USA 84: 6452–6456

O'Farrell PH (1975) High resolution two-dimensional electrophoresis of proteins. J Biol Chem 250: 4007–4021

O'Farrell PZ, Goodman HM, O'Farrell PH (1977) High resolution two-dimensional electrophoresis of basic as well as acidic proteins. Cell 12: 1133–1142

Patterson RJ, Laing JG, Wang JL, Werner EA (1991) Evidence for carbohydrate recognition in the splicing of messenger RNA precursors in a cell free assay. Glycoconjugate J 8: 229

Raz A, Meromsky L, Svibel I, Lotan R (1987) Transformation-related changes in the expression of endogenous lectins. Int J Cancer 39: 353–360

Roff CF, Wang JL (1983) Endogenous lectins from cultured cells. Isolation and characterization of carbohyrate-binding proteins from 3T3 fibroblasts. J Biol Chem 258: 10657–10663

Todaro GJ, Green H (1963) Quantitative studies of the growth of mouse embryo cells in culture and their development into established lines. J Cell Biol 17: 299–313

Towbin H, Staehelin J, Gordon J (1979) Electrophoretic transfer of proteins from polyacrylamide gels to nitrocellulose sheets: procedure and some applications. Proc Natl Acad Sci USA 76: 4350–4354

Wang JL, Laing JG, Anderson RL (1991) Lectins in the cell nucleus. Glycobiol 1: 243–252

Wang JL, Werner EA, Laing JG, Patterson RJ (1992) Nuclear and cytoplasmic localization of a lectin-ribonucleoprotein complex. Biochem Soc Trans 20: 269–274

Wray W, Boulikas T, Wray V, Hancock R (1981) Silver staining of proteins in polyacrylamide gels. Anal Biochem 118: 197–203

10 Chemical Modification of Lectins by Group-Specific Reagents

F.-Y. ZENG and H.-J. GABIUS

Modification of certain groups of amino acid side chains serves two experimental purposes. It makes it possible to infer which groups can be target for labeling reagents without adversely affecting the ligand-binding properties. When modification is carried out in the absence or presence of a suitable ligand, any differential effect hints at the involvement of certain groups in ligand binding. Impairment of ligand binding only in the absence of the ligand is suggestive evidence for a spatial association of the respective type of side chain to the binding site. Access to such groups should be blocked by the specific glycomolecule. A selection of common reactions to achieve selective side chain modification is given from the possible panel (Glazer et al. 1976). It is necessary to run control reactions for each individual type of modification in the absence of reagents under identical conditions to reliably assess the impact of actual modification on the lectin's binding capacity. The experimental part illustrates the conditions used for a fetuin-binding protein, calcyclin.

10.1 Experimental Part

Materials

Saturated NaHCO$_3$ solution

Glycine-NaOH buffer: 0.2 M, pH 10.5

Sodium borate buffer: 0.2 M pH 9.0

Phosphate buffer 1: 0.02 M pH 8.0; 2: 0.02 M, pH 7.2

Tris-HCl buffer: 0.01 M, pH 7.5

Tris-buffered saline (TBS): 0.01 M Tris-HCl, pH 7.5, and 0.9% NaCl

Sodium acetate–acetic acid buffer: 0.05 M, pH 4.0

Glycine–ethylester: 1 M in H$_2$O, pH 4.75

Citraconic anhydride

O-methylisourea: 0.6 M in glycine-NaOH buffer, adjusted to pH 10.5 with NaOH

Cyclohexane-1,2-dione: 0.05 M in ethanol

Phenylglyoxal: 5 mg/ml in dimethylsulfoxide (DMSO)

N-Acetylimidazole: 10 mg/ml in Tris-HCl buffer

N-Bromosuccinimide (NBS): 0.25 mM in acetate buffer

1-Ethyl-3-(3-dimethyl-aminopropyl)-carbodiimide (EDAC): 1.1 M in glycine-ethylester buffer

Diethylpyrocarbonate

Microdialysis bag (molecular mass cut off: 6000–8000 Da)

1a. Modification of Lysine Residues by Citraconic Anhydride (Dixon and Perham, 1968)

Eighty μg calcyclin is dissolved in 0.5 ml of saturated NaHCO₃ solution. 10 μl of citraconic anhydride is slowly added. The reaction is carried out at room temperature (RT) for 2 h under gentle shaking. The mixture is then dialyzed exhaustively against 10 mM TBS, pH 7.5.

1b. Modification of Lysine Residues by O-Methylisourea

Eighty μg calcyclin is dissolved in 0.5 ml of glycine-NaOH buffer. The reaction begins by addition of 0.4 ml of O-methylisourea solution. After gentle shaking at 4 °C for 24 h, the mixture is then dialyzed exhaustively against 10 mM TBS, pH 7.5.

2a. Modification of Arginine Residues by Cyclohexane-1,2-Dione (Patthy and Smith, 1975)

Eighty μg calcyclin is dissolved in 0.5 ml of borate buffer and mixed with 0.5 ml of cyclohexane-1,2-dione solution. The reaction mixture is flushed with N₂ and incubated at 37 °C in the dark for 3 h. The reaction is terminated by addition of 0.1 ml of 5% acetic acid in H₂O and extensively dialyzed against 10 mM TBS, pH 7.5.

2b. Modification of Arginine Residues by Phenylglyoxal (Mukherji and Bhaduri, 1986)

Eighty μg calcyclin is dissolved in 0.7 ml of phosphate buffer 1 and mixed with 0.1 ml of phenylglyoxal solution. After reaction at RT for 4 h, the mixture is dialyzed exhaustively against 10 mM TBS, pH 7.5.

3. Modification of Tyrosine Residues by N-Acetylimidazole (Riordan et al. 1965)

Eighty μg calcyclin is dissolved in 0.7 ml of Tris-HCl buffer and 0.5 ml of N-acetylimidazole solution. After gentle shaking at RT for 1 h, the mixture is dialyzed exhaustively against 10 mM TBS, pH 7.5.

4. Modification of Tryptophan Residues by NBS (Spande and Witkop, 1967)

Eighty μg calcyclin is dissolved in 0.7 ml of borate buffer and 0.5 ml of NBS solution is added. After gentle shaking at RT for 2 h, the mixture is dialyzed exhaustively against 10 mM TBS, pH 7.5.

5. Modification of Carboxyl Group Residues by EDAC (Kundu et al. 1987)

Eighty μg calcyclin is dissolved in 1.0 ml of glycine-ethylester solution. Two 5 μl portions of EDAC solution are added at 30- min intervals. The mixture is shaken and pH is maintained at 4.7 by the addition of HCl. After incubation at RT for 2 h, the mixture is dialyzed exhaustively against 10 mM TBS, pH 7.5.

6. Modification of Histidine Residues by Diethylpyrocarbonate (Church et al. 1985)

Eighty μg calcyclin is dissolved in 1.2 ml of phosphate buffer 2; 100 μl of diethylpyrocarbonate solution is added. The reaction is carried out at RT for 1 h and then dialyzed extensively against 10 mM TBS, pH 7.5.

The dialyzed fractions are lyophilized, and then redissolved in a small volume of TBS. The protein content is determined by the dye-binding assay, adapted for

Table 1. Effect of amino acid modification by group-specific reagents on binding of biotinylated fetuin to calcyclin (first row) and of biotinylated heparin to the heparin-binding lectin in the absence of (second row) and the presence of heparin (third row)

Chemical treatment	Residues modified	Binding activity (%)		
		1	2[a]	3[a]
Native protein		100	100	100
O-methylisourea	Lysine	40	80	100
Citraconic anhydride	Lysine, N-terminal amino group	40	10	60
Cyclohexane-1,2-dione	Arginine	40	30	100
Phenylglyoxal	Arginine	60	40	30
N-Acetylimidazole	Tyrosine	100	20	60
Diethylpyrocarbonate	Histidine	100	0	10
Ester-carbodiimide	Carboxyl group	100	100	80
N-Bromosuccinimide	Tryptophan	—	0	60

[a]From Kohnke-Godt and Gabius (1991)

microtiter plates, with BSA as standard (Redinbaugh and Campbell 1985). The fetuin-binding activity is assessed in a solid-phase assay, as described in Chapter 8. To exclude any further influence of the reaction conditions on the activity, respective controls are performed under identical conditions in the absence of modifying reagent.

10.2 Results and Discussion

It is evident that the fetuin-binding capacity of calcyclin is primarily affected by chemical modification of lysine and arginine residues (Table 1). Similarly, such groups also appear to play a role for the heparin-binding lectin (Table 1). In order to obtain labeled probes to characterize lectin-specific ligands biochemically and histochemically, hydrazide derivatives of respective reagents offer a suitable choice, as shown for the heparin-binding lectin (Gabius et al. 1991).

References

Church FC, Lundblad RL, Noyes CM (1985) Modification of histidine in human prothrombin. J Biol Chem 260: 4936–4940.

Dixon HBF, Perham RN (1968) Reversible blocking of amino groups with citraconic anhydride. Biochem J 109: 224–228.

Gabius HJ, Kohnke-Godt B, Leichsenring M, Bardosi A (1991) Heparin-binding lectin of human placenta as a tool for histochemical ligand localization and isolation. J Histochem Cytochem 39: 1249-1256.

Glazer AN, Delange RJ, Sigman DS (1976) Chemical modification of proteins. In: Work TS, Work S (eds) Laboratory techniques in biochemistry and molecular biology, vol. 4, part 1, North Holland Publ. Co., Amsterdam.

Kohnke-Godt B, Gabius HJ (1991) Heparin-binding lectin from human placenta: further characterization of ligand binding and structural properties and its relationship to histones and heparin-binding growth factors. Biochemistry 30: 55–65.

Kundu M, Basu J, Ghosh A, Chakrabarti P (1987) Chemical modification studies on a lectin from *Saccharomyces cerevisiae* (baker's yeast). Biochem J 244: 579–584.

Mukherji S, Bhaduri A (1986) UDP-glucose-4-epimerase from *Saccharomyces fragilis*: presence of an essential arginine residue at substrate-binding site of the enzyme. J Biol Chem 261: 4519–4524.

Patthy L, Smith EL (1975) Reversible modification of arginine residues. J Biol Chem 250: 557–564.

Redinbaugh MG, Campbell WH (1985) Adaptation of the dye-binding protein assay to microtiter plates. Anal Biochem 147: 144–147.

Riordan JF, Wacker WEC, Vallee BL (1965) N-acetylimidazole: a reagent for determination of "free" tyrosyl residues of proteins. Biochemistry 4: 1758–1765.

Spande TF, Witkop B (1967) Determination of the tryptophan content of proteins with N-bromosuccinimide. Methods Enzymol 11: 498–505.

Biophysical Methods for the Characterization of Lectin-Ligand Interactions

11 Study of Oligosaccharide-Lectin Interaction by Various Nuclear Magnetic Resonance (NMR) Techniques and Computational Methods

H. C. Siebert, R. Kaptein, and J. F. G. Vliegenthart

Various Nuclear Magnetic Resonance (NMR) techniques in combination with computational methods are successfully used for elucidation of the structure of biomolecules in solution. These powerful methods are necessary if one wants to study and understand biomolecular interaction processes on an atomic level. In this chapter we introduce methods which are suitable for a study of lectin-oligo-saccharide interaction on this level. The first step is assignment of the proton resonances, which can be done with multidimensional methods if the spectra have a high resolution (Fig. 1a). Nuclear Overhauser Effect (NOE) intensities are the most important source of structural information in NMR because of their distance dependence. The dihedral angles between vicinal coupled protons can be determined from scalar coupling constants with the help of the Karplus curve (Karplus 1959; Bystrow 1976). The Transfer NOE (TrNOE) experiment is a special nuclear Overhauser experiment, which gives information about the conformation of a small ligand bound to a receptor. This experiment is especially useful if the receptor molecule is very large and only broad, nonresolved NMR signals (Fig. 1b) can be obtained. Computer modeling, especially computer calculations like Molecular Dynamics (MD), are complementary methods used to test the results of the NMR study.

11.1 Nuclear Magnetic Resonance Study of the Oligosaccharide-Lectin Interaction

Usually the samples were repeatedly treated with ^2H_2O, finally using 99.96 atom% ^2H_2O (Merck) at p^2H 7 and room temperature. Resolution-enhanced ^{1}H-NMR spectra were measured at 360, 500, and 600 MHz on a Bruker AM-360 or AMX-500 spectrometer (Bijvoet Center, Utrecht University) or AM-600 spectrometer (SON hf-NMR facility, Nijmegen University). The sample temperatures were 30 °C, and chemical shifts are expressed in ppm relative to internal acetone in ^2H_2O (δ 2.225). The sample concentration was about 1 mM and the buffer concentration (e.g. K_2HPO_4) about 50 mM. Two-dimensional spectra were obtained using Bruker software for the pulse programs, recording 512 measurements of 2 K data points with 32 scans per t_1 value.

Equipment and Chemicals

11.1.1 Multidimensional Measurements

For the structural elucidation of biomolecules in solution (under physiological conditions) NMR is the most powerful method (Kaptein et al. 1988). NMR structures are primarily based on a set of short proton-proton distances obtained from NOEs (Noggle and Schirmer 1971; Neuhaus and Williamsen 1989). The origin

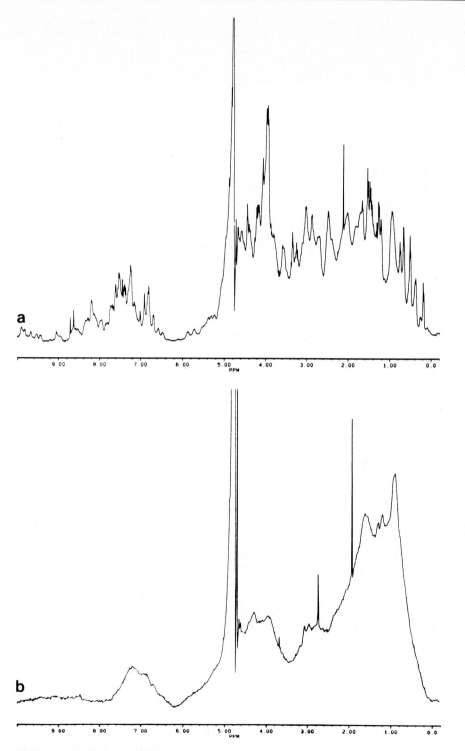

Fig. 1. a. One-dimensional ¹H-NMR spectrum of *Urtica dioica* agglutinin (320 scans). **b.** One-dimensional ¹H-NMR spectrum of Concanavalin A (320 scans)

of the NOE is dipolar cross-relaxation between protons. Because of the r^{-6} distance dependence of the effect, NOEs can only be measured between protons at relatively short distances (<5 Å). Since the NOE is a spin-relaxation phenomenon, it depends upon the dynamic behavior of the molecule in solution. Therefore, both the structure and dynamics of a biomolecule are seen by NMR.

A requirement for the analysis of a protein structure by NMR is the assignment of the great majority of its proton resonances. For small proteins (MW <10 kDa) that do not aggregate at millimolar concentrations, this can be accomplished using a combination of various 2D experiments. The procedure for so-called sequential assignment of protein ^{1}H NMR spectra has been extensively described (Wüthrich 1986). An analogous procedure for sequential assignment of oligosaccharide ^{1}H NMR spectra has also been described (Dabrowski 1987, 1989). Briefly, two main classes of 2D experiments are used (Ernst et al. 1987). In the first, off-diagonal cross-peaks arise only between protons through J-coupling networks, with COSY (COrrelated SpectroscopY) as the prime example. Another very useful experiment in this class is TOCSY (TOtal Correlation SpectroscopY) or 2D HOHAHA (HOmonuclear HArtmann-HAhn). Patterns of cross-peaks can be traced between pairs of J-coupled protons in COSY spectra or between several J-coupled protons within an amino acid chain or saccharide ring in HOHAHA/TOCSY spectra. In the second class of 2D NMR experiments, cross-peaks connect protons that are spatially in close proximity (distance <5 Å). The 2D NOE or NOESY (Nuclear OverhausEr SpectroscopY) experiment and its rotating frame counterpart ROESY (Rotating frame OverhausEr SpectroscopY), which is used for medium-sized molecules (e.g., oligosaccharides and peptides), fall into this class.

Procedure

A 2D NOE spectrum is recorded in the three-pulse experiment,

$$90° - t_1 - 90° - t_m - 90° - t_2 \text{ (acq).}$$

In this sequence, 90° stands for a 90° radio frequency pulse, t_1 and t_2 are the variable times, which after double Fourier transformation yield the ω_1 and ω_2 frequency domains of a 2D spectrum, and t_m is a fixed mixing time, which allows exchange of magentization between nuclei. The proton-proton distance constraints are most conveniently derived from cross-peak intensities in 2D NOE spectra. The initial build-up rate of these cross-peaks in spectra taken as short mixing times (e.g. 50 ms) is proportional to the cross-relaxation rate σ_{ij} between the protons i and j. For a rigid isotropically tumbling molecule σ_{ij} is simply related to the distance d_{ij} and the correlation time τ_c:

$$\sigma_{ij} \sim \tau_c d_{ij}^{-6} . \tag{1}$$

Therefore, using a known calibration distance (d_{cal}) the proton-proton distances follow from the relation

$$d_{ij} = d_{cal}(\sigma_{cal}/\sigma_{ij})^{1/6} . \tag{2}$$

In practice, these two equations are only approximately valid, because there are two main problems associated with accurate determination of proton-proton distances. The first is that of indirect magnetization transfer or "spin diffusion". In reality the NOE cross-peaks are the result of multispin relaxation and only in the limit of extremely short mixing times (where the signal-to-noise ratio is poor) is the two spin approximation of (1) valid. As discussed by Boelens et al. (1989) the effect of spin diffusion can be calculated and procedures based on a full relaxation matrix treatment are being developed to solve the problem. The second problem is that proteins and oligosaccharides are not rigid bodies and intramolecular mobility leads to nonlinear averaging of distances and to different effective correlation times for different interproton vectors in the molecule. Internal motions occur over a wide range of time-scales and only the fast fluctuations (up to a few hundred pico-seconds) can presently be simulated by molecular dynamics calculations (see sect. 11.2 below Computational Methods). The slower motions are difficult to handle and therefore constitute the most serious source of error in the determination of proton–proton distances from NOEs. For this reason, the distance card method has been invented for oligosaccharides (Poppe et al. 1990 a, b). For proteins, the approach usually taken in that of translating the NOE information into distance ranges (e.g., 2–3 Å, 2–4 Å, 2–5 Å for strong, medium and weak NOEs, respectively) rather than attempting to obtain precise distances. The distances can be calculated from the NOE values with relaxation matrix methods. A direct NOE structure refinement can be done with the DINOSAUR program (Bonvin et al. 1991). Amide protons exchange against ^2H_2O but they are also important for assignment and structure elucidation. Therefore, it is necessary to obtain some spectra from the lectin in ^1H_2O (with 10% ^2H_2O for field locking). These measurements are much more difficult, because the huge proton signal of H_2O has to be suppressed and it is necessary that the magnetic field is very homogeneous. In some cases, 2D NMR experiments do not provide enough information for complete structural determination. An overlap of resonances in a 2D spectrum can prohibit the measurement of certain interactions of a spin system. As molecules become larger, the probability of overlap of 2D cross-peaks increases, which leads to ambiguities in the analysis of 2D spectra. The introduction of a third frequency domain displaying chemical shifts then makes it possible to spread the overlapping 2D peaks and thus identify the peaks and their chemical shifts unambiguously and derive the structural information (Griesinger et al. 1989; Vuister et al. 1989). Isotopic labeling of the lectin and/or oligosaccharide ligands with ^{13}C and/or ^{15}N-atoms is a convenient method to make the spectra of bigger proteins (10–40 kDa) much easier to analyse.

Lectins with a molecular weight over 50 kDa like Concanavalin A (Con A) (MW = 102 kDa, 4 subunits) (Fig. 1b), succinyl-Con A (51 kDa, 2 subunits) and *Erythrina corallodendron* lectin (ECorL, MW = 60 kDa, 2 subunits) show non-resolved NMR spectra, which means that it is not possible to assign the proton resonances of the different amino acids. Likewise, lectins which are in principle small enough for highly resolved NMR techniques but with a tendency to aggregation at millimolar concentrations like the *Pseudomonas aeruginosa* agglutinin (PsA) (Gilboa-Garber 1982) provide nonresolved spectra. By contrast, wheat germ agglutinin (WGA) (molecular mass 36 kDa, 2 subunits) and *Urtica dioica* agglutinin (UDA) provide highly resolved 1D-NMR spectra. As a representative of the highly conserved group of lectins of the Graminae family, WGA possesses several properties distinct from other plant lectins: (a) high stability due to the presence of 16 disulfide bridges distributed over four homologous isostructural domains, (b) specificity for two different types of carbohydrates (GlcNAc oligomers and NeuNAc, α-anomer), in contrast to most lectins that display specificity for only a single type, (c) requirement of both subunits for carbohydrate binding, (d) the presence of two independent, non cooperative binding sites per subunit, located in domains that represent contact regions of opposite subunits and are related by two fold symmetry. The crystal structures of two WGA isolectins have been described and compared (Wright 1984, 1990). It is therefore possible to compare the structure of these two isolectins with and without bound ligand to study the influence of the different amino acids at the

1D spectrum

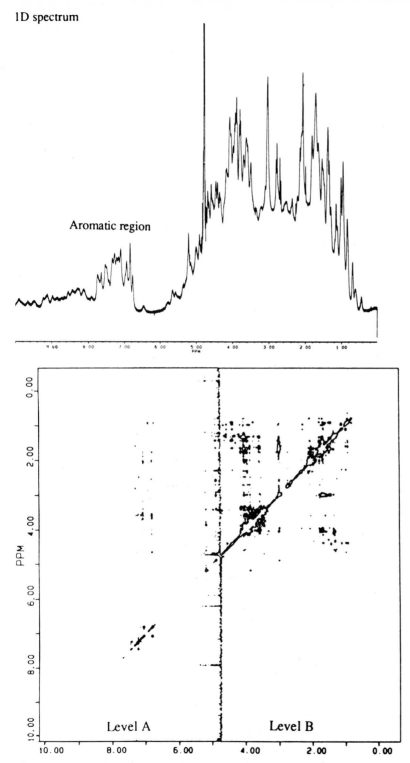

Fig. 2. One-dimensional ^{1}H-NMR spectrum and two-dimensional ^{1}H-NMR NOESY spectrum of a Wheat Germ Agglutinin-N,N′-diacetylchitobiose complex in ^{2}H$_2$O

binding site. Further, the influence of different carbohydrate ligands can be studied, e.g., several sialic acids (N-glycolyl and O-acetyl) which form a complex with WGA. NMR spectroscopic measurements probing the environment of bound carbohydrate have not yet allowed the assignment of effects specifically to either of the two independent binding sites. It has been possible to characterize WGA-N,N'-diacetylchitobiose and WGA-N-acetylneuraminic acid complexes by X-ray structure analysis. A 600 MHz 2D NOESY spectrum together with the 1D spectrum of a WGA-N,N'-diacetylchitobiose complex is shown in Fig. 2. The 2D spectrum is plotted on two levels. The lower level (A) shows contacts between protons of aromatic amino acids and protons of N,N'-diacetylchitobiose. For an analysis of the protein carboydrate contacts it is, of course, necessary to compare this spectrum with a 2D NOESY spectrum of native WGA recorded under the same experimental conditions.

11.1.2 Transfer NOE Spectroscopy

Transfer NOE (TrNOE) experiments can provide information about the conformation of a small ligand bound to a larger receptor molecule (Albrand et al. 1979; Clore and Gronenborn 1983). These experiments take advantage of chemical exchange to transfer information about the bound state to the free state and involve measurement of the negative NOE of a ligand spin following irradiation of the free, bound or averaged resonances of another ligand spin. The transfer of information allows observation of effects developed in the bound state at the resonance of an abundant nucleus with negligible effects of line broadening and interference from the protein.

Methods In principle, any lectin can be used for TrNOE experiments, even those that show nonresolved NMR spectra, because only the sharp signals of the carbohydrate ligand are irradiated. These experiments have proven to be an extremely effective means of exploring the conformation and exchange dynamics of carbohydrates bound to lectins, as has been described for the ricin B-chain (Bevilacqua et al. 1990). Lectin-carbohydrate interactions are usually assumed to be second-order reactions in the formation of the complex and first-order in the dissociation of the complex. In TrNOE studies, it is the equilibrium between free ligand and complex that is monitored. These studies depend on the possibility of acquiring data in an appropriate "time window" for accurate distance evaluation. This means that a suitable compromise between carbohydrate/lectin concentration and mixing time – depending on the binding constant – has to be chosen in order to determine distances in a sufficiently accurate manner. In the case of ligands with positive NOE, the molar fractions of bound and free ligand must be adjusted adequately to emphazise the strong negative NOE arising from the bound state. Consequently, only a ten fold molar excess of the carbohydrate ligand over lectin was used. Three conditions should be fulfilled to ensure a successful study: (1) The positive NOE of the free ligand has to be suppressed, (2) the effects of "spin diffusion" have to be minimized, and (3) it must be possible to identify signals originating from protein-protein interactions or to avoid contributions of such signals to the intraligand NOEs. Regarding point (2), a straightforward analysis of fully protonated ligands using equations based on the independent spin pair approximation proves to be misleading. However, selective deuteration and/or a complete relaxation matrix analysis allow determination of a bound structure. A program for the analysis of the relaxation matrix has been developed (Boelens et al. 1989). For TrNOE experiments it is also important to have information about the thermodynamic properties of ligand-receptor interaction. Exchange is described by the rate constants k'_1 and k_{-1}, which are reciprocal lifetimes for molecules in bound and free states. The forward rate is explicitly dependent on the free protein concentration; $k'_1 = k_1[P]$. Free protein concentration is in turn related to an equilibrium constant, $K_a = k_1/k_{-1}$, and a free ligand concentration that is taken to be approximately equal to the total ligand concentration in our experiments. ^{13}C-TrNOE experiments have recently been described (Oschkinat et al. 1992). The advantages of TrNOE measurements using ω_1-^{13}C-filtered NOESY and ^{13}C-^1H-relayed NOESY are the removal of residual protein signals, the simplification of zero quantum suppression, and the improvement of water suppression, which enabled the semi-quantitative study of ^1H-^1H distances of the ligand in the bound state.

11.2 Computational Methods

Molecular Mechanics and Molecular Dynamics (van Gunsteren 1988) calculations are of great importance in assessing the conformations of flexible oligosaccharides and proteins. A molecule can be energetically minimized by using an energy function or interaction potential and searching for the local minima.

MD simulations were carried out on a Local Area VAX Cluster using the GROMOS (GROningen MOlecule Simulation) computer program package. The system consisted of the oligosaccharide and 300 water molecules in a cubic periodic box, and was simulated at a constant temperature (300 K) and pressure (1 atm).

Equipment

The potential energy, $V(r)$, of the molecular system to be analysed is the basic function of the Molecular Mechanics and Molecular Dynamics calculations:

Procedure

$$V(r_1, r_2, \ldots, r_N) = \sum_{\text{bonds}} 1/2K_b[b - b_0]^2 + \sum_{\text{angles}} 1/2K_\theta[\theta - \theta_0]^2$$

$$+ \sum_{\text{torsions}} 1/2K_\xi[\xi - \xi_0]^2 + \sum_{\text{dihedrals}} 1/2K_\phi[1 + \cos(n\phi - \delta)]$$

$$+ \sum_{\text{pairs } (i,j)} [C_{12}(i,j)/r_{ij}^{12} - C_6(i,j)/r_{ij}^6 + q_iq_j/(4\pi\varepsilon_0\varepsilon_r r_{ij})]. \qquad (3)$$

$V(r)$ describes the potential energy of the molecular system as a function of the positions r_i of the N atoms labelled by the index i. The first term represents the covalent bond-stretching interaction along bond b. It is a harmonic potential in which the minimum energy bond b_0 and the force constant K_b vary with the particular type of bond. The second term describes the bond-angle bending inter-action in a similar form. Two forms are used for the torsional or dihedral-angle interactions: a harmonic term for dihedrals (torsion ξ) that are not allowed to make transitions, e.g., dihedral angles within aromatic rings, and a sinusoidal term for the other dihedrals (ϕ), which may make 360° rotations. The last term is a sum over all pairs of atoms and represents the effective nonbonded interaction, composed of van der Waals and the Coulombic interactions between atoms i and j with charges q_i and q_j at a distance r_{ij}. When applying Energy Minimization (EM) algorithms, one searches for a minimum energy configuration by moving along the gradient of the potential energy through configuration space

$$\Delta r_i \; \mathrm{d}/\mathrm{d}r_i \; V(r_1, r_2, \ldots, r_3).$$

These force fields contain hundreds of parameters that have been derived from ab initio quantum calculations on small molecules, or from experimental data (spectroscopic, crystallographic, neutron scattering etc.). Since in this way one moves basically only downhill over the energy hypersurface in configuration space, EM moves the system to a minimum energy configuration that is generally not far from the initial one. The combination of systematic search with EM searches a larger part of the conformation space than the application of EM alone.

The search of configuration space can be expanded by applying the MD computer simulation technique. The classical Newtonian equations of motions of all atoms in the system are solved by the application of MD. These are for a set of N atoms with masses m_i and Cartesian position vectors r_i ($i = 1, \ldots, N$):

$$\mathrm{d}^2 r(t)/\mathrm{d}t = m_i^{-1} F_i[r_1(t), r_2(t), \ldots, r_N(t)], \tag{4}$$

where the force exerted on atom i is found from

$$F_i(t) = \partial/\partial r_i(t) V[r_1(t), r_2(t), \ldots, r_N(t)]. \tag{5}$$

These equations can be numerically integrated using small ($\sim 10^{-15}$ s) time-steps Δt producing a trajectory (atomic positions as a function of time t) of the system. Due to the inertia of the atoms, MD enables the system to surmount energy barriers which are of the order of kT, since the average kinetic energy per degree of freedom is $1/2kT$. The relation between velocity and temperature according to the Maxwell equation means that it is possible to calculate the dynamics of the molecule at various temperatures. Even when the flexibility of a system is taken into account by minimization of the potential energy with respect to the atomic positions, a description of its properties in terms of static (energy minimized) structures is inadequate. A molecular system at room temperature is by no means static. It traverses multiple minima of the potential energy surface. In principle, one would like to know the multidimensional distribution function of all atomic coordinates and the development of this function over time. In practice, computer simulation techniques like MD generate an ensemble of configurations which is assumed to be an adequate representation of the complete distribution function. In a number of cases, a description of the system in terms of an ensemble of configurations is essential.

1. When molecular properties are to be obtained as a function of temperature, a static model is inadequate.
2. When the molecular system is characterised by a dynamic equilibrium between different configurations, e.g., different mutually exclusive patterns of hydrogen bonds, static modeling techniques are insufficient.
3. The solvent around a protein, a DNA fragment or a complex carbohydrate cannot be described by one or a few configurations, but can only be represented by an ensemble of configurations such as generated in a MD simulation.

Oligosaccharides are considered to be flexible about their glycosidic linkage (Cumming and Carver 1987). Analyzing the conformations of several GM3 gangliosides, it has been shown (Siebert et al. 1992a) that MD calculations at elevated temperature and in water as solvent generate the conformations predicted by NMR measurements. The possible conformations of the Neu5Ac(α2-3)Gal linkage have been determined with the help of NMR methods and molecular modeling (e.g., distance geometry and molecular mechanics). MD calculations with the Consistence Valence Force Field (CVFF) (Hagler et al. 1974) show that transitions occur between the two predicted conformations (Dabrowski et al. 1990; Siebert 1990). It should be emphasised that the conformations of the glycosidic linkage are independent of the different functional groups (N-acetyl, N-glycolyl and O-acetyl) on the sialic acid. This is remarkable, because the biological influence of the different functional groups on sialic acid is significant (Schauer 1982; Reuter and Schauer 1987). A 500 ps MD simulation of a Neu5Ac(α2-3)Gal disaccharide and Neu5Ac(α2-3)Gal(β1-4)GlcNAc trisaccharide with the GROMOS program also shows transitions between the two conformations of the sialic acid (α2-3) galactose linkage. A trajectory of the glycosidic angles Φ and Ψ is shown in Fig. 3.

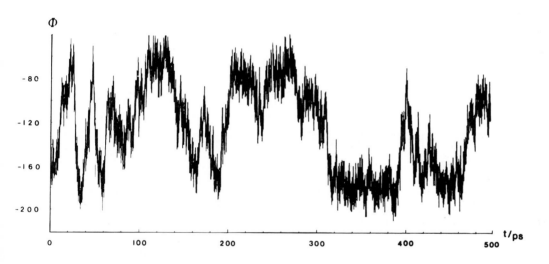

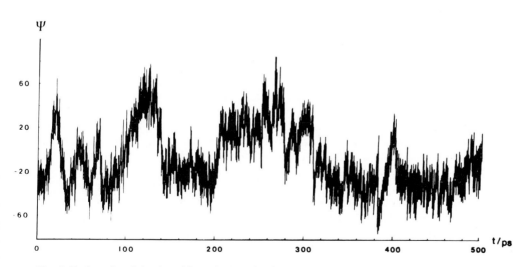

Fig. 3. Trajectories of the glycosidic angles Φ and Ψ from the Neu5Acα2-3Gal disaccharide in a 500 ps Molecular Dynamic calculation in water

Monte Carlo calculations of oligosaccharides (Stuike-Prill and Maier 1990) can also be used as an other computational technique to explore the conformational space.

11.3 Discussion

The methods introduced in this chapter are complementary to each other and suitable to study the lectin-oligosaccharide interaction under physiological conditions. With the help of these methods, much important information about carbohydrate-lectin interaction can be delineated. Flexibility of oligosaccharides is important in that it allows them a fit into binding sites. The binding site might also undergo changes in order to optimize the fit. Therefore a conformational analysis of both the oligosaccharide and protein as well as the structural analysis of the complex are necessary for oligosaccharide-lectin interaction studies. The importance of hydrogen bonding and van der Waals forces have long been appreciated in determining the specificity and affinity of carbohydrate-protein interactions, while the stacking of aromatic residues against the faces of carbohydrate residues has only recently been appreciated (Quiocho 1989). This is not surprising, as there is a conformation of a sugar ring ($^4C_1$) with partially exposed hydrophobic clusters of C–H bonds on each face that can interact with suitable aromatic residues. This fact is very important for laser photo CIDNP (Chemically Induced Nuclear Polarisation) studies (Kaptein et al. 1978; Kaptein 1982) and makes this method a proper tool for oligosaccharide-lectin interactions. Laser photo CIDNP signals can only be obtained from surface-exposed aromatic amino acids. If these surface-exposed aromatic amino acids are part of the binding domain of the lectin, the laser photo CIDNP signal can be influenced by the bound oligosaccharide. If aromatic amino acid protons of the binding site can be identified by this method, one has important additional information for the assignment of the proton signals belonging to amino acids at the binding site (Siebert et al. 1992b).

Local changes in lectin and lectin/oligosaccharide structure are relatively easy to predict with MD calculations if reliable NMR and/or X-ray data are available. MD simulation is an excellent tool for searching local configuration space for low energy conformations if proper conformational data are available. Static modeling of a protein ignores a possible structural response of the protein to an amino acid residue substitution. In these calculations it is only the absence or presence of the charge which influences the field, not its precise location at a distance from the active site. Structure-function relationships have been difficult to establish for carbohydrates, and the synthesis of glycopeptides and glycolipids holds great promise for addressing the biological role of oligosaccharides. The sequence of the carbohydrate chains, their linkage type and branching mode, and particular modifications such as phosphorylation, sulfation, O-acetylation etc. establish a vocabulary, fulfilling the requirements for a role in biosignaling (Rademacher et al. 1988). However, it is only possible to understand these vocabularies if their complementary counterpart (the lectins) is partly understood on a molecular level. Since the oligosaccharide interacts with the protein surface, preferential stabilization of one conformation presumably arises.

Lectins from plant origin sometimes have a toxic domain in addition to the carbohydrate-binding domain. Structural information can be used to modify the toxic part as well as the binding part. Drug design and drug targeting are the goals of the structural analysis of lectin-carbohydrate complexes with NMR and computational methods.

References

Albrand JP, Birdsall B, Feeney J, Roberts GCK, Burgen ASV (1979) The use of transferred nuclear Overhauser effects in the study of the conformations of small molecules bound to proteins. Int J Biol Macromol 1: 37–41

Bevilacqua VL, Thomson DS, Prestegard JH (1990) Conformation of methyl β-lactoside bound to ricin B-Chain: interpretation of transferred nuclear Overhauser effects facilitated by spin simulation and selective deuteration. Biochemistry 29: 5529–5537

Boelens R, Koning TMG, van der Marel GA, van Boom JH, Kaptein R (1989) Iterative procedure for structure determination from proton-proton NOEs using a full relaxation matrix approach. Application to a DNA octamer. J Magn Reson 82: 290–308

Bonvin AMJJ, Boelens R, Kaptein R (1991) Direct NOE refinement of biomolecular structures using 2D NMR data. J Biomol NMR 1: 305–309

Bystrow VF (1976) Spin-spin coupling and the conformational states of peptide systems. Progress in NMR Spectroscopy 10: 41–81

Clore GM, Gronenborn AM (1983) Theory of time-dependent transferred nuclear Overhauser effect: applications to structural analysis of ligand-protein complexes in solution. J Magn Reson 53: 423–442

Cumming DA, Carver JP (1987) Virtual and solution conformations of oligosaccharides. Biochemistry 26: 6664–6676

Dabrowski J (1987) Application of two-dimensional NMR methods in the structural analysis of oligosaccharides and other complex carbohydrates. In: Croasmun WR and Carlson RMK (eds) Two-dimensional NMR spectroscopy, Verlag Chemie, Weinheim, pp 349–386 (Methods in stereochemical analysis, Vol. 9)

Dabrowski J (1989) Two-dimensional proton magnetic resonance spectroscopy. Meth Enzymol 179: 122–156

Dabrowski J, Poppe L, Siebert HC, v d Lieth CW (1990) Conformational equilibria of oligosaccharides as determined by rotating frame ^{1}H-NMR spectroscopy and confirmed by molecular dynamics simulations. In:Bethge K (ed) Structure and conformational dynamics of biomacromolecules. Europhysics Conference Abstracts 14H, European Physical Society, High Tatras, p 45

Ernst RR, Bodenhausen, G, Wokaun A (1987) Principles of nuclear magnetic resonance in one and two dimensions. Clarendon Press, Oxford

Gilboa-Garber N (1982) Pseudomonas aeruginosa lectins. Meth Enzymol 83: 378–385

Griesinger C, Sørensen OW, Ernst RR (1989) Three-dimensional Fourier spectroscopy. Application to high resolution NMR. J Magn Reson 84:14–63

van Gunsteren WF (1988) The role of computer simulation techniques in protein engineering. Protein Engineering 2, 1: 5–13

Hagler AT, Huber E, Lifson S (1974) Energy functions for peptides and proteins. Deviation of a consistent force field including the hydrogen bond for amide crystals. J Am Chem Soc 96: 5316–5327

Kaptein R, Dijkstra K, Nicolay K (1978) Laser photo-CIDNP as a surface probe for proteins in solution. Nature 274: 293–294

Kaptein R (1982) Photo CIDNP studies of proteins. In: Berliner LJ (ed) Biological magnetic resonance, Vol. 4, Plenum, New York, pp. 145–191

Kaptein R, Boelens R, Scheek RM, van Gunsteren WF (1988) Protein structures from NMR. Biochemistry 27: 5389–5395

Karplus M (1959) Contact electron-spin coupling of nuclear magnetic moments. J Chem Phys 30: 11–15

Neuhaus D, Williamsen MP (1989) The nuclear Overhauser effect in structural and conformational analysis. VCH Publishers, New York

Noggle JH, Schirmer RE (1971) The nuclear Overhauser effect – chemical applications. Academic Press, New York

Oschkinat H, Schott K, Bacher A (1992) Conformation of 6,7-dimethyl-8-ribityllumazine bound to β-subunits of heavy riboflavin synthase: transferred nuclear Overhauser effect studies employing ω_1-^{13}C-filtered NOESY including a novel technique for zero quantum suppression. J Biomol NMR 2: 19–32

Poppe L, Dabrowski J, v d Lieth CW, Koike K, Ogawa T (1990a) Three-dimensional structure of the oligosaccharide terminus of globotriaosylceramide and isoglobotriaosylceramide in solution. Eur J Biochem 189: 313–325

Poppe L, v d Lieth CW, Dabrowski J (1990b) Conformation of the glycolipid globoside head group in various solvents and in micelle-bound state. J Am Chem Soc 112: 7762–7771

Quiocho FA (1989) Protein-carbohydrate interactions: basic molecular features. Pure & Appl Chem 61, 7: 1293–1306

Rademacher TW, Parekh RB, Dwek RA (1988) Glycobiology. Annu Rev Biochem 57: 785–838

Reuter G, Schauer R (1987) Isolation and analysis of gangliosides with O-acetylated sialic acids. In: Rahmann, H (ed) Gangliosides and modulation of neuronal functions, NATO ASI Series H, Springer Verlag, Berlin Heidelberg New York, pp 155–167 (Cell biology, Vol. 7)

Schauer R (1982) Chemistry, metabolism and function of sialic acids. Adva Carbohydr Chem Biochem 40: 131–234

Siebert HC (1990) Konformationsanalyse verschiedener Ganglioside mit Hilfe von ^{1}H-Kernersonanzmethoden und Computerberechnungen. Thesis, Heidelberg University

Siebert HC, Reuter G, Schauer R, v d Lieth CW, Dabrowski J (1992a) Solution conformation of GM3 gangliosides containing different sialic acid residues as revealed by NOE-based distance-mapping, and molecular mechanics and molecular dynamics calculations. Biochemistry 31: 6962–6971

Siebert HC, Pouwels PJW, Kaptein R, Kamerling JP, Vliegenthart JFG (1992b) Study of the wheat germ agglutinin-oligosaccharid interaction by laser photo CIDNP experiments. Abstracts of the XVIth International Carbohydrate Symposium, 5–10 July 1992, Paris, France p 534

Stuike-Prill R, Maier B (1990) A new force-field program for the calculation of glycopeptides and its application to a heptacosapeptide-decasaccharide of immunoglobolin G_1. Eur J Biochem 194: 903–913

Vuister GW, de Waard P, Boelens R, Vliegenthart JFG, Kaptein R (1989) The use of 3D NMR in structural studies of oligosaccharides. J Am Chem Soc 111: 772–774

Wright CS (1984) Structural comparison of the two distinct sugar binding sites in wheat germ agglutinin isolectin II. J Mol Biol 178: 91–104

Wright CS (1990) 2.2 Å resolution structure analysis of two refined N-acetylneuraminyl–lactose–wheat germ agglutinin isolectin complexes. J Mol Biol 215: 635–651

Wüthrich K (1986) NMR of proteins and nucleic acids. Wiley, New York

12 Lectin-Glycoconjugate Cross-Linking Interactions

D.K. MANDAL and C.F. BREWER

Lectin binding to the surface of cells often leads to cross-linking and aggregation of specific glycoconjugate receptors including glycoproteins and glycolipids which, in many cases, is associated with a variety of biological responses. For example, cross-linking of glycoconjugates on the surface of cells has been implicated in the mitogenic activities of lectins such as Concanavalin A (Con A) and soybean agglutinin (SBA) (cf. Nicolson 1976); in the molecular sorting of glycoproteins in the secretory pathway of cells (Chung et al. 1989); and in signal transduction mechanism of certain glycoprotein hormones (Sairam 1989). Furthermore, lectin-induced cross-linking of transmembrane glycoproteins result in changes in their interactions with cytoskeletal proteins and alterations in the mobility and aggregation of other surface receptors (cf. Edelman 1976; Carraway and Carraway 1989).

Asparagine-linked (N-linked) oligosaccharides represent one class of cell surface carbohydrates which are generally classified into three subtypes: oligomannose, complex and hybrid (Lennarz 1980). Complex type oligosaccharides are often the final products of N-glycosylation pathways and are generally present on the surface of cells in greater amounts than the other two subtypes (Kornfeld and Kornfeld 1985). Our investigations of the molecular recognition properties of N-linked oligosaccharides and glycopeptides and their interactions with lectins have revealed that many of these carbohydrates are multivalent and can bind, cross-link, and precipitate with lectins of various carbohydrate binding specificities (cf. Bhattacharyya et al. 1990). These cross-linking interactions lead to a new dimension of specificity in carbohydrate-protein interactions: namely, the formation of a unique, homogeneous cross-linked complex between each carbohydrate and lectin, even in the presence of mixtures of the molecules.

The stoichiometry and the specificity of the cross-linking interactions between lectins and carbohydrates are analyzed by quantitative precipitation analyses. Furthermore, several lectin-carbohydrates precipitates are crystalline and can be investigated by electron microscopy and X-ray diffraction techniques. Quantitative precipitation studies can also be extended to examining glycoprotein-lectin cross-linking interactions. The results indicate that certain plant lectins as well as an S-type animal lectin form specific cross-linked complex(es) with certain glycoproteins by carbohydrate-mediated cross-linking (Khan et al. 1991; Mandal and Brewer 1992). In the present article, we describe the techniques of quantitative precipitation analyses and electron microscopy to study the cross-linking interactions between lectins and glycoconjugates.

12.1 Experimental Part

Lectins and glycoproteins (prepared according to published methods)

Materials

N-dimethylated and ^{3}H-/^{14}C-radiolabeled proteins [prepared by reductive methylation (Khan et al. 1991; Mandal and Brewer 1992)]

Oligosaccharides and glycopeptides of appropriate structures [(can be isolated from glycoproteins according to published methods or obtained by synthesis (BioCarb Chemicals, Sweden); the purity of the oligosaccharide and glycopeptide is checked by high resolution ^{1}H NMR at 500 MHz (Vliegenthart et al. 1983)]

N-dimethylated and ^{3}H-/^{14}C-radiolabeled glycopeptide [prepared by reductive methylation (Bhattacharyya et al. 1988a); the oligosaccharides can be radiolabeled by reduction to the corresponding alcohols with sodium borotride in NaOH (Bhattacharyya and Brewer 1992)]

Borosilicate glass culture tubes (13 × 100 mm)

Centrifuge (Sorvall RC 5B; rotor SS 34)

Spectrophotometer

Scintillation counter (LKB Rackbetta)

All reagents are of analytical grade

Concentrations

Determination of Protein and Carbohydrate Concentrations
Protein concentrations are usually determined spectrophotometrically using published $A1\%$, 1 cm values at 280 nm and expressed in terms of monomer. The concentrations of oligosaccharides and glycopeptides are measured by the phenol-sulfuric acid method (Dubois et al. 1956) using appropriate mixtures of monosaccharides as standards.

Precipitation

Quantitative Precipitation Assays
The experiments are set up in borosilicate glass culture tubes (13 × 100 mm) in a final volume of 400 or 200 μl at room temperature or 4 °C (when precipitation occurs at 4 °C but not at room temperature) in appropriate buffer (generally, 0.1 M Tris-HCl or 0.1 M HEPES buffer, pH 7.2 containing 0.9 M NaCl, 1 mM MnCl$_2$ and 1 mM CaCl$_2$ is used as a high salt containing buffer and 10 mM sodium phosphate buffer, pH 7.2 containing 0.15 M NaCl, 0.1 mM MnCl$_2$ and 0.1 mM CaCl$_2$ is used as a low salt containing buffer for assays with plant lectins).

12.1.1 Assays with Individual Oligosaccharide and Glycopeptide

Increasing amounts of carbohydrate are taken in a series of tubes containing buffer in a total volume of 200 μl. In the same buffer 200 μl of lectin solution was added in each tube, mixed immediately in a Vortex mixer and allowed to stand for about 20 h at room temperature (or at 4 °C). The mixture is centrifuged at 6000 rpm for 3 min at the same temperature. Supernatants are drawn off carefully by a Pasteur pipet. The concentrations of lectin are measured in the supernatants (after dilution of an aliquot with buffer to an appropriate volume) in a spectrophotometer at 280 nm to obtain percent lectin precipitated. Control experiments are performed with a competing sugar added to the solutions. Plot of % lectin precipitated as a function of oligosaccharide or glycopeptide concentration gives the profile for the quantitative precipitation of lectin with the carbohydrate. The stoichiometry of the precipitation reaction is obtained from the ratio of the concentration of carbohydrate to lectin at the equivalence point (region of maximum precipitation).

12.1.2 Assays with Binary Mixture of Carbohydrates or Lectins

Increasing amounts of a mixture of two oligosaccharides or glycopeptides in the appropriate ratio (one carbohydrate is ^{3}H-labeled and the other ^{14}C-labeled) are taken in a series of tubes containing buffer to a total volume of 200 μl. Two hundred μl of lectin solution is added in each tube, mixed in a Vortex mixer and allowed to stand for about 20 h at room temperature (or at 4 °C). The mixture is centrifuged at 6000 rpm for 3 min at the same temperature and the supernatants are drawn off carefully and rejected. The precipitates are resuspended in 100 μl of ice-cold buffer by vortexing, centrifuged at 4 °C and the supernatants rejected. The washing procedure is repeated and the precipitates are dissolved in 0.1 M solution of a competing sugar (e.g, methyl α-D-mannopyranoside for D-Glc/D-Man-specific plant lectin Concanavalin A) to a final volume of 2 ml. The resulting solutions are analyzed for protein concentration in a spectrophotometer at 280 nm and an aliquot (1 ml) of the solutions is used for measuring radioactivity in a liquid scintillation counter to monitor the amounts of the carbohydrate in the precipitates. A plot of % lectin precipitated and cpm of each carbohydrate as a function of total concentration of carbohydrates gives the mixed precipitation profile. The presence of two protein peaks corresponding to the precipitation maxima of the carbohydrates indicates the formation of homogeneous cross-linked lattice between the lectin and each carbohydrate.

Assays with binary mixture of lectins are performed in a similar way by titrating a constant amount of carbohydrate with increasing concentration of a mixture of lectins in appropriate ratio.

12.1.3 Assays with Individual Glycoproteins

In these assays, lectin and glycoprotein are radiolabeled with ^{3}H and ^{14}C or vice versa. Increasing amounts of radiolabeled glycoprotein are taken in a series of tubes containing buffer and a constant amount of radiolabeled lectin is added to each tube to a final volume of 200 μl. The precipitates are treated in the same way as described in Section 12.1.2 above. However, in some cases, higher concentration of sugar solution is used to speed up the dissolution process of the precipitates. The specific activity of individual proteins is determined by measuring the cpm values for solutions of known concentrations of each protein and expressed in cpm/nmol. From the cpm of lectin and glycoprotein in the precipitates and their respective specific activity values, mol/mol ratio (per monomer) of glycoprotein to lectin is calculated. Plot of absorbance (total protein), cpm of each protein as well as mol/mol ratio (glycoprotein:lectin) as a function of glycoprotein concentration yields the quantitative precipitation profile.

Precipitates are formed either by using equivalent concentrations (in terms of binding sites) of lectin and the oligosaccharides or from different data points across the precipitation profile. Precipitates are negatively stained by placing the samples on 300 mesh carbon-coated Parlodion grids that had been freshly glow-discharged, touched to filter paper, floated on a drop of 1% phosphotungstic acid, pH 7.0, and blotted immediately. Samples are observed at 80 kV in a JEOL 1200EX electron microscope. For freeze fracture studies, samples are placed in a gold double-replica device, frozen in liquid Freon, and fractured in a Balzers BAF301 freeze-fracture unit

Electron Microscopy

at −115 °C. The fracture face was shadowed at a 45° angle with platinum and stabilized with carbon. Samples are observed as described above.

12.2 Remarks

1. In all precipitation assays, control experiments should be performed in the presence of competing as well as noncompeting sugars. Precipitates are inhibited from forming, or dissolved by competing sugars but not by non-competing ones.
2. Reductive methylation is found to be a suitable method to introduce a radiolabel in a glycopeptide, lectin, or glycoprotein as the structure, affinity, and precipitating activity of the derivatized protein/glycopeptide are similar to the parent ones. However, in some cases, the affinity and the precipitating activity are drastically reduced by N-methylation of lectins (e.g., RCA-I). In such cases, the amount of protein in the precipitate can be monitored by alternative procedure such as quenching of fluorescence of fluorescein mercuric acetate by disulfide bonds in 1 M NaoH at excitation and emission wavelengths of 490 and 520 nm, respectively (Karush et al. 1964), if one of the two proteins in a mixture possesses disulfide bonds.
3. While several lectin-carbohydrate precipitates show observable lattice pattern in electron microscopy, others do not. This may be due to several factors which are not well understood at present.
4. Structural analyses of the highly organized carbohydrate-lectin lattices can be performed by combined X-ray diffraction, electron microscopy, image analysis, and computer modeling techniques.

12.3 Results

12.3.1 Quantitative Precipitation Profile of Lectins with Individual Oligosaccharides or Glycopeptides

A typical quantitaive precipitation profile D-Glc/D-Man specific plant lectin Concanavalin A (Con A) (76 µM) in the presence of bisected hybrid type glycopeptide **2** (Fig. 1) is shown in Fig. 2 (Bhattacharyya et al. 1988b). The curve is bell-shaped, which indicates tight binding of the glycopeptide to Con A (Kabat 1976). Figure 2 also shows the cpm of the glycopeptide in the precipitate, which is similar to the profile of the percent Con A precipitated when an equimolar mixture of N-dimethylated and ^{14}C-labeled glycopeptide is used. The ratio of the concentration of glycopeptide to Con A monomer at the equivalence zone (maximum precipitation) is 1:1.8 which gives the stoichiometry of the reaction. Since each monomer of Con A possesses one sugar-binding site, this result indicates that the glycopeptide is bivalent for Con A binding. Detailed structure-activity studies have shown that the two Con A binding sites are located on the α(1-6) and α(1-3) arms of the core β-mannosyl residue of the glycopeptide. The site on the α(1-6) arm (the trimannosyl moiety) is identified as the high affinity or primary site, and the site on the α(1-3) arm as the low affinity or secondary site.

Man —α1,2— Man ⟍α1,6
Man —α1,2— Man ⟋α1,3 ⟍Man⟍α1,6 —β1,4— ⟍β1,4⟍
Man —GlcNAc —β1,4— GlcNAc —β1— Asn   **1**
Man —α1,2— Man —α1,2— Man ⟋α1,3

Man ⟍α1,6
Man —α1,3— ⟍Man⟍α1,6 —β1,4
GlcNAc —β1,4⟋ Man —GlcNAc —β1,4— GlcNAc —β1— Asn   **2**
GlcNAc —β1,2— Man ⟋α1,3

Fuc —α1,3— GlcNAc —β1,2— Man ⟍α1,6
⟍Man - OH   **3**
Fuc —α1,3— GlcNAc —β1,2— Man ⟋α1,3

Fuc ⟍α1,3
GlcNAc —β1,2— Man ⟍α1,6
Gal ⟋β1,4
Gal ⟍β1,4 ⟍Man - OH   **4**
GlcNAc —β1,2— Man ⟋α1,3
Fuc ⟋α1,3

Fuc ⟍α1,3
GlcNAc ⟍β1,6
Gal ⟋β1,4
⟍Gal —β1,4— Glc - OH   **5**
Gal ⟍β1,4 ⟋β1,3
GlcNAc ⟋
Fuc ⟋α1,3

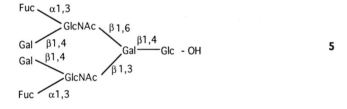

Gal —β1,4— GlcNAc —β1,2⟍
⟍Man ⟍α1,6
Gal —β1,4— GlcNAc ⟋β1,6
⟍Man-OH   **6**
Gal —β1,4— GlcNAc —β1,4— Man ⟋α1,3

Gal —β1,4— GlcNAc —β1,2— Man ⟍α1,6
⟍Man -OH   **7**
Gal —β1,4— GlcNAc —β1,2⟍
⟍Man ⟋α1,3
Gal —β1,4— GlcNAc ⟋β1,4

Fig. 1. Structures of the oligomannose type glycopeptide **1**, bisected hybrid type glycopeptide **2**, fucosyl biantennary oligosaccharides **3–5**, triantennary complex type oligosaccharides **6–7**, and tetraantennary complex type oligosaccharides **8–9**. *Fuc, Gal, Man*, and *GlcNAc* represent fucose, galactose, mannose, and N-acetylglucosamine residues, respectively. Fucose is in the L configuration; all other sugars in the D configuration

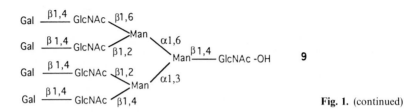

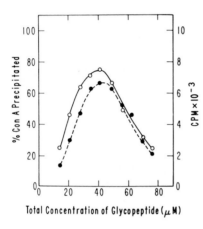

Fig. 1. (continued)

Fig. 2. Precipitation profile of Con A (○) in the presence of an equimolar mixture of glycopeptide **2** and the corresponding ^{14}C-dimethylated glycopeptide (●) in the precipitate. The buffer was 0.1 M Tris-HCl containing 0.9 M NaCl, 1 mM $MnCl_2$, and 1 mM $CaCl_2$ at pH 7.2. The protein concentration was 76 μM in a final volume of 400 μl. (Bhattacharyya et al. 1988b)

Quantitative precipitation studies of several galactose-specific plant lectins including those from *Erythrina indica* (EIL), *Ricinus communis* (agglutinin I) (RCA-I) and *Glycine max* (soybean) (SBA) with the complex type tri- and tetraanatennary oligosaccharides **6,7,8** (Fig. 1), which contain nonreducing terminal galactose residues show that these carbohydrates precipitate the lectins (Bhattacharyya et al. 1988a). The equivalence points of the precipitin curves indicate that the tri- and tetraantennary oligosaccharides are tri- and tetravalent, respectively, for EIL and SBA binding (Table 1). However, the oligosaccharides are all trivalent for RCA-I binding, due apparently to the larger size of the monomeric subunit of the lectin. The results indicate that, in general, each arm of the branched chain oligosaccharides can bind individual lectin molecules, which leads to cross-linking and precipitation. The extent of precipitation is dependent not only on the valency of the oligosaccharides but, as in the case of the triantennary carbohydrates, on their branching pattern as well. For example, at a protein concentration of 44 μM, **7** precipitates only 15% of EIL at the equivalence zone, compared to about 32% by **6** and 49% by **8**, whereas at a protein concentration of 92 μM, **7** precipitates 65% of RCA-I at the eqivalence zone, compared to 57% and 75% by **6** and **8**, respectively.

Table 1. Stoichiometries of precipitation reactions of the galactose-specific plant lectins with tri- and tetraantennary oligosaccharides

Oligosaccharide	Protein conc. (μM)	Oligosaccharide conc. at equivalence point (μM)	Stoichiometry
		EIL	
6	44	15	1:2.9
7	94	35	1:2.7
8	44	11	1:4.0
		RCA-I	
6	92	31	1:3.0
7	92	33	1:2.8
8	92	32	1:2.9
		SBA	
6	280	120	1:2.3
7	260	92	1:2.8
8	260	69	1:3.8

Data for EIL and RCA-I were obtained at 22 °C and for SBA at 4 °C. Stoichiometry refers to number of lectin monomers bound to one carbohydrate molecule.

12.3.2 Quantitative Precipitation Profiles with Binary Mixture of Carbohydrates or Lectins

A typical quantitative precipitation profile of Con A (77 μM) in the presence of an equimolar mixture of an oligomannose type Man9 glycopeptide **1** (Fig. 1) and bisected hybrid type glycopeptide **2** in Fig. 3, shows two protein peaks for the percent protein precipitated. The radioactive profiles show that the precipitation maximum of **2** occurs at a total glycopeptide concentration of 56 μM which corresponds with the first protein peak and that the precipitation maximum for **1** occurs at 90 μM, which corresponds with the second protein peak. These findings indicate that each glycopeptide forms a unique, homogeneous cross-linked lattice

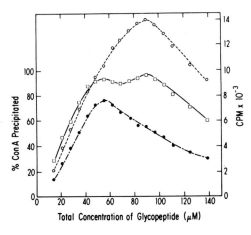

Fig. 3. Precipitation profile of Con A (□) in the presence of a 50:50 mixture of ^{3}H-dimethylated glycopeptide **1** and ^{14}C-dimethylated glycopeptide **2** at 22 °C. The cpm of **1** (○) and (●) in the precipitate are also shown. The buffer was the same as that in Fig. 2. The protein concentration was 77 μM in a final volume of 400 μl

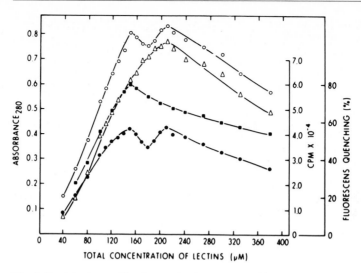

Fig. 4. Precipitation profile of an equimolar mixture of ^{14}C-labelled EIL and RCA-I in the presence of ^{3}H-labelled **9** at 22 °C. (○) Total protein ; (●) ^{3}H; (△) ^{14}C; (■) fluorescence quenching (RCA-I in the precipitates). The buffer was 10 mM sodium phosphate containing 0.15 M NaCl, 0.1 mM MnCl$_2$ and 0.1 mM CaCl$_2$ at pH 7.2. The oligosaccharide concentration was 41 µM in a final volume of 200 µl. (Bhattacharyya and Brewer 1992)

with the lectin which excludes the lattice of another glycopeptide in a mixture (for detailed discussion, see Bhattacharyya et al. 1988b).

Figure 4 shows the results of the quantitative precipitation of ^{3}H-labeled tetra-antennary oligosaccharide **9** in the presence of an equimolar mixture of lectins, ^{14}C-labeled EIL and RCA-I (Bhattacharyya and Brewer 1992). The profile of the total amount of the protein in the precipitates shows two protein peaks which run essentially parallel with the two peaks in the oligosaccharide profile (^{3}H data). The ^{14}C data in the precipitates show that the second peak in the protein (and in the oligosaccharide) profile corresponds to the precipitation maximum of EIL, whereas the fluorescence quenching data show that the precipitation maximum of RCA-I corresponds to the first peak in the protein profile. These results indicate the formation of homogeneous lectin–carbohydrate cross-linked lattices in a mixture of lectins with similar physicochemical and carbohydrate binding properties in the presence of an oligosaccharide.

12.3.3 Formation of Unique Homogeneous Oligosaccharide-Lectin Cross-Linked Lattices Observed by Electron Microscopy

The negative-stain and freeze-fracture electron micrographs of the precipitates of L-fucose specific *Lotus tetragonolobus* isolectin A (LTL-A) in the presence of oligo-saccharides **3, 4, 5** (Fig. 1) are shown in Fig. 5 (Bhattacharyya et al. 1990). The freeze-fracture and the negative-stain patterns of the precipitates of LTL-A with **3** are similar (Fig. 5a,b). The patterns of the negatively stained precipitates of LTL-A in the presence of the oligosaccharides **4** and **5** are shown in Fig. 5c and d. The results show a distinct lattice pattern for each oligosaccharide, indicating the presence of long-range order and well-defined geometry in each cross-linked complex. The

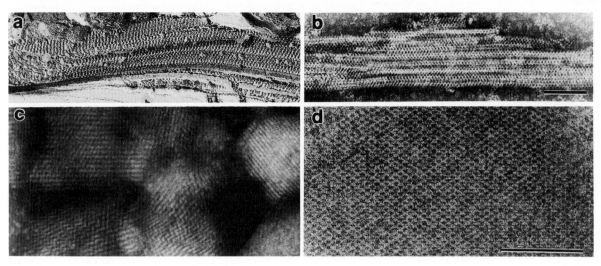

Fig. 5a–d. Electron micrographs of the precipitates of LTL-A (100 μM) with fucosyl biantennary oligosaccharides. **a** Freeze-fracture pattern with **3** (60 μM), **b** Negative-stain pattern with **3** (60 μM), **c** Negative-stain pattern with **4** (60 μM), **d** Negative-stain pattern with **5** (90 μM). The *bars* in **b** and **d** represent 0.1 μm. Magnification in **a, b,** and **c** is the same. (Bhattacharyya et al. 1990)

Table 2. Electron microscopic patterns of the precipitates of LTL-A in mixed oligosaccharide precipitation systems

Mixture of	Conc. of LTL-A (μM)	Oligosacch. total conc. (μM)	Molar ratio of oligosacch.	Incubation time (h)	Pattern[a]
3 and **4**	130	70	50:50	24	c
	100	60	70:30	24	c
	100	60	80:20	24	b
3 and **5**	130	60	50:50	96	b
	130	130	50:50	0.5	b
	140	120	20:80	96	b
	100	90	5:95	96	d

[a]The letters b, c, and d represent patterns b, c, and d in Fig. 5, respectively.

precipitates of LTL-A formed in the presence of a mixture of two oligosaccharides show patterns characteristic of one or the other oligosaccharide depending on the relative concentration of the carbohydrates (Table 2). These findings show the formation of distinct homogeneous cross-linked lattices between LTL-A and three bivalent oligosaccharides.

12.3.4 Quantitative Precipitation Profiles of Lectins with Glycoproteins

The quantitative precipitation profiles of lectins with glycoproteins are, in general, more complex than those observed with oligosaccharides and glycopeptides. However, in a few cases, the profile is a simple, bell-shaped one (as observed with multiantennary carbohydrates) shown in Fig. 6 which describes the precipitation of

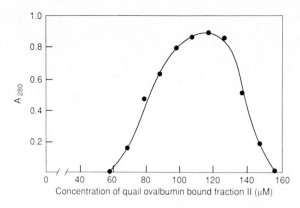

Fig. 6. Precipitation profile of Con A in presence of quail ovalbumin bound fraction II at 22 °C. The buffer was the same as that in Fig. 2. Con A concentration was fixed at 210 µM in a final volume of 200 µl

Con A (210 µM) in presence of an increasing amount of a preparation of quail ovalbumin (designated as quail ovalbumin bound fraction II), monomeric glycoprotein that possesses only one N-linked oligomannose type chain consisting of either Man7 or Man8 oligosaccharide. The equivalence point (point of maximum precipitation) shows that the oligomannose chain on quail ovalbumin is bivalent for Con A binding, which is similar to that observed for the corresponding free oligomannose type glycopeptides. This results shows that quail ovalbumin forms a 1:2 cross-linked complex with tetrameric Con A.

The quantitative precipitation profile of ^{3}H-labeled Con A (33 µM) in presence of ^{14}C-labeled SBA, a tetrameric glycoprotein possessing a single Man9 oligomannose

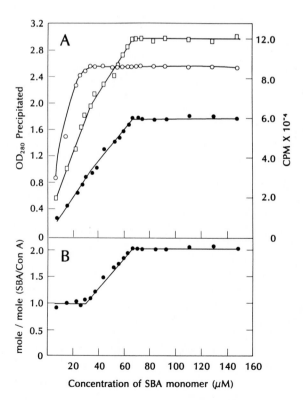

Fig. 7A, B. Precipitin curves for the quantitative precipitation of Con A by SBA at 22 °C. **A** Profile of the total protein (A_{280}) precipitated (□) and cpm of ^{3}H-Con A (○) and ^{14}C-SBA (●) in the precipitate. **B** Ratio (●) of mole of SBA precipitated per mole of Con A. The buffer was the same as that in Fig. 2. ^{3}H-Con A concentration was fixed at 33 µM in a final volume of 1 ml. (Khan et al. 1991)

chain per monomer, is shown in Fig. 7 (Khan et al. 1991). The profile has several charateristic features. The curve for the total protein precipitated (A_{280}) shows three different slopes with two break points at SBA concentrations of $33\,\mu M$ (one equivalent) and $66\,\mu M$ (two equivalents), at which point the curve becomes flat. These break points correspond with two break points in the radioactive profile for SBA (^{14}C data) (Fig. 7A). Moreover, at first break point, maximum precipitation (98%) of Con A takes place. Thus, the increase in the total amount of protein precipitated after the first break point is due only to increasing amounts of SBA precipitated. The mole ratio (per monomer) of SBA to Con A in the precipitates shows that at a concentration of one equivalent or less, SBA forms only $1:1$ cross-linked complex with Con A. This ratio increases with further additions of SBA until it reaches a value of $2:1$ at SBA concentration of $66\,\mu M$ (two equivalents), beyond which no further change in the ratio takes place. These results indicate that SBA forms two different types viz, $1:1$ and $2:1$ cross-linked complexes with tetrameric Con A, depending on the relative ratios of the two molecules in the solution. The total carbohydrate valency of SBA is four in the $1:1$ complex and two in the $2:1$ complex. Thus, the individual Man9 oligosaccharide, which as a glycopeptide is bivalent for binding to Con A, expresses univalency or one-half valency when present on the protein matrix of SBA.

References

Bhattacharyya L, Brewer·CF (1992) Formation of homogeneous carbohydrate–lectin cross-linked precipitates from mixtures of D-galactose/N-acetyl-D-galactosamine-specific lectins and multiantennary galactosyl carbohydrates. Eur J Biochem 208: 179–185

Bhattacharyya L, Fant J, Lonn H, Brewer CF (1990) Binding and precipitating activities of *Lotus tetragonolobus* isolectins with L-fucosyl oligosaccharides. Formation of unique homogeneous cross-linked lattices observed by electron microscopy. Biochemistry 29: 7523–7530

Bhattacharyya L, Haraldsson M, Brewer CF (1988a) Precipitation of galactose-specific lectins by complex-type oligosaccharides and glycopeptides: Studies with lectins from *Ricinus communis* (agglutinin I), *Erythrina indica*, *Erythrina arborescens*, *Abrus precatorius* (agglutinin), and *Glycine max* (soybean). Biochemistry 27: 1034–1041

Bhattacharyya L, Khan MI, Brewer CF (1988b) Interactions of concanavalin A with asparagine-linked glycopeptides: formation of homogeneous cross-linked lattices in mixed precipitation systems. Biochemistry 27: 8762–8767

Carraway KL, Carraway CAC (1989) Membrane-cytoskeleton interactions in animal cells. Biochim Biophys Acta 988: 147–171

Chung K-N, Walter P, Aponte GW, Moore H-PH (1989) Molecular sorting in the secretory pathway. Science 243: 192–197

Dubois M, Gilles KA, Hamilton JK, Rebers PA, Smith F (1956) Colorimetric method for determination of sugars and related substances. Anal Chem 28: 350–356

Edelman GM (1976) Surface modulation in cell recognition and cell growth. Science 192: 218–226

Kabat EA (1976) Structural concepts in immunology and immunochemistry, 2nd edn, Holt, Rinehart and Winston, New York

Karush F, Klinman NR, Marks R (1964) An assay method for disulfide groups by fluorescence quenching. Anal Biochem 9: 100–114

Khan MI, Mandal DK, Brewer CF (1991) Interactions of concanavalin A with glycoproteins. A quantitative precipitation study of concanavalin A with the soybean agglutinin. Carbohydr Res 213: 69–77

Kornfeld R, Kornfeld S (1985) Assembly of asparagine-linked oligosaccharides. Annu Rev Biochem 54: 631–664

Lennarz WJ (ed) (1980) The biochemistry of glycoproteins and proteoglycans. Plenum, New York

Mandal DK, Brewer CF (1992) Cross-linking activity of the 14 kDa β-galactoside specific vertebrate lectin with asialofetuin: comparison with several galactose-specific plant lectins. Biochemistry 31: 8465–8472

Nicolson GL (1976) Trans-membrane control of the receptors on normal and tumor cells. II Surface changes associated with transformation and malignancy. Biochim Biophys Acta 458: 1–72

Sairam MR (1989) Role of carbohydrates in glycoprotein hormone signal transduction. FASEB J 3: 1915–1926

Vliegenthart JFG, Dorland L, Van Halbeek H (1983) High-resolution, ^{1}H-nuclear magnetic resonance spectroscopy as a tool in the structural analysis of carbohydrates related to glycoproteins. Adv Carbohydr Chem Biochem 41: 209–374

13 Lectins as Analytical Probes to Define the Physical Characteristics of Binding Events

P.L. MANN, C. HANOSH, C.M. TELLEZ, R. WENK, and R. DIAZ

The lectins' remarkable molecular homogeneity and adequate affinity constants make them excellent candidates for development as analytical reagents. If one makes the assumption that the binding target is carbohydrate (C) and the binder is lectin (L), then the simplest bimolecular reaction can be represented by:

$$C_j + L_i \leftrightarrows CL_{ij}, \quad K_{ij} = \frac{[CL_{ij}]}{[C_j] [L_i]}.$$

Under conditions of equilibrium, non co-operativity, and simple single stage interactions the estimate of the equilibrium constant (K_{ij}) is represented by the equation given above. If a synthetic ligand (I), identical to cell surface entity C, is introduced into the system, a competitive inhibition reaction for the original binding event occurs. The analysis of this competition, using mathematical models, permits estimates of the identity and quantitative aspects of the binding event through the prediction of K_{ij} (equilibrium constant) and R_{ij} (binding capacity).

13.1 Objectives, Limitations, and the Outline of the Procedure

As an example of the analytical capacity of lectins, we will investigate the structural nature of the carbohydrate component of H-Y antigen (Shapiro and Goldberg 1984). H-Y (histocompatibility Y, or as used here H-Ys its serologically defined counterpart) is a minor histocompatibility antigen which defines "maleness" in mammalian species (Heslop et al. 1989). H-Ys antigen represents a very low incidence protein antigen of approximately 20 kDa molecular weight with fundamental biological importance. It is important in the primary determination of sex, as a growth regulator, and as an integrator of cellular differentiation. In this study we are not concerned with the antigen's peptidic properties but rather its oligosaccharide's conformational arrangement on the cell's surface. The H-Ys model, in this context, permits the necessary verification and validation procedures of the lectin interaction. These validation procedures utilize monoclonal antibody reagents against specific carbohydrate structures to verify the lectin binding entity, as well as specific antisero H-Ys reagents to cross-reference the biological structure and function of the cellular binding epitope. The purpose of this study is to illustrate the potential utility of lectin reagents in the experimental documentation of the physical properties of lectin-specific binding events. As in any other new application of technology, great care must be taken not to overinterpret the experimental data. The most striking feature of the interpretation of these data is that there is an apparent coherent structure consisting of monosaccharide units on cellular surfaces. Secondly, these purported structures are related to known biological functions, and thirdly, specific changes in these structures are an apparent prelude to specific cellular differentiation patterns (Mann et al. 1991).

Construction of binding saturation curves permit the determination of lectin concentration, L_s (epitope saturation concentration). L_s is then mixed with a range of specific carbohydrates at various concentrations ($I_1 \ldots I_{15}$). The resultant lectin binding function (B/F, ratio of the bound to free fraction) is then plotted versus the concentration of I. Analysis of these data permits modeling of the original inter-action of the lectin with its specific epitope in terms of analytical parameters, i.e., number of binding sites which best fits the data (from 0 to 5), the equilibrium constant (K_{ij}), and binding capacities (R_{ij}) for each model (see Munson and Rodbard 1980 for a discussion of the mathematical models).

Controls for these studies include: carbohydrate specificity of the binding event; lectin specificity; independent verification of H-Ys identity; independent verification of the carbohydrate identity; as well as internal statistically based validation procedures for the modeling programs.

13.2 Preparations and Materials

Equipment   Ninety six-well flat bottom, tissue culture trays (Corning, Corning, NY); 50 and 100 µl octapettes (Costar, Cambridge, MA) for dispensing reagents; an eight-place "rake" for washing the 96-well plates; Enzyme Immuno Assay (EIA) wash solution (PBS supplemented with 0.02% Tween 20 to lower the nonspecific adsorption of reagents); an EIA plate reader with 405 nm interference filter (Multiscan, Flow, McLean, VA); a convenient PC- or MAC-based program for Scatchard-type analyses of the plotted data (ALLFIT by DeLean, Munson, and Rodbard for the logistic analysis and LIGAND by Munson for the Scatchard analysis, both pro-grams available from the Laboratory of Theoretical and Physical Biology, NICHHD, NIH, Bethesda, MD).

Chemicals and Materials   *Cells.* Two sources of cells are used for these studies. Lymph node (LN) cells from age-matched male and female C57BL/6J mice (Jackson Labs, Bar Harbor, ME) representing a nonadherent cellular population, and an adherent fibroblast (F) cell line from neonate male and female C57BL/6J mice.

The LN cells are fixed to the polystyrene culture dish surface by first coating the surface of the 96-well trays with 90 kDa poly L-ornithine (PLO, Sigma, St. Louis, MO). A solution of PLO (1 mg/ml in PBS) is incubated on the surface for 90 min at 37 °C, removed and followed by 3 washes (5 min each) with PBS. The cell suspension (200 µl) at 2×10^6/ml in PBS is aliquoted onto the PLO surface and incubated at room temperature for 60 min. After two washes (all washes from this point on must be gentle; the maintenance of cell density checked by microscopic examination after each procedure) the cells are fixed (described below) and stored in PBS/BSA at 4 °C until use. If the plates are properly sealed and stored, they are stable for months.

The F cells are maintained in log phase growth for at least three passages before use. Population Doubling Levels (PDLs) are calculated and together with the generation times (T_g) are used to standardize the cell line's performance in culture. The F cells used for these studies have a T_g of 28.7 ± 2.1 h and all experiments are performed between PDL 19 and 25. The cells are aliquoted into the 96-well culture trays in complete growth medium at 2×10^5/ml in 200 µl volumes. After 16–24 h of culture at 37 °C in a 5% CO_2 atmosphere, the culture surface coverage is estimated microscopically (no less than 55% and no more than 85%). The cells are then washed three times in PBS, and fixed with 2.5% formaldehyde + 5% dextrose (Tousimis, Rockville, MD) in PBS for 20 min at room temperature. After three further washes in PBS, the plates are blocked in PBS/BSA and stored at 4 °C until use. These preparations are also stable for months if sealed and hydrated properly. The growth conditions, surface confluence, PDL, and fixation technique are all crucial, as they have been shown to affect the cell surface oligosaccharide display.

Lectins. All lectins used in these studies are purchased from Vector Labs (Burlingame, CA). It is important when purchasing lectins to obtain quality control data on their performance in terms of specificity, purity, and stability.

Carbohydrates. All carbohydrates used for competitive inhibition analyses are purchased from BioCarb (Lund, Sweden) and are accompanied with NMR spectral data on purity. In some cases authentic β 1-3 Gal-GalNAc (as the ortho nitrophenyl salt) is used, the kind gift of Dr. K.L. Matta (Roswell Park Memorial Institute Buffalo, NY).

Immunological Reagents. The monoclonal reagents are prepared and characterized by us and show a high level of selectivity between the α and β conformers of the Gal-GalNAc carbohydrate. The anti-H-Ys reagent is prepared by published methods (Goldberg et al. 1971).

Preparation

Peanut agglutinin (PNA) used in these studies is diluted to 1 mg/ml in phosphate buffered saline (PBS) pH = 7.5 containing 0.1 mM Ca^{2+}, and 10 mM HEPES. The PNA solution is aliquoted in 250 µl volumes and frozen at -20 °C. Any lectin remaining after 4 months of storage is discarded.

PBS/BSA bovine albumin (V, Sigma, St.Louis, MO) at 1 mg/ml in PBS

Avidin-D (Vector Labs), stock solutions in 0.1 M $NaHCO_2$ at pH 8.5 and 10 mg/ml, the working solution is prepared fresh at 0.1 µg/ml in PBS/BSA.

Biotinylated Alkaline-Phosphatase (Vector Labs) is diluted in PBS (supplemented with Mg^{2+} at 1.0 mM, 10 mM HEPES, and BSA at 1 mg/ml) at 10 U/ml stored at -20 °C, it is thawed and diluted in PBS/BSA to 0.01 U/ml immediately before use.

Para nitro-phenyl-phosphate PNPP (Behringer-Mannheim, Indianapolis, IN) is stored at 4 °C until use and then diluted to 1.0 mg/ml in alkaline buffer (96 g diethanolamine + 800 ml water, pH to 10.2 then add a mixture of 100 mg $MgCl_2 \cdot 6H_2O$ + 136 mg $ZnCl_2$ in 10 ml of water, make to 1000 ml) store at 4 °C and replace monthly.

The antibody reagents are stored either as lyophilized powders at 4 °C or at 10 mg/ml at -20 °C, and diluted in PBS/BSA to 1 mg/ml and stored at 4 °C until use.

13.3 Detailed Procedure

13.3.1 Saturating Lectin Concentration (L$_s$)

Procedure

– Aliquot 100 µl of PBS/BSA into columns 1, and 3–12 with octapipet
– Aliquot 200 µl of lectin at 20 µg/ml (PBS/BSA) into column 2
– Remove 100 µl of column 2 and mix well with column 3

– Repeat this procedure for rest of columns
– Discard excess 100 µl from column 12

Recommendations a) Keep all solutions at 4 °C during this procedure to minimize binding until after the entire experiment is constructed
b) Use square petri dishes for the solutions as this facilitates use of the octapipets
c) Remove the very fine ends of the disposable tips for the octapipets to minimize damage to the cells
d) Rows A to D are used for L_1 and E to H for L_2
e) Column 1 is the reagent background (this should be similar to no-cell background)

Procedure – Incubate the trays for 90 min at 37 °C
– After three washes (EIA) incubate with 100 µl/well Avidin-D at 37 °C
– After 90 min, wash three times as above
– Add 100 µl B-alkaline phosphatase and incubate for 90 min at 37 °C
– After three washes, incubate with 200 µl of PNPP solution
– Quantitate the reaction at 405 nm (OD 405)
– Subtract the OD 405 values for column 1 from columns 2–12
– Plot OD 405 as a function of added lectin concentration
– Determine the saturating concentration (L_s) from the curves (inflection point)

Recommendations a) Maintain all solutions for this phase of the procedure at room temperature
b) Use the 8-channel "rake" for EIA washes
c) Carefully "flick" the wash solutions out of the trays in a single action
d) After the final wash lightly dry the trays by tapping them on a paper towel (do not allow cells to desiccate)
e) (L_s) = the inflection point from a four parameter logistic analysis (ALLFIT)

13.3.2 Carbohydrate Competition Analysis

Procedure – Add 100 µl of PBS/BSA to column 1 of the 96-well trays of cells
– Add 100 µl of lectin (L_s) to column 2 and 4–12
– Add 200 µl of lectin and carbohydrate I at 10 mM to column 3
– Remove 100 µl of solution from column 3 and mix with column 4
– Repeat as above to construct a doubling dilution curve of I (10–0.02 mM)
– Repeat from § above
– Calculate B as the OD 405 at concentration I_1 to I_{11}
– Calculate T as the maximal OD 405 (column 2); $F = T - B$
– Calculate B/F and plot versus concentration I_1 to I_{11}
– Analyze data using Scatchard-type analysis (Ligand)

Recommendations a) Maintain the initial solutions at 4 °C to minimize unwanted reactions
b) Starting at 10 mM carbohydrate concentrations insures a complete competition curve (10–0.02 mM)
c) It is very important to be able to accurately estimate the total bound fraction [B_T] and baseline binding (B_0)

13.3.3 Immunological Competition Analysis

– Substitute the appropriate antibody reagent for the lectin (L_s) as above
– Repeat the carbohydrate competition analysis as described above

a) It is very important to accurately determine the saturating concentration of the immunological reagents
b) Matching the specific and control reagents with respect to affinity is important because large differences in affinity result in large changes in baseline binding and therefore, present difficulties in estimating the true baseline (B_0)

Recommendations

13.4 Results

Under the conditions outlined herein the saturating concentration of the majority of lectins used (L_s) is typically between 5 to 10 µg/ml. The L_s for PNA binding used in these experiments is 5 µg/ml and is determined separately from the inflection point of the saturation curve. We determine L_s visually from these S-shaped saturation curves, if more analytic precision is needed the four parameter logistical analysis capabilities of ALLFIT could be used to determine the exact concentration at which 85% binding saturation occurs. ALLFIT is also very useful as a quality control tool to compare the shapes of the various binding curves. We have tested L_s values from 50 to 100% of the total cellular binding capacity and as would be expected, at low values of L_s the sensitivity of the assay is severely reduced and above 95% the reproducibility of the high B/F values becomes less reliable.

The B/F values are calculated directly from the binding data and $(B/F, I)_{1-15}$ datasets are analyzed by LIGAND. Figure 1 shows the results of a competition analysis of PNP-β-galactose for PNA binding on C57BL/6J male and female cells. The lines in Fig. 1 are simple visual representations and only the analytical data, generated from the mathematical modeling of these datasets, are used for comparison. The visual representation indicates clearly that there is a sex-dependent difference in the β 1–3 Gal-GalNAc (nominal specificity of PNA) cellular epitope. Table 1 shows the results of the LIGAND analysis of these datasets. These analyses, in agreement with the visual representations, indicate that the male cells have a sex-specific high affinity β 1–3 Gal-GalNAc family of sites as well as the lower affinity family shared with the female cells. The LIGAND algorithm is a nonlinear model fitting program which makes no assumptions or approximations. It exactly represents the physical binding interaction, handles nonspecific background binding as a parameter, allows for multiple ligands, and permits inter-experimental comparisons. The program models the data for 0–5 binding sites and assesses the "best" fit on a solid statistical basis, providing estimates of K_{ij} and R_{ij}. This analysis relies heavily on the accurate estimation of B_T and B_0, as these two parameters define the extremities of the binding.

The data shown here suggest that male cells differ from female cells by the appearance of an additional high affinity β 1–3 Gal-GalNAc site on male cells. The total number of binding sites (R_{ij}) does not appear to be significantly different, thus the "difference" appears to be one of conformational arrangement of an endogenous oligosaccharide residue. All other lectin studies indicate no other lectin-dependent

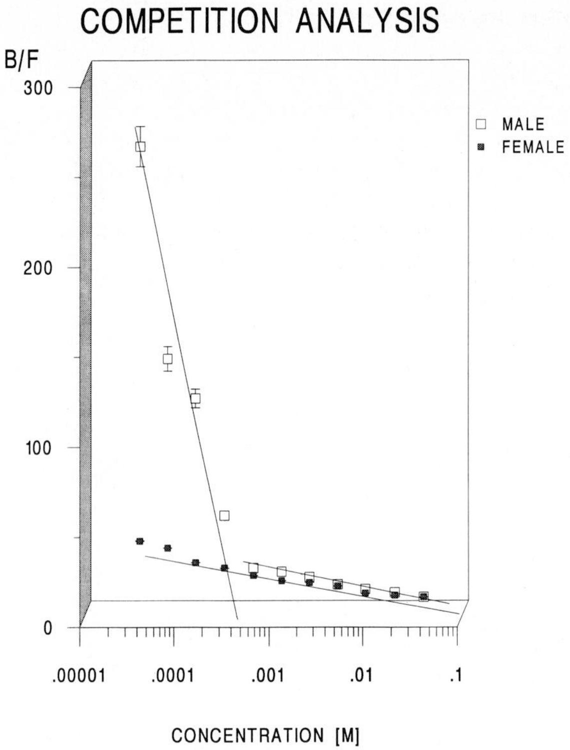

COMPETITION ANALYSIS

Fig. 1. The results of a competition of para-nitrophenyl β-D-galactose carbohydrate for PNA binding on male and female murine cells. The bound/free parameter (B/F) is plotted against the concentration of ligand added (M). The data $\pm$ SEM for a typical experiment, representative of four separate experiments, is plotted here. The *lines* are drawn visually as these are complex mathematical functions. The ligand algorithm calculates the physical data which most nearly approximates the experimental data. The data presented show four points on the high affinity portion of the curve. This is the minimum number of points required for the analysis. The addition of three more points in this concentration range is advisable

Table 1. The results of the ligand analysis of the data presented in Fig. 1. Ligand attempted to fit the data to zero through five binding site families, in this case a two-site model is the best for the male cells and a single-site is best for the female cells. Other lectin studies (RCA-120, Con A, and WGA) indicate that the male and female cells are identical for these carbohydrate specificities. The values given are from a single experiment (typical of five replicates) and suggest that there is s specific β 1–3 Gal-GalNAc conformer difference between the male and female cells

Binding parameter	Male	Female
K_{11}	8.3E06	—
K_{12}	6.0E03	6.0E03
R_1	1.3E-04	—
R_2	7.0E-03	1.8E-03

PNA binding to C57BL fibroblasts.

"difference" between the cell surfaces of male and female cells. Similar results are observed for both lymph node and fibroblast cells. It appears that the conformation of this specific carbohydrate residue represents a significant component of the male/female cellular differential.

As part of the extensive quality control required for these experiments, we tested the competitive nature of at least 3 specific and 2 nonspecific carbohydrates. In this case β 1–3 Gal-GalNAc, PNP-β-D-Gal, and 1-O-Me-β-D-Gal gave very similar data, suggesting that the actual analytical results (the numbers) may be significant, in that they appear to be cell surface specific rather than ligand specific.

13.5 Trouble shooting

The data presented above suggest that the sole oligosaccharide sex-dependent "difference" is the presence of an additional high affinity β 1–3 Gal-GalNAc binding site on male cells. It is important to ascertain the quality of that observation and to develop an independent verification of the observation. In many cases these are difficult criteria to meet. In the case of H-Ys antigen we have reagents for both procedures.

We made and characterized monoclonal reagents to the alpha and beta conformation of 1–3 Gal-GalNAc. Specificity of these reagents and all such reagents is not absolute but rather a case of relative differences, for instance, dependent on the difference in the concentration of the reagent required to bind 50% (ED_{50}) of the specific reagent relative to the nonspecific. This ED_{50} parameter can be derived from the four parameter logistic analysis (ALLFIT) discussed above. The greater the "difference" in ED_{50}, the greater the specificity of these reagents. The other factor which severely affects the quality of these reagents is the absolute concentration of the ED_{50}. Nanomolar values yield reagents which possess both specificity and selectivity.

Table 2 shows the results of a LIGAND analysis of $(B/F, I)_{1-15}$ datasets where the competitor an anti-β 1–3 Gal-GalNAc monoclonal antibody (MoAb) preparation. This reagent was characterized much like the lectin reagents discussed above. This characterization includes; specificity (comparison with the anti α reagent), concentration required for saturation, and the ability of authentic α and β 1–3 Gal-GalNAc to compete for its binding to cells and/or authentic carbohydrate epitopes. This reagent shows that it competed with PNA for its binding epitope on both lymph node and murine fibroblast cells in a sex-dependent fashion. The male cells show a distinctive two-site arrangement analogous to the data with the PNA and specific sugars. This monoclonal reagent does not compete with other lectin specificities.

Table 2. The results of the ligand analysis of the competition of a specific anti-β 1–3 Gal-GalNAc MoAb for PNA binding on male and female murine cells. This is typical of four experiments, and indicates that there are two binding sites on the male cells and a single family on the female

Sex	K_{11}	K_{12}	R_1	R_2
Male	8E04	6E01	9E-04	5E-02
Female	—	5E02	—	5E-02

PNA binding to C57BL lymph node cells.

Table 3. The results of the ligand analysis of the competition of a specific anti-H-Ys MoAb for PNA binding on male and female murine cells. This is typical of three experiments, and indicates that there are two specific binding sites on the male cells and a single family on the female

Sex	K_{11}	K_{12}	R_1	R_2
Male	5E04	1E02	1E-05	5E-03
Female	—	2E03	—	5E-03

PNA binding to C57BL lymph node cells.

Other lectins also do not compete with this reagent for binding. The alpha specificity MoAb does not compete for the PNA binding in this manner.

Finally, the use of an authentic anti H-Ys reagent is useful to connect the observations given above to a functional activity. This reagent (both monoclonal and polyclonal reagents were tested) has demonstrated specificity for male fibroblasts and lymph node cells from the C57BL/6J model. When used as a competitor for the PNA binding it shows a sex-dependent bifurcation of the binding event. Table 3 shows the results of an experiment where a polyclonal anti-H-Ys antibody is used to compete for PNA binding onto male and female lymph node cells. Here again there is evidence of a sex-specific high affinity β 1–3 Gal-GalNAc epitope.

13.6 Summary

This study demonstrates that lectins can be used for detailed analyses of cell surface structures. FITC conjugated PNA studies revealed no demonstrable difference in staining between the male and female cells used in these studies. Thus, the use of lectins in these competition assays permits a unique perspective on the binding epitope. The analysis reveals that male murine fibroblasts and lymph node cells possess a sex-specific high affinity β 1–3 Gal-GalNAc epitope on their surface. To the first approximation it appears that this unique site is the result of the re-arrangement of existing material as opposed to the insertion of new material. Furthermore, a similar or identical sex-specific β 1–3 Gal-GalNAc site can be identified with monoclonal reagents directed towards authentic carbohydrate and H-Ys antigen.

The final stage in the process of the use of lectin in the identification and characterization of H-Ys antigen structures would be the purification of the binding epitope and its subsequent biochemical characterization.

This study has demonstrated the utility of lectin reagents as analytical reagents much in the same way that immunological reagents have been used. Great care is needed when using these reagents, and especially in the interpretation of the data. Specifically, the following points of caution are made:

- The competition must be similar for at least three carbohydrates with the same specificity
- The data sets must be large enough to permit statistical reliability within the 1–5% range
- Independent verification of both the biochemical and biological identity should be attempted.

References

Goldberg EH, Boyse EA, Bennett D, Scheid M, Carswell EA (1971) Serological demonstration of H-Y (male) antigen on mouse sperm. Nature 232: 478–480

Heslop BF, Bradley MP, Baird MA (1989) A proposed growth regulatory function for the serologically detectable sex-specific antigen H-Ys. Hum Genet 81: 90–104

Mann PL, Eshima D, Bitner DM, Griffey RH, Wenk R, Born JL, Matwiyoff NA (1991) Biomodulation: an integrated approach to access and manipulate biological information. In Lectins and cancer (H.-J. Gabius and S. Gabius eds.) Springer-Verlag Berlin. pp 179–206

Munson PJ, Rodbard D (1980) LIGAND: a versatile computerized approach for characterization of ligand binding systems. Anal Biochem 107: 220–239

Shapiro M, Goldberg EH (1984) Analysis of a serological determinant of H-Y antigen: evidence for carbohydrate specificity using an H-Y specific monoclonal anitbody. J Immunogen 11: 209–218

Lectins as Tools
for the Characterization of Glycoconjugates

14 Glycoprotein-Lectin-Immunosorbent Assay (GLIA)

E. KÖTTGEN, A. KAGE, B. HELL, C. MÜLLER, and R. TAUBER

Much reported work in recent years points to the essential physiological and pathobiochemical importance of the regulated expression of the variable glycosylation of proteins and lipids. (Schachter 1984; Martinez and Barsigian 1987). The structural analysis of the glycans of single purified glycoproteins requires, however, a range of extraordinarily complicated and costly methods. For this reason, the investigation of changes in protein glycosylation in body fluids or tissue homogenates is often not performed on individually purified glycoproteins, but on glycoprotein mixtures. An important contribution to such studies has been made by lectin-affinity chromatography (Osawa and Tsuji 1987), but this technique presents certain difficulties when applied to the analysis of very small samples, or large sample numbers. As an alternative approach – usually with limited success – attempts have been made to relate the biochemical or pathobiochemical regulatory role of glycoproteins to the total content of distinct sugar residues, e.g., of Neu5Ac or fucose in glycoprotein mixtures. With the technique described here it is possible to selectively analyze the glycans of defined glycoproteins without prior purification. The method displays high sensivity, and requires only small sample quantities; it is quick to perform, and the operational costs are favorable (Köttgen et al. 1988, 1989). In addition, the necessary apparatus is available in practically all modern biochemical laboratories.

14.1 Test Principle

The test principle of the GLIA is derived essentially from that of the ELISA sandwich technique. Thus, the glycoprotein under investigation is absorbed from the protein mixture by specific binding to a deglycosylated antibody coupled to the wall of a microtiter plate. Other proteins are removed by washing, and the specifically bound glycoprotein is incubated with biotin- or peroxidase-labeled lectin. Unbound lectin is removed by a further washing cycle, and the bound lectin is quantified enzymatically. Alternatively, in particular for the analysis of purified glycoproteins, the lectin may also be coupled to the solid phase of the microtiter plates, and the bound glycoprotein determined by the use of labeled antibodies. Both test principles are summarized in the following scheme (Fig. 1).

14.2 Materials and Apparatus

Deglycosylation of the Solid Phase Antibody:

Buffers and Solutions

- SDS-solution: 1.0% sodium dodecyl sulfate (m/v in H_2O)
- NP-40 solution: 7.5% NP-40 (v/v in H_2O)
- Phosphate buffer: 0.5 M K/Na phosphate, pH 8.6
- Dialysis buffer: 10 mM Tris, 1 mM $CaCl_2$, 1 mM $MgCl_2$, pH 7.4

A

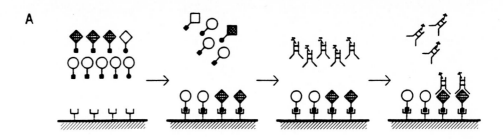

B

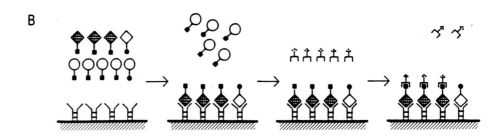

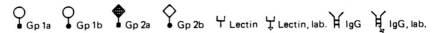

Gp 1a   Gp 1b   Gp 2a   Gp 2b   Lectin   Lectin, lab.   IgG   IgG, lab.

Fig. 1A.B. Versions of the GLIA method. **A** Carrier-bound lectins bind defined glycan structures, which are present on certain components of glycoprotein mixtures. The lectin-bound glycoproteins are recognized by peroxidase-coupled antibodies, thus permiting their quantitation. Identical glycans of different glycoproteins compete for binding to the lectin, and they may decrease or even prevent the binding of the glycoprotein under investigation (*gp 2a*). **B** The glycoprotein under investigation (*gp 2a*) is absorbed from the glycoprotein mixture by binding to the solid phase, via the carrier-bound antibody. The nonbound accompanying glycoproteins, which could compete later for lectin binding, are removed by washing. Finally, peroxidase-labeled lectins are added, which bind specifically and exclusively to the defined glycans of the previously bound ligands. This procedure was adopted in the present manual. *Gp 1a* glycoprotein with antigenic determinant "1" and glycan type "a"; *Gp 1b* glycoprotein with antigenic determinant "1" and glycan type "b"; *Gp 2a* glycoprotein with antigenic determinant "2" and glycan type "a"; *Gp 2b* glycoprotein with antigenic determinant "2" and glycan type "b"; *Lectin* lectin with binding specificity for glycan type "a"; *lab* labeled with peroxidase; *IgG* specific antibody for antigenic determinant "2"; *lab* labeled with biotin. (Köttgen et al. 1988)

Biotinylation of Lectins:

– Biotin solution: dissolve 4 mg NHS-LC-biotin in 40 µl dimethyl sulfoxide (DMSO), and add 960 µl H$_2$O (prepare immediately before use)
– Lectin stock solution: dissolve 1 mg lectin in 500 µl Tris (10 mM, pH 7.4), 500 µl ethyleneglycol. The stability of the solution depends on the lectin used (see supplier's data)

Glycoprotein-Lectin-Immunosorbent Assay (GLIA):

- Antibody coupling buffer: see dialysis buffer (above)
- Washing buffer: 10 mM Tris, 1 mM $CaCl_2$, 1 mM $MgCl_2$, 0.05% (v/v) Tween 20, pH 7.4
- HSA-blocking buffer: 5 g/l human serum albumin (HSA), 10 mM Tris, 1 mM $CaCl_2$, 0.1% (v/v) Tween 20, pH 7.4. To remove impurities, the protein is previously chromatographed over ConA-Sepharose, and only the protein fraction eluted in the exclusion volume is used
- Standard buffer: 10 mM Tris, 1 mM $CaCl_2$, 1 mM $MgCl_2$, 0.5% (v/v) glycerol, 0.1% (v/v) Tween 20, pH 7.4
- Lectin stock solution: see above
- Substrate-buffer: (1) TMB solution: dissolve 240 mg tetramethylbenzidine (TMB) in 5.0 ml DMSO and 5.0 ml ethanol (stable for about 4 weeks at 4 °C in the dark); (2) add 68 µl H_2O_2 (30%) to 200 ml citrate buffer (stable for about 2 weeks at 4 °C in the dark)

Immediately before use, mix 10 ml citrate buffer/H_2O_2 (in the dark at 25 °C) and 100 µl TMB solution

In addition to the usual provisions of a biochemical laboratory, the following equipment is also necessary or recommended: **Apparatus**

- Work station for microtiter plates: photometer with printer, dilutor, vibrator, plate washer, 8-channel pipet and reagent baths
- Water bath or incubator

14.3 Performance of the Test

Deglycosylation of the Antibody (the given volumes are for coating one microtiter plate):   **Preparatory Stages**

	Deglycosylation	Control
	(µl)	(µl)
Antibody[a]	2.0	2.0
Distilled water	3.0	3.0
SDS-solution	5.0	5.0

[a]Protein concentration of the antibody solution 10 g/l.

Incubate sample mixture for 25 min in ultrasonic bath

Phosphate buffer	10.8	10.8

Incubate sample mixture for 20 min in ultrasonic bath

NP-40 solution	5.0	5.0
N-glycanase[b]	1.2	—
Distilled water	—	1.2

[b]Enzyme activity 250 U/ml.

Incubate sample mixture for 5 h at 25°C, followed by 16 h dialysis at 4°C against buffer.

The control mixture serves as a test for the successful deglycosylation of the antibody. The success of the deglycosylation is also tested electrophoretically, assessing the shift in the apparent molecular mass of the antibody polypeptides in the gel. After deglycosylation the antibody solution is stable for about 2 weeks.

Biotinylation of the Lectins (Müller et al. 1986):

Incubate 1 ml lectin stock solution with 25 µl biotin solution for 1 h at 25°C, followed by storage for 16 h at 4°C. The material is stable for about 4 weeks.

Coupling of the Deglycosylated Antibody to the Microtiter Plates:

- Dilute 27 µl antibody solution (after the deglycosylation step, see above) with 10 ml antibody coupling buffer (corresponds to a final antibody dilution of 1 : 5 000)
- Add 100 µl diluted antibody to each well and incubate for 1 h at 25°C
- Perform three washing cycles, each with 200 µl washing buffer per well. The plate can then be used immediately. Alternatively, the plate may be washed once more with distilled water then lyophilized. Dried plates can be stored dry for 1 week at 4°C in the absence of air

GLIA Test	*Pipeting schemes*	Explanation
	Preparation of reagents and microtiter plates (see Sect. Preparatory Stages)	(1)
	↓	
	150 µl diluted sample or standard	(2)
	2 h incubation at 25°C	
	Wash three times	
	↓	
	125 µl lectin solution	(3)
	1 h incubation at 25°C	
	Wash three times	
	↓	
	125 µl streptavidin peroxidase	(4)

30 min incubation at 25 °C

Wash three times   (5)
↓
125 µl substrate solution   (6)

Incubate 10 min at 25 °C (protected from light)

50 ml H_2SO_4 (2 M)   (7)
↓
Photometric measurement at 450 nm

Explanations to the pipeting scheme

1. All quoted volumes are for a single well. The following blank values must be determined for each microtiter plate: (a) sample blank (standard buffer in place of sample); (b) unspecific binding of lectins (normal test mixture in wells not coated with antibody). If the blank value due to unspecific binding is too high (absorbance > 0.1), HSA-blocking buffer should be used in place of standard buffer for the entire test run.

   The protein concentration of the investigated glycoprotein is first determined immunologically, then adjusted with standard buffer to 1.0, 0.5, and 0.2 mg/l (Köttgen et al. 1986). If possible, a purified sample of the investigated glycoprotein is also run as a standard; standard plasma may be run as an alternative. Standards are also run in a series of dilutions for the subsequent preparation of a calibration curve. When certain lectins are used (e.g., those with galactose specificity), the standard material must first be desialylated.

2. Solutions of standards and samples are diluted 1:15 (10 µl sample plus 140 µl standard buffer) then added directly to the wells. If required, highly concentrated samples were additionally diluted before. All samples and standards are analyzed in triplicate.

   Before each subsequent incubation step, the plates are packaged in foil and briefly vibrated on the vibrator.

3. The lectin stock solution is diluted 1:2000 with standard buffer before use.

4. The streptavidin-peroxidase solution (1 mg/ml) is diluted 1:10000 with standard buffer before use.

   At this stage, the substrate buffer can be brought to 25 °C in the dark.

5. After the third addition of washing buffer, the plate must be incubated for 5 min on the vibrator, to ensure the release of unbound residues of streptavidin-peroxidase.

6. Preparation of substrate solution: immediately before use, 100 µl TMB solution are added to 10 ml citrate buffer,

7. H_2SO_4 stops the enzymatic substrate conversion. The photometric measurement must be made within the next 30 min.

14.4 Evaluation and Results

The average absorbance value is calculated for each triplicate determination. The difference between the test and the blank absorbance values represents the primary analytical value. The specific binding activity of the glycoprotein-lectin interaction is

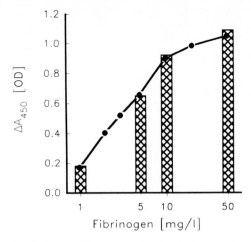

Fig. 2. Assay of fibrinogen using ConA. A calibration curve for the fibrinogen-concanavalian A (Con A) interaction was prepared with a dilution series of human standard plasma (●—●). Different concentrations of purified fibrinogen were analyzed in the same test run (*columns*). Equal concentrations of purified fibrinogen and fibrinogen in human standard plasma always give rise to approximately the same measurement signal (ΔA_{450}). This means that fibrinogen is the only protein of standard human plasma that becomes specifically bound to the solid phase antibody, indicating that the method is accurate at this level. For calculation of the "specific binding activity" of the Con A-fibrinogen interaction, see Section 14.3. The values shown represent the average of five replicate determinations. The variation coefficient for individual measurements is < 5%.

derived from this analytical result. First, the lectin-glycoprotein reactivity of the sample material is calculated in arbitrary units/l. One unit defines the binding capacity of all glycoproteins bound to the solid phase for a lectin, relative to the binding capacity of glycoprotein standard. The lectin binding activity of the sample material, however, is determined not only by the nature and quantity of the lectin-reactive glycan residues of one individual glycoprotein, but also by the number of glycoprotein molecules bound to the solid phase. A conventional protein determination must therefore be performed in parallel (e.g., immunologically using ELISA procedures), in order to determine the lectin reactivity in relation to the existing ligand concentration. The binding capacity of the glycoprotein for a lectin, expressed in U/l, can thus be converted to the concentration-independent glycoprotein-lectin-specific binding activity (U/mg protein). An example of the calculation is taken from the results in Fig. 2:

ΔA_{450} of the sample = 0.4; on the calibration curve, this absorbance corresponds to a fibrinogen concentration of 2.0 mg/l. In accordance with the above definition, the fibrinogen-ConA reactivity of the sample is therefore 2.0 U/l. The fibrinogen protein concentration of the sample (determined previously and independently in the ELISA) must now be included in the calculation; in the present example this concentration is 5.0 mg/l. The "specific binding activity" of the fibrinogen in this sample for ConA is therefore 0.4 U/mg.

The results in Table 1 provide a further example of the analytical spectrum of the GLIA, i.e., an efficient quantitative differentiation of fibronectin isoforms from the plasma of different animal species by the determination of their specific binding activities.

Table 1. Specific binding activity of plasma fibronectin to different lectin species

Lectin	Specific binding activity of p-FN [U/mg]			
	Human	Calf	Pig	Dog
ConA	1.35	1.30	2.19	1.33
WGA	0.14	0.15	0.17	0.19
LCA	0.14	0.67	2.21	0.09
LTA	0.11	0.27	0.42	0.02

Note. Calculation of the specific binding activities (see Sect. 14.3) for the interaction of glycoproteins and lectins represents a new method for quantifying the interaction and for the characterization of the ligands. Rabbit-antihuman fibronectin was used as the solid phase antibody. This antibody shows complete cross-reactivity with plasma fibronectin (p-FN) of the animal species used here. The calibration curve was prepared with human desialylated p-FN.
ConA, concanavalin A; WGA, wheat germ agglutinin; LCA, *Lens culinaris* agglutinin; LTA, *Lotus tetragonobolus* agglutinin. The values represent the average of five replicate determinations. The variation coefficient for individual measurements is < 5%.

Test of Binding Specificity by Inhibitor Studies

The glycan-dependent nature of the binding reaction is demonstrated by inclusion in the test incubation mixture of mono-, di- or oligosaccharides, which are known inhibitors of glycan binding by the respective lectins. These sugars are added together with the lectin solution (see above) to the test mixture. A control incubation must be performed in parallel, in which the sugar solution is replaced by an equimolar NaCl solution. In analogy with the analysis of enzyme inhibition, the inhibitor constant can be determined from a double reciprocal transformation of the results (percentage inhibition of binding) obtained with different concentrations of inhibitor (Fig. 3).

Quality Control

To test the imprecision in series and the imprecision from day to day, standard material is included as described under Section 14.3. In our experience, the variation coefficient (CV) of the individual values of the triplicate determination is < 5%, and the CV for imprecision in series and from day to day is < 10%. Since these are relative measurements, and there is no defined glycan standard, accuracy can only be monitored at the level of specific protein binding to the solid-phase bound antibody (see Fig. 2).

14.5 Appraisal of the GLIA

The GLIA provides a new approach to the quantitative analysis of glycans. The method is very sensitive and shows good reproducibility. Important advantages are the small quantity of sample required, and the use of various mechanized analytical steps. Nevertheless, the following limitations must be taken into account: (a) the

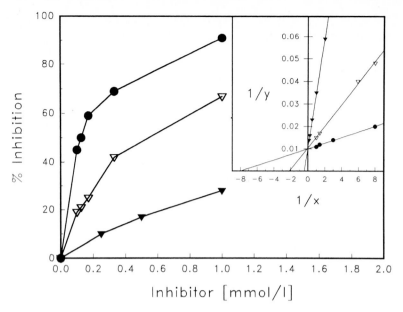

Fig. 3. Inhibition of fibrinogen–ConA binding by monosaccharides. The interaction of fibrinogen and Con A is inhibited by the addition of "Con A-specific" monosaccharides glucose (▼), mannose (▽), or methylmannoside (●) to the GLIA test incubations (see Sect. 14.3). This concentration-dependent inhibition of binding is not observed in the presence of other sugars (e.g., galactose, fucose, Neu5Ac; not shown). It should be noted, however, that a "NaCl control" must be analysed in parallel in all inhibitor studies, because the binding reaction is influenced by the molarity of the buffer system. The inhibitor constant (K_i) for each sugar can be determined from the double reciprocal transformation of the values on the abscissa and ordinate (insert).

An inhibitor constant is valid only for the ligand system in which it is determined. The following K_i values apply to the present test system: methylmannoside 0.12 mmol/l, mannose 0.45 mmol/l, glucose 3.7 mmol/l. On the other hand it is also possible to characterize the isoforms of a defined glycoprotein (e.g., fibrinogen from adult plasma and umbilical cord blood) by the different inhibitor constants for a single sugar (not shown). The values shown represent the average of triplicate determinations. The variation coefficient for individual measurements is < 5%.

method does not provide information on the heterogeneity of glycosylation of a glycoprotein in the glycoprotein mixture; differential analysis of carbohydrate chains is possible only after purification of the individual glycoprotein, and analysis of the glycan-carrying peptides; (b) in its present stage of development, the method cannot be used to quantify the total carbohydrate content of a glycoprotein.

14.6 List of Suppliers

Reagents   Lectins: Boehringer Mannheim, Sigma, Medac. Antisera and purified antibodies: Dakopatts and Behring. Standard human plasma and purified proteins: Behring. N-Glycanase (EC 3.5.1.52): Genzyme. Sialidase from *Clostridium perfringens* (EC 3.2.1.18): Sigma. NHS-LC-Biotin: Pierce. Streptavidin-peroxidase: Calbiochem. Tetramethylbenzidine: Fluka. All other reagents were from Merck, Sigma, or Boehringer Mannheim.

MTP-photometer: Dynatech MR 7000(Flow Labs). Microtiter plates: NUNC.   **Equipment**
 Multichannel pipets and washing comb: Flow Labs
Vibrator: Labinstruments (Austria)

References

Köttgen E, Hoeft S, Müller Ch, Hell B (1986) Functional analysis of plasma fibronectin with special consideration of binding interferences. J Clin Chem Clin Biochem 24: 541–549

Köttgen E, Hell B, Müller Ch, Tauber R (1988) Demonstration of glycosylation variants of human fibrinogen, using the new technique of glycoprotein lectin immunosorbent assay (GLIA). Biol Chem Hoppe Seyler 369: 1157–1166

Köttgen E, Hell B, Müller Ch, Kainer F, Tauber R (1989) Developmental changes in the glycosylation and binding properties of human fibronectins. Biol Chem Hoppe Seyler 370: 1285–1294

Martinez J, Barsigian C (1987) Carbohydrate abnormalities of N-linked plasma glycoproteins in liver disease. Lab Invest 57: 240–257

Müller Ch, Moritz R, Köttgen E (1986) Immunological analysis. A general micro-ELISA design utilizing monoclonal antibodies: application to assays for AFP and CEA. Fresenius J Anal Chem 324: 246

Osawa T, Tsuji T (1987) Fractionation and structural assessment of oligosaccharides and glycopeptides by use of immobilized lectins. Annu Rev Biochem 56: 21–42

Schachter H (1984) Glycoproteins: their structure, biosynthesis and possible clinical implications. Clin. Biochem 17: 3–14

15 Use of Lectins in Quantification and Characterization of Soluble Glycoproteins

J.M. RHODES, J.D. MILTON, N. PARKER, and P.J. MULLEN

There are two reasons why we sought to develop lectin binding assays for soluble glycoproteins rather than relying on conventional immuno-assays. Firstly, we have a research interest in mucus glycoproteins, which can be difficult to assay using conventional antibody techniques, because antibodies generally have narrow specificity. Most antibodies to mucins are directed against carbohydrate determinants, presumably because the protein core of the mucin is so heavily glycosylated that it is less immunogenic than the oligosaccharide side chains. There is, however, a considerable potential for heterogeneity between different oligosaccharide chains on the same mucin molecule. This implies that an antibody which recognizes a complex carbohydrate may show considerable variation in binding between mucins. Whilst this may be useful in characterization of mucins it is inconvenient when quantification is the aim. Although little is known about the precise spatial epitope requirements of many anti-carbohydrate antibodies, it is assumed that these generally bind to a larger and therefore more complex epitope than most lectins. This generally broader specificity makes lectins particularly useful for quantification of mucins, provided that one can be certain that the mucin preparation is free from other glycoproteins bearing similar carbohydrate structures. Enzyme-linked lectin binding assays are highly sensitive and can be applied to fractions of mucin extracted from small biopsy samples in situations that are well beyond the limits of detection of conventional protein or carbohydrate assays.

Our second reason for wishing to develop lectin binding assays was the recognition that many tumor marker assays used in the serological diagnosis of epithelial tumors depend on the detection of mucins in serum samples (Magnani et al. 1983). In previous studies (Ching and Rhodes 1988), we had shown that these tumor-related mucins could be detected, but not easily quantified, using lectin blotting of material transferred from polyacrylamide gels and that in many cases they seemed to account for much of the binding activity of the serum to one or more lectins. Lectin blotting, the use of enzyme-tagged lectins to identify carbohydrates on nitrocellulose blots from polyacrylamide gels in manner similar to Western blotting, may be combined with sequential chemical or enzymatic treatment of the blotted glycoprotein to obtain information about oligosaccharide structure. This technique is described in Ching and Rhodes (1990), Rhodes and Ching (1993). This chapter will concentrate on the use of lectins for quantification rather than characterisation.

Lectins may be used in glycoprotein assays either as a detector system where the target glycoprotein is attached to the solid phase or as a capture system where the lectin is attached to the solid phase, with an antibody, or some other recognition system, used for detection of the captured glycoprotein. We have found both these systems useful and they will be described separately. It should be realized that when a lectin is used as a detector system, the assay will give quantification of the number of specific binding sites (usually a monosaccharide or a disaccharide) and that as the number of these sites may vary between otherwise similar glycoproteins (e.g., due to differing degrees of degradation), the result obtained may not equate directly with

Tris/HCl 0.1 M pH 8.0) were diluted 1:1 in carbonate buffer pH 9.5 and the following procedure used:

ELISA plate well coated with 100 µl sample in coating buffer
– incubate 16 h at 4 °C
– wash ×2 with PBS/Tween

Add 120 µl blocking buffer (PBS/Tween)
– incubate 2 h at 37 °C
– wash ×2 with PBS/Tween

Add 100 µl peroxidase-lectin conjugate (2 µg/ml in PBS/Tween)
– incubate 4 h at 37 °C
– wash ×5 with PBS/Tween

Add 100 µl OPD substrate
– incubate 6 min 20 °C
add 100 µl 4 M sulfuric acid
– Read optical density at 429 nm.

Plates should be read within 30 min if at room temperature, but may be left overnight at 4 °C before reading with no deleterious effect.

Remarks   Any peroxidase-labeled lectin can, in principle be used as the detector in this assay. WGA (which binds N acetyl glucosamine and sialic acid) is very useful as a nonspecific probe for detection and quantification of many glycoproteins. Other lectins can be used to give more specific information about carbohydrate constituents of glycoproteins, e.g., LFA for sialic acids and Con A for mannose (or glucose). Some lectins are more fastidious than others in their requirements for metal

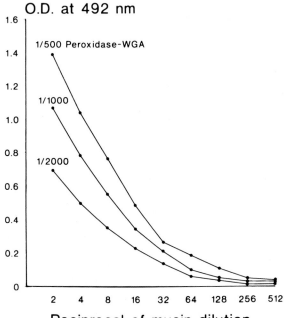

Fig. 1. Wheat germ agglutinin (*WGA*) binding assay applied to quantification of mucin extracted from human colonic biopsy samples: effect of varying concentrations of WGA. (Raouf et al. 1991)

ions and care needs to be taken that the appropriate ions are incorporated in the lectin buffer; Con A requires Ca and Mg ions but WGA and LFA do not require metal ions. Preliminary experiments may be needed to determine the optimum concentrations of any new lectin. (Fig. 1). We have found that lectin concentrations of around 2–5 µg/ml are usually satisfactory. A commonly used blocking buffer for ELISA procedures is 0.5% Tween/0.5% bovine serum albumin (BSA) in PBS but we have found that this decreases the sensitivity of the assay by about four to eight fold compared with the more gentle blocking buffer of 0.1% Tween in PBS. This reduced sensitivity seems to be due to the BSA rather than the altered Tween concentration. We do not use the outside wells of the plates, as these tend to give variable results. The nature of the coating buffer is not critical as we have found that diluting our column samples, which are in Tris/HCl pH 8.0, with carbonate coating buffer, Tris/HCl or PBS pH 7.4 does not affect the result. Extending the period of incubation of the plate with the lectin from 4 h at 37 °C to 16–20 h at 4 °C, while it increases the optical density readings at higher concentrations of mucin does not improve the overall sensitivity of the assay (Fig. 2).

The same assay system can also be used to quantify specific carbohydrate structures carried on serum glycoproteins. Its use in this way to quantify total peanut agglutinin (*Arachis hypogea*) binding activity in serum has been shown to be effective in pancreatic cancer diagnosis (Ching and Rhodes 1989). There are, however, peanut agglutinin binding glycoproteins present in normal sera, and serum samples have to be diluted 1 : 20 000 prior to application to the plate. We now use the combined lectin/antibody binding assay as described in the following section, as the latter assay gives lower background readings with normal sera.

When tested on intestinal mucus glycoprotein fractions all the lectin binding assays with the exception of Con A (mannose is almost totally lacking from purified mucus glycoprotein) were able to detect 1–2 µg (Lowry protein)/ml of mucus glycoprotein. Using wheat germ agglutinin, the assay has a coefficient of variation of 8.7% between plates ($n = 6$). Figure 3 shows the application of these assays to mucus glycoprotein fractions obtained after ion-exchange chromatography of purified mucus glycoprotein extracted from a single intestinal biopsy sample.

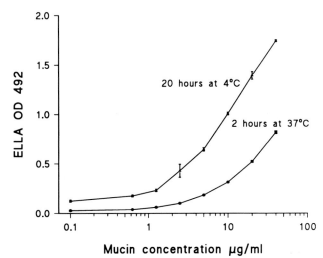

Fig. 2. Effect of prolonged incubation of lectin with mucin in WGA binding assay

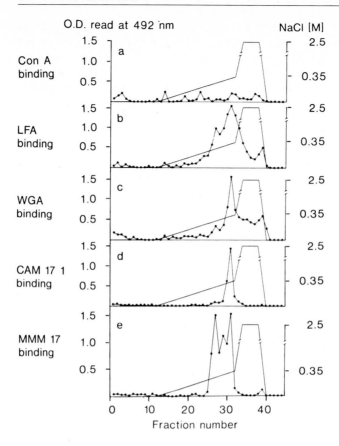

Fig. 3. Application of lectin binding assays (combined with immunoassays using the monoclonal anti-mucin antibodies CAM17.1 and MMM17) in the profiling of mucus glycoprotein fractions obtained after ion-exchange chromatography of a mucin preparation purified from a single colonoscopic biopsy sample. (Raouf et al. 1991)

15.2.2 Lectin/Mucin Antibody Sandwich ELISA for Quantification of Mucus Glycoproteins in Cancer Sera

Procedure   This assay (Parker et al. 1992) uses wheat germ agglutinin (WGA) attached to the solid phase to trap glycoproteins containing sialic acid or N-acetyl-glycosamine and as a detector uses a mouse IgM monoclonal antibody with specificity for sialomucin (CAM17.1), binding of which is in turn detected by peroxidase-labeled anti-mouse immunoglobulin. Checkerboard studies have been performed to determine optimal concentrations of lectin, antibodies, and test sera (Fig. 4).

– ELISA plate well coated with 100 µl WGA (3.3 µg/ml in carbonate coating buffer)
– incubate 16 h at 4 °C
– wash × 3 with PBS

– Add 120 µl blocking buffer (PBS/Tween)
– incubate 1 h at room temperature
– wash × 2 with PBS/Tween

– Add 100 µl of test sera diluted 1/20 in PBS/Tween
– incubate 2 h at 37 °C
– wash × 3 with PBS/Tween

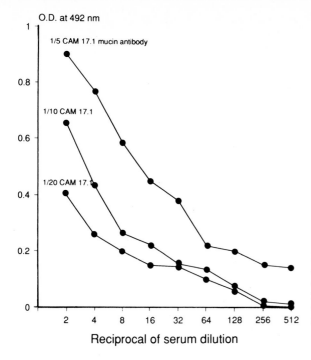

O.D. at 492 nm

1/5 CAM 17.1 mucin antibody

1/10 CAM 17.1

1/20 CAM 17.1

Reciprocal of serum dilution

Fig. 4. Effects of varying anti-mucin antibody concentration and serial dilution of a pancreatic cancer mucin-containing serum sample – CAM17.1/WGA assay. (Parker et al. 1992)

- Add 100 µl CAM17.1 culture supernatant diluted 1/20 in PBS
- incubate 2 h at 37 °C
- wash × 3 with PBS/Tween

- Add 100 µl peroxidase-goat anti-mouse-IgM conjugate, diluted 1/600 in PBS/Tween
- incubate for 2 h at 37 °C
- wash × 4 with PBS/Tween 20

- Add 100 µl OPD substrate
- incubate 6 min at 20 °C
- add 100 µl 4 M sulfuric acid

- Read optical density at 492 nm.

Remarks

The monoclonal antibody CAM17.1 was produced after immunization of a mouse with a mixture of Coll 2-23 colorectal cancer cells and PC/AA polyposis coli cells and a lipid extract of human meconium (Makin 1986). Evaluation of this antibody on slot blotted mucins has shown that sialic acid is an essential part of its epitope. Red cell agglutination studies have shown that it agglutinates all adult erythrocytes regardless of blood group, but has no activity against cord blood, findings that are compatible with activity against sialylated I antigen. Immunohistochemical studies have shown strong reactivity against intestinal mucus, particularly in the colon, small intestine, biliary tract, and pancreas, and we have used it to quantify purified colonic mucus glycoprotein fractions in a procedure similar to that described above with lectins (Raouf et al. 1991).

Attempts to biotin label CAM17.1 led to loss of affinity making an antibody–antibody sandwich system unworkable, and this was the initial reason for the development of the lectin-antibody sandwich ELISA. This system should be readily applicable to many other glycoprotein assays.

The assay has subsequently been refined further using a biotin/avidin-labeled anti-immunoglobulin system prior to marketing in kit form.

Results The assay has proved very robust with consistently low background activity and a coefficient of variation within assay of 4.7% and between assay of 10.3%. Early assays were standardized using aliquots of a pancreatic cancer serum sample, but the assay is now calibrated using tissue culture supernatant from a mucin-secreting cell

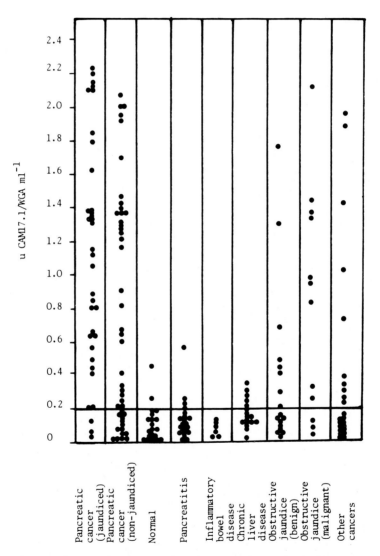

Fig. 5. Application of CAM17.1/WGA assay to sera from patients with pancreatic cancer and controls (Parker et al. 1992). At the time these assays were performed 1 unit/ml was the activity of a pancreatic cancer standard serum. The units have since been redefined with reference to a supernatant from a mucin-secreting cell line but sensitivity and specificity remain as shown

line. The assay is not affected significantly by repeated freezing and thawing of the serum samples.

The assay has been tested in sera from 79 patients with pancreatic cancer and 120 controls. As with the CA19-9 assay (Ching and Rhodes 1989), some false positive results are found in sera from patients with benign obstructive jaundice. This is presumably because biliary mucus has carbohydrate structures similar to pancreatic mucus. The overall results are encouraging, with a sensitivity of 78% and specificity of 76% for pancreatic cancer is non-jaundiced patients (Fig. 5). This test should be useful in the diagnosis of patients with unexplained pain or weight loss and in the monitoring of patients with pancreatic cancer following resection or chemotherapy.

References

Ching CK, Rhodes JM (1988) Identification and partial characterisation of a new pancreatic cancer-related serum glycoprotein by sodium dodecyl sulfate electrophoresis and lectin blotting. Gastroenterology 95: 137–142

Ching CK, Rhodes JM (1989) Enzyme-linked PNA-binding assay compared with CA19-9 and CEA radioimmunoassay as a diagnostic test for pancreatic cancer. Br J Cancer 59: 949–953

Ching CK, Rhodes JM (1990) Purification and characterisation of a peanut-agglutinin-binding pancreatic-cancer-related serum mucus glycoprotein. Int J Cancer 45: 1022–1027

Magnani JL, Steplewski Z, Mitchell K, Herlyn M, Fuhner P. (1983) Identification of the gastrointestinal and pancreatic cancer-associated antigen detected by monoclonal antibody 19-9 in the sera of patients as a mucin. Cancer Res 43: 5489–5492

Makin CA (1986) Monoclonal antibodies raised to colorectal carcinoma antigens. Annals R Coll Surg Eng 68: 298–301

Parker N, Makin CA, Ching CK, Eccleston D, Taylor OM, Milton JD, Rhodes JM (1992). A new enzyme-linked lectin/mucin antibody sandwich assay (CAM 171./WGA) assessed in combination with CA 19-9 and peanut lectin binding assay for the diagnosis of pancreatic cancer. Cancer 70: 1062–1068

Raouf A, Parker N, Iddon D, Ryder S, Langdon-Brown B, Milton JD, Walker R, Rhodes JM (1991) Ion-exchange chromatography of purified colonic mucus glycoproteins in inflammatory bowel disease: absence of a selective subclass defect. Gut 32: 1139–1146

Rhodes JM, Ching CK (1991) Serum diagnostic tests for pancreatic cancer. In: Neoptolemos J (ed.) Cancer of the pancreas. Clinical gastroenterology. Vol 4, No 4. London: Bailliere Tindall, 833–852

Rhodes JM, Ching CK (1993) The use of lectins in the characterisation of mucosal glycoproteins. In: Methods in molecular biology. ed E Hounsell Humana Press, Totowa, NJ. pp 247–262

16 Lectins as Indicators of Disease-Associated Glycoforms

N. Sumar, K.B. Bodman, and P.M. Rudd

16.1 Glycosylation and Disease

Alterations in protein glycosylation are associated with many diseases; these may be useful markers of a disease state, reflecting changes in the environment in which proteins are glycosylated. Both the carbohydrate structure and the concentration of glycoproteins may change in disease, providing diagnostic information. Examples of glycoproteins where such modifications have been described include the tumor products alpha-fetoprotein (AFP), chorionic gonadotropin (HCG), and alpha antitrypsin and the plasma proteins IgG, transferrin, α-1 acid glycoprotein, and fibrinogen.

Abnormal glycosylation may also contribute to pathogenesis, particularly in view of the fact that many important biological functions such as cell adhesion and endocytosis depend on the recognition of specific oligosaccharides on target ligands. The expression of specific oligosaccharide structures on cells may, for example, contribute to the metastatic behavior of tumor cells: transformation of hamster kidney (BHK) cells with polyoma virus leads to an increase in the level of expression of GlcNAc of transferase V (Yamashita et al. 1985). The effect of this is that abnormally large amounts of tetra-antennary oligosaccharides are synthesized from the tri-antennary isomers, substantially changing the glycosylation of the cell surface.

The extensive use of lectins to investigate glycosylation changes in development and disease is currently increasing our understanding of the biological and pathological significance of glycosylation events and allowing the possibility of new diagnostic approaches. Table 1 lists various lectins together with their oligosaccharide specificity.

Table 1. Some lectins and their sugar specificities

Lectin	Common name	Specificity
Abrus precatorius agglutinin	Abrin	βGal
Abrin precatorius toxin	Abrin A and Abrin C	βGal
Agaricus bisporos	Mushroom	βDGal(1-3) DGalNAc
Allomyrina dichotoma	Beetle	NeuNAc
Androctonus australis	Saharan scorpion	NeuNAc
Anguilla anguilla	Eel	αL-Fuc
Arachis hypogaea (PNA)	Peanut	βDGal(1-3) DGalNAc
Bandeiraea/(BS)	—	
Griffonia simplicifolia (GS)		
BSI		αDGal, αDGalNAc
BSI-A$_4$		αDGalNAc
BSI-B$_4$		αDGal
BSII		α and βDGlcNAc
Bauhania purpurea	Camies foot tree	βDGal(1-3) DgalNAc
Biomphalaria glabarata	Fresh water snail	NeuNAc

Table 1. (continued)

Lectin	Common name	Specificity
Canavalia ensiformis (Con A)	Concanavalin A (Con A) (Jack bean)	αDMan, αDGlc
Cancer antennarius	Marine crab	NeuNAc
Caragana arborescens	Siberian pea tree	DGalNAc
Carcinoscorpius rotunda cauda	Indian horsehoe crab	Neu5Acα2-6GalOH
Cicer arietinum	Chick pea	Fetuin
Codium fragile	Green marine algae	DGalNAc
Datura stramoniun (DSA)	Thorn apple	(DGlcNAc)$_2$
Dolichos biflorus (DBA)	Horsegram	αDGalNAc(1-3) GalNAc
Erythrina corallodendron	Coral tree	βDGal(1-4) DglcNAc
Euonymus europaeus	Spindle tree	αDGal(1-3) Dgal
Glycine max (SBA)	Soybean	DGalNAc
Helix aspersa	Garden snail	DGalNAc
Helix pomatia (HPA)	Roman snail (edible snail)	DGalNac
Heterometrus granulomanus	Indian scorpion	NeuNAc, GalNAc
Hadrurus arizonensis	Hairy scorpion	NeuNAc
Lathyrus odoratus	Sweet pea	αDMan
Lens culinaris (LCA)	Lentil lectin	αDMan
Limax flavus	Slug	NeuNAc
Limulus polyphemus	Limulin	NeuNAcα(2-6) GalNAc
Lotus tetragonolobus (LTG)	Winged pea	αLFuc
Lycospersicon esculentem	Tomato	DGlcNAc(β1-4GlcNAc) 1-3
Malcura pomifera	Osage orange	αDGal, αDGalNAc
Mytilus edulis	Sea mussel	NeuNAc
Oryza sativa	Rice	GlcNAc(β1-4GlcNAc) 1-2
Paruroctonus mesaensis	Scorpion	NeuNAc
Phaseolus limensis	Lima bean	DGalNAc
Phaseolus vulgaris (PHA)	PHA-E (erythroagglutinin)	Bisected bi- and tri-antenate -N linked sequences
	PHA-L (leucoagglutinin)	Branched non bisected -sequences
Phytolacca americana (PWM)	Pokeweed mitogen	(GlcNAcβ1-4GlcNAc) 1-3
Pisum sativum	Garden pea	αDMan
Pseudomonas aeruginosa	PA-1	Dgal
Psophocarpus tetragonolobus	Winged bean	DGalNAc
Ptilota plumosa	Red marine algae	αDGal
Ricinus communis (RCA)	Castor bean	
RCA toxin (RCA 60)		β and α Gal
RCA agglutinin (RCA120)		βGal
Sambucus nigra	Elderberry	NeuNAc(α2-6)-DGal/DGalNAc
Solanum tuberosum	Potato	(DGlcNAc) 3
Sophora japonica	Pagoda tree	βGalNAc
Triticum vulgaris (WGA)	Wheat germ	GlcNAc(β1-4GlcNAC) 1-2, NeuNAc
Ulex europaeus (UEA)	Gorse/furze	
UEA I		αLFuc
UEA II		LFuc α1-2 Galβ1-4 GlcNAc
Vicia faba	Broad bean	αMan
Vicia villosa (VVA)	Hairy vetch	
A4		DGalNAc
B4		DGalNAc
Vigna radiata	Mung bean	αDGal
Viscum album	Mistletoe	βDGal
Wisteria floribunda	Japanese wisteria	DGalNAC

Refs. Goldstein and Poretz (1986) and Mandal and Mandal (1990).

16.2 Lectin-Carbohydrate Interactions

Recent studies have established some basic concepts relating to lectin-carbohydrate interactions. Such insights may provide a basis for understanding the mechanisms by which lectins distinguish normal from abnormal ligands in diagnostic tests.

The following specific examples of lectin interactions illustrate a number of general principles:

- A multivalent sugar or a multiply glycosylated protein can bind several large protein ligands, allowing the possibility of cross-linking.
- Oligosaccharides can respond to accommodate the spatial requirements of lectin-protein ligands.
- Lectins can flexibly respond to stabilize a lectin-carbohydrate complex.
- Lectins can be exquisitely sensitive to small alterations in the presentation of their carbohydrate/protein ligands.
- Some carbohydrate residues in an oligosaccharide may interact with the protein to which they are attached, directing the presentation of the remainder of the sugar chain to the receptor.

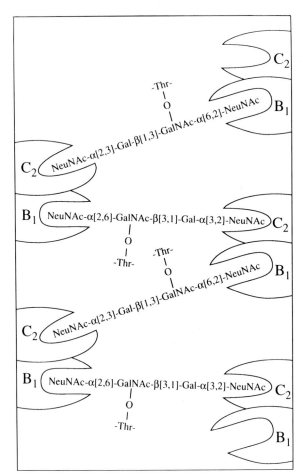

Fig. 1. Schematic representation of the T-5 oligosaccharide of glycophorin A cross-linking WGA dimers. The B1-C2 domains of five WGA dimers form an array, with tetrasaccharides O-linked to glycophorin bound in anti-polar fashion to each of the unique sites [B1 (primary binding site) /C2 and C2 (primary binding site) /B1] (Wright 1992)

– Carbohydrates attached to glycoproteins with lectin domains may interact internally with these domains, protecting both the lectin site and the carbohydrate during transport.

16.2.1 Wheat Germ Agglutinin

The definition of the crystal structure of wheat germ agglutinin (WGA)–glycophorin sialoglycopeptide receptor complex (Wright 1992) has demonstrated that a single multivalent tetrasaccharide can bind several large protein ligands (Fig. 1). In this case, the well-observed ability of lectins to form agglutinates depends on the fact that the lectin carbohydrate recognition domains, which contain equivalent ligand binding sites, are able to interact asymmetrically with complex cell surface receptors. The tetrasaccharide, T-5, was found to adopt different conformations in the different lattice environments, illustrating the fact that oligosaccharides can respond to accommodate the spatial requirements of lectin protein receptors.

16.2.2 *Lathyrus ochrus*

The importance of oligosaccharide structure and flexibility has also been demonstrated in a study of the binding of a biantennary oligosaccharide to clefts in a lectin, LOL-1, isolated from the seeds of *L. ochrus* (Bourne et al. 1992). In this case, the sugar adopts an extended S-shape conformation when bound to the lectin (Fig. 2). The Man 4 and GlcNAc 5 residues interact on either side of the phenyl 123 ring, gripping it as a clamp. The residues 5 and 6 fit into a partly hydrophobic cleft and

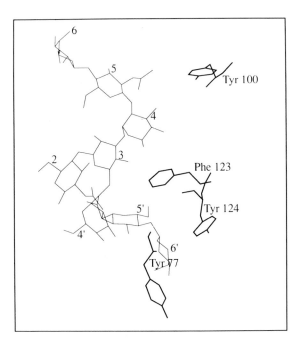

Fig. 2. View of the Van der Waals interactions of four important nonpolar residues with octasaccharide. Phe-123 stacks between Man 4 and GlcNAc5' and partly Man 3, Tyr- 124 to Gal 6' and Tyr 100 to GlcNAc5, Tyr-77 is shielded by the octasaccharide, but is not in direct Van der Waals contact (Bourne et al. 1992)

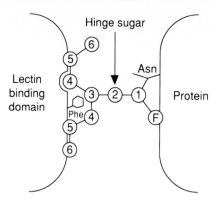

Fig. 3. Possible scheme for the interaction between a lectin and a protein with an N-linked oligosaccharide containing fucose (*F*) substituted in residue l. (Bourne et al. 1992)

participate in numerous interactions with the lectin. The X-ray data also show that the lectin adapts to the sugar; the protein backbone of the lectin was observed to move towards mannose in the formation of the complex, which was stabilized by multiple hydrogen bonds.

Bourne et al. (1992) have proposed that the binding of LOL-1 to a glycopeptide from human lactotransferrin, which contains a core fucose (F) (Fig. 3), might depend on a mechanism which has been described in a high resolution X-ray study by Bode et al. (1989) in which the presentation of the human leucocyte elastase was probed. This showed that most of the noncovalent contacts between the polysaccharide and the peptide backbone involve a fucose ring F, (Fig. 3) suggesting that the sugars 1 and F are firmly attached to the protein while the remainder of the sugar chains extend into the solution. In this way the fucose residue keeps residue 1 ordered, restraining the hinge region to residue 2.

16.2.3 Murine *Zona Pellucida*

The fine specificity of lectin interactions and the manner in which cross-linking can be achieved by a multiply glycosylated protein is illustrated by a study of murine ZP3 (Miller et al. 1992). All three glycoproteins in murine zona pellucida (ZP) are recognized by bovine Gal-transferase (Gal-T) in a lectin-like interaction. This depends on the presence of terminal N-acetyl-glucosamine on the ZP proteins; this is the monosaccharide recognized by the Gal-T. In contrast, only ZP3 is recognised by murine sperm Gal-T, demonstrating a more stringent substrate specificity for the biologically active enzyme which may involve the protein as well as sugar. ZP3 induces the acrosome reaction by cross-linking its sperm receptor; this is achieved through the presence of several binding sites for Gal-T on ZP3. After the acrosome reaction, Gal-T is re-distributed to the lateral sperm head, where it loses affinity for ZP3. Following fertilization, ZP3 is no longer recognized by Gal-T, although it is still recognized by non-sperm Gal-T. The factors controlling such exquisite specificity are still being explored.

16.2.4 Concanavalin A

An intriguing idea that glycosylated lectins may bind their own carbohydrate is explored in a description of the activation of the precursor of the legume lectin

Concanavalin A (Con A). The precursor is converted to the mature form through a complex sequence of events (deglycosylation, proteolytic cleavage and transpeptidation or ligation). Sheldon and Bowles (1992) showed that the glycoprotein precursor of Con A, which is inactive as a lectin, could be converted into a protein possessing carbohydrate binding activity by deglycosylation alone; this is the first step of processing in vivo and suggests that the other steps are not required for activation. It is possible that the lectin is "self-neutralized" by its own N-linked oligosaccharide locating within the binding site of the precursor, and that this may prevent interaction of the lectin with glycosylated molecules during its transport through subcellular compartments.

16.2.5 Immunoglobulin G

An interesting possibility arises in the case of IgG which contains a protein region able to bind the Fc carbohydrate, thereby fixing the conformation of the oligosaccharide. High resolution NMR of the free biantennary sugars associated with the Fc fragment of IgG contain regions of defined secondary structure, and all but one of the glycosidic linkages exist in a single preferred conformation, giving rigidity to the structure (Holmans et al. 1987). The α(1-6) linkage, however, can assume two conformations giving a flexible arm to the oligosaccharide. In the Fc, part of the carbohydrate interacts with the domain surface while the remainder extends into the intestitial space. The α(1-6) antennae interact with the hydrophobic and polar residues Phe 243, Pro 245 and Thr 260 in the domain surface resulting in the loss of flexibility since the antennae are now constrained by the protein (Fig. 4). In rheumatoid arthritis (RA) there is an increase in the population of glycoforms devoid of terminal galactose and sialic acid in both arms (Parekh et al. 1985). The lack of these monosaccharides therefore uncovers lectin-like sites on the protein. The consequences of abnormal glycosylation of IgG observed in RA have not yet been

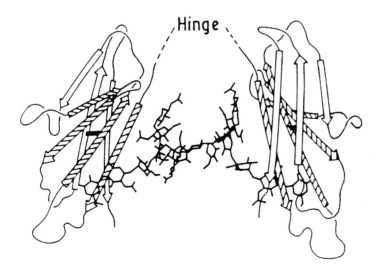

Fig. 4. In the IgG molecule the branched oligosaccharide chains cover part of the four stranded faces of the molecule

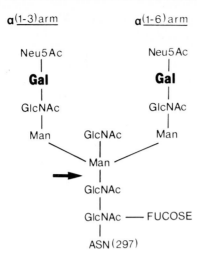

Fig. 5. The biantennary complex oligosaccharide structure found in the Fc region of IgG showing galactose moieties present in both the α(1-3) and α(1-6) arms thus depicting the galactosylated form of IgG (G2). In the absence of both these terminal galactose residues, the resulting structure is referred to as agalactosyl IgG or the GO glycoform completely lacking galactose

established. However, an increased proportion of agalactosyl glycoforms (termed IgG GO) (Fig. 5) in the total IgG population compared with normal levels, correlates with increased disease activity (Parekh et al. 1988), while high levels of agalactosyl IgG detected early in the course of the disease correlate with a poor prognosis (Young et al. 1991). The determination of IgG GO levels is therefore valuable both as a disease marker and as a prognostic indicator for the disease.

16.3 Protocols

Recently, a number of techniques have been developed which use lectins as a means of probing glycosylation changes. These include lectin column chromatography, lectin affino-diffusion, crossed affino-immuno-electrophoresis, sodium dodecyl sulfate (SDS)-PAGE followed by probing with lectins, radiolabeled or enzyme-labeled lectins in histochemical localization, radiolabeled lectin immunoassays and enzyme-linked lectins immuno assays. Table 1 summarizes some diseases where lectins are used as diagnostic reagents; the protocols which follow, describe assays which use different techniques, Western and dot blotting, immunohistochemical staining and Enzyme Linked Lectin Assays (ELLA) to diagnose and predict the course of disease.

Protocol 1. Lectins in Dot Blotting and Western Blots
Analysis of Glycosylation Changes in IgG (Sumar et al. 1990)

Principle   Glycosylation changes in IgG from rheumatoid patients have been described earlier. Age-corrected IgG GO levels in the serum of patients with RA are significantly higher when compared with levels in osteoarthritis or normal serum. The results are expressed as percentage oligosaccharide chains lacking galactose (%GO).

A method suitable for diagnostic testing has been developed whereby glycosylation changes in IgG are analyzed using the lectins *Ricinus communis agglutinin*

(RCAI) and *Bandeiraea simplicifolia II* (BSII). These lectins bind specifically to galactose and N-acetyl glucosamine residues respectively, allowing the percentage of GO to be determined.

All chemicals were Analar grade from BDH and Sigma. Nitrocellulose and Bradford reagent were from Schleicher and Schull and Biorad respectively.
Biotinylated BSII and RCAI were obtained from Vector Labs and Streptavidin-HRP conjugate from Dakopatts.

Materials

Purified IgG (10 or 20 µg/ml)
↓
Dot blot 100 µl IgG standards of known galactose deficiency and samples onto nitrocellulose (2 blots) and dry blots for 15 min.
↓
Boil blots at 90 °C in PBS for 15 min (this denatures the IgG and exposes the sugars)
↓
Block in PBS-T (0.05%)-BSA (1%)
↓
Treat with ricin-biotin (1 mg/ml, 1/250 dilution) 2 h RT or O/N at 4 °C/*Bandeiraea* biotin (1 mg/ml, 1/50 dilution) O/N at 4 °C
↓
5 × 10-min washes in PBS-T-BSA
↓
Treat with streptavidin – HRP conjugate (1/1000 dilution), 2 h RT
↓
5 × 10-min washes in PBS-T
↓
Develop blots with chloronaphthol (0.3 mg/ml methanol)
incubation mix: 1 vol substrate solution, 5 vols 50 mM Tris buffer pH 7.6 and 0.01% H_2O_2
↓
Wash, dry and read blots on a Biorad video densitometer.
↓
Plot OD ratio of *Bandeiraea* binding/ricin binding for the IgG standards against % GO values (oligosaccharide chains lacking galactose). Use this standard curve to read the %GO of samples.

Method

Correct results for age using an age-related %GO distribution obtained for the normal population (Sumar et al. 1990). The %GO results are expressed as standard deviations (SD) about the mean for the control group at that age.

IgG GO values obtained by this method have been shown to correlate with those obtained by the more rigorous biochemical method using exoglycosidase sequencing; hence this simpler method can be used with confidence. This technique uses lectins to detect terminal sugars on serum IgG samples applied on nitrocellulose blots and provides valid data for the analysis of glycosylation changes in IgG. The method is also suitable for analysing the relative abundance of accessible terminal sugars in any given glycoprotein of known carbohydrate structure and clearly has wider applications. This method has also been applied to transferrin (Sumar et al. unpubl.).

Conclusion

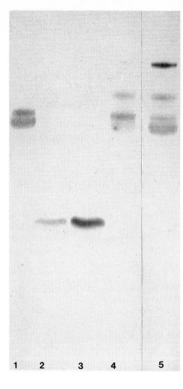

Fig. 6. Western blot of IgG, Fc and ammonium sulphate fractions of serum from a control and a rheumatoid patient probed with ricin agglutinin.
Lane 1 Purified IgG from Sigma (10 µg); *Lane 2* Purified Fc from a rheumatoid patient A (8 µg); *Lane 3* Purified Fc from a control subject B (8 µg); *Lane 4* Ammonium sulfate fraction of serum from rheumatoid patient A (20 µg); *Lane 5* Ammonium sulfate fraction of serum from control subject B (20 µg)

The same methodology can be applied on Western blots. Once samples have been run on SDS-PAGE gels and transferred onto nitrocellulose the blots can be treated as described for dot-blots.

Example. Probing normal and rheumatoid IgG and Fc with ricin agglutinin on Western blots.

Method   10 µg of purified commerical IgG, 8 µg of Fc and 20 µg of 33% ammonium sulfate fractions of serum from a patient and a control were run on an SDS-PAGE gel (10%) under reducing conditions by the method of Laemmli (1970). The gel was Western blotted onto nitrocellulose and boiled, blocked and probed with biotinylated ricin agglutinin and treated as described for dot-blots.

Results   There is a decrease in the extent of ricin agglutinin binding to rheumatoid IgG and Fc compared with normal; (Fig. 6), this can be quantified by densitometry. When using crude samples, it is necessary to monitor the amount of protein that is being probed to obtain comparable data.

Protocol 2. Lectins in Immunohistochemistry
The binding of *Helix pomatia* agglutinin (HPA) to N-acetyl glucosaminyl oligosaccharides in breast tissue in the prediction of long-term survival of breast cancer patients (Leathem and Brooks 1987; Brooks and Leathem 1991).

Positive tumors are more likely to have metastasized locally and to distant sites and patients therefore have a shorter survival time. It has been shown that HPA binds to a population of breast cells associated with metastasis to local lymph nodes. The clinical progress of 179 patients followed for 15–20 years was related to staining of paraffin sections of their primary breast cancers by HPA. This recognizes N-acetylgalactosamine residues which are associated with metastasis and poor prognosis in breast cancer.

Principle

Paraffin sections from blocks of primary cancers; HPA (Pharmacia, Sigma and made by the authors); rabbit antiserum to HPA (Laclab or produced by the authors); biotinylated swine anti-rabbit immunoglobulin (IgG) (Dako); avidin peroxidase, hydrogen peroxide (Sigma) and diaminobenzidinetetrahydrochloride (DAB) (Sigma).

Materials

Blocks of primary cancers removed by mastectomy, fixed in formalin and embedded in paraffin wax

↓

5-mm sections cut by microtone and dewaxed

↓

Sections incubated with HPA at 10 µg/ml for 1 h

↓

Rabbit anti-HPA 1/200 for 1 h

↓

Biotinylated swine anti-rabbit IgG 1/400 for 30 min

↓

Avidin-HRP at 5 µg/ml for 30 min

↓

Stain for peroxidase activity using H_2O_2 and DAB

Method

The tissue is classified as "stainer" or "nonstainer", irrespective of staining intensity, since there is no correlation between intensity of staining and prognosis, only with expression or nonexpression of the HPA binding ligand. The dividing line between the two groups was:

"Stainers" were cases in which 5% or more of cells were clearly stained (+) or 30% or more of the cancer cells were very weakly positive (±)
"Nonstainers" were cases in which 5% of the cancer cells were scored negative (−) or less than 50% were very weakly positive (±).

In practice, most cases were either intensely positive or completely negative.

HPA shows nominal specificity for the monosaccharide GalNAc. In breast cancers, HPA recognizes a GalNAc-bearing glycoconjugate associated with metastases. This HPA binding ligand has not yet been identified; however, it has been suggested (Springer 1989) that it might be related to the Tn blood group precursor substance absent in normal tissues but readily detectable in a high proportion of breast cancers and other cancers (see Table 2). In this paper, the HPA binding ligand is absent from normal breast tissue and early (nonmetastatic) breast cancer but is expressed by breast cancer cells as they become metastatically competent.

The value of HPA lies in its simple application to routinely fixed paraffin-embedded sections.

Conclusion

Table 2. Lectins as probes

Disease	Lectin probe/altered glycoprotein	Diagnostic application	Reference
1. Liver diseases 1.1.	Alpha feto protein (AFP) in serum Con A and LCA and affinity chromatography In chronic active hepatitis (BLD), < 20% AFP bound to LCA In primary liver cancer (PLC) and metastatic liver disease (MLD), > 20% AFP bound to LCA In BLD and PLC > 70% AFP bound to Con A In MLD < 70% AFP bound to Con A	A simple diagnostic test for differential diagnosis of chronic active hepatitis, primary liver cancer and metastatic liver disease	Baumah et al. (1987) Nikolic et al. (1990)
1.2.	Serum concentrations of AFP are increased in both yolk sac and hepatocellular carcinoma (HCC) and non-hepatic neoplasms. Determination of Con A and LCA nonreactive AFP variants by affinity chromatography on Con A and LCA columns differentiates between HCC and yolksac tumors and HCC from benign tumors	Distinguishes HCC from other diseases associated with increased AFP levels. LCA binding differentiates HCC from benign tumors	Wu (1990) Du (1991)
1.3.	Glycoforms of serum AFP were separated by electrophoresis, Western blotted and probed with: LCA (L), PHA (P), ConA (C), Allomyrina Bands were labelled 1–5. Distinct differences in lectin binding in bands 3 and 4 were detected with LCA and PHA	AFP bands L3 and P4 binding in HCC increased compared to chronic hepatitis and liver cirrhosis, therefore combined evaluation of L3 and P4 can be used as a discriminatory test	Taketa et al. (1990b)

1.4.	AFP bands as above, probed with DSA
	Four bands D1–D4 observed
	AFP D1 showed no affinity for DSA
	  D2 binding increased marginally in hepatitis and liver cirrhosis with or without HCC
	  D3 binding increased in HCC
	  D4 showed the highest affinity for DSA in tumors of gastrointestinal origin
		AFP-D3 is a highly specific marker for HCC
		AFP-D4 bound to DSA in other tumors
			Taketa et al. (1990a)

1.5.	Alpha-1-antitrypsin (AAT) analysis using crossed immuno electrophoresis and LCA. LCA reactive and nonreactive forms were detected in gels containing LCA a) The percentage of LCA reactive AAT in neoplastic diseases of liver increased compared with benign liver disease and normal controls b) Monitoring of AAT glycoforms before and after HCC showed that LCA reactive AAT increases after the development of HCC	The measurement of reactivity of AAT with LCA can be used as a diagnostic marker of HCC	Sekine et al. (1987)
2. Lung Carcinoma	Binding of lectins to cells in benign and malignant effusions were examined by lectin-immuno cytochemistry using ConA, DBA, PHA, RCA, SBA, UEA and WGA. 60% adenocarcinomas stained with UEA, while cells from benign effusions did not	Several patterns were observed but UEA gave positive indication of carcinoma	Rosen-Levin et al. (1989)
3. Oral cavity 3.1. Squamous cell carcinoma	The binding of a panel of 15 lectins to saccharides associated with the epithelial component of normal and leukoplakic mucous and squamous cell carcinoma from a variety of locations in the oral cavity was examined by lectin-immunocytochemistry	Neoplasia in oral squamous epithelia was found to be associated with alterations in terminal GalNAc, GlcNAc, mannose, fucose residues in O- and N-linked glycoconjugates	May and Sloan (1991)

Table 2. (continued)

Disease	Lectin probe/altered glycoprotein	Diagnostic application	Reference
3.2. Nasopharyngeal carcinoma	Cell surface carbohydrate profiles were examined in normal and neoplastic epithelium using a panel of nine lectins on formalin fixed paraffin embedded tissue sections. Alterations in terminal sialic acid, fucose, GalNAc residues present in outer parts of glyconjugates were detected. A decrease in staining was seen in undifferentiated nasopharyngeal carcinoma. Significant differences in intensity and distribution seen in UEA and PNA staining (after neuraminidase treatment) of normal nasopharyngeal epithelium and nasopharyngeal-neoplasia	Evaluation of lectin-binding to distinguish pre-invasive from invasive lesions of the nasopharynx	Chew et al. (1991)
4. Prostatic cancer	The binding of UEA 1 to benign and malignant prostatic epithelium was examined by immunoperoxidase staining of formalin fixed (FFPE) and cryostat sections. In benign epithelium 10% of cells expressed UEA 1 binding sites on FFPE sections while in malignant epithelium 90% cells bound UEA 1	Distinguishes benign from malignant prostate epithelium	Abel et al. (1989)
5. Colon cancer	The expression of T antigens (oncodevelopmental cancer-associated antigens) was studied in fetal, normal adult and malignant chlororectal tissues by immunocytochemistry using *Vicia villosa* lectin. Tn antigen = GalNAc to ser/thr; T antigen = Gal to Tn. Sialosyl-Tn = sialyl to Tn	Cells of normal colonic mucosa did not express Tn, sialosyl Tn or T-anitgen while malignant tissues did	Itzkowitz et al. (1989)
6. Pancreatic cancer 6.1	A high mw PNA binding glycoprotein in serum which expresses blood group H and sialylated Lewis antigen was studied using PNA lectin blotting	Increased binding of high mw glycoprotein to PNA in pancreatic cancer	Ching and Rhodes (1990)
6.2.	PNA and enzyme-linked lectin assays probed a mucin glycoprotein that expresses oncofetal serum antigen Gal β1-3 GalNAc (T blood group antigen and binding site for PNA). Increased binding and elevated levels of this antigen in serum are associated with pancreatic cancer	Serological test for pancreatic cancer	Ching and Rhodes (1989)
7. Breast cancer 7.1.	Malignant transformation of murine and human epithelial cells have increased – GlcNAc β 1-6 Man α 1-6 Man β branching in N-linked oligosaccharides. The level of this oligosaccharide was studied by binding of PHA to paraffin embedded tissue sections	(a) Intensity of L-PHA staining correlated with disease pathology. (b) L-PHA reactive β 1-6 branched N-linked glycoforms are increased in neoplasias of human breast cancer	Fernandes et al. (1991)
7.2.	Con A binding to oligosaccharides on normal and breast tissue was examined by immunoperoxidase staining on paraffin sections	Con A binds cancer cells but not normal epithelium	Leathem et al. (1983)
7.3.1.	HPA binding to N-acetyl galactosaminyl oligosaccharides on sections of primary tumors; immunoperoxidase staining on paraffin sections	Long-term prognosis. Cancers with postivie HPA staining were associated with lower survival rate	Fukutomi et al. (1989) Leathem and Brooks (1987)

Table 2. (continued)

Disease	Lectin probe/altered glycoprotein	Diagnostic application	Reference
7.3.2.	Changes in glycosylation of primary breast cancer cells using the above technique	Increased HPA binding correlates with the presence of lymph node metastasis. Distinguishes between cancers that have metastasized and those that have not	Brooks and Leathem (1991) Leathem (1990)
8. Cystic fibrosis	Abnormal fucosylation of mucin-ileal mucus in ileal goblet cells detected by histochemistry and stained with peroxidase conjugated LTG and UEA	Strong staining of illeal goblet cells with LTG showed increased fucosylation of mucus in cystic fibrosis compared with controls. UEA did not differentiate between CF and controls	Thiru et al. (1990)
9. Rheumatoid arthritis (RA) Juvenile arthritis (JRA) Tuberculosis (TB)	Reduced galactosylation of the biantennary oligosaccharide chains in serum IgG was studied using dot blotting and probing with BSII and Ricin agglutinin	Increased BSII binding in serum IgG in RA, JRA, and TB patients. IgG GO results used in conjunction with Rheumatoid Factor as diagnostic test	Sumar et al. (1990, 1991) Roitt et al. (1988) Young et al. (1991)
10. Carcinomas	Accute phase proteins. a) α-1 acid glycoprotein in plasma. The variation in the five highly branched N-linked glycans was examined by crossed immuno-affinity electrophoresis	Changes in di/tri antennary glycan ratios were detected in diseased states	Hanson et al. (1984)
11. Severe burns	b) Variation in the glycosylation of α-1 acid glycoprotein detected by Con A affinity chromatography	A major shift towards diantennary glycans indicates severe burns	Mallet et al. (1987)
12. Chorio-carcinoma	Human chorionic gonadotropin from urine quantitated by lectin-radiometric assays in normal and choriocarcinoma urine with RCA and PNA	Elevated levels of desialylated forms of HCG associated with choriocarcinoma	Imamura et al. (1987)

Protocol 3. Lectins in Enzyme-Linked Lectin Assays (ELLA)

A serological test for pancreatic cancer (Ching CK and Rhodes JM 1989) (For more extensive details of ELLA, the reader is referred to a thesis dedicated to ELLA (Pekalharing 1989)

Principle   Pancreatic cancer can be difficult to diagnose, particularly when the patient is non-jaundiced. A simple diagnostic test has been developed to allow diagnosis without the need for invasive tests. Tumor antigens which have been described in pancreatic cancer sera include carcinoembryonic antigen (CEA) and CA 19-9. Lectin immunoblotting studies (Ching and Rhodes 1988) have shown that there is increased binding of peanut agglutinin (PNA) and increased levels of CA 19-9 in pancreatic cancer sera.

PNA binds to exposed Thomson Friedenreich antigen (galactose 1-3, N-acetyl galactosamine) on desialylated cell surface and mucin glycoproteins, while CA 19-9 recognizes the sialylated blood group Lewis antigen expressed on mucin secreted by the pancreatic tumor. The high level of PNA binding activity and CA 19-9 levels in pancreatic cancer sera reflect mucin which is probably structurally immature, i.e., incompletely glycosylated and which has been secreted into the serum.

This method detects abnormally glycosylated mucin in serum by enzyme linked lectin assay (ELLA) using PNA.

Materials

Sera from pancreatic cancer patients, controls (which included other patients with inflammatory disease and other chronic diseases) and from normal individuals were stored at $-70\,°C$.

Micro ELISA plates (M12A B) were obtained from Dynatech. Peroxidase-tagged PNA from Sigma, CA 19-9 RIA and CEA RIA Kits from CIS, UK.

Method

Apply 100 µl (1/20 000 diluted) serum samples and serially diluted positive controls in duplicate on ELISA plates and incubate for 16 h at 4 °C. Samples are randomly distributed on the plate and assayed blind.

↓

Wash plates three times and block with PBS/Tween 20 (PBS-T) buffer

↓

Add 100 µl peroxidase-PNA (12.5 µg/ml) to each well; incubate 16 h at 4 °C

↓

Wash three times, PBS-T

↓

Bound lectin is detected with O-phenyldiamine (10 mg) and H_2O_2 (40 µl) in phosphate citrate buffer (25 ml, pH 5), 100 µl/well. The reaction is allowed to proceed for 10 min at room temperature and then terminated by addition of 4 M H_2SO_4, 100 µl/well.

Optical density is read at 492 nm, the O.D. of the test sera converted into units of peanut agglutinin binding activity (PLBA) per ml with reference to the positive control serum. The same positive control serum must be used throughout the assay. One unit PLBA per ml can be arbitrarily defined as peanut agglutinin binding activity equivalent to a 1:20 000 dilution of the positive control serum.

Results

This assay is highly reproducible; it has a sensitivity of 77% (0.6 µg/ml taken as the normal cut-off limit) and specificity of 83% for pancreatic cancer. It does not distinguish between jaundiced and nonjaundiced patients with pancreatic cancer; however, patients with metastatic disease have significantly higher PLBA than those with localized disease. The PNA ELLA method gave very similar results to CA 19-9 radioimmunoassays; CEA assays were found to be less useful (47% specificity). Combination of CA 19-9 assay with PNA ELLA improved the sensitivity to 85% with only a small decrease in specificity to 85%. Anomalous positive results may arise in some patients with epithelial cancers.

Conclusion

The enzyme-linked peanut lectin assay is cheap, easy to perform, and reproducible and its use in conjunction with CA 19-9 RIA is a valuable advance in the diagnosis of pancreatic cancer.

16.4 Summary

In this chapter, we have described three different techniques which use lectins in the diagnosis of diseases which involve glycosylation changes. The ELLA method is more convenient when compared to SDS-PAGE followed by blotting and probing; however, the electrophoretic step can be valuable, since it allows the analysis of a single glycoprotein in a mixture or crude sample. When using SDS PAGE/lectin blotting techniques, it is essential to load equal amounts of each sample protein in order to obtain comparable data. Another powerful technique not discussed, but which may be used in the differential diagnosis of tumors is lectin-affinity electrophoresis (Taketa and Hirai 1989, Taketa et al. 1990a, b).

Lectins can successfully be used to localize carbohydrate in both frozen and paraffin embedded, formalin-fixed tissues. However, the preparation of tissue for such staining must be done carefully and consistently or the sensitivity of the technique may be altered (Walker 1985). Several factors limit the precise definition of the lectin receptor ligands in tissue sections. Most lectins react with terminal, nonreducing sugar ligands, but some recognize internal oligosaccharide sequences. Moreover, adjacent nonspecific sugars can either markedly enhance or severely diminish binding to the specific sugar sequence. Lectins used in histochemistry can be conjugated to fluorochromes (lectin-fluorescence isothiocyanate) or chromogens (lectin-horseradish peroxidase) and applied directly to tissue sections. The sensitivity of such reagents can be further increased by the use of anti-lectin antibodies and peroxidase-antiperoxidase amplification or biotin conjugated lectins and avidin–biotin peroxidase complex amplification. Colloidal gold is another popular electron-dense marker employed in immunocytochemistry (Versura et al. 1989).

The advantages of using a panel of lectins with different fine specificities is illustrated by a study of abnormal fucosylation in ileal mucous in cystic fibrosis. Thiru et al. (1990) used peroxidase conjugated lectins to analyze the glycoproteins of small intestinal mucins in normal infants and those with cystic fibrosis to ascertain whether there are any detectable histochemical differences in saccharide composition. A significant decrease in *Lotus tetragonolobus* (LTG) binding was found in normal small intestinal mucin, whereas patients with cystic fibrosis showed persistent and intense LTG binding. *Ulex europaeus* (UEA), which also binds αL-fucose, did not show any difference between controls and patients. Such studies can be complemented by tests with monoclonal antibodies if these are available (King et al. 1990).

The dot-blotting technique (Sumar et al. 1990) describes how a glycoprotein was purified and subsequently probed with the appropriate lectins. This is a straightforward method to use, although the initial protein purification step may be time consuming. If the nitrocellulose can be pre-coated with appropriate reagents to specifically pull out the glycoprotein in question, purification may not be necessary. The dot-blotting method for IgG analysis, for example, has been modified by coating with protein A, which binds IgG allowing the majority of the other serum proteins to be washed away.

Lectins have proved to be powerful tools for the study of glycosylation changes in many disease states. They are very important diagnostic reagents, especially in tumor biology, where detection of early changes in glycosylation may lead to effective treatment. Most of these methods are much less time-consuming than conventional biochemical techniques which, although more accurate, are not suitable for routine diagnostic tests. The use of lectins to study glycosylation changes may be expected to result in novel diagnostics while allowing new insights into disease pathogenesis.

Acknowledgements. The Glycobiology Institute is funded by Monsanto Company. The authors wish to thank Dr. Philip Sheldon for helpful discussions on Concanavalin A and Professor R.A. Dwek for his valuable suggestions and critical reading of the manuscript.

References

Abel PD, Keane P, Leathem A, Tebbutt S, Williams G (1989) Change in glycoconjugate for the binding site of the lectin *Ulex europaeus* 1 following malignant transformation of prostatic epithelium. Br J Urol 63: 183–185

Baumah PK, Cornell C, Cassells-Smith AJ, Skillen AW (1987) Differential reactivity of alpha-fetoprotein with lectins and its usefulness in the diagnosis of various liver disease. Clin Chim Acta 168: 69–73

Bode W, Meyer E, Powers JC (1989) Human leukocyte and porcine pancreatic elastase: X-ray crystal structures, mechanism, substrate specificity and mechanism-based inhibitors. Biochemistry 28: 1951–1963

Bourne Y, Rouge P Cambillau C (1992) X-ray structure of a biantennary octasaccharide-lectin complex refined at 2.3-Å resolution. JBC 267: 197–203

Brooks SA, Leathem AJ (1991) Prediction of lymph node involvement in breast cancer by detection of altered glycosylation in the primary tumour. Lancet 338: 71–74

Chew EC, Yuen KE, Lee JC (1991) Lectin histochemistry of normal and neoplastic nasopharyngeal epithelium. Anti Cancer Res 11: 697–704

Ching CK, Rhodes JM (1988) Identification and partial characterisation of a new pancreatic cancer related serum glycoprotein by SDS-PAGE and lectin blotting. Gasteroenterology 95: 137–140

Ching CK, Rhodes JM (1989) Enzyme-linked PNA lectin binding assay compared with CA 19-9 and CEA radioimmunoassay as a diagnostic blood test for pancreatic cancer. Br. J Cancer 59: 949–953

Ching CK, Rhodes JM (1990) Purification and characterisation of a peanut-agglutinin-binding-pancreatic-cancer-related serum mucus glycoprotein. Int J Cancer 45: 1022–1027

Du MQ, Hutchinson WL, Johnson PJ, Williams R (1991) Differential alpha-fetoprotein lectin binding in hepatocellular carcinoma. Diagnostic utility at low serum levels. Cancer 67: 476–480

Fernandes B, Sagman N Auger M, Demetrio M, Dennis JW (1991) Beta 1-6 branched oligosaccharides as a marker of tumour progression in human breast and colon neoplasia. Cancer Res 51: 718–723

Fukotomi T, Habishi M, Tsugane S, Yamamoto Nanasawa T, Hirota T (1989) Prognostic contributions of *Helix pomatia* and carcinoembryonic antigen staining using histochemical techniques in breast carcinomas. Jpn J Clin Oncol 19: 127–134

Goldstein IJ, Poretz RD (1986) Isolation, Physicochemical characterisation and carbohydrate-binding specificity of lectins. In: The lectins eds Liener IE, Sharon N, Goldstein IJ Academic press pp 35–248

Hanson JES, Larsen VA, Bog Hansen TC (1984) The microheterogeneity of α-1 acid glycoprotein in inflammatory lung disease, cancer of the lung and normal health. Clin Chim Acta 138: 41–48

Holmans SW, Dwek RA, Rademacher TW (1987) Tertiary structure in N-linked oligosaccharides. Biochemistry 26: 6553–6560

Imamura S Armstrong EG Birken S Cole LA- Canfield RE (1987) Detection of desialylated forms of human chorionic gonadotropin Clin Chim Acta 163: 339–349

Itzkowitz S, Yuan M, Montgomery CK, Kjeldsen T, Takahashi HK, Bigbee WL, Kim YS (1989) Expression of Tn, sialosyl-Tn and T antigens in human colon cancer. Cancer Res 49: 197–204

King A, McLeish M, Thiru S (1990) Abnormal fucosylation of ileal mucus in cystic fibrosis: II A histochemical study using monoclonal antibodies to fucosyl oligosaccharides. J Clin Pathol 43: 1019–1022

Laemmli UK (1970) Cleavage of structural proteins during the assembly of the head of bacteriophage T4. Nature 227: 680–685

Leathem A, Dokal I, Atkins N (1983) Lectin binding to normal and malignant breast tissue. Diagn Histol Pathol 6: 171–180

Leathem AJ (1990) Biological, biochemical and morphological markers of breast disorders and of breast cancer. Acta Histochem Suppl 40: 51–58

Leathem AJ, Brooks SA (1987) Predictive value of lectin binding on breast-cancer recurrence and survival. Lancet i: 1054–1056

Mallet B, Franc JL, Miquel M, Arnaud C (1987) Effects of severe burns on glycan microheterogeniety of four acute phase proteins. Clin Chim Acta 167: 247–257

Mandal C, Mandal C (1990) Sialic acid binding lectins. Experientia 46: 433–441

May DPJ, Sloan P (1991) Lectin binding to normal mucosa, leucoplakic and squamous cell carcinoma of the oral cavity. Med Lab Sci 48: 6–18

Miller D, Macek MB, Shur BD (1992) Complementarity between sperm surface β-1, 4-galactosyltransferase and egg-coat ZP3 mediates sperm-egg binding. Nature 357: 589–594

Nikolic JA, Stajic M, Cuperlovic M, Hajdukovic L, Golubovic G (1990) Serum alpha fetoprotein levels and microheterogeneity in patients with different liver diseases. J Hepatol 11: 252–256

Parekh RB, Dwek RA, Sutton DJ, Fernandes DL, Leung A, Stanworth D, Rademacher TW, Mizuochi T, Tanaguchi T, Matsuta K, Takeuchi F, Nagano Y, Miyamoto T, Kobata A (1985) Association of rheumatoid arthritis and primary osteoarthritis with changes in the glycosylation pattern of serum IgG. Nature 316: 452–457

Parekh R, Isenberg D, Ansell B, Roitt IM, Dwek RA, Rademacher TW (1988) Galactosylation of IgG associated oligosaccharides: reduction in patients with adult and juvenile arthritis and relation to disease activity. Lancet i: 966–969

Pekalharing JM (1989) Lectin enzyme binding assays. Thesis, Rotterdam

Roitt IM, Dwek RA, Parekh RB, Rademacher TW, Alavi A, Axford JS, Bodman KB, Bond A, Cooke A, Hay FC, Isenberg DA, Lydyard PM, Mackenzie-Rook G, Smith M, Sumar N (1988) The role of antigen in autoimmune response with special reference to changes in carbohydrate structure of IgG in rheumatoid arthiritis. J Autoimm 1: 499–506

Rosen-Levin E, Patil JR, Watson CW, Jagirdar J (1989) Distinguishing benign from malignant pleural effusions by lectin immunocytochemistry. Acta-Cytol 33: 499–504

Sekine C, Aoyagi Y, Suzuki Y, Ichida F (1987) The reactivity of alpha-1-antitrypsin with Lens culinaris agglutinin and its usefulness in the diagnosis of neoplastic disease of the liver. Br J Cancer 56: 371–375

Sheldon PS, Bowles DJ (1992) The glycoprotein precursor of Concanavalin A is converted to an active lectin by deglycosylation. EMBO 11: 1297–1301

Springer GF (1989) Tn epitope (N-acetyl-D-galactosamine-O-serine/threonine) density in primary breast carcinoma. A functional predictor of aggressiveness. Mol Immunol 26: 1–5

Sumar N, Bodman KB, Rademacher TW, Dwek R, Williams P, Parekh RB, Edge J, Rook GAW, Isenberg DA, Hay FC, Roitt IM (1990) Analysis of glycosylation changes in IgG using lectins. J Immunol Meths 131: 127–136

Sumar N, Isenberg DA, Bodman KB, Soltys A, Young A, Leak AM, Round J, Hay FC, Roitt IM (1991) Reduction in IgG galactose in juvenile and adult onset rheumatoid arthritis measured by a lectin binding method and its relation to rheumatoid factor. Annal Rheum Dis 50: 607–610

Taketa K, Hirai H (1989) Lectin affinity electrophoresis of alpha-fetoprotein in cancer diagnosis. Electrophoresis 10: 562–567

Taketa K, Ichikawa E, Yamamoto T. Matsuura S. Taga H, Hirai H (1990) *Datura stramonium* agglutinin reactive alpha-fetoprotein isoforms in hepatocellular carcinoma and other tumours, Tumour Biol 11: 220–228

Taketa K, Sekiya C, Namiki M, Akamatsu K, Ohta Y, Endo Y, Kosata K (1990b) Lectin reactive profiles of alpha-fetoprotein characterising hepatocellular carcinoma and related conditions. Gastroenterology 99: 508–518

Thiru, S, Devereux G, King A (1990) Abnormal fucosylation of ileal mucus in cystic fibrosis: 1 A histochemical study using peroxidase labelled lectins. J. Clin Pathol 43: 1014–1018

Versura P, Maltarello MC, Bonvicini F, Laschi R (1989) The lectin–gold technique: an overview of applications to pathological problems. Scanning Microsc 3: 605–620

Walker RA (1985) The use of lectins in histopathology. Histopathology 9: 1121–1124

Wright CS (1992) Crystal structure of a wheat germ agglutinin/glycophorin-sialoglycopeptide receptor complex: structural basis for cooperative lectin binding. 267: 14345–14352

Wu JT (1990) Serum alpha-fetoprotein and its lectin reactivity in liver diseases: a review. Ann Clin Lab Sci 20: 98–105

Yamashita K, Tachibana Y, Ohkura T, Kobata A (1985) Enzymic basis for the structural changes of asparagine- linked sugar chains of membrane glycoproteins of baby hamster kidney cells induced by polyoma transformation. J Biol Chem 260: 3963–3969

Young A, Sumar N, Bodman KB, Goyal S, Sinclair H, Roitt I, Isenberg D (1991) Agalactosyl IgG: an aid to differential diagnosis in early synovitis. Arth Rheum 34: 1425–1429

17 Viral Lectins for the Detection of 9-O-Acetylated Sialic Acid on Glycoproteins and Glycolipids

B. SCHULTZE, G. ZIMMER, G. HERRLER

A number of viruses are able to recognize specific carbohydrate structures and to use these structures for the attachment to the cell surface. Influenza C virus and bovine coronavirus specifically attach to receptors containing N-acetyl-9-O-acetyl-neuraminic acid. Therefore, they can be used as lectins for the detection of glycoconjugates containing this type of sialic acid. These viruses also contain an acetylesterase on their surface which can be exploited for the detection of virions bound to immobilized glycoconjugates (Zimmer et al. 1992). As shown in Fig. 1, in the case of influenza C virus the acetylesterase activity is a function of the surface glycoprotein HEF that also mediates the binding to N-acetyl-9-O-acetylneuraminic acid. In the case of bovine coronavirus the acetylesterase is a function of the HE protein. Though this protein also recognizes 9-O-acetylated sialic acid, the binding of virus is primarily due to a another surface glycoprotein designated S (Schultze et al. 1991). The esterases of both viruses are able to cleave the synthetic substrate α-naphthyl acetate giving rise to naphthol. The latter compound reacts with a diazonium ion such as Fast Red resulting in a coloured insoluble complex, which reveals the presence of bound virus.

Influenza A and B viruses and paramyxoviruses have been used for the detection of gangliosides containing N-acetylneuraminic acid (i.e. sialic acid lacking an O-acetyl group) in defined linkage types. These viruses contain a neuraminidase rather than an acetylesterase. As no commercial detection system is available that makes use of neuraminidases to yield a coloured insoluble complex, binding of these viruses has to be determined by other means, e.g. enzyme-linked immunoreagents.

Here we describe how bovine coronavirus and influenza C virus can be used to detect 9-O-acetylated sialic acid on glycoproteins bound to nitrocellulose as well as on gangliosides bound to thin-layer plates. In addition a microtiter assay is desribed, which is useful for quantitative studies.

17.1 Experimental Part

17.1.1 Detection of Glycoproteins Containing N-Acetyl-9-O-Acetylneuraminic Acid by Bovine Coronavirus

- Nitrocellulose (0.45 μm, Schleicher & Schuell, Dassel, Germany)   **Materials**
- Phosphate-buffered saline (PBS): 8.0 g NaCl, 0.2 g KCl, 1.15 g $Na_2PO_4 \times 2\,H_2O$, 0.2 g KH_2PO_4 in 1 l H_2O, pH 7.2
- PBS/BSA: 1% bovine serum albumin (Serva, Heidelberg, Germany) in PBS
- Nonfat dry milk: 10% suspension in PBS
- PBS/Tween: 0.1% Tween 20 (Serva, Heidelberg, Germany) in PBS
- Rat serum

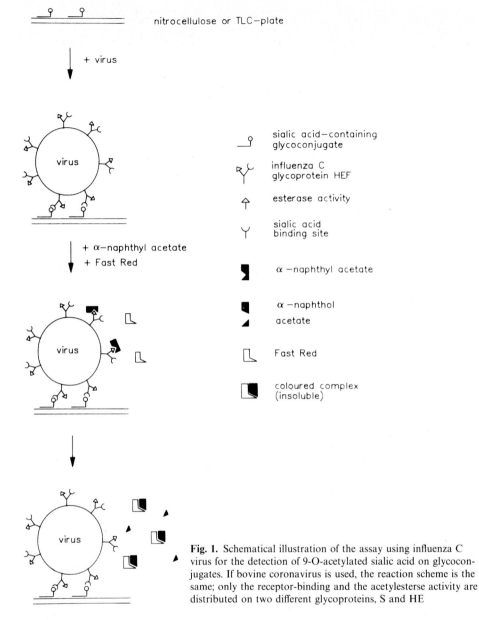

nitrocellulose or TLC–plate

+ virus

virus

sialic acid–containing
glycoconjugate

influenza C
glycoprotein HEF

esterase activity

sialic acid
binding site

+ α–naphthyl acetate
+ Fast Red

virus

α –naphthyl acetate

α –naphthol

acetate

Fast Red

coloured complex
(insoluble)

virus

Fig. 1. Schematical illustration of the assay using influenza C
virus for the detection of 9-O-acetylated sialic acid on glycoconjugates. If bovine coronavirus is used, the reaction scheme is the
same; only the receptor-binding and the acetylesterse activity are
distributed on two different glycoproteins, S and HE

- α-Naphthyl acetate (Sigma, Deisenhofen, Germany): 50 mM stock solution in
  acetone
- Esterase substrate A: 1 mM α-naphthyl acetate in PBS, 0.1% Fast-Red TR-salt
  (Sigma, Deisenhofen, Germany), filtered (e.g., folded filters, Schleicher & Schuell,
  Dassel, Germany)
- Biotinylated anti-rabbit donkey immunoglobulin (Amersham, Braunschweig,
  Germany)

- Streptavidin-biotinylated horseradish peroxidase complex (Amersham, Braun-schweig, Germany)
- 4-Chloro-1-naphthol (Sigma, Deisenhofen, Germany): 3 mg/ml in DMSO
- H_2O_2 38%, (Fluka, Neu-Ulm, Germany)
- Bovine coronavirus (BCV) is grown in MDCK I cells (Schultze et al. 1990). After clarification of the supernatant by low speed centrifugation, the virus is used for the binding assay

Procedure

1a. Different dilutions of the glycoproteins to be analyzed are applied to nitrocellu-lose in a volume of 2 µl and air-dried for about 20 min
1b. Proteins are separated by SDS-polyacrylamide gel electrophoresis and blotted on a nitrocellulose sheet
1c. Control samples were incubated with 0.1 N NaOH for 1 h at room temperature in order to release O-acetyl groups. After three washes with PBS, the blocking reagent was added
2. Nonspecific binding sites are blocked with nonfat milk overnight at 4 °C (alternatively, PBS/BSA may be used)
3. The nitrocellulose strips are washed three times for 5 min with PBS/Tween and incubated for 1 h with BCV (256 HA-units/ml) at 4 °C
4. After washing three times with PBS–Tween at 4 °C, bound BCV is detected by immersing the nitrocellulose in esterase substrate A. After incubation for 5–10 min at room temperature, the reaction can be stopped by replacing the substrate solution with distilled water

Alternative

If the assay is applied to viruses lacking an acetylesterase, a conventional immunolo-gical detection system has to be used to detect bound virus.

1–3. As above
4. The following steps (4–6) are performed at 4 °C. After three washes with PBS/Tween, the nitrocellulose strips are incubated for 1 h with a rabbit antiserum directed against BCV (1:1000 dilution)
5. The strips are washed three times with PBS/Tween and then incubated for 1 h with a 1:1000 dilution of biotinylated anti-rabbit donkey immunoglobulins
6. Following three washes with PBS/Tween, the nitrocellulose is incubated for 1 h with streptavidin-biotinylated horseradish peroxidase complex (1:1000) and then washed again
7. Bound BCV is detected by incubation of the strips with PBS, 4-chloro-1-naphthol, H_2O_2 (500:100:1) for 10–20 min at 20 °C. The color development is stopped by washing with distilled water

17.1.2 Detection of Gangliosides Containing 9-O-Acetylated Sialic Acid by Influenza C Virus

Materials

- PBS, PBS/BSA, PBS/Tween, α-naphthyl acetate, esterase substrate A: see above
- Glass-backed high-performance thin-layer chromatography (HPTLC) plates (10 × 10 cm or 5 × 10 cm), silica gel 60 (Merck, Darmstadt, Germany)
- Polyisobutylmethacrylate (Plexigum P28; Röhm, Darmstadt, Germany)

- Bovine brain gangliosides (BBG), prepared as described (Svennerholm and Fredman 1980)
- Orcinol spray reagent: 40.7 ml 37% HCl, 0.1 g orcinol (Sigma, Deisenhofen, Germany), 1 ml 1% aqueous $FeCl_3$, 25 ml H_2O
- Influenza C virus (strain Johannesburg/1/66) is grown in embryonated chicken eggs as described (Herrler et al. 1979). The allantoic fluid is clarified by low speed centrifugation (4000 g, 10 min) and stored in small aliquots at $- 80\,°C$. The virus is quantitated by a hemagglutination assay using 0.5% chicken erythrocytes. The hemagglutination titer (HA-units/ml) indicates the reciprocal value of the maximum dilution that causes agglutination of the erythrocytes.

Procedure

1a. Gangliosides containing 1–5 µg sialic acid are applied to each lane of the glass-backed HPTLC-plates and chromatographed. A suitable solvent system is chloroform/methanol/0.2% aqueous $CaCl_2$ (60:40:9)

1b. Control samples are exposed to ammonia vapors at room temperature for about 12 h in order to release O-acetyl groups. After thorough drying, the plates are chromatographed as described above

1c. For chemical detection of gangliosides after chromatography, a dried control plate is sprayed with the orcinol reagent, covered by a second glass plate and heated at 120 °C for 20 min. Gangliosides are indicated by a characteristic violet color

2. The developed chromatogram is dried thoroughly under a stream of air for at least 10 min at room temperature

3. The plate is dipped for 2 min in diethylether containing 0.5% polyisobutyl-methacrylate and dried as above

4. After spraying with PBS, the plate is immersed in PBS/BSA for 60 min at room temperature

5. Some drops of allantoic fluid containing influenza C virus with a hemagglutinating activity of at least 256 HA-units/ml are added and spread over the whole chromatogram by covering it with a piece of parafilm. Virus is allowed to bind for 60 min at 4°C

6. The plate is washed three times with PBS/Tween at 4 °C, 5 min each

7. Bound virus is visualized by immersing the plate in esterase substrate A for about 30 min at room temperature. The reaction is stopped by rinsing the chromatogram with H_2O and drying it as above.

17.1.3 Microtiter Assay for the Detection of 9-O-Acetylated Sialic Acid

Materials

- PBS, PBS/BSA, PBS/Tween, and influenza C virus: see above
- 96-well, flat-bottom polystyrene microtiter plates (Immuno-Module MaxiSorp F8, Nunc, Wiesbaden, Germany)
- Fluorescence spectrophotometer (Perkin-Elmer, Offenbach, Germany) with plate reader
- 4-Methylumbelliferyl acetate: 2 mM stock solution in acetone/H_2O (1:1, by vol)
- p-Nitrophenyl acetate (Sigma, Deisenhofen, Germany): 100 mM stock solution in acetone
- Esterase substrate B: 60 µM 4-methylumbelliferyl acetate in PBS
- Esterase substrate C: 1 mM p-nitrophenyl acetate in PBS.

Procedure

1a. Glycoproteins to be analyzed are dissolved in PBS (up to 10 μg of bound sialic acid/ml); 100 μl are added per well of the microtiter plate and incubated overnight at 4 °C

1b. Gangliosides are applied in a volume of 100 μl methanol containing up to 1 μg of bound sialic acid. The solvent is allowed to evaporate completely at room temperature. For the analysis of glycolipids the BSA concentration of PBS/BSA is raised to 2% and the wells are washed with PBS rather than with PBS/Tween (see steps 2, 3, and 5)

1c. In control wells, immobilized glycoconjugates are saponified with 0.1 N NaOH at room temperature to release O-acetyl groups. After 30 min, cells are washed three times with PBS

2. Remaining binding sites are blocked by incubation with PBS/BSA (200 μl/well) for 1 h at room temperature (2 h for glycolipids)

3. The wells are washed twice with PBS/Tween (200 μl/well)

4. After dilution with PBS, influenza C virus is added to the wells in a volume of 100 μl containing an acetylesterase activity of about 1–2 mU. Virus is allowed to bind for 1 h at 4 °C

5. The wells are washed three times with PBS/Tween at 4 °C

6. Each well is incubated with 100 μl esterase substrate B for 10–30 min at 37 °C. The reaction is stopped by the addition of 100 μl ethanol. The amount of 4-methylumbelliferone released is determined using a fluorescence spectro-photometer operating at wavelengths 365 nm for excitation and 450 nm for emission. If a fluorimeter is not available, esterase substrate C can be used. In this case the reaction is monitored at 405 nm.

17.2 Results

The detection of glycoproteins containing 9-O-acetylated sialic acid by bovine coronavirus is shown in Fig. 2. Different dilutions of rat serum have been used for a

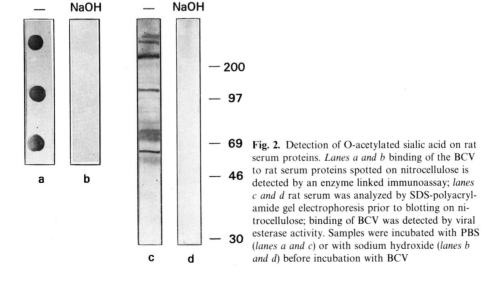

Fig. 2. Detection of O-acetylated sialic acid on rat serum proteins. *Lanes a and b* binding of the BCV to rat serum proteins spotted on nitrocellulose is detected by an enzyme linked immunoassay; *lanes c and d* rat serum was analyzed by SDS-polyacryl-amide gel electrophoresis prior to blotting on ni-trocellulose; binding of BCV was detected by viral esterase activity. Samples were incubated with PBS (*lanes a and c*) or with sodium hydroxide (*lanes b and d*) before incubation with BCV

spot assay (1:100, 1:500, 1:1000, lanes a and b, from top to bottom) using the immunological reagents to visualize bound virus. In lanes c and d, the serum proteins have been separated by SDS-polyacrylamide gel electrophoresis prior to blotting on nitrocellulose; the viral esterase has been used to visualize bovine coronavirus bound to individual glycoproteins. Virus binding is abolished if the proteins on the nitrocellulose strips are pretreated with sodium hydroxide (lanes b and d).

A typical example for the detection of gangliosides containing 9-O-acetylated sialic acid by influenza C virus is shown in Fig. 3. After thin-layer chromatography of bovine brain gangliosides, several bands are stained by the virus binding assay (lane 3). Alkaline pretreatment of the sample results in the disappearance of the three slower migrating bands (lane 4) indicating that they represent gangliosides containing N-acetyl-9-O-acetylneuraminic acid. The bands at the top of the chromatogram may be due to acid phospholipids and sulfatides, which migrate more rapidly than most gangliosides. These compounds appear to interfere with the polyisobutylmethacrylate coating and are often not well covered by the plastic film, so that nonspecific

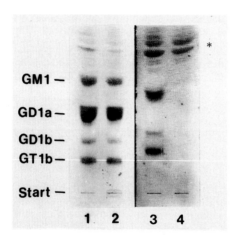

Fig. 3. Detection of 9-O-acetylated sialic acid on gangliosides from bovine brain separated by HPTLC. Prior to development, *lanes 2 and 4* have been exposed to ammonia vapors. *Lanes 1 and 2* were sprayed with the orcinol reagent staining the main gangliosides *GM1, GD1a, GD1b,* and *GT1b*. *Lanes 3 and 4* were overlayed with influenza C virus for the detection of 9-O-acetylated gangliosides. The *asterisk* indicates nonspecific staining

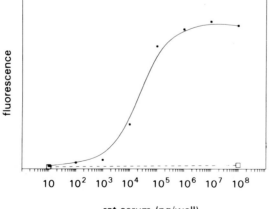

Fig. 4. Binding of influenza C virus to serial dilutions of rat serum immobilized in microtiter wells. Binding of influenza C virus was assayed using 4-methylumbelliferyl acetate as substrate. (———) untreated serum; (----) alkali-treated serum

staining in these areas may be observed. Gangliosides containing 9-O-acetylated sialic acid constitute only a minor fraction. They are not stained by orcinol (detection limit about 0.5 µg sialic acids). This reagent only reveals the four major ganglioside species (lane 1) and is not affected by alkaline pretreatment of the sample (lane 2).

In Fig. 4, a virus binding assay with microtiter plates is shown. This assay is useful for quantitative studies.

17.3 Remarks

Because of the instability of the 9-O-acetyl group of sialic acid under alkaline conditions, care was taken to avoid pH values above 9.0.

The alkaline pH values of the electrophoresis buffers (Laemmli 1970) do not interfere with the binding assay.

For electroblotting, the method described by Kyhse-Andersen (1984) was modified by lowering the pH value of the two buffers at the anode site from 10.4 to 9.0 (facing the anode) and 7.4 (facing the nitrocellulose), respectively.

Treatment with alkaline methanol, which is commonly used in the preparation of gangliosides to destroy contaminating phospholipids, is not practicable, if O-acetylation of gangliosides is to be preserved.

References

Herrler G, Compans RW, Meier-Ewert H (1979) A precursor glycoprotein in influenza C virus. Virology 99: 49–56

Kyhse-Andersen J (1984) Elektroblotting of multiple gels: a simple apparatus without buffer tank for rapid transfer of proteins from polyacrylamide to nitrocellulose. J. Biochem Biophys Methods 10: 203–209

Laemmli UK (1970) Cleavage of structural proteins during the assembly of the head of bacteriophage T4. Nature 227: 680–685

Schultze B, Gross H-J, Brossmer R, Klenk H-D, Herrler G (1990) Hemagglutinating encephalomyelitis virus attaches to N-acetyl-9-O-acetylneuraminic acid-containing receptors on erythrocytes: comparison with bovine coronavirus and influenza C virus. Virus Res 16: 185–194

Schultze B, Groß H-J, Brossmer R, Herrler G (1991) The S-protein of bovine coronavirus is a hemagglutinin recognizing 9-O-acetylated sialic acid as a receptor determinant. J Virol 65: 6232–6237

Svennerholm L, Fredman P (1980) A procedure for the quantitative isolation of brain gangliosides. Biochem Biophys Acta 617: 97–109

Zimmer G, Reuter G, Schauer R (1992) Use of influenza C virus for detection of 9-O-acetylated sialic acids on immobilized glycoconjugates by esterase activity. Eur J Biochem 204: 209–215

18 Affinoblotting Detection of the Glycoprotein Ligands of the Endogenous Cerebellar Soluble Lectin CSL and of Concanavalin A

J.P. Zanetta, S. Maschke, and A. Badache

The endogenous cerebellar soluble lectin CSL, first isolated from the cerebellum of young rats (Zanetta et al. 1987), plays an important role in several tissues including the central and peripheral nervous system. Its role in cell adhesion is mediated through its interaction with glycoprotein glycans. It displays a strong affinity for some oligomannosidic N-glycans, and a very strong affinity for rare hybrid-type N-glycans. Using active immunopurified CSL, it was possible to detect endogenous ligands of high affinity. This chapter reports the methodology used for detection on blots of the glycoprotein ligands of CSL and of Concanavalin A.

18.1 Chemicals

Materials The various chemicals were obtained from following sources: bovine serum albumin (BSA, fraction V), avidin-alkaline phosphatase (AKP), extravidin-AKP, horseradish peroxidase (HRP; fraction VI), nitro blue tetrazolium (NBT) and 5-bromo-4-chloro-3-indolyl-phosphate (BCIP), 4-chloro-1-naphthol tablets, p-nitrophenyl phosphate tablets from Sigma Chem. Co (St. Louis, Mo, USA); biotinylation reagent (NHS-biotin) was from Biorad (Richmond, USA); Ultrogel AcA 202 and Concanavalin A (Con A) was from IBF (Villeneuve-la-Garenne, France); nitrocellulose (0.22 and 0.45 µm pore size) from Schleicher and Schuell (Dassel, Germany) and Triton-X-100 from Aldrich SA (Strasbourg, France). Products for electrophoresis (molecular biology grade) were from Appligen (Strasbourg, France). Periodate-treated BSA (pBSA) was prepared according to Glass et al. (1981). This treatment is performed to destroy the carbohydrate moiety of contaminant globulins in BSA preparations. After treatment, pBSA is exhaustively dialyzed (twice 24 h at 4 °C) against 20 vol of PBS buffer (25 mM sodium phosphate buffer, pH 7.2 containing 0.15 M sodium chloride), then centrifuged to eliminate denatured material. This treatment of BSA has been found to be very efficient in reducing background lectin staining on blots.

18.2 Experimental Procedure

Preparation of Biotinylated CSL

Biotinylation For biotinylation, CSL purified by immunoaffinity chromatography (Zanetta et al. 1987; Kuchler et al. 1988) was exhaustively dialyzed against PBS buffer. The lectin solution was adjusted to pH 9.0 by addition of 0.1 M NaOH (30 µl/500 µl) and was reacted with biotin N-hydroxy-succinimide ester for 1 h at 4 °C with intermittent agitation. The excess reagent was neutralized by the addition of 100 µl of 0.2 M glycine in 0.1 M disodium phosphate and 100 µl of a 10% solution of Triton-X-100 was added. The biotinylated CSL was separated from low molecular weight biotinylated compounds by gel filtration onto a 7.5×1 cm column of Ultrogel AcA 202 equilibrated in PBS buffer containing 0.1% Triton-X-100. One-ml fractions were recovered.

Isolation and Test of Biotinylated CSL

Twenty-µl aliquots of each fraction were placed in wells of an ELISA microtiter plate and dried in an oven at 37 °C. The wells were saturated by incubation for 1 h at room temperature with 100 µl of a 3% solution of pBSA in PBS. After removal of the liquid and three washes with PBS, avidin-AKP [1/1500 dilution (0.7 µg/ml) in PBS] was added and incubated for 1 h at room temperature. After repetitive washes in PBS over a 1 h period, the wells were washed twice with water and incubated with the p-nitrophenyl phosphate reagent at 37 °C. The color intensity was measured using an ELISA scanner. The colored material corresponding to the void volume of the column was recovered, made 0.1% in pBSA and stored in small aliquots at -70 °C until used. At various intervals of time, the agglutinating activity of the biotinylated CSL preparation was checked (Marschal et al. 1989) in order to be sure of the carbohydrate-binding activity. Generally, preparations are still active after 6 months.

Isolation Test

Treatments of Samples for Electrophoretic Studies

Cultures were carefully washed three times with large volume of 25 mM sodium phosphate buffer at pH 7.2 containing 150 mM NaCl (PBS), cells were suspended at a concentration of 2 mg/ml protein (Lowry et al. 1951) in 1% sodium dodecyl sulfate (SDS) and homogenized with a Potter Elvejheim homogenizer. For animal tissues, rats (wistar albino 45 day- or 90 day-old) were killed by decapitation. The samples were directly homogenized in 1% SDS and stored at -80 °C until used. Aliquots of the samples were brought to 1 mg/ml protein by addition of 1 vol of the Laemmli (1970) dissociating buffer and boiled for 5 min before electrophoresis.

Samples

Electrophoretic Techniques

Proteins were separated by sodium dodecyl sulfate (SDS) polyacrylamide gel electrophoresis (10 or 13% polyacrylamide) in the buffer system of Laemmli (1970). The slab gels (0.75 mm thick) were transferred onto nitrocellulose according to Towbin et al. (1979). Proteins were revealed by incubation during 30 s in a solution of 0.2% Ponceau red in 3% TCA and rinsed with water. After photography, the dye was eliminated by repetitive washes in PBS. The destained blots were revealed using staining with biotinylated-CSL and avidin-AKP or the Con A-HRP technique.

Electrophoresis

Staining of CSL Ligands on Blots with Biotinylated CSL

Nitrocellulose filters were saturated by incubation with 3% pBSA in PBS buffer, under slow agitation, for 1 h at room temperature, followed by the addition of

Staining with CSL

biotinylated CSL (final concentration 5 ng/ml) and incubation for 3 h at room temperature. The filters were washed every 10 min for 1 h with PBS, then for 5 min in PBS containing 0.1% Triton-X-100, and washed again in PBS for 1 h. The blots were incubated for 1 h in 3% pBSA in PBS, then avidin-alkaline phosphatase (1/1500 final dilution) was added and incubation was proceeded overnight at 4 °C. After repetitive washes, as above, the blots were washed twice for 10 min in TBS buffer (10 mM Tris-HCl at pH 7.2 containing 0.15 M NaCl) and twice for 5 min in water. The fixed avidin-AKP was revealed using the NBT-BCIP reagent (Leary et al. 1983). Blanks were performed by omiting biotinylated CSL or by including mannose (0.3 M) or heparin (10 mg/ml) during incubation with biotinylated CSL and the subsequent three first washes, or using denatured biotinylated CSL. In the last case, aliquots of biotinylated CSL were boiled for 15 min before incubations. In all experiments, a sample consisting of total adult sciatic nerve was used as an internal standard. It allows verifying the specificity of the staining and to stop the NBT-BCIP reaction at the same intensity from one experiment to the other.

Staining of Con A-Binding Glycoproteins on Blots Using the Con A-HRP Technique

Con A-HRP Technique   Filters were saturated for 1 h at room temperature with a 3% solution of pBSA in PBS. During this period, a solution of Con A was prepared (10 mg/ml) in 10 mM Tris-HCl buffer at pH 7.2 containing $CaCl_2$, $MnCl_2$, and $MgCl_2$ (100 mM each). After saturation, Con A was added (final concentration 200 µg/ml) and incubated for 3 h at room temperature. Unbound Con A was eliminated by repetitive washes as described above for biotinylated CSL. The blots were transferred to a 3% solution of pBSA in PBS, incubated during a 1 h period and HRP (200 µg/ml final concentration) was added. After incubation overnight at 4 °C, unbound HRP was eliminated by repeated washes as above, the last two washes being in TBS buffer. Bound HRP was revealed using the chloro-naphthol technique of Hawkes et al. (1982). A sample containing rat sciatic nerve material was used as a positive control. Blanks were performed by including mannose (0.3 M) during incubation and the first three washes after incubation with Con A or by omitting Con A.

18.3 Results and Discussion

As shown in Fig. 1, the technique using biotinylated CSL and avidin-AKP allowed the easy detection of CSL ligands. The positive control, the rat sciatic nerve (Fig. 1A, lane 3), showed the classical profile of CSL ligands, including the major glycoprotein of the peripheral nervous system myelin, P0 (Mr 29 000), and the myelin-associated glycoprotein, MAG (Mr 100 000). The minor component at 31 kDa corresponded to an axonal glycoprotein identified as the P31 or the 31 kDa glycoprotein. The low Mr components (which remain unknown) were present in young animals and absent in 90 day-old rats. The CSL ligands are relatively rare when compared to the protein profile of the sciatic nerve or Con A-binding ligands (Fig. 1B, lane 3). The rat sciatic nerve sample constitutes a very good internal standard. The absence of staining of these specific bands indicates the inactivation of biotinylated CSL (Fig. 1A, lane 4). Reversely, staining of other bands indicates the degradation of avidin- or extravidin-

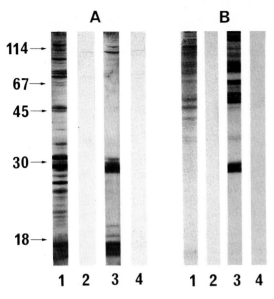

Fig. 1. Detection of CSL-binding glycoproteins (**A**) and Con A-binding glycoprotein (**B**) on blots. Fifteen µg protein were loaded in each lane. *Lanes 1* and *2* total extract of C6 glioma cells; *lane 3* and *lane 4* 60 day-old rat sciatic nerve. In **A**, *lanes 1 and 3* correspond to total incubations with active biotinylated CSL followed by avidin-AKP detection and *lanes 2 and 4* to the same experiment performed in the presence of denatured (boiled) CSL. Note the almost total inhibition of the staining after boiling the lectin and the presence in controls of two faint bands at 110 and 70 kDa also detected in incubations with avidin-AKP alone. In *lane 3*, the component around 110 kDa corresponds to MAG and the major component at 29 kDa corresponds to P0. Note the presence of a faint band at 31 kDa corresponding to the axonal glycoprotein P31. In **B**, *lanes 1 and 3* correspond to the Con A-HRP staining and *lanes 2 and 4* to controls obtained by including 0.3 M mannose during incubations with Con A and the three subsequent washes. Note the total inhibition of the binding of Con A in the presence of mannose. Note that, with the exception of P0, large differences in the intensity and numbers of bands are revealed using CSL instead of Con A

AKP. In our hands, these avidin conjugates are not very stable with time: due to the loss of activity by freezing, they have to be stored at 4 °C. But after 3 months at 4 °C, the conjugates give more and more unspecific staining. Blanks performed in identical conditions using heat denatured biotinylated CSL were not totally unstained. The same pattern was obtained using avidin- or extravidin-AKP without prior incubation with biotinylated CSL. Using good reagents, two faint bands with Mr 110 000 and 70 000 were detected in most samples, which correspond probably to endogenous biotinylated proteins (like acetyl CoA carboxylase). But this pattern varies from one tissue to the other. For example, the pattern in rat liver is very complex. Thus, when unknown samples are analyzed, it is imperative to perform these blanks. Other controls have been performed for many samples (not shown). They include incubations (and three washes) in the presence of 0.3 M mannose or heparin. We were not successful in inhibiting completely the binding of biotinylated CSL to the endogenous ligands with mannose, although the binding is significantly decreased relative to incubations and washes in the presence of 0.3 M galactose. Using heparin, the specific staining of CSL ligands was inhibited, but extraneous bands appeared. Since these extraneous bands are observed in the presence of heparin (in the presence or not of biotinylated CSL), it is likely that they correspond to formation of strong ionic interactions between proteins on the blots, heparin, and AKP-conjugates.

Provided that controls are performed, this method allows a specific and very sensitive detection of CSL glycoprotein ligands. Since the sensitivity is as high as that obtained using iodinated CSL, the method presents the advantages of easier application and of a greater stability of labeled CSL (CSL is very sensitive to radiolysis). The application of this method for quantitative determinations is questionable. As for other methods of staining on blots, the staining is proportional to the quantity of ligands only in a short range of concentrations. Below this level, the ligands are not detected, and above, the signal is saturated when measured densitometrically. Thus, generally only semi-quantitative data are obtained, unless various dilutions of the same sample are used. In these conditions and using the sciatic nerve internal standard to stop the staining reaction at equivalent levels, quantitative data can be obtained. Another problem for quantitative determinations is related to the efficiency of the transfer of protein to nitrocellulose. High Mr components ($Mr > 100\,000$) are not quantitatively transferred to nitrocellulose from a 13% acrylamide gel, necessary to separate low Mr compounds. Furthermore, these low Mr compounds, in contrast with high Mr compounds, are not strongly retained on 0.45 μm pore size nitrocellulose (they are well retained on 0.22 μm pore size). Since the efficiency of the electrotransfer step is difficult to control, variations are observed from one experiment to the other. Thus, when quantitative determinations are expected, all samples to be compared have to be analyzed on the same gel and on the same blot.

This technique has been applied to C6 glioma cells cultured or not in the presence of inhibitors of N-glycosylation. The number and quantity of CSL ligands was high (Fig. 1A, lane 1), although most of them correspond to very minor cell components as revealed by the CBB profile or the Con A-HRP profile (Fig. 1B, lane 1). When the cells are treated with tunicamycin, castanospermine and deoxynojirimycin, the CSL ligands disappear (not shown). In contrast, when cells are treated with deoxymannojirimycin or swainsonine the level of the CSL ligands was unchanged. This confirms that most of the CSL ligands are N-glycans of the oligomannosidic type (Marschal et al. 1989) or hybrid type (one of the best ligand of CSL is the glycan of glycoprotein P0 endowed with a short oligomannosidic branch and a N-acetyllactosaminic branch terminated by the glucuronic acid-3-sulfate group (HNK-1 epitope)).

For the Con A-binding ligands (Fig. 1B, lane 3), the presence of the sciatic nerve internal standard is also useful because P0 glycoprotein constitutes a well defined standard. The controls are generally easy to perform, including 0.3 M mannose during incubation and the three first washes. In these conditions (Fig. 1B, lane 4), an almost total elimination of the staining occurs, equivalent to incubations with HRP alone. The sensitivity of the method is by far lower than the previous one, but generally Con A ligands are important cell constituents and, consequently, their detection is easy. However, the same problems are encountered for the quantitation of these compounds, for the reasons mentioned above. A more specific problem is concerned with the low Mr ligands of Con A (between 12 000 and 30 000), related (but not totally) to problems of the efficiency of the transfer. One explanation for the experimental variations is that these ligands have only one N-glycosylation site and that the affinity of Con A is lower for these mono-glycosylated compounds than for higher Mr poly-glycosylated components. Here also, when comparison of samples has to be performed, this has to be done on the same gel and the same blot.

18.4 Conclusions and Perspectives

Provided care is taken to overcome technical problems using multiple controls, the techniques described above are very powerful and secure, although quantitative data are difficult to obtain reliably. The biotinylated CSL technique allows a very specific and sensitive detection of generally minor cell components, which were not identified before and seem to play important roles in cell adhesion mechanisms (in relationship with CSL as a molecule forming bridges between the cell surface ligands). This technique, as the Con A-HRP technique, can be extrapolated to the detection of CSL ligands on cells, using either the avidin-AKP detection, or, better, the avidin-HRP technique. Recently, the experiment on the effects of the glycosylation inhibitors on C6 glioma cells was followed also histochemically using biotinylated CSL and avidin-HRP. CSL ligands are at the cell surface and disappear in the presence of the drugs inhibiting the synthesis of the CSL ligands. In the same experiment, it was shown that monensin does not inhibit the synthesis of the CSL ligands but provokes an intracellular accumulation of these ligands. Thus, this method allows the biological interpretation of biochemical data, with few modifications.

References

Glass II WF, Briggs RC, Hnilica LS (1981) Use of lectins for detection of electrophoretically separated glycoproteins transferred onto nitrocellulose sheets. Anal Biochem 115: 219–224

Graham RC, Karnovsky MJ (1966) The early stages of absorption of injected horseradish peroxidase in the proximal tubules of mouse kidney: ultrastructural cytochemistry by a new technique. J Histochem Cytochem 14: 291–302

Hawkes R, Niday E, Gordon J (1982) A dot immunobinding assay for monoclonal and other antibodies. Anal Biochem 119: 142–147

Kuchler S, Fressinaud C, Sarlieve LL, Vincendon G, Zanetta J-P (1988) Cerebellar soluble lectin is responsible for cell adhesion and participates in myelin compaction in cultured rat oligodendrocytes. Dev Neurosci 10: 199–212

Leary JJ, Brigati DJ, Ward DC (1983) Rapid and sensitive colorimetric method for visualizing biotin-labeled DNA probes hybridized to DNA or RNA immobilized on nitrocellulose: bio-blots. Proc Natl Acad Sci USA 80: 4045–4049

Laemmli UK (1970) Cleavage of structural proteins during assembly of the bacteriophage T4. Nature 227: 680–685

Lowry OH, Rosebrough NJ, Farr AL, Randall RJ (1951) Protein measurement with Folin phenol reagent. J Biol Chem 193: 265–275

Marschal P, Reeber A, Neeser JR, Vincendon G, Zanetta J-P (1989) Carbohydrate and glycoprotein specificity of two endogenous cerebellar lectins. Biochimie 71: 645–653

Towbin H, Staehelin T, Gordon J (1979) Electrophoretic transfer of proteins from polyacrylamide gels to nitrocellulose sheets: procedure and somme applications. Proc Natl Acad Sci USA 76: 4350–4354

Zanetta J-P, Dontenwill M, Reeber A, Vincendon G, Legrand C, Clos J, Legrand J (1985) Con A binding glycoproteins in the developing cerebellum of control and hypothyroid rats. Dev Brain Res 21: 1–6

Zanetta J-P, Meyer A, Kuchler S, Vincendon G (1987) Isolation and immunochemical study of a soluble cerebellar lectin delineating its structure and function. J Neurochem 49: 1250–1257

Determination of Expression
of Lectins and Their Ligands

19 An Enzyme Immunoassay (EIA) for Human Endogenous β-Galactoside-Binding Lectin

M. CARON

The S-type (thiol-dependent) or S-Lac (soluble lactose-binding) β-galactoside-specific lectins are found in a wide variety of tissue and cells (Caron et al. 1990). Human brain contains a lectin of this family that has been isolated (Bladier et al. 1989) and subsequently has been well characterized (Bladier et al. 1991; Joubert et al. 1992). The subunit molecular weight (14.5 K) and amino acid composition of the human brain lectin (HBL) were similar to those for human placenta lectin (Hirabayashi and Kasai 1988).

An important prerequisite in the study of endogenous lectin structure and function is the availability of sensitive determinations in protein mixtures. The enzyme immunoassay (EIA) detailed below was developed to determine the concentration of HBL-immunoreactive material (molecules reacting with anti-HBL antibodies) in biological fluids and tissue extracts. It uses two polyclonal antibodies and both biotinylated and unlabeled HBL.

19.1 Outline of Procedure

The principle of the assay is summarized in Fig. 1.

Coating	Coating of microtiter plate with anti-rabbit IgG
Saturation	Wells are blocked with bovine albumin
Immunological reaction	Absorption of anti-HBL polyclonal antibody on anti-rabbit IgG. Simultaneous application of biotinylated and unlabeled HBL (or sample)
Biotin–streptavidin reaction	Biotinylated HBL is reacted with a streptavidin-peroxidase conjugate
Enzyme reaction	Bound peroxidase is quantified with the substrate orthophenylenediamine

19.2 Materials

Microplate reader (e.g., Metertech Model 960)
96-Well microtiter plates (Nunc-Immuno Plate Maxisorp, Denmark)

Equipment

Goat anti-rabbit IgG antiserum
Rabbit anti-HBL polyclonal antiserum (Bladier et al. 1989)
HBL (Bladier et al. 1989)
Biotinylated derivative of carboxamidomethylated-HBL (CAM HBL-Biot; Avellana-Adalid et al. 1990)

Reagents, Solutions

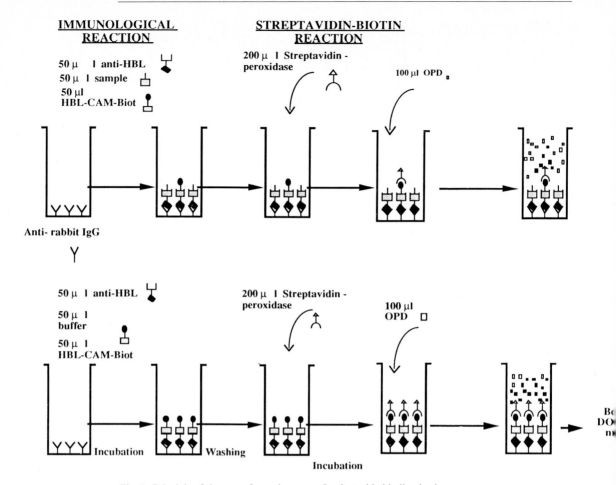

Fig. 1. Principle of the assay for endogenous β-galactoside binding lectin

Glutaraldehyde 25% (Sigma)
Bovine serum albumin (BSA) (Sigma, Fraction V, no. A-4503)
Phosphate buffer (PB): 0.5 M K_2HPO_4/KH_2PO_4, pH 7.4, NaN_3 0.002%
EIA buffer (EIAB): 1:5 dilution of PB containing 0.4 M NaCl, 1 mM EDTA and
    0.01% BSA
Washing buffer (WB): 1:50 dilution of PB containing 0.05% Tween (Tween 20, no.
    822184, Merck)
Saturation buffer: EIAB containing 3% BSA
Streptavidin coupled to horseradish peroxidase (Strep-HRP, Sigma)
o-phenylenediamine (OPD, Roche or Sigma)

19.3 Detailed Procedure

Procedure   – 30 µl of anti-rabbit IgG antiserum and 50 µl of glutaraldehyde are mixed in 15 ml
    of EIAB.

- Each well of a microtiter plate is coated with 100 µl of the mixture for one night at 4 °C. Wells are washed twice with 300 µl of WB, then blocked with 300 µl of saturation buffer for 2 h at 37 °C, followed by two further washes.
- Six dilutions of purified HBL are used for the calibration curve. One unit of HBL, or of immunoreactive material, corresponds to the antigenic activity of a nanomolar HBL solution (30 ng/ml) at pH 7.4 and 37 °C.

Calibration curve:

Standards HBL:	ng/ml	Units
B_0	0	0
1	6	0.2
2	12	0.4
3	24	0.8
4	48	1.6
5	96	3.2
6	192	6.4

- Sample dilutions need to be optimized according to their immunoreactive material concentration. Assays are made in triplicate and the final incubation volume of 150 µl is comprised of three aliquots:
- 50 µl of rabbit anti-HBL at a final concentration of 1:1000 in EIAB (50 µl of EIAB instead of antiserum in blank wells)
- 50 µl of standard or sample dilution (50 µl of EIAB in B_0 and blank wells)
- 50 µl of CAM-HBL-Biot at a constant concentration of 75 ng/ml.
- After incubation at 37 °C for 2 h and at 4 °C for one night, the wells are washed three times and 200 µl of Strep-HRP (0.2 µg/ml in PBS containing 1% BSA) are added to each well.

After incubation for 1 h at 37 °C, the wells are washed four times. The enzymatic reaction starts by adding 100 µl of the substrate OPD (40 µM in 0.1 M citrate buffer, pH 5.2, containing 6 µmol/ml hydrogen peroxidase). After a visible color development, the reaction is stopped with 100 µl of a 0.5 M H_2SO_4 solution. The color change in each well is read spectrophotometrically at 490/492 nm against blank wells.

19.4 Parameters

The optimal conditions for the determination of HBL were established from the examination of two parameters, i.e., the concentration of both CAM-HBL-Biot and the specific anti-HBL antibody (Eloumami et al. 1991). When a new lot of antibody is used, it is necessary to test several dilutions (e.g., from 1:500 to 1:4000). To transpose the assay to another biotinylated lectin, it would be necessary to study the effect of different concentration of this derivative used without competitor on the absorbance at 490/492 nm. The optimal concentration is the lowest concentration which saturates the antibody.

19.5 Results

Figure 2 shows two representations of the calibration curve:

- the ratio B/B_0 (absorbance of the standard/absorbance without competitor $\times 10^2$) vs. the concentration (in ng/ml or in units);
- the concentration vs. the absorbance. In this case a relation Concentration $= f$ (Absorbance) can be determined using a computer.

Figure 3 demonstrates how the assay can be used to study antigenic cross-reactions between lectins and other proteins, here the absence of cross-reactions between HBL and the myelin basic protein (MBP).

19.6 Troubleshooting

As with any microtiter plate immunoassay, one must take care that an immunological cross-reactivity does not necessary mean molecular identity. Consequently, in many cases, and particularly for complex protein mixtures, the material detected must be considered as an "immunoreactive material" and not as a definite lectin. If

A

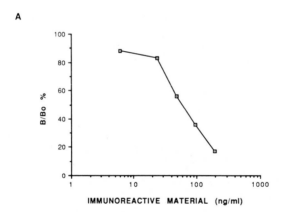

B

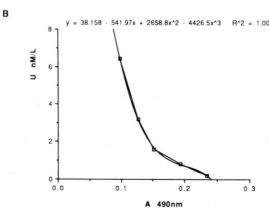

Fig. 2A,B. Representations of a standard curve computed from the results of EIA of HBL. In the theorical equation of the curve, y represents the concentration in immunoreactive material, and x the absorbance at 490 nm

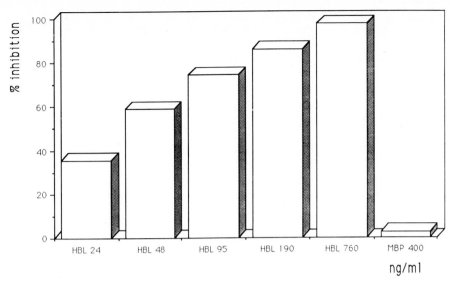

Fig. 3. Inhibition of the binding of CAM-HBL-Biot, by purified HBL and MBP, tested by EIA on anti-HBL. % inhibition: $100\text{-}B/B_0$, HBL concentrations: 24, 48, 95, 190, 760 ng/ml. MBP concentration: 400 ng/ml

the amount of material is sufficient for an immuno-blotting detection, both of these complementary assays can be performed to obtain quantitative and qualitative information.

References

Avellana-Adalid V, Joubert R, Bladier D, Caron M (1990) Biotinylated derivative of a human brain lectin: synthesis and use in affinoblotting for endogenous ligand studies. Anal Biochem 190: 26–31

Bladier D, Joubert R, Avellana-Adalid V, Kemeny JL, Doinel C, Amouroux J, Caron M (1989) Purification and characterization of a galactoside-binding lectin from human brain. Arch Biochem Biophys 269: 433–439

Bladier D, Le Caer JP, Joubert R, Caron M, Rossier J (1991) β-Galactoside soluble lectin from human brain. Complete amino acid sequence. Neurochem Int 18: 275–281

Caron M, Bladier D, Joubert R (1990) Soluble galactoside-binding vertebrate lectins: a protein family with common properties. Int J. Biochem 22: 1379–1385

Eloumami H, Caron M, Joubert R, Doinel C, Bladier D (1991) Human brain lectin immunoreactive material in cerebrospinal fluids determined by enzyme immunoassay (EIA). J Neurol Sci 105: 6–11

Hirabayashi J, Kasai K1 (1988) Complete amino-acid sequence of a β-galactoside-binding lectin from human placenta. J Biochem 104: 1–4

Joubert R, Caron M, Avellana-Adalid V, Mornet D, Bladier D (1992) Human brain lectin: a soluble lectin that binds actin. J Neurochem 58: 200–203

20 Analysis of Lectin Expression by Immunoblotting

R. LOTAN

Two types of soluble galactoside-binding lectins of 14.5 and 29–35 kDa, respectively, have been studied in different laboratories and found to be distributed widely in species ranging from electric eel to man. Within each species, these lectins exhibit distinct tissue distribution and their levels are regulated during development. The lower and higher molecular weight lectins have been cloned and sequenced and found to share some sequence homology. Their physiological function(s) is not known; however, their ability to bind galactosides and the changes in their level during embryogenesis and development, processes which are also accompanied by changes in glycoconjugates, suggest that the lectins play a role in recognition and cellular interactions (reviewed by Barondes 1988; Lotan 1992). Galactoside-specific lectins have been purified from various tissues by affinity chromatography using different immobilized galactosides or galactoglycoproteins (e.g. asialofetuin). The molecular weights reported for the lectins purified in different laboratories and from different tissues and species vary from 12–15 kDa for the low molecular weight lectin and 29–35 kDa for the higher molecular weight lectin. These will be designated here as L-14.5 and L-34, respectively. Additional galactoside-binding lectins of 6.5, 16, 17, 18, 22, 67 and 70 kDa have been detected, but only limited information is available on these lectins. The presence of lactoside-binding lectins in tissues and cells can be screened rapidly by immunoblotting using the procedures described below.

20.1 Objectives and Limitations

The immunoblotting method is useful for several objectives:

a) An initial characterization of the presence and relative amount of lactoside-binding lectins in cells and tissues.
b) Analyzing the presence of lectins in eluates of affinity chromatography columns during lectin isolation.
c) Assessment of the specificity of anti-lectin antibodies (i.e., to determine whether the antibodies bind only to the antigen that they are expected to bind) before their application to immunohistochemical analyses.

The main advantage of the immunoblotting analysis is its ability to detect lectins using small amounts of tissue or cell extracts compared to affinity chromatography on immobilized saccharides or glycoproteins. In addition, it provides a more accurate estimate of the total amount of lectin in the solubilized tissue analyzed without losses of lectins incurred during purification by affinity chromatography. Such losses are expected due to the use of nondenaturing extraction buffers that may not dissociate lectins from complexes with endogenous glycoconjugates. These complexes presumably will not bind to affinity columns. In addition, affinity chromatography involves several steps of dialysis during which lectins may be lost due to non-specific binding to the dialysis membranes.

The limitations of the immunoblotting method are:

a) It does not provide information on the activity of the lectins because they are detected by their antigenicity.
b) It does not indicate which cell type in a heterogeneous population of cells expresses lectins;
c) If it detects proteins with molecular weights different from 14.5 and 29–35 kDa, it is not known whether these are lectins or cross-reacting nonlectin proteins.

These limitations may be overcome by complementary analyses using affinity chromatography and immunofluorescence or immunohistochemical methods some of which are described in other chapters of this book.

20.2 Equipment and Materials

- Microfuge (Beckman model 12, Beckman Instruments, Fullerton, CA)   **Equipment**
- Benchtop clinical centrifuge (e.g., IEC model HN-SII; Damon/IEC Division, Needham Heights, MA)
- Tissue homogenizer (Tissue-Tearor, Biospec Products, Bartsville, OK)
- Mini-PROTEAN II vertical gel electrophoresis and blotting system (Bio-Rad, Richmond, CA)
- Reciprocating shaker (e.g., Bellco Biotechnology, Vineland, NJ)
- Beckman DU 70 scanning densitometer (Beckman Instruments)

These are examples for the equipment used in the author's laboratory, similar equipment from other sources should be equally useful.

Cells or tissues to be analyzed: **Materials**

- Extraction solution I = 0.9% NaCl, 1% sodium deoxycholate, 1% Triton X-100, 0.1% sodium dodecyl sulfate (SDS), 0.05% iodoacetamide, and 1 mM phenyl-methyl sulfonyl fluoride (PMSF) in 10 mM Tris-HCL buffer, pH 7.2 (Crittenden et al. 1984)
- Extraction solution II = 1% SDS, 4% β-mercaptoethanol, and 60 mM Tris buffer pH 6.8
- Extraction solution III = 0.5% Nonidet P-40, 0.25 M sucrose, 0.05 mM $CaCl_2$, 0.1 mM PMSF, 0.3 M lactose in 5 mM Tris-HCl buffer, pH 7.4
- CMF-PBS = calcium-free and magnesium-free phosphate-buffered saline, pH 7.2 3× SDS:PAGE sample buffer = 187.5 mM Tris-HCl buffer pH 6.8, containing 6% SDS, 3% β-mercaptoethanol, 1.5 mM EDTA, and 30% glycerol
- Supplies for polyacrylamide gel electrophoresis of proteins (Bio-Rad, Richmond, CA)
- Nitrocellulose filter membrane (0.1 μm, Micron Separation Inc., Westboro, MA) Ponceau S protein stain (Sigma Chemical Co., St. Louis, MO; 0.1% [w:v] solution in 1% acetic acid)
- Blocking buffer = 15% skim milk, 0.15 M NaCl, and 0.02% sodium azide in 10 mM Tris-HCl, pH 7.4

Antilectin antibodies:

- Monoclonal (M3/38) rat anti-mouse Mac-2 antigen (L-34 lectin) (Boehringer - Mannheim Chemicals, Indianapolis, IN)
- Polyclonal anti-L-14.5 lectin antibodies prepared as described by Couraud et al. (1989) and Lotan et al. (1991)
- ^{125}I-labeled goat anti-rabbit IgG antibodies (12.7 μCi/μg, ICN Radiochemicals, Irvine, CA)
- ^{125}I-labeled sheep anti-rat IgG (2–15 μCi/μg, Amersham, Arlignton Heights, IL)
- Wash buffer = 0.15 M NaCl and 0.15% Tween-20 in 10 mM Tris: HCl, pH 7.4
  X-ray film (XAR-5, Kodak, Rochester, NY)

20.3 Experimental Procedure

Preparation of Tissue Extracts

Tissue Extracts   The method is a modification of that described by Crittenden et al. (1984).

Fresh tissues or thawed frozen specimens (at least 5 mg weight) are minced with scalpels into small fragments (< 0.5 mm^3) and suspended in 50–200 μl tissue extraction solution I and homogenized.

The homogenate is centrifuged at 1500 g for 10 min at 4 °C and the supernatant fraction is collected and diluted 1:1 (v:v) in extraction solution II.

The suspension is heated at 95 °C for 20 min and recentrifuged as above.

The supernatant fraction is then analyzed for protein content (Bradford 1976), mixed 2:1 (v:v) with 3 × SDS:PAGE sample buffer and adjusted to 3 mg/ml with 1 × sample buffer. The samples can be stored frozen at − 70 °C.

An alternative extraction of tissues can be achieved by mixing 50-mg samples with 0.5 ml extraction solution III for 18 h at 4 °C.

The mixtures are centrifuged at 12 000 g for 5 min and the supernatant fractions are collected and prepared for SDS:PAGE as above.

Preparation of Cell Extracts

Cell Extracts   Cells grown as adherent monolayers are washed three times in CMF-PBS and harvested by repeated pipeting after a brief (5–10 min) incubation in CMF-PBS containing 2 mM EDTA. Cells that fail to detach under such conditions can be harvested by scarping with a rubber policeman into CMF-PBS.

The cells are pelleted by centrifugation at 2000 g for 10 min at 4 °C, resuspended in extraction solution III, and incubated for 30 min at 4 °C.

The insoluble material is then pelleted by centrifugation in a microfuge at 12 000 g for 5 min at 4 °C and the supernatant fraction is mixed 2:1 (v:v) with 3 × SDS:PAGE sample buffer, heated at 95 °C for 5 min, and frozen at − 70 °C.

Gel Electrophoresis

Electrophoresis   Samples of tissue or cell extracts containing about 100 μg protein in 30–35 μl sample buffer are subjected to gel electrophoresis in 14% polyacrylamide minislab gels in the presence of SDS as described previously (Lotan et al. 1980).

The proteins are then transferred electrophoretically (150 V, 2 h) to a nitrocellulose membrane as described by Burnette (1981). After the transfer, the nitrocellulose membranes are immersed in a staining solution containing Ponceau S and washed once in water before observation of the protein transfer results.

Immunoblotting

The membrane is immersed overnight in 10-ml blocking buffer in a closed plastic container to saturate nonspecific binding sites on the nitrocellulose. The next steps are carried out by mixing on an reciprocating shaker

Immunoblotting

The membrane is incubated for 2 h at 23 °C with antilectin antibodies diluted into in blocking buffer supplemented with 0.1% Tween-20

The membrane is washed four times (15 min each) with wash buffer

The membrane is incubated with 1.5×10^6 cpm of ^{125}I-labeled goat anti-rabbit IgG or ^{125}I-labeled sheep anti-rat IgG in 10 ml of blocking buffer with 0.1% Tween-20 for 1 h.

After an additional four washes, the membrane is dried and placed against an X-ray film between two intensifying screens and kept in the dark at -70 °C for 1 or more days, after which the film is developed and the ^{125}I-IgG binding to the lectin is detected as a dark band.

Quantitation of the Immunoblotting Results

The autoradiograms are placed on top of the nitrocellulose filters, and the areas where lectin bands are present on the film are excised from the nitrocellulose using a razor blade or an Exacto knife. The amount of radioactivity associated with each band is determined using a gamma counter. Each slab gel includes a lane with an extract prepared from cells which contain L-14.5 and L-34 lactose-binding lectins (e.g., 1×10^5 murine RAW117 lymphosarcoma cells); these lanes serve as standards for comparison among different gels. Immunoblotting experiments with serially diluted lymphosarcoma extracts indicated that the bound radioactivity is proportional to the amount of lactose-binding lectins in the range that includes an extract from 1×10^5 cells. The amount of radioactivity bound to lectin bands on the nitrocellulose membrane is divided by the radioactivity bound to the L-14.5 and the L-34 lectins of the lymphosarcoma cells, respectively. The amount of radioactivity bound to these standard lectin bands from 100 µg cell extract is designated as 1 unit.

Quantitation

An alternative method is to quantitate the lectin from the density of the bands on the autoradiograms using a scanning densitometer. The areas under the peaks corresponding to lectins L-14.5 and L-34 are calculated and the values are divided by values obtained by scanning the lanes that contain the standard lectins when more than one gel is used. Scanning of autoradiograms exposed for different times may be required to ensure that saturation is not reached on the more intense bands.

20.4 Recommendations

The use of Minigels is advantageous over the use of larger gels in that the duration of electrophoresis is shorter, and an experienced investigator can run four slab gels concurrently and transfer the proteins to nitrocellulose in 5–6 h.

It is useful to include prestained low molecular weight protein standards in each gel to follow the progress of electrophoresis and to identify the molecular weight of the antigens detected by the antilectin antibodies.

The staining of the nitrocellulose membranes with Ponceau S after the electrophoretic transfer of proteins is important for assessment of transfer results. For example, this staining allows one to determine whether similar amounts of proteins are present on the membrane in lanes where presumably similar amounts of protein were loaded on the gel. The staining also indicates whether there are any irregularities in the transfer due to the inadvertent presence of air bubbles between the membrane and the polyacrylamide gel during the transfer.

It is possible to use ^{125}I-protein A from *Staphylococcus* instead of ^{125}I-labeled second antibody (Burnette 1981) if the latter exhibits nonspecific binding.

It is possible to use the same nitrocellulose membrane blot to analyze the two lectins by first performing the procedure with one anti-lectin antibody and, after the autoradiograms are developed and the exposure is found to be optimal, with the antibodies against the other lectin.

20.5 Results

Figure 1 shows typical results of immunoblotting of extracts from cultured cells (lanes 1–5, and 11–13) and tissues (lanes 6–10). For example, lanes 1–5 show that the expression of the lectins is variable in several clones of cultured mouse melanoma cells. Lanes 6–10 demonstrate different levels of lectins in tissue extracts of primary colon cancer tumor (lane 6) and liver metastases (lanes 7–10). Lane 9 shows an additional band of 22 kDa, which is presumed to be a proteolytic fragment of the L-34 lectin because it is detected by the anti-Mac-2 antibody but not by the anti-L-14.5 antibodies. Lanes 11–13 show the expression of lectins in mouse RAW117 lymphosarcoma cells (lane 11), in mouse PYS parietal endodermal cells (lane 12), and in PSA-5E mouse visceral endodermal cells (lane 13). Note that lane 13 shows an antigen of about 45 kDa is addition to L-34 and L-14.5. This antigen is cross-reactive with the anti-L-14.5 antibodies but not with the anti-L-34 antibodies. These results

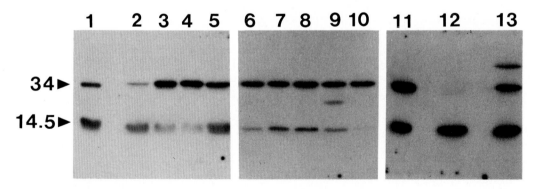

Fig. 1. Typical results of immunoblotting of extracts from cultured cells (*lanes 1–5*, and *11–13*) and tissues (*lanes 6–10*). For details see text

indicate that the blotting method can provide qualitative and quantitative information on the expression of lectins in cells and tissues.

References

Barondes SH (1988) Bifunctional properties of lectins: lectins redefined. Trends Biochem Sci 13: 480–482

Bradford MM (1976) A rapid and sensitive method for the quantitation of microgram quantities of protein utilizing the principle of dye binding. Ann Biochem 72: 248–254

Burnette WN (1981) "Western blotting": electrophoretic transfer of proteins from sodium dodecyl sulfate-polyacrylamide gels to unmodified nitrocellulose and radiographic detection with antibody and radioiodinated protein A. Anal Biochem 112: 195–203

Couraud PO, Casentini-Borocz D, Bringman TS, Griffith J, McGrogan M, Nedwin GE (1989) Molecular cloning, characterization, and expression of a human 14-kDa lectin. J Biol Chem 264: 1310–1316

Crittenden SL, Roff CF, Wang JL (1984) Carbohydrate binding protein 35: identification of the galactose-specific lectin in various tissues of mice. Mol Cell Biochem 4: 1252–1259

Lotan R (1992) β-Galactoside-binding vertebrate lectins: synthesis, molecular biology, function. In: Allen H, Kisailus E (eds) Glycoconjugates: composition, structure, and function. marcel dekker, Inc., New York, pp. 635–671

Lotan R, Kramer RH, Neumann G, Lotan D, Nicolson GL (1980) Retinoic acid-induced modification in cell growth and cell surface components of a human carcinoma (HeLa) cell line. Exp. Cell Res. 130: 401–414

Lotan R, Matsushita Y, Ohannesian D, Carralero D, Ota DM, Cleary KR, Nicolson GL, Irimura T (1991) Lactose-binding lectin expression in human colorectal carcinomas. Relation to tumor progression. Carbohyd Res 213: 47–57

21 Application of Immuno- and Affinocytochemistry to the Detection of an Endogenous β-Galactoside-Specific 14 kDa Lectin and Its Ligands

R. JOUBERT-CARON

In recent years, the importance of β-galactoside-binding soluble animal lectins and of the different biological processes in which they may participate through inter-action with their ligands has been better appreciated (Leffler et al. 1989; Caron et al. 1990). As a consequence, a growing number of studies implicating the cellular localization of these molecules (Joubert et al. 1989; Cooper and Barondes 1990; Kuchler et al. 1990; Gabius 1991), and more recently the visualization of their ligands (Avellana-Adalid et al. 1990; Gabius et al. 1991; Joubert et al. 1992) have been published. Some properties of these lectins, however (Barondes 1984; Caron et al. 1990; Bladier et al. 1991), pose several specific methodological and interpretative problems. Thus, the choice of (a) tissue or cell preparation and (b) detection methods, may be guided by these considerations.

The present chapter will focus (a) on the immunocytochemical detection of a β-galactoside soluble lectin expressed in murin Neuroblastoma cell line N1E115 (MBL 14) using an antibody (anti-HBL 14) directed against an homologous protein (HBL 14) purified from human brain, (b) on the affinocytochemical visualization of complementary lectin ligands on the same model using a biotinylated derivative of the human brain – 14 kDa lectin (CAM-HBL Biot) – as probe.

21.1 Immunocytochemical Detection of 14 kDa Soluble Lectin in Cell Cultures

21.1.1 Fixation Method

Objectives Although fixation is necessary to avoid artifactual diffusion of soluble cellular components, it constitutes in itself a major artifact whose denaturing effects may be minimized. The fixative should be adapted to the immunoreactive material under investigation, i.e., endogenous soluble lectins. Thus, for these molecules, several considerations should be kept in mind, particularly the fact that they could be found in the cytosol rather than in the membrane-bound compartment. Consequently, to immunostain only extracellular antigens, some precautions have to be taken. Although a good preservation of intracellular lectin is gained by strong fixation, cell surface determinants can be best detected under mild fixation conditions. To bring these contradictory demands into equilibrium, several fixatives and parameters have been tested.

Materials 7.5 mM Na_2HPO_4, 12 H_2O; 2.6 mM NaH_2PO_4,H_2O; 150 mM NaCl pH 7.2 = PBS (Phosphate-buffered saline)
Fixative of Carnoy: 60 ml ethanol 95% (Merck), 30 ml chloroform (Merck); 10 ml acetic acid (Carlo Erba)

Fixative of Lillie: 100 ml formaldehyde 36–40% (Merck); 900 ml distilled water; 4 g
    Na_2HPO_4 (Merck); 6.5 g anhydrous Na_2HPO_4 (Merck)
Fixative of Carson: 10 ml formaldehyde 36–40% (Merck); 17 ml NaOH 2.56% (w/v);
    83 ml Na_2HPO_4, H_2O (2.26% w/v)
Methanol/formaldehyde: 90 ml methanol (Carlo Erba); 10 ml formaldehyde 36–40%
Methanol/acetone: (mixed v/v)
Formaldehyde/acetic acid: 30 ml formaldehyde 36–40%; 10 ml acetic acid

Neuroblastoma cells were cultivated on glass coverslips and carefully washed with **Detailed**
PBS before fixation with 500 µl of the solutions described under *Materials* for 15 or **Procedure**
30 min at two different temperatures, 0° and 20 °C. Cells were rinsed 4 × 5 min with
PBS before immunological reaction.

Carson fixative used for 15 min at 20 °C was selected for subsequent lectin immuno- **Parameters**
detection. This fixative permits the detection of intracellular antigen by permeabili-
zing the plasma membrane to allow the passage of antibody to the cell interior. To
immunostain extracellular antigens only, cells were fixed *after* the specific immunological reaction as described below. The described conditions gave the best results
both in cell preservation and in conservation of the antigenic properties of the
lectins.

21.1.2 Immunodetection of Lectin by Two-Step Antibody Reaction

The two-step technique has the advantage of increasing sensitivity and as only one **Objectives**
conjugated antibody is required, it is a convenient and versatile method for studying
endogenous lectins. The anti-lectin primary antibody is used unconjugated.

3-3′-Diaminobenzidine tetrahydrochloride (BDH) = DAB   **Materials**
Oxygen peroxide 30% (Sigma) = H_2O_2
Bovine serum albumin (Sigma) = BSA; Tween 20 (Merck)

Detection of cytosolic lectin: **Outline of**
                                                           procedure
1. Fixation with Carson fixative
2. Blocking of endogenous peroxidase activity
3. Blocking of nonspecific interactions with blocking solution
4. Incubation with anti-HBL 14 lectin antiserum (primary antibody)
5. HRP-swine anti-rabbit IgG (Dako) (second antibody)
6. Visualization of HRP-activity by DAB

Detection of extracellular lectin:

1. Incubation of living cells with sterile anti-lectin antiserum at 37 °C for 60 min
2. Fixation and subsequent steps are performed as described above

– After fixation, wash coverslips 3 × 5 min with PBS   **Detailed**
– Block endogenous peroxidase activity with a fresh 1% solution of hydrogen **Procedure**
   peroxide in methanol PBS (1:1) for 15 min at 20 °C
– Wash coverslips 2 × 5 min with PBS

- Expose coverslips to blocking solution [3% BSA (w/v), 0.3% Tween 20 (v/v) in PBS] for 60 min at 20 °C
- Excess of blocking solution is removed without washing, prior incubation in primary rabbit anti-lectin antiserum optimally diluted in blocking solution, at 20–22 °C for 60 min
- Wash carefully with PBS
- Apply swine anti-rabbit IgG conjugated to HRP (1 : 250 in blocking solution) for 60 min at 20–22 °C.
- Wash thoroughly with PBS.

The end product is revealed with a freshly made solution of 0.05% DAB in 0.01%
- H_2O_2 in PBS
- Wash in PBS

Controls The cross-reactivity of antihuman brain lectin antibody with the murine soluble β-galactoside 14 kDa lectin (MBL14) was checked using classical immunoblotting technique (Joubert et al. 1989)
Controls were performed replacing anti-HBL14 serum dilution by PBS or by preimmune serum (1 : 10)

Troubleshooting Another type of S-Lac lectins has been found in different cells or tissues (Caron et al. 1990). These molecules showed an apparent molecular weight in range of 29–35 kDa. The amino-acid sequence of the C-terminal portion of these lectins is homologous to the 14 kDa lectins (Robertson et al. 1990)

These 29–35 kDa lectins could be detected with anti-14 kDa lectin antibody. Consequently, it is necessary to verify, by immunoblotting, what type of lectin is expressed by the cells to be studied before performing immunocytodetection. Moreover, the major constituent of the myelin sheaths, the myelin basic protein (MBP), presents a common tetrapeptide (WGAE) with 14 kDa lectin from bovine origin, and a cross-reaction has been found with monoclonal (Abbott et al. 1989) or polyclonal antibodies directed against bovine lectins but not with antibodies directed against human or rodent 14 kDa lectins (data not shown in this study).

Results Figure 1 shows the immunoblotting detection of mouse lectin using anti-HBL14 antibody. Only one band corresponding to 14 kDa lectin is immunodetected. Figure

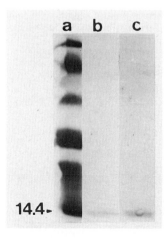

Fig. 1. Immunoblot detection of mouse antigens using anti-human brain lectin antibody. *Lane a* biotinylated calibration proteins used as molecular mass markers, *from top to bottom* phosphorylase B (94 000); bovine serum albumin (67 000); ovalbumin (43 000); carbonic anhydrase (30 000), soybean trypsin inhibitor (20 100); alpha lactalbumin (14 400); *lane b* mouse brain crude extract; *lane c* purified mouse brain lectin.

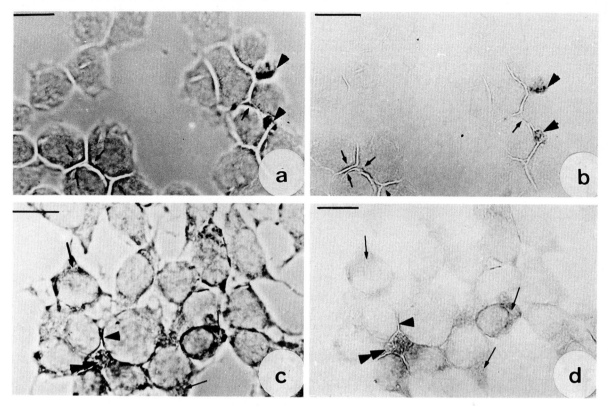

Fig. 2a–d. Immunocytochemical localization of mouse lectin in undifferentiated Neuroblastoma cells. **a** and **c** are phase-contrast micrographs of the fields **b** and **d**, respectively. Extracellular detection is shown in **a** and **b**. Note the presence of extracellular antigen in externalized vesicles (**a**, **b**, *arrowheads*) and associated with plasma membrane at contact sites between cells (**a**, **b**, *arrows*). Intracytosolic immunoreactive material is highly concentrated into patches close to the plasma membrane, presumably immediately before its externalization in vesicles. It is also detected diffusely in the cytoplasm of the cells in clusters (**c**, **d**, *arrows*) and expressed extracellularly by the same cells either on the membrane (**c**, **d**, *arrowheads*) or in deposits between cells (**c**, **d**, *double arrowheads*). *Bars* 10 μm

2a and b shows the immunodetection of extracellular antigen before cell permeabilization while Fig. 2c and d represents the results obained after cell permeabilization. In this case extra- and intracellular antigen is seen.

21.2 Affinocytochemical Detection of Potential Ligands of Soluble Lectin in Cell Cultures

Until now, investigations of cellular glycoproteins have been performed using plant lectins as probes and moreover, these studies focused only on the membrane glycoconjugates. Important findings are that endogenous lectins are often soluble intracellular proteins and may contain a second type of binding site that interacts with noncarbohydrate ligand (Avellana-Adalid et al. 1990; Barondes 1988; Joubert et al. 1992). For this reason, the use of plant lectins that can recognize only glycoconjugates cannot give sufficient information concerning the lectin ligands. We

Objectives

developed a probe to investigate the presence of soluble and/or membrane lectin ligands in Neuroblastoma cells.

Probe   The probe consists of a carboxamidomethylated derivative of the brain lectin conjugated to biotinyl-amidocaproicacid (CAM-HBL Biot) (Avellana-Adalid et al. 1990).

Materials   These are identical to those used for immunocytochemistry

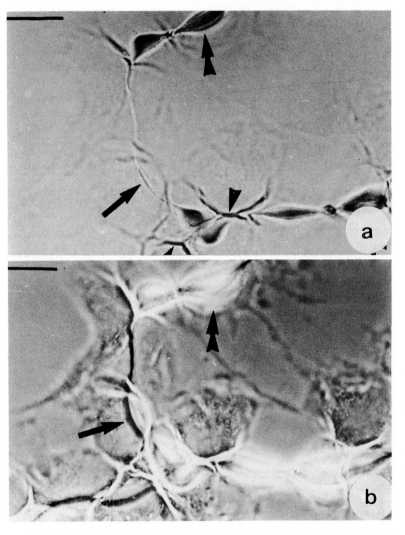

Fig. 3a,b. Extracellular affinocytochemistry showing lectin-binding sites. Lectin-binding sites are observed at the external face of the plasma membrane of undifferentiated cells as shown in **a** and **b** (*arrowheads*), positive network (*arrow*) and swells appear localized under the attached cells (**a, b**) (*double arrowheads*). Phast contrast micrograph of the same field (**b**) clearly shows that the cells are positioned above this network. *Bars* 10 μm

Detection of cytosolic ligands

Outline of procedure

1. Fixation with Carson fixative
2. Blocking of endogenous peroxidase activity
3. Blocking of nonspecific interactions with blocking solution
4. Incubation with CAM-HBL Biot
5. Treatment with Strep-HRP
6. Visualization of HRP-activity by DAB

Detection of extracellular ligands

1. Incubation of living cells with sterile CAM-HBL Biot at 20 °C for 60 min
2. Fixation and subsequent steps are performed as described above

Culture, fixation, and elimination of endogenous peroxidase activity are performed as described for immunocytochemistry.

Detailed Procedure

Wash coverslips 2 × 5 min with PBS

Expose coverslips to blocking solution [3% BSA (w/v), 0.3% Tween 20 (v/v) in PBS] for 60 min at 20 °C.

Excess of blocking solution is removed without washing prior to incubation in CAM-HBL Biot diluted in blocking solution, at a final concentration of 10 µg/ml at 20–22 °C for 60 min.

Wash carefully with PBS.

Apply Strep-HRP (0.26 µg/ml in PBS containing 1% BSA) for 60 min at 25 °C.

Wash thoroughly with PBS.

The end product is revealed as described for immunocytochemistry.

Wash in PBS.

Intra- and extracellular ligands are detected by the biotinylated probe. Extracellular ligands are found as a positive network and swellings localized under attached cells

Results

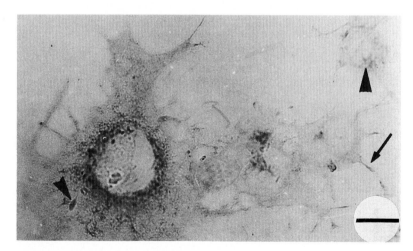

Fig. 4. Intracellular lectin affinocytochemistry of permeabilized N1E115 cells. Lectin-binding sites are shown as diffuse deposits of peroxidase positive materials as well as inside undifferentiated cells (*arrowheads*). Lectin-binding sites are shown in neurite extensions of differentiated cells (**a, b** *arrows*). *Bars* 10 µm

(Fig. 3c). Phase contrast micrograph of the same field (Fig. 3d) shows clearly that cells are positioned above this network. Figure 4 shows intracellular lectin binding sites as diffuse deposits of peroxidase-positive materials in cell somata and in neurite-like extensions. As cells were permealized during fixation, the extracellular ligands were no longer detected.

References

Abbott MW, Mellor A, Edwards Y, Feizi (1989) Soluble bovine galactose-binding lectin, cDNA cloning reveals the complete amino acid sequence and an antigenic relationship with the major encephalitogenic domain of myelin basic protein. Biochem J 259: 283–290

Avellana-Adalid V, Joubert R, Bladier D, Caron M (1990) Biotinylated derivative of a human brain lectin: synthesis and use in affinoblotting for endogenous ligands studies. Anal Biochem 190: 26–31

Barondes SH (1984) Soluble lectins: a new class of extracellular proteins. Science 233: 1259–1264

Barondes SH (1988) Bifunctional properties of lectins: lectins redifined. TIBS 13: 480–482

Bladier D, Le Caer JP, Joubert R, Caron M, Rossier J (1991) β-galactoside-soluble lectin from human brain; complete amino acid sequence. Neurochem Int 18: 275–281

Caron M, Bladier D, Joubert R (1990) Soluble galactoside-binding lectins: a protein family with common properties. Int J Biochem 22: 1379–1385

Cooper DNW, Barondes SH (1990) Evidence for export of a muscle lectin from cytosol to extracellular matrix and for a novel secretory mechanism. J Cell Biol 110: 1681–1691

Gabius HJ (1991) Detection and functions of mammalian lectins — with emphasis on membrane lectins. Biochim Biophys Acta 1071: 1–18

Gabius HJ, Wosgien B, Hendrys M, Bardosi A (1991) Lectin localization in human nerve by biochemically defined lectin-binding glycoproteins, neoglycoproteins and lectin-specific antibody. Histochem 95: 269–277

Joubert R, Kuchler S, Zanetta JP, Bladier D, Avellana-Adalid V, Caron M, Doinel C, Vincendon G (1989) Immunohistological localization of a β-galactoside-binding lectin in rat central nervous system. I Light and electron microscopical studies on developing cerebral cortex and corpus callosum. Dev Neurosci 11: 397–413

Joubert R, Caron M, Avellana-Adalid V, Mornet D, Bladier D (1992) Human brain lectin: a soluble lectin that binds actin. J Neurochem 58: 200–203

Kuchler S, Joubert R, Avellana-Adalid V, Caron M, Bladier D, Vincendon G, Zanetta JP (1989) Immunohistological localization of a β-galactoside-binding lectin in rat central nervous system. II Light and electron microscopical studies on developing cerebellum. Dev Neurosci 11: 414–427

Kuchler S, Zanetta JP, Vincendon G, Gabius HJ (1990) Detection of binding sites for biotinylated neoglycoproteins and heparin (endogenous lectins) during cerebellar ontogenesis in the rat. Eur J Cell Biol 52: 87–97

Leffler H, Masiarz F, Barondes SH (1989) Soluble lactose-binding vertabrate lectins: a growing family. Biochemistry 28: 9222–9229

Robertson MW, Albrandt K, Keller D, Liu FT (1990) Human IgE binding protein: a soluble lectin that exhibiting a highly conserved interspecies sequence and differential recognition of IgE glycoforms. Biochemistry 29: 8093–8100

Lectins and Neoglycoconjugates
in Histochemical and Cytochemical Analysis

22 Tissue Lectins in Histopathology – Markers in Search of Their Physiological Ligands

K. Kayser, H.-J. Gabius, and S. Gabius

Lectins have proven their value to localize defined carbohydrate structures in tissue sections. It should, however, not be over-looked that such binding sites, recognized by plant or invertebrate lectins, will not necessarily be ligands to tissue lectins when mammalian organs are monitored, unless their specificities have rigorously been shown to be identical. To draw meaningful conclusions on the presence of histo-chemically accessible ligands for endogenous lectins in tissue sections, it is thus required to work with the tissue lectins, as described for cell cultures by Joubert–Caron (Chap. 21, this Vol.). Prior chemical modification studies with group-specific reagents, as described by Zeng and Gabius (Chap. 8, this Vol.), serve as a guideline, which amino acid side chains can be modified by label incorporation such as a fluorescent dye, biotin or digoxigenin, without seriously harming the activity.

Human lung has been shown to contain several β-galactoside-binding lectins (Sparrow et al. 1987). We have purified the two lectins HL-14 and HL-29 and employed them as histochemical tools to localize accessible ligands in sections of 75 lung cancer cases. Carboxyamidomethylation with 40 mM iodoacetamide during affinity elution from lactose-Sepharose prevents oxidative inactivation (Powell and Whitney 1984; Hirabayashi and Kasai 1991; Tracey et al. 1992). The histochemical properties of the stabilized lectins from human lung are compared to the binding pattern of the immunomodulatory effective galactoside-binding lectin from mistletoe.

22.1 Experimental Part

Biotinyl-N-hydroxysuccinimide ester **Chemicals**
Dimethylformamide
Biotin-ε-aminocaproic acid N-hydroxysuccinimide ester
Biotin-amidocaproyl hydrazide
Pyridine-HCl (pH 4.8)
1-Ethyl-3-(3-dimethylaminopropyl)-carbodiimide
Avidin
ABC-kit reagents

Biotinylation of the Lectins

Two ml lectin-containing solution were dialyzed at 4 °C against five mM phosphate-buffered saline (pH 8.0) containing 20 mM lactose to protect the carbohydrate-binding site. Three mg biotinyl-N-hydroxysuccinimide ester, dissolved in 0.6 ml dimethylformamide, were slowly added and the reaction was kept at 4 °C for 14 h, followed by extensive dialysis against 2 mM phosphate-buffered saline (pH 7.2) and **Biotinylation**

4 mM β-mercaptoethanol. Alternatively, biotin-ϵ-aminocaproic acid N-hydroxysuccinimide ester can be used.

When modification of carboxyl groups is preferable, as has been done for the heparin-binding lectin (Gabius et al. 1991a), the lectin-containing solution is dialyzed against 10 mM phosphate-buffered saline (pH 4.8). Three mg lectin, dissolved in 2 ml of this buffer, were treated in 20 ml solution, obtained by adding 18 ml 10 mM pyridine-HCl (pH 4.8), with 100 mg biotin-amidocaproyl hydrazide, dissolved in 20 ml dimethylsulfoxide: H_2O (1:1), 50 mg 1-ethyl-3-(3-dimethylaminopropyl)-carbodiimide were added. The solution was kept at room temperature for 20 h under gentle stirring. Following concentration by ultrafiltration, the solution is extensively dialyzed to remove residual reagents.

Determination of Accessible Biotin, Conjugated to the Lectin

Determination The biotin content can be measured by a solid-phase assay, employing a streptavidin-conjugate as signal-generating reagent, as described by Zeng and Gabius (Chap. 8. this Vol.). Since avidin exhibits an increased absorbance at 233 nm ($\Delta E_{233} = 2.4 \times 10^4\,M^{-1}$/mol biotin) upon complex formation with biotin, titrations can be performed in solution to calculate the biotin content of biotinylated lectins (Powers et al. 1989). Starting with 2 ml of a $1-5 \times 10^{-6}$ M solution of avidin in phosphate-buffered saline (pH 7.2), 20 µl aliquots of the solution containing biotinylated lectin are successively added during titration up to saturation of the increase in absorbance. It is necessary to wait 40–70 min at each step for a reliable reading.

Histochemical Procedure

Histochemistry Three to five µm sections from formalin-fixed, paraffin-embedded specimen of 75 cases of lung cancer were treated, as described by Danguy and Gabius (Chap. 25 this Vol.). Biotinylated lectins from human lung were then incubated with the sections for 60 min at room temperature at a concentration of 10 µg/ml. The galactoside-binding lectin from mistletoe, purified by affinity chromatography (Gabius 1990), was used at a concentration of 5 µg/ml. Controls were performed by inhibition with 0.2 M lasctose + 1 mg/ml asialofetuin to prove the dependence of lectin binding on the carbohydrate-binding site and by omission of the biotinylated probe to exclude any binding by reagents such as the glycoproteins avidin or peroxidase. The cases were judged to be positive when dark brown color was seen in all or in clusters of the analyzed cells.

22.2 Results and Discussion

The two lectins from human lung bound to accessible binding sites in the tumor cells and the inflammatory host cells, the extent of binding of the 14 kDa lectin being higher than that of the 29 kDa lectin (Table 1). The galactoside-binding lectin from mistletoe stained tumor cells at an intermediate level, whereas labeling of inflammatory host cells was rather similar to that of the 14 kDa lectin. These results emphasize

Table 1. Binding of β-galactoside-specific lectins from human lung (14 kDa, 29 kDa) and of galactoside-binding lectin from mistletoe (ML-l) to tumor and inflammatory host cells in human lung carcinomas ($N = 75$)

Type of cell Lectin	Tumor cells[a]	Inflammatory host cells
14 kDa	72	65
29 kDa	29	23
ML-l	45	57

[a]The number of positive cases from a total of 75 cases is given.

that the three lectins of related, but definitely not identical carbohydrate specificity recognize accessible ligands to a different degree. As a next step to elucidate the physiological functions of individual lectins colocalization studies with lectin-specific antibodies and labeled tissue lectins, as described by Joubert–Caron for cells (Chap. 21. this Vol.), are attractive.

In addition to covalent labeling by low molecular weight-reagents, exemplary generation of an adequate tool to localize lectin ligands has been reported, employing genetic engineering. A chimeric protein containing a lectin (L-selectin) as well as the hinge and constant regions of human immunoglobulin heavy chains was constructed on the level of the gene and expressed (Watson et al. 1990). Any lectin whose cDNA is cloned could thus in principle be converted into such a monoclonal antibody-like tool.

Having ascertained the presence of lectin-binding sites by the histochemical technique, respective glycoproteins can be isolated by lectin-affinity chromatography. Their labeling gives access to markers that will interact with the tissue lectin with high affinity (Gabius et al. 1991b). Any other ligand properties of these purified lectin ligands have to be excluded, when the labeled lectin-binding glycoproteins are to be applied in histochemistry. Overall, the described probes not only augment the panel of substances for thorough glycohistochemical analysis. Notably, their application will be helpful to provide clues to solve the pertinent problem on the description of the physiological functions of diverse tissue lectins.

References

Gabius HJ (1990) Influence of type of linkage and spacer on the interaction of β-galactoside-binding proteins with immobilized affinity ligands. Anal Biochem 189: 91–94

Gabius HJ, Kohnke-Godt B, Leichsenring M, Bardosi A (1991a) Heparin-binding lectin of human placenta as a tool for histochemical ligand localization and isolation. J Histochem Cytochem 39: 1249–1256

Gabius HJ, Wosgien B, Brinck U, Schauer A (1991b) Localization of endogenous β-galactoside-specific lectins by neoglycoproteins, lectin-binding tissue glycoproteins and antibodies and of accessible lectin-specific ligands by mammalian lectin in human breast carcinomas. Path Res Pract 187: 839–847

Hirabayashi J, Kasai K (1991) Effect of amino acid substitution by site-directed mutagenesis on the carbohydrate recognition and stability of human 14 kDa β-galactoside-binding lectin. J Biol Chem 266: 23648–23653

Powell JT, Whitney PL (1984) Endogenous ligands of rat lung β-galactoside-binding protein (galaptin) isolated by affinity chromatography on carboxyamidomethylated galaptin-Sepharose. Biochem J 223: 769–776

Powers JD, Kilpatrick PK, Carbonell RK (1989) Protein purification by affinity binding to unilamellar vesicles. Biotechnol Bioeng 33: 173–182

Sparrow CP, Leffler H, Barondes SH (1987) Multiple soluble β-galactoside-binding lectins from human lung. J Biol Chem 262: 7383–7390

Tracey BM, Feizi T, Abbott WM, Carruthers RA, Green BN, Lawson AM (1992) Subunit molecular mass assignment of 14,654 Da to the soluble β-galactoside-binding lectin from bovine heart muscle and demonstration of intramolecular disulfide bonding associated with oxidative inactivation. J Biol Chem 267: 10342–10347

Watson SR, Imai Y, Fennie C, Geoffroy JS, Rosen SD, Lasky LA (1990) A homing receptor-IgG chimera as a probe for adhesive ligands of lymph node high endothelial venules. J Cell Biol 110: 2221–2229

23 Lectin Cytochemistry Using Colloidal Gold Methodology

G. EGEA

The capability of lectins to recognize and bind sugar residues makes them useful tools to investigate the cellular and subcellular distribution of glycoconjugates in situ (Nicolson 1978; Roth 1978). In order to be used as cytochemical reagents, lectins must be coupled to appropriate markers. In addition to fluorescein or rhodamine isothiocyanate conjugation (Roth et al. 1978) and horseradish peroxidase coupling (Roth and Binder 1978; Spicer and Schulte 1992), colloidal gold particles can be used either at light or electron microscopy (Roth 1978, 1987; Benhamou 1989). Colloidal gold labeling techniques are versatile and can be combined with a variety of techniques for specimen preparation. This chapter describes the preparation of lectin-gold and glycoprotein-gold complexes and their application in diverse labeling strategies at microscopical and nonmicroscopical techniques.

23.1 Preparation of Lectin- and Glycoprotein-Gold Probes

Ultracentrifuge and rotors (Beckman Ti-50, Ti-70, SW-41 or equivalents), pH meter, table centrifuge, spectrophotometer. **Equipment**

Chemicals from MERCK: absolute ethanol, acetic acid, ammonium chloride (NH$_4$Cl), chlorhidric acid (HCl), formalin, glycerol, gold trichloride acid trihydrate (HAuCl$_4$·3H$_2$O), hydrogen peroxide (H$_2$O$_2$), hydroquinon, paraformaldehyde, polyethylene glycol 20 000 (PEG 20 000), potassium monocarbonate (K$_2$CO$_3$), sodium azide (Na$_3$N), sodium hydroxide (NaOH), sodium metaperiodate (NaJO$_4$), trisodium citrate 2-hydrate (C$_6$H$_5$Na$_3$O$_7$·2H$_2$O) **Chemicals**
Chemicals from SIGMA: lysozyme, poly-L-lysine (M_r 560 000), polyvinylpyrrolidone, Triton X-100, Tween-20 (polyoxyethylene-sorbitan monolaurate)
Chemicals from FLUKA: bovine serum albumin (BSA), glycine, silver acetate, silver lactate
Others: Aleppo tannic acid (MALLINCKRODT), Agfafix (AGFA)

Preparation of Gold Particles

Colloidal gold has an orange, red, or red-violet color, depending on the preparation method. Colloidal gold is prepared by condensation of metallic gold produced by reducing gold salts with numerous reagents, the most common being phosphorus, tannic acid, ascorbic acid, and sodium citrate. Depending on the method used, colloidal gold preparations vary in particle size [expressed as the average particle diameter (APD)] and size variability, i.e., the coefficient of variation (CV) of the gold particle diameter in a sol [monodisperse (when CV is smaller 15%) and polydisperse **Gold Particles**

Table 1. Different gold particle size obtained by the sodium citrate method. (Frens 1973)

Average particle diameter (APD) (in nm)	Amount of 1% trisodium citrate added (in ml)
14	5
24	2
30	1.6
46	0.8
65	0.6

(when CV exceeds 15%)]. For cytochemistry purposes, colloidal gold is prepared using two main methods.

Method I. Sodium Citrate (Frens 1973). This method provides monodisperse colloidal gold with a size range from 14 to 150 nm. To prepare a 14 nm colloidal gold solution, 1 ml of a solution of 1% $HAuCl_4$ (in double distilled water) is added to 100 ml of double distilled water in a very clean Erlenmeyer flask covered with aluminum foil with a small hole in the center. The solution is heated to boiling. As soon as it starts to boil, 5 ml of filtered 1% trisodium citrate (in double distilled water) is quickly added (i.e., blow the solution out of 5 ml glass pipet). The solution is gently boiled until an orange-red color develops (usually, a color changes from translucid to blue to red within minutes). Some variations using the sodium citrate method have been reported. For instance, 100 ml double distilled water are brought to boil alone, and then 1% trisodium citrate is added. Subsequently, 1% $HAuCl_4$ solution is added. Depending on the desired size of gold particles, the amount of 1% trisodium citrate solution added should be changed (Table 1).

Method II. Tannic Acid Method (Slot and Geuze 1985). To perform high resolution studies, smaller gold markers are generally necessary (1–10 nm). Slot and Geuze (1985) developed a method that combines tannic acid (TA) with trisodium citrate as reducing agents. They described the optimal conditions to obtain monodisperse sols of different size (from 3 to 15 nm), varying the amount of tannic acid added. This method is adequate for preparing gold particles of different size to be used in multiple labeling protocols.

To make 100 ml of a sol, two solutions are made:

Solution A: Gold chloride solution (250 ml beaker with a stirring bar)
– 79 ml double distilled water
– 1 ml 1% $HAuCl_4$ in double distilled water

Solution B: Reducing mixture (50 ml beaker)
– 4 ml of filtered 1% trisodium citrate in distilled water
– a variable volume (x) of 1% tannic acid (Table 2)
– 25 nm K_2CO_3 (to correct the pH of the reducing mixture) in the same volume as tannic acid if the tannic acid is above 1 ml.
– double distilled water to make 20 ml

Solutions A and B are heated to 60 °C. Then B is quickly added to A under vigorous

Table 2. Different gold particle size obtained by the tannic acid method. (Slot and Geuze 1985)

Average particle diameter (APD) (in nm)	Amount of 1% tannic acid added (in μl)
3	3000
4	2000
5	1000
˙6	500
8	200
10	100
14	15

stirring. Red sols are formed very rapidly, depending on the amount of tannic acid added. This colloid is heated until boiling.

Preparation of Direct Lectin-Gold Complex

Colloidal gold sols are unstable in the presence of electrolytes. When electrolytes are added, the color turns blue, and gold precipitates. However, when proteins are added under proper conditions, these bind spontaneously to the gold particles, and sols are rendered hydrophylic and stable in the presence of added electrolytes (Geoghegan and Ackerman 1977). The binding of proteins to gold is practically irreversible, and the proteins maintain their biological activities. This basic principle is applied to lectins. The binding of protein to gold is pH-dependent. In general, stable complexes are achieved at pH equal to, or 0.5 pH units higher than the iso-electric point of the lectin involved (Table 3). The pH of the colloidal gold solution can be adjusted by the addition of 0.1 M K_2CO_3 or 0.1 N NaOH or by addition of 0.1 N HCl or 0.1 N acetic acid. The pH is directly measured with pH paper or with a gel-filled electrode. When a normal electrode is used, 0.5 ml of 1% polyethylene glycol 20 000 to 9.5 ml of colloid are added prior to the measurement to avoid plugging the pores of the electrode.

Lectin-Gold Complex

Estimation of the Minimal Amount of Lectin Needed to Stabilize a Volume of Gold Sol. To provide optimal conditions for complex formation, the ionic strength a lectin solutions must be very low, since salts facilitate particle aggregation. When a critical distance is reached, the particles aggregate and flocculation occurs (red gold sol turns blue). Macromolecules adsorbed on the surface of the gold inhibit such aggregation induced by electrolytes (Roth 1983a). This characteristic is utilized to determine the minimum amount of protein needed to stabilize a certain volume of colloidal gold. Thus, whenever possible, dissolve the proteins in distilled water. This is not a major problem in the case of lectins, since most of them are water-soluble. If this is not the case, methods to complex proteins in the presence of salt have been described (Lucocq and Baschong 1986). The easiest method is the salt flocculation test where the color change of the gold sol is observed (Roth and Binder 1978). This method is performed as follows:

1. Lectin is dissolved in distilled water at 1 mg/ml
2. Serial dilutions of the lectin in distilled water (100 μl) are prepared in glass tubes

Table 3. Optimal Conditions for the Preparation and Use of Lectin-Gold Complex

Lectin	pH of gold sol	Amount of lectin (µg) needed to stabilize 10 ml of colloidal gold		Lectin–gold complex optical density at 525 nm (OD_{525}) or concentration for post-embedding staining of Lowicryl K4M ultrathin sections.
		Citrate method (15 nm)	Tannic acid method (8–10 nm)	
Amaranthus caudatus	7.4	60	120	0.1 in PBS + + + ([a]) (1)
Aplysia depilans	9.5	250		Dilution 1:4 in PBS/PEG pH 6.0 (2)
Arachis hypogaea (peanut lectin)	6.3	130	250	25 µg/ml in PBS (3)
Concanavalin A	8.0	275		
Dolichos biflorus	5.6	250	500	25 µg/ml in PBS (3)
Glycine max (soybean lectin)	6.1	65		25–50 µg/ml in PBS (3)
Helix pomatia	7.4	65	125	0.7 in PBS + + + (4)
Lens culinaris	7.0	130		100 µg/ml in PBS (3)
Lotus tetragonolobus	6.3	130		100 µg/ml in PBS(3)
Maackia aurensis	6.0	125	250	0.5 in PBS/1% BSA (5)
Phaseolus vulgaris	5.3	130		100 µg/ml in PBS (3)
Ricinus communis-I	8.0	130		50–200 µg/ml in PBS (3)
Ricinus communis-II				
Sambucus nigra	6.0-6.2	400 (ascorbic acid method)		0.5–0.7 in PBS + + + (6)
Ulex europaeus	6.3	250		50 µg/ml in PBS (3)
Viscum albium	9	200		

[a] PBS (0.01 M phosphate buffer, 0.15 M NaCl, pH 7.2) containing 1% BSA, 0.075% Triton X-100 and 0.075 % Tween 20. (1) Sata et al. (1990); (2) Benhamou et al. (1988); (3) Roth (1983[b]); (4) Taatjes et al. (1987); (5) Sata et al. (1989); (6) Taatjes et al. (1988).

3. pH-Adjusted gold sol (0.5 ml) is added (with stirring) to each tube (Fig. 1A)
4. After 2 min, 100 µl of 10% NaCl (in distilled water) is added
5. Tubes that contain enough amount of lectin to stabilize gold sol are maintained red even in the presence of electrolytes

Flocculation takes place immediately where the gold solution is unstable and the red color turns to blue/violet (Fig. 1B) In any case, wait 30 m in to conclude results, and a short centrifugation (1000 rpm for 10 min) of the tubes provides clearer results. The amount of lectin to be considered corresponds to the second tube containing less protein whose sol color does not change (Fig. 1B). Conjugation can also be estimated spectrophotometrically at the maximum of absorbance of the colloid (518–550 nm, depending on the gold size -518 nm for Au_5; 525 nm for Au_{26}; 539 nm for Au_{47}; Horrisberger 1985).

Formation of Direct Gold Complex. Once the optimal pH for adsorption and the lectin stabilizing amount have been determined, large-scale lectin-gold complex preparations (5–100 ml) is carried out (Table 3). The optimal stabilizing amount of lectin plus a ten per cent in excess is dissolved in distilled water (0.5–1 ml) and then filtered through a Millipore filter (0.2–0.4 µm pore size) and placed at the bottom of a clean plastic centrifuge tube (Fig. 1C) (in some cases this tube is soaked overnight at 4 °C with a solution of 1% BSA in distilled water and rinsed ten times with double

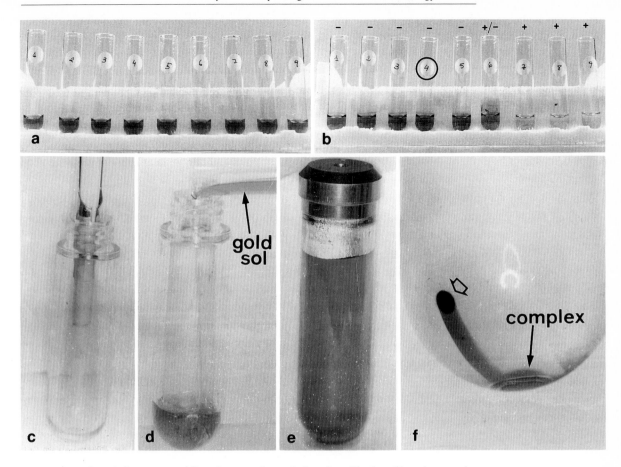

Fig. 1a–f. Salt flocculation test (**a** and **b**) and preparation and obtention of lectin-gold or glycoprotein-gold complexes (**c, d, e,** and **f**). Serial dilutions of lectin or glycoprotein (0.1 ml) in the presence of pH-adjusted colloidal gold (0.5 ml) before (**a**) and after (**b**) the addition of sodium chloride (10%). Note the tubes where gold sol is severely (+), averagely (+ / −), or not (−) flocculated. The appropriate glass tube with the amount of lectin or glycoprotein sufficient for the estabilization of gold sol is *encircled*. To prepare lectin or glycoprotein-gold probes, protein is firstly located in the bottom of the centrifuge tube (**c**) and, subsequently, an adequate volume of pH-adjusted gold sol is added (**d**). The crude lectin-gold or glycoprotein-gold complex (**e**) is centrifuged. Mobile lectin-gold or glycoprotein-gold complex is situated at the bottom (**f**), while uncomplexed, unstable, and/or aggregated gold particles form a tight pellet in a side (**f**, *empty arrow*)

distilled water before the large-scale gold complexing of the lectin). Colloidal gold (adjusted to the adequate pH) is then added to the lectin solution (Fig. 1D) and mixed rapidly. After 10 min, filtered 1% PEG 20 000 in distilled water is added (0.5–5 ml). Some authors suggest that the pH has to be brought to 7–8 with 0.1 M K_2CO_3. In my experience, this step is not essential. The colloid (Fig. 1E) is then centrifuged to remove excess of unbound lectin and to concentrate the lectin-gold complex. Centrifugation is performed in a Sorvall of Beckman ultracentrifuges at 4°C for 45–60 min with Ti-50 or Ti-70 rotors at 23 000 rpm for $Au_{10-15\ nm}$ or 29 000 rpm for $Au_{5-8\ nm}$. The pellet is composed of a large and very mobile sediment of lectin-gold complex at the bottom, and a small tight pellet on the side of the tube

(Fig. 1F). The translucid supernatant is carefully aspired and the loose sediment can be resuspended in 1–2 ml in the appropriate buffer containing 0.2 mg/ml of PEG 20 000. The recovered lectin-gold complex is placed in a clean glass tube, further centrifuged at 3 000 rpm for 15–20 min (to remove gold aggregates) and supernatant is stored at 4 °C. Also, the lectin-gold complex may be centrifuged over a 10–30% continuous glycerol (or sucrose) gradient in the appropriate buffer. Briefly, the lectin-gold complex sediment is first resuspended in a small volume (0.5 ml) of buffer and then placed over the continuous glycerol gradient carefully. This is centrifuged in a SW41 or SW40 rotors (Beckman) for 30 min at 14 000 rpm. The lectin-gold complex is situated 1–2 cm below the top of the gradient. When the tannic acid gold preparation method is used, this procedure is recommended to eliminate tannic acid, uncomplexed lectin, and gold aggregates. Alternatively, tannic acid is destroyed when crude gold sols are treated with 0.2% (final concentration) H_2O_2 for 2 h at room temperature and then boiled for 10 min.

Preparation of Glycoprotein-Gold Probes

Glycoprotein-Gold Probes   Certain lectins present difficulties for performing direct complex formation with colloidal gold due to their low molecular weight, high isoelectric point, or very low solubility in distilled water. However, a glycoprotein bearing sugar specific for the lectin (with two or more sugar binding sites – one recognizing the saccharide(s) of the tissue preparation and other recognizing the saccharide(s) of the glycoprotein – is directly conjugated to colloidal gold and used as a two-step technique. The preparation of glycoprotein-gold complex is identical to the preparation of lectin-gold complex. Lectins with their corresponding glycoprotein-gold complex are summarized in Table 4.

Storage of Gold Reagents

Storage   Most lectin-gold and glycoprotein-gold complexes retain their biological activity for months in the refrigerator at 4 °C with 0.02% sodium azide added to the buffer solution. Good bioactivity preservation for a longer period is also obtained with storage at −20 °C in the presence of 45% glycerol (solution remains fluid).

23.2 Application of Lectin Cytochemistry for Electron Microscopy

Fixation Conditions

Fixation   Carbohydrates are rather stable under fixation conditions that are generally used. The following fixatives are more or less routinely used and provide a good compromise between morphological preservation and lectin recognition capability for saccharide structures:

- 2–4% paraformaldehyde in 0.1 M phosphate buffer or PBS.
- 2–4% paraformaldehyde with 0.01–0.5% glutaraldehyde in 0.1 M phosphate buffer or PBS.
- 0.5–1% glutaraldehyde in 0.1 M phosphate buffer or PBS.

Table 4. Optimal Conditions for the Preparation and Use of Glycoprotein-Gold Complex for the Two-Step Lectin Cytochemical Post-Embedding Technique

Lectin	Glycoprotein	pH of gold sol	Amount of glycoprotein (µg) to stablizie 10 ml of gold		Lectin (µg/ml)	Glycoprotein–gold (OD$_{525}$ or µg/ml)
			Citrate method (15 nm)	Tannic acid method (8–10 nm)		
Amaranthin caudatus	Asialoglycoprotein	6		250	10	0.1 in PBS (1)
	Asialomucin	6		30	10	0.2 in PBS (1)
Concanavalin A	Horseradish peroxidase	8	65	130	10–20	5–10 µg/ml (2)
Datura stramonium	Asialofetuin	7.4	250		75–100	0.15 in PBS + + + [a] (3)
	Asialorosomucoid	5.0	65			0.2 in PBS + + + (3)
	Fetuin	5.4	70	200		0.35 in PBS + + + + (3,5)
	Ovomucoid	4.8	16	150		0.2 in PBS + + + (3)
Erythrinia cristagalhi	Asialofetuin	7.4	250		100	0.3 in PBS + + + (4)
Limax flavus	Fetuin	5.4	70	200	50–100	0.35 in PBS + + + + (5)
Maackia amurensis	Fetuin	5.4	70	200	100	0.3 in PBS + + + (6)
Ricinus communis I	Asialofetuin	7.4	250		50–75	0.3 in PBS + + + + (4)
	Asialoorosomucoid	5.0	65			0.2 in PBS + + + (7)
Sambucus nigra	Fetuin	5.4	70	200	100–250	0.3 in PBS + + + (8)
Wheat germ	Ovomucoid	4.8	16	150	10–20	0.2 in PBS + + + (3)

[a] PBS (0.01 M phosphate buffer, 0.15 M NaCl, pH 7.2) containing 1% BSA, 0.075% Triton X-100 and 0.075% Tween-20.
(1) Sata et al. (1990); (2) Roth (1983); (3) Egea et al. (1989); (4) Taatjes et al. (1990); (5) Roth et al. (1984); (6) Sata et al. (1989); (7) Egea and Marsal (1992); (8) Taatjes et al. (1988).

Prior to processing the fixed material further, free aldehyde groups should be blocked by treatment with 50 mM NH_4Cl or 0.15 M glycine in phosphate buffer or PBS for 1 h at room temperature.

Strategies for Lectin Labeling

Lectin Labeling   There are diverse approaches to demonstrate lectin-binding sites, the most frequent being:

Direct Methods. This method is based on lectin conjugation to the label (gold particles in our case) and direct incubation on the tissue sections (Table 3).

Indirect Method. The two-step cytochemical technique consists in incubating first with an unlabeled lectin and, subsequently visualizing the sugar-lectin interaction with a gold-complexed glycoprotein or polysaccharide that has a high affinity for the lectin (Table 4). Other indirect methods are the antibody method, where anti-lectin immunoglobulins are used and they are visualized with protein-A gold (Perruzo and Rodriguez 1989); the biotinylated-lectin followed by an avidin-gold or streptavidin-gold complex (Skutelsky and Bayer 1979); the lectin-digoxigenin conjugates (Sata et al. 1990b); and the enzyme-gold technique, where glycohydrolases are directly complexed to gold particles detecting the monosaccharide susceptible to be detached (Bendayan 1985). However, these approaches will not be described in this chapter.

Lectin Cytochemistry Applications

Cytochemistry   *Pre-Embedding Technique.* Pre-embedding techniques imply the performance of the cytochemical lectin reaction on the cells or tissues before they are processed for microscopic observation. The reactions are performed either after a mild pre-fixation or in the absence of fixation. This approach eliminates the interference of dehydrating agents and embedding media. This technique is used for detecting plasmalemmal carbohydrates, although it presents problems for localization of intracellular saccharides because of the dimensions of the gold spheres. In this case, the plasma membranes need to be permeabilized, leading to altered morphology, and target retention of the cell is diminished. Usually, this technique is applied for the study of outer oligosaccharides on cultured cells and on isolated cells and membranes. A current pre-embedding protocol (Versura et al. 1989) using lectin-gold complexes is the following:

1. Rinse cultured cells, isolated membranes, or tissues blocks briefly in washing buffer (PBS, cacodylate, etc).
2. Lightly pre-fix cells or tissue blocks with an appropriate fixative (see Sect. 23.2.1)
3. Rinse several times (5–10 min each) with washing buffer and block free-adehydes groups (see Sect. 23.2.1).
4. In the case of tissues, thick sections (15–40 μm) are obtained through a Vibratome. Sections are rinsed in the buffer to be applied with the lectin-gold complexes.
5. Incubate the cells or tissues (in constant agitation) with the optimal concentration or OD_{525} of the lectin-gold complexes for 45–90 min at room temperature, or 3–12 h at 4 °C.

6. Rinse several times (10 min each) in PBS or TBS (Tris buffer saline).
7. Rinse three times in washing solution.
8. Fix cells or tissue sections with 1–2.5% glutaraldehyde in PBS at room temperature for 15–30 min.
9. Rinse briefly in water or overnight in PBS.
10. Post-fix with osmium tetraoxide for 1 h.
11. Rinse and dehydrate in a graded series of ethanol.
12. Process for embedding in Epon or Araldite according to standard protocols.

This basic protocol can also be applied with the indirect procedure.

Post-Embedding Technique. The post-embedding technique is the most widely used lectin cytochemical technique for light and electron microscopy. Basically, cells or tissues are embedded and cytochemical reactions are performed on ultrathin cryosections or on sections from diverse embedding media (mainly Epon, Araldite, LR White and Lowicryl K4M) can be used for postembedding procedures. At EM level, the most immunocytochemical routine methodology utilizes the low-temperature (− 35 °C) embedding of the sample in the low cross-linked hydrophylic resin Lowicryl K4M (Carlemalm et al. 1982) and the epoxy resin Epon. Whichever embedding medium is used, ultrathin sections mounted on Formvar-coated nickel grids (no copper grids!) are processed for lectin-gold cytochemistry at room temperature as follows:

One-step methodology

1. Grids carrying sections are floated on a droplet of PBS for 5–10 min
2. Grids are directly transferred (Fig. 2A) onto a droplet (10–20 µl) of diluted lectin-gold complex for 30–60 min (Table 3; Fig 2B) in a moist chamber
3. Grids are "jet" washed by a mild spray of PBS (Fig. 2C), and then they are immersed in PBS for 2–5 min. This process is repeated once more. Alternatively,

 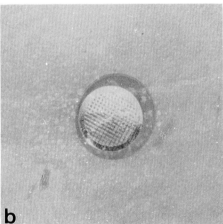 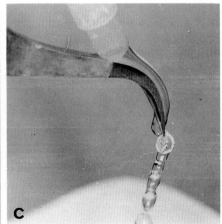

a   b   c

Fig. 2a–c. Sequences for lectin-gold staining of Lowicryl-K4M ultrathin sections. Formvar-coated nickel grids containing thin sections are carefully faced down onto a small droplet of lectin-gold probe (**a, b**). After the incubation time, the grid is submitted to a continuous flow of rinsing buffer ("jet washing") (**c**) for 10–20 s

grids are floated onto the surface of a large volume of PBS (for example), the microwells of an ELISA plaque) and submitted to constant stirring with a magnetic vortex for 3×20 s

4. Finally, grids are quickly "jet" rinsed with distilled water and allowed to air-dry before counterstaining with uranyl acetate (2% in distilled water) for 5–7 min, and lead citrate or lead acetate (1%) for 45 s (Fig. 3A)

Two-step methodology

1. Grids carrying sections are floated on a drop of PBS for 5–10 min.
2. Grids are directly placed onto a droplet (10–20 µl) of uncomplexed lectin in PBS (Table 4), and incubaed for 45 min in a moist chamber.
3. Grids are rinsed as described above (direct method).
4. Subsequently, grids are quickly dried with filter paper from the side that does not contain the sections (do not let sections dry!), and grids are transferred onto a droplet of the glycoprotein-gold complex (Table 4) for 30 min.
5. Finally, grids are "jet" washed with PBS and distilled water, air-dried, and counterstained with uranyl acetate and lead citrate or lead acetate as described above.

Total immersion of the grid during the incubation reactions should be avoided since this will increase the background staining. The above protocols are representative when ultrathin sections of Lowicryl-K4M embedded cells and tissues are used. These strategies are equivalent when thin sections are obtained from Epon-embedded material. However, when epoxy resins are used, is necessary an etching step before the cytochemical reactions. Diverse procedures can be used:

– 1–10% H_2O_2 for 5–10 min at room temperature
– 3% NaOH dissolved in absolute ethanol for 10 s–30 min (depending on the thickness of the section)
– Saturated aqueous solution of sodium metaperdiodate for 10–60 min. This method is very useful when epoxy-embedded tissues were previously post-fixed with osmium tetraoxide (Bendayan and Zollinger 1983)

When cryosections are used, the procedure is as described above, but without using detergents in the lectin–gold or glycoprotein–gold solutions. Usually, the incubation times can be shortened.

Fig. 3a–f. Examples of lectin-gold cytochemistry at electron (**a, b, e, f**) and light microscopy (**c, d**) techniques. **a** Lowicryl K4M ultrathin section directly stained with *Dolichos biflorus* (DBA)-gold complex (revealing terminal and nonreduced GalNAc) at the ovine submaxillary gland. Gold particles are restricted to the Golgi apparatus (*G*). The rough endoplasmic reticulum (*RER*) and the mucin droplets (*M*) are devoid of labeling. **b** Replica-staining label-fracture procedure of quick-frozen and freeze-fractured Torpedo electric organ synaptosomes stained with *Datura stramonium* (DSA) (recognizes Galβ1, 4GlcNAcβ sequences) followed by ovomucoid-gold complex. Gold particles are seen superimposed on the Pt/C replicated extracellular hemimembrane leaflet (*EF*). Most of gold particles are not associated with intramembrane particles (IMPs) (*arrowheads*). **c,d** soybean (SBA)-gold staining pattern (recognizes terminal and nonreduced GalNAc) of normal human colon visualized at light microscopy with the silver-enhancement technique. Black spots are viewed over the mucin droplets (*M*) of goblet cells by phase contrast (c) or bright field (d) imaging. **e, f** Lowicryl-K4M ultrathin sections of cultured human colon carcinoma cells stained for *Limax flavus* (LFA)-fetuin gold complex (to visualize neuraminic acid) before **e** and after **f** the action of *Clostridium perfingens* neuraminidase type X. As a consequence of the enzymatic digestion, gold particles are virtually absent from the mucin droplets (*M*) (**f**). Moreover, note the disappearance of the fuzzy structures (*small arrows*) located inside the mucin droplets

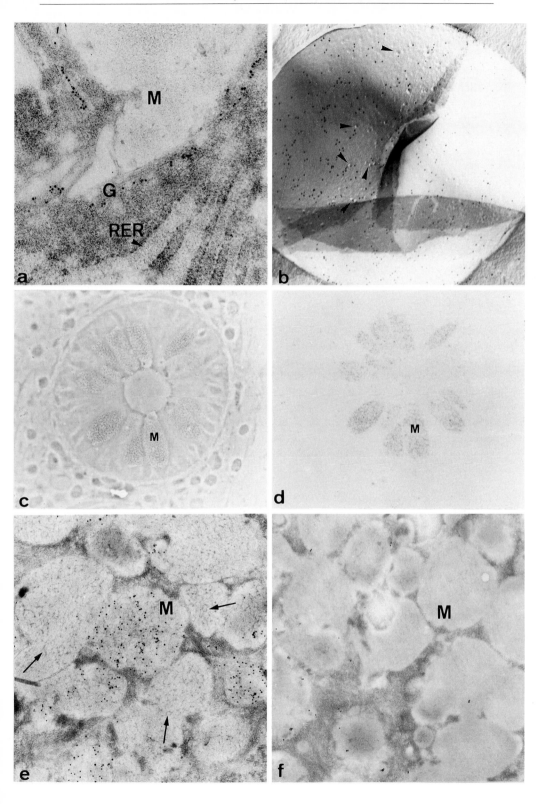

Scanning Electron Microscope Technique. Horrisberger and collaborators have shown that lectin-gold complexes can be succesfully used with the scanning electron microscope (SEM) (Horrisberger 1985). This is a sensitive technique to study tridimensionally the external distribution of carbohydrates. The only limitation is the gold particle size, which should be larger than 20 nm. The cytochemical procedure is very similar to that of the pre-embedding technique:

1. Cells are prefixed with aldehydes.
2. Incubation with an excess of lectin-gold complex for 4 h (with occasional handly shaking) in the presence of 0.15 M NaCl–0.02 M Tris, pH 7.4 containing 0.5 mg/ml Carbowax 20 M or PEG.
3. Light centrifugation and rinse with the buffer until the supernatant is colorless. Finally rinse with water.
4. Dehydration in graded series of ethanol (or acetone), critical point dry and direct examination at SEM (without deposit of metal).

When tissues are used, the visualization requires techniques that do not contain metals. In this case, tissues are impregnated with osmium tetroxide and thiocarbohydrazide (noncoating technique) or by evaporation of carbon (10–15 nm). For details see Horrisberger (1985, 1989).

Freeze-Fracture Technique. Pinto da Silva and collaborators have developed several strategies that combine freeze-fracture with the in situ gold labeling techniques (Kan and Pinto da Silva 1989 as a general review). The most frequently used freeze-fracture cytochemical approach is the label fracture. Label fracture is a cell surface labeling method (Pinto da Silva and Kan 1984). In summary, cell surface gold labeling is followed by conventional freeze fracturing but platinum/carbon replicas are rinsed with distilled water instead of chemical digestion with acids or bases. This procedure allows the outer half of the membrane to remain attached to the replica. Gold particles are seen superimposed on the freeze-fractured exoplasmic hemimembrane face (EF). A common procedure is as follows:

1. Cells are prefixed with glutaraldehyde (1–2.5% in PBS) for 1 h at 4 °C or at room temperature.
2. Cells are rinsed two times in PBS by successive centrifugation (1500 rpm for 5 min) and resuspension in PBS.
3. Aldehyde-free groups are quenched with 0.1 M glycine or 50 mM amonium chloride both in PBS for 1 h at room temperature.
4. Cells are washed three times in PBS.
5. Cells are labeled with an excess of lectin-gold complex or nonlabeled lectin, followed by glycoprotein-gold complex diluted in PBS alone or in the presence of 1% BSA. It is important not to use detergents in the lectin-gold or glycoprotein-gold complex solutions.
6. Impregnate gradually in glycerol (20–35% in PBS).
7. Freeze in liquid propane of Freon 22 and freeze fracture and replicate as standard conditions.
8. Wash replicas solely in distilled water (6 × 30 min).
9. Collect the replicas in Formvar-coated nickel grids.

Recently, a modification of this technique has been developed (replica-staining label-fracture technique; Andersson-Forsman and Pinto da Silva 1988). Basically, the samples are first freeze fractured and replicated, rinsed in distilled water, and subsequently processed for lectin-staining as described above (Fig. 3B). Other strategies have been developed by Pinto da Silva and collaborators (fracture-label, fracture-flip) and the reader should address to original sources for more details.

23.3 Application of Lectin Cytochemistry for Light Microscopy

Although extensively used in electron microscopy, colloidal gold complexes have also proven to be suitable for light microscopy (Lucocq and Roth 1985). The gold particle label is visualized by the colloidal gold's own red color. However, when gold labeling is low, no color is apparent. Under these circumstances, the gold signal is evidenced by the application of the photochemical silver amplification method (Holgate et al. 1983; Danscher and Nörgaard 1983). The limit of detection of this technique has been described as being superior to the peroxidase labeled reagents (Springal et al. 1984). The labeling techniques can be performed on frozen sections (cryostat or semithin) of prefixed and unfixed samples, on sections of fixed paraffin or semithin resin-embedded specimens (Epon, Araldyte, Lowicryl K4M) (Roth 1983b,c; Lucocq and Roth 1984, 1985; Springal et al. 1984; Tokuyasu, 1986; Taatjes et al. 1987).

Fixation Conditions

The general conditions described for electron microscopy can also be applied for light microscopy. When retrospective paraffin embedded pathological material has to be studied, carbohydrates are "resistant" to routine fixators as formol but give a low and/or nonspecific staining when Bouin's (18 h) or Carnoy's solution (4 h) were used (Roth 1983c).

Fixation

Handling and Lectin Labeling of Tissues Sections

Cryostat, vibratome, or semithin cryosections are mounted on glass slides according to routine procedures. The glass slides have to be very clean (use ethanol-HCl) and coated with poly-L-lysine (1 mg/ml in distilled water, MW 560 000) before use. Paraffin sections are obtained and mounted also on poly-L-lysine-treated glass slides according to standard procedures. Deparaffination is performed in xylene (2 × 5 min) followed by rehydration in decreasing grades of ethanol (5 min each) and storage in PBS. Semithin Epon sections are collected on a droplet of distilled water and mounted on glass slides treated with poly-L-lysine and allowed to dry overnight at 40 °C. Before labeling, Epon resin should be removed. In this case, semithin sections are covered for 2–5 min with a freshly prepared and filtered solution that contains 2 g KOH in 10 ml of absolute methanol plus 5 ml propylene oxide. Then, this solution is discarded and sections are covered with PBS/methanol (50:50 v/v) and PBS. Semithin Lowicryl K4M sections require neither resin removing procedure. Sections are directly placed on glass slides treated with poly-L-lysine and dried overnight at 40 °C. Lectin cytochemistry is performed according the protocols described previously for electron microscopy techniques and are common for frozen,

Lectin Labeling

paraffin, or semithin resin sections. Differences are only in rinsing times with PBS (2 × 5 min each) and distilled water (2 × 2 min). Moreover, before the final rinsing with distilled water and the signal amplification procedure, it is highly recommended to fix the sections with glutaraldehyde (1% in PBS) for 15 min in order to avoid that gold reagents detach from sections due to the low pH in which silver amplification reaction occurs.

Silver-Enhancement of Lectin-Gold Cytochemistry

Silver-Enhancement

With the photochemical method, invisible but latent gold particles are visualized because of the reduction on the gold particle surface of silver ions to methallic silver by the action of hydroquinon. The photochemical silver staining is performed as follows (Taatjes et al. 1987) :

Solution I
12 vol distilled water
 2 vol citrate buffer (0.5 M, pH 3.5)
 3 vol hydroquinon (0.85 g in 15 ml of distiled water)
 2 vol distilled water.

Solution II
12 vol distilled water
 2 vol citrate buffer (0.5 M, pH 3.5)
 3 vol hydroquinon
 2 vol silver lactate (0.11 g in 15 ml distilled water)

Hydroquinon is dissolved in a very clean light-protected recipient. Since silver lactate solution is photosensible, it should be dissolved only in the dark room. In the silver staining procedure, all the steps are performed in the dark room with the safe light on. Glass slides containing cytochemically labelled sections are equilibrated for 5 min in solution I. Meanwhile, previously weighted silver lactate is dissolved in distilled water in a very clean glass recipient. The soltuion should be translucid. The corresponding part of the dissolved silver lactate is added to the solution II and, immediately, sections are immersed in such a solution for 2–4 min. Sometimes, a black spot around the sections develops in a time-dependent manner. Discard the solution when it becomes dark and rinse quickly with distilled water (2–3 times). Finally, sections are placed on a photographic fixative (diluted 1:10) for 2 min, and washed with distilled water before dehydration and mounting. After the silver amplification, gold labeling is visualized as a black stain (Fig. 3C and D). Recently, a modification of this technique has been described by Roth (1989), where the complete procedure can be performed at daylight as follows:

Solution I
Dissolve 250 mg hydroquinon in 50 ml of citrate buffer (0.5 M, pH 3.5)

Solution II
Dissolve 100 mg silver acetate in 50 ml of double-distilled water

Glass slides containing sections are incubated for 5 min in a solution which contains equal volumes of solution I and double-distilled water. Subsequently, the sections are transferred to a freshly prepared solution which contains equal volumes of solutions I and II. A time-dependent developing reaction occurs (15–20 min), which is controlled under the microscope. Finally, rinse the sections with double-distilled water, fix for 2–5 min, rinse with tap water, dehydrate, and mount.

23.4 Application of Lectin Cytochemistry in Blotting Techniques

Overlay techniques for proteins separated electrophoretically and blotted onto nitrocellulose sheets are useful for the identification and analysis of binding activities of proteins with antibodies and lectins (Rohringer and Holden 1985). Analysis is also performed after spotting material onto nitrocellulose (Brada and Roth 1984). This technical approach is frequently utilized to evaluate the effects of diverse fixation protocols with respect to carbohydrate-binding properties of the lectin(s). Gold probes have been successfully applied besides more classical approaches as radio-activity- and enzyme-conjugated reagents. The following protocol applies for analysis of spot blots:

1. Apply protein to nitrocellulose (or other membrane).
2. Wash blot with 0.1–0.5% Tween-20 in PBS or TBS for 1–2 h at room temperature under constant agitation. Polyvinylpirrolidone (1–2%) and lysozyme (2% in 0.15 M NaCl) can also be used as blocking agents.
3. First determine whether the blotted protein binds uncomplexed colloidal gold nonspecifically.
4. If no inert gold binding is observed, incubate with the lectin-gold complex or uncomplexed lectin diluted in PBS/Tween-20 or blocking agent according the optimal conditions described in Tables 3 and 4.
5. Wash blot 3×10 min with PBS/Tween-20 under constant agitation.
6. Wash 2×10 min with PBS and fix with glutaraldehyde (1% in PBS) for 15–30 min.
7. Wash 2×5 min with PBS and, finally, 2×3 min with distilled water. Blot is air-dried.

Usually, pink-red stained bands or spots appear. However, when the gold signal is weak or virtually absent, the silver amplification technique (under the same conditions as described under Light Microscopy) is recommended to obtain more sensitivity (Brada and Roth 1984; Moeremans et al. 1984).

23.5 Control Reactions

The specificity of lectin histo- and cytochemistry should be carefully controlled, since lectins are not exclusive in their carbohydrate-binding specificities. A list of the main control reactions is given.

a) The addition of inhibitory and noninhibitory sugars or oligosaccharides (0.01–0.5 M) to the lectin-gold complex 30–45 min before the incubation of sections.   **Direct Method**
b) Incubation of the sections with an excess of the unlabeled lectin (30–45 min) before the application of the lectin-gold complex.

Table 5. Optimal Conditions for the Glycohydrolase Treatments of Semithin and U-trathin Tissue Sections

Enzyme	Enzyme source	Concentration (U/ml)	Incubation buffer	Time (h) and temperature (°C)
A. Exoglycohydrolases				
α-N-Acetylgalactosaminidase	Chicken liver	1	0.1 M citrate, pH 4.5	18, 37 (1)
β-N-Acetylhexosaminidase	Jack bean	5	0.05 M citrate, pH 5	20, 37 (2)
α-L-Fucosidase	Bovine kidney	1	0.1 M citrate, pH 5	20, 37 (2)
β-Galactosidase	Green coffee beans	2–2.5	3.2 M (NH_4) SO_4, pH 6	20, 37 (2)
	Recombinant *E. coli*	50	PBS	6, 22 (3)
β-Galactosidase	*Aspergilus cryzae*	150	PBS	6, 30 (3)
	Charonia lampas	0.5–1	0.2 M citrate, 0.5 M NaCl	20, 37 (2)
	Diplococcus pneumoniae	0.5	0.1 M acetate, pH 4.5	20, 37 (4)
Neuraminidase	*Arthrobacter ureafaciens*	0.25	0.1 M acetate, pH 5.2	20, 37 (2)
	Clostridium perfingens	1–4	0.1 M acetate, pH 4.2–5	2–20, 37 (5)
B. Endoglycohydrolases				
Endo-α-N-acetylgalactosa-minidase	*Diplococcus pneuminae*	0.17	0.05 M phosphate buffer, pH 7	20, 37 (2)
Endo-α-N-acetylglucosami-nidase	*Streptomyces griseus*	0.001–0.2	0.5 M citrate, pH 5.5	18, 37 (4,6)
PNGase F	*Flavobacterium meningosepticum*	60	0.2 M TRIS-HCl, 2.5 mM EDTA pH 8.6	18, 37 (4)

(1) Egea et al. (1993); (2) Ito et al. (1989); (3) Taatjes et al. (1990); (4) Egea et al. (1989); (5) Roth et al. (1984); (6) Lucocq et al. (1987)

Indirect Method   a) Incubation of the sections with the unlabeled lectin absorbed previously with inhibitory and noninhibitory sugars followed by the glycoprotein-gold complex.
b) Incubation with the glycoprotein-gold complex alone.
c) Incubation with unlabeled glycoprotein between the unlabeled lectin and the glycoprotein-gold complex.

Enzymetic Digestions   It is possible to dissect the glycoconjugate structure by determination of lectin binding before and after the action of glycohydrolases over sections (paraffin, frozen, semithin and ultrathin). This methodology also allows one to identify the next terminal sugar. However, the amount of enzyme needed to cleave saccharides moieties is higher than established for biochemical uses (Table 5).

1. Sections are incubated with the appropriate enzyme buffer for 10 min at room temperature or 37 °C.
2. Incubation with the glycohydrolase according the manufacture's instructions or original biochemical conditions with respect to buffer, pH, temperature and cation(s) requirements. Usually, ten fold higher enzyme concentration is used at histological level (Table 5).
3. Wash the sections extensively with distilled water (4 × 5 min), and finally with PBS before the direct or indirect lectin staining method (Fig. 3E and F).

Chemical Treatments   Oligosaccharides can also be removed by chemical treatments as periodic acid (1%) for 15 min (Versura et al. 1989), and 0.1–0.2 N NaOH for 12–36 h at 37 °C

(β-elimination) to release and O-linked glycans from Lowicryl K4M semithin sections (Egea et al. 1989).

Acknowledgments. I apologize to those contributors whose work is referred to indirectly due to constraints in the number of references. I thank F.X. Real (IMIM) for critical reading of the manuscript and for his continuous support, and Roser Badía for help in the preparation of the manuscript. Part of the work described here was done while the author was recipient of an EMBO Long Term Fellowship in the laboratory of Jürgen Roth at the Biozentrum of Basel University (Basel, Switzerland). The work in Spain was supported by CICYT grant to G.E. (PM 89-0059) and CICYT (88/1795) and FIS (89/456) grants to F.X.R.

References

Andersson-Forsman C, Pinto da Silva P (1988) Label-fracture of cell surfaces by replica staining. J Histochem Cytochem 36: 1413–1418

Bendayan M, Zollinger M (1983) Ultrastructural localization of antigenic sites on osmium-fixed tissues applying the protein-A-gold technique. J Histochem Cytochem 31: 101–109

Bendayan M (1985) The enzyme-gold technique: a new cytochemical approach for the ultrastructural localization of macromolecules. In: Bullock GR, Petrusz P (eds), Techniques in Immunocytochemistry, vol 3, Academic Press, London, pp 179–201

Benhamou N (1989) Preparation and application of lectin-gold complexes. In: Hayat MA (ed), Colloidal gold: principles, methods, and applications, vol 1. Academic Press, San Diego, pp 96–145

Benhamou N, Gilbou-Garbes N, Trudel J Asselin A (1988) A new lectin-gold complex for utlrastructural localization of galacturomic acid. J Histochem Cytochem 36: 1403–1411

Brada D, Roth J (1984) "Golden blot" – Detection of polyclonal and monoclonal antibodies bound to antigens on nitrocellulose by protein-A-gold complexes. Anal Biochem 142: 79–83

Carlemalm E, Garavito RM, Villiger W (1982) Resin development for electron microscopy and analysis of embedding at low temperature. J Microsc (Oxf) 126: 123–143

Danscher G, Nötgaard RJO (1983) Light microscopic visualization of colloidal gold on resin-embedded tissue. J Histochem Cytochem 31: 1394–1399

Egea G, Goldstein IJ, Roth J (1989) Light and electron microscopic detection of (3Galβ1, 4GlcNAcβ1) sequences in asparagine-linked oligosaccharides with the *Datura stramonium* lectin. Histochem 92: 515–522

Egea G, Marsal J (1992) Carbohydrate patterns of the pure cholinergic synapse of Torpedo electric organ: a cytochemical and immunocytochemical electron microscopic approach. J Histochem Cytochem 40: 513–521

Egea G, Franci C, Gambius G, Lessufleur T, Zwaibaum A, Real FX (1993) as-Golgi verident motens and O-glycans are abnormally compartmentalized in the RER of human colon cancer cells. J Cell Sci (in press)

Frens G (1973) Preparation of gold dispersions of varying particle size: controlled nucleation for the regulation of the particle size in monodisperse gold suspensions. Nature (Phys Sci) 241: 20–22

Geophegan WD, Ackerman GA (1977) Adsorption of horseradish peroxidase, ovomucoid and anti-immunoglobulin to colloidal gold for the indirect detection of concanavalin A, wheat germ agglutinin and goat anti-human immunoglobulin G on cell surfaces at the electron microscopie level: A new method, theory and application. J. Histochem. Cytochem 25: 1187–1200

Goldstein IJ, Hughes RC, Monsigny M, Osawa T, Sharon N (1980) What should be called a lectin? Nature 285: 66

Holgate CS, Jackson, P, Cowen PN, Bird CC (1983) Immunogold-silver staining: a new immunostaining with enhanced sensitivity. J Histochem Cytochem 31: 938–944

Horrisberger M (1985) Gold method applied to lectin cytochemistry in transmission and scanning electron microscopy. In: Bullock GR, Petrusz P (eds), Techniques in immunocytochemistry, vol 3, Academic Press, London, pp 155–178

Horrisberger M (1989) Colloidal gold for scanning electron microscopy. In: Hayat MA (ed) Colloidal gold: principles, methods, and applications, vol 1. Academic Press, San Diego, pp 217–229

N, Nishi K, Nakajima M, Okamura Y, Hirota T (1989) Histochemical analysis of the chemical structure of blood group-related carbohydrate chains in serous cells of submandibular glands using lectin staining and glycoxidase digestion. J Histochem Cytochem 37: 1115–1124

Kan FWK, Pinto da Silva P (1989) Label fracture cytochemistry. In: Hayat MA (ed) Colloidal gold: principles, methods, and applications, vol 2. Academic Press, San Diego, pp 175–201

Lucocq JM, Baschong W (1986) Preparation of protein colloidal gold complexes in the presence of commonly used buffers. Eur J Cell Biol 42: 332–337

Lucocq JM, Roth J (1984) Applications of immunocolloids in light microscopy. III. Demonstration of antigenic and lectin-binding sites in semithin resin sections. J Histochem Cytochem 32: 1075–1083

Lucocq JM, Roth J (1985) Colloidal gold and colloidal silver-metallic markers for light microscopic histochemistry. In: Bullock (GR, Petrusz P (eds), Techniques in immunocytochemistry, vol 3, Academic Press, London, pp 203–236

Lucocq JM, Berger EG, Roth J (1987) Detection of terminal N-linked GlcNAc-residues in the Golgi apparatus using galactosyltransferase and endoglucosaminidase F/peptide N-glycosidase F: adaptation of a biochemical approach to electron microscopy. J Histochem Cytochem 35: 67–74

Moeremans M, Daneels G, van Dijk A, Langanger G, de Mey J (1984) Sensitive visualization of antigen antibody reaction in dot and blot immune-overlay. J Immunol Methods 74: 353–360

Nicolson GL (1978) Ultrastructural localization of lectin receptors. In: Koehler JK (ed) Advanced techniques in biological electron microscopy, vol 2, Springer-Verlag, New York, pp 1–38

Perruzo B, Rodríguez, EM (1989). Light and electron microscopical demonstration of concanavalin A and wheat-germ agglutinin binding sites by use of antibodies against lectin or its label. Histochemistry 92: 505–513

Pinto da Silva P, Kan FWK (1984) Label-fracture: a method for high resolution labelling of cell surfaces. J Cell Biol 99: 1156–1161

Rohringer R, Holden DW (1985) Protein blotting: detection of proteins with colloidal gold, and of glycoproteins and lectins with biotin-conjugated and enzyme probes. Anal Biochem 144: 118–124

Roth J (1978) The lectins: molecular probes in cell biology and membrane research. Exp Pathol Suppl 3: 5–186

Roth J (1983a) The colloidal gold marker system for light and electron microscopic cytochemistry. In: Bullock GR, Petrusz P (eds), Techniques in immunochemistry, vol 2, Academic Press, London, pp 216–284

Roth J (1983b) Application of lectin-gold complexes for electron microscopic localisation of glycoconjugates on thin sections. J Histochem Cytochem 31: 987–999

Roth J (1983c) Applications of immunocoll0ids in light microscopy. II Demonstration of lectin-binding sites in paraffin sections by the use of lectin-gold and glycoprotein-gold complex. J Histochem Cytochem 31: 547–555

Roth J (1987) Light and electron microscopic localization of glcycoconjugates with gold-labeled reagents. Scanning Microsc 2: 695–703

Roth J (1989) Postembedding labeling on Lowicryl K4M tissue sections: detection and modification of cellular components. In: Tartakoff AM (ed) Methods in cell biology, vol 31. Academic Press, San Diego, pp 514–553

Roth J, Binder M, Gerhard UJ (1978) Conjugation of lectins with fluorochromes: an approach to hiotochemical double labelling of carbohydrate components. Histochem 56: 265–273

Roth J, Lucocq JM, Charest PM (1984) Light and electron microscopic demonstration of sialic acid residues with the lectin from Limax flavus: a cytochemical affinity technique with the use of fetuin-gold complexes. J Histochem Cytochem 32: 1167–1176

Sata T, Lackie PM, Taatjes DJ, Peumans W, Roth J (1989) Detection of the NeusAc (α2,3) Gal (β1,4) GlcNAc sequence with the leukoagglutinin from Maackia amurensis: light and electron microscopic demonstration of differential tissue expression of terminal sialic acid in α2,3-and α2,6-linkage. J Histochem Cytochem 37: 1577–1588

Sata T, Zuber C, Runderle SJ, Goldstern IJ, Roth J (1990) Expression patterns of the T antigen and the cryptic T antigen in rat fetuses: detection with the lectin amaranthin. J Histochem Cytochem 38: 763–774

Sata T, Zuber C, Roth J (1990b) Lectin-digoxigenin conjugates: a new hapten system per glycoconjugate cytochemistry. Histochem 94: 1–11

Skutelsky E, Bayer EA (1979) The ultrastructural localization of cell surface glycoconjugates: affinity cytochemistry via the avidin-biotin complex. Bio Cell 36: 237–252

Slot JW, Geuze HJ (1985) A new method of preparing gold probes for multiple-labelling cytochemistry. Eur J Cell Biol 38: 87–93

Spicer SS, Schulte BA (1992) Diversity ofl cell glycoconjugates shown histochemically: a perspective. J Histochem Cytochem 40: 1–38

Springal RD, Hacker GW, Grimelius L, Polak JM (1984). The potential of the immunogold-silver staining method for paraffin sections. Histochem 81: 603–612

Taatjes DJ, Schaub U, Roth J (1987) Light microscopical detection of antigens and lectin binding sites with gold-labelled reagents on semi-thin Lowicryl K4M sections: usefulness of the photochemical silver reaction for signal amplification. Histochem J 19: 235–245

Taatjes DJ, Roth J, Peumans W, Goldstein IJ (1988) Elderberry bark lectin-gold techniques for the detection of Neu5Ac(α2,6) Gal/GalNAc sequences: applications and limitations. Histochem J 20: 478–490

Taatjes DJ, Barcomb LA, Leslie KO, Low RB (1990) Lectin binding patterns to terminal sugars of rat lung alveolar epithelial cells. J Histochem Cytochem 38: 233–244

Tokuyasu KT (1986) Application of cryoultramicrotomy to immunohistochemistry. J Microsc (Oxf) 143: 139

Versura P, Maltarello MC, Bonvicini F, Laschi R (1989) The lectin-gold technique: an overview of applications to pathological problems. Scanning Microsc 3: 605–621

24 Neoglycoprotein-Gold Complexes as a Tool in the Study of Carbohydrate-Specific Binding Proteins

V. Kolb-Bachofen and C. Egenhofer

Labeling of ligands or antibodies by using colloidal gold particles has become a widely used technique. Initially, this technique was introduced into electron microscopy exploiting the electron density of these particles (Faulk and Taylor 1971; Horrisberger and von Lanthen 1979). Since then, gold labeling has in addition been applied in Western or dot blots (Brada and Roth 1984), in photometric measurements (Teradeira et al. 1983), and in light microscopy (Geoghegan et al. 1978; Holgate et al. 1983) also in connection with in situ hybridization experiments (Wolber and Beals 1989). The latter applications exploit the dark red color of the colloids and the possibility to increase the signal by the silver enhancement technique (Moeremans et al. 1984), which has led to a very high sensitivity, so that in Western blots detection levels are in the picogram range (Holgate et al. 1983; Moeremans et al. 1984; Springall et al. 1984).

24.1 Preparation of Colloidal Gold Particles

Colloidal gold particles can be purchased from an increasing number of companies. Since most of the generally used sizes are easy and inexpensive to prepare, recipes are given below (for review, see also the article of J. Roth in *Techniques in Immunocytochemistry*, 1983b).

Colloidal gold preparations can be stored for very long times, sterile conditions provided. We usually autoclave the freshly prepared batches and remove aliquots under a clean bench with sterile pipets. Otherwise, growth of fungi may occur, which appears as a black "precipitate" in the flasks.

Neither uncoated nor coated gold colloids can be frozen. The resulting loss of its colloidal nature is easily recognized as a complete loss of its dark red color.

Take care when measuring the pH in an uncoated (stabilized) gold colloid: the colloid will immediately destroy the electrode. Therefore use pH-paper strips or sticks only.

Materials   Gold sol batches are prepared in glass vessels. These should be clean and siliconized prior to use.

- Tetrachloroauric acid $[H(AuCl_4) \cdot 4H_2O]$ for example from Merck, Darmstadt, is sealed by the manufacturer in a glass ampule; prepare a 5% (w/v) stock solution with double distilled water.
- Potassium carbonate (K_2CO_3, water-free), 0.2 M should be freshly prepared.
- A saturated ether solution of white phosphorus (for the Au_5) is prepared by giving small pieces of white phosphorus sticks (cut under water) to absolutely water-free diethyl ether (p.a. grade). A saturated solution is ready after overnight incubation at room temperature. From this a 1:5 dilution with diethyl ether is finally used.

– Trisodium citrate ($C_6H_5Na_3O_7 \cdot H_2O$) is stored as a 1% solution. Where no manufacturers are indicated, chemicals were obtained from Merck, Darmstadt or Sigma, Taufkirchen, both FRG.

Preparation of Gold Particles 15 to 17 nm in Diameter (Au$_{17}$)
(Frens et al. 1973)

1. Mix together

100.0 ml	double distilled water
0.2 ml	5% $H(AuCl_4)$

Heat under constant stirring to boiling

2. Add

4.0 ml	1% sodium citrate solution

Heat and stir until the solution has become deep red (about the color of a good burgundy).

Particles with a narrow size distribution of 15 to 18 nm in diameter result from this procedure.

With exactly the same procedure, but adding less sodium citrate, larger particles are prepared: for particles 50 nm in diameter add 0.5 ml instead of 4 ml 1% sodium citrate.

Gold Particles (Au$_{17}$)

Preparation of Gold Particles 5 nm in Diameter (Au$_5$)
(Faulk and Taylor 1971)

1. Mix together

120.0 ml	double distilled water
0.3 ml	5% $H(AuCl_4)$ solution
0.7–0.8 ml	0.2 M K_2CO_3 (to adjust pH $\approx$ 7.2, use strip or stick, see above)

2. Add rapidly

1 ml	1:5 diluted white phosphorus in ethyl ether (see above)

Stir ($\approx$ 15 min) at room temperature and heat the solution slowly to boiling with constant stirring. Boil for app. 10 min until procedure results in the development of the wine-red color as an indication of the reduction process being complete. Particles of sizes 3 to 12 nm are produced.

Gold Particles (Au$_5$)

Preparation of Gold Particles 2 nm in Diameter (Au$_2$)
(Baschong et al. 1985)

1. Mix together under stirring

97.5 ml	double distilled water
0.2 ml	5% $H(AuCl_4)$
1.5 ml	0.2 M K_2CO_3
0.6 ml	1 M thiocyanate solution.

Gold Particles (Au$_2$)

When this mixture develops a yellowish color over the next 15–30 min, the solution is left for 15 h at 22 °C in the dark to complete the reaction, as indicated by a brownish color.

The final colloid is reported to be remarkably uniform in size (2.5 nm mean diameter; coefficient of variation is 15%).

24.2 Protein Adsorption (Coating) to the Colloidal Particles

The resulting colloids consist of electronegative particles, which is responsible for the colloid stability due to particle repulsion. Addition of electrolytes will cause flocculation (particle aggregation) and this process is accompanied by a change in color from red to blue, or in turbidity in the case of the Au_2.

The electronegativity can be exploited to adsorb biological macromolecules to the particles, thereby coating their surface. Since the protein-gold complex is formed mainly by van der Waals forces, formation of the complex is pH-dependent, with stable complexes formed best around the isoelectric point of the respective macromolecule (Geoghegan and Ackerman 1977).

Successful coating of the colloidal particles stabilizes the particles against flocculation by electrolytes (Zsigmondy 1901). Thus, addition of NaCl after molecule adsorption and monitoring for changes in color are a useful control to ensure complete and stable coating of the gold colloid.

The coupling procedure for using neoglycoproteins, e.g., bovine serum albumin with chemically introduced carbohydrate groups (Lee and Lee 1980), will be described in detail. With variations depending mostly on molecular weight and isoelectric points of the respective molecule, the recipe given also holds true for most (glyco) proteins and polysaccharides as further useful tools in the study of carbohydrate-specific binding.

24.2.1 Preparation of Neoglycoprotein-Gold Complexes

The following procedure describes in detail the preparation of glycated BSA-Au_{17} complexes as this is done in our laboratory. Variations needed for the preparation of other complex sizes or other coating substances with the various sizes will then be described.

Procedure

1. Fill into siliconized polycarbonate centrifuge tubes (45 ml sizes)

   20 ml Au_{17} sol (prepared as described above)

   Adjust the pH to around 6.0 (attention, use strips or sticks) by addition of freshly prepared K_2CO_3 (0, 2 M)

2. Add to each tube

   30 µl   glycated BSA stock solution (5 mg/ml)

   Mix the solution

3. Check saturation of coating and stable complex formation by NaCl test. Fill into a 1.5 ml micro sample tube (Eppendorf, Hamburg)

   50 μl   aliquot of the probe

   Place the centrifuge tubes in the cold

   Add

   5 μl   10% NaCl solution (can be stored)

   Monitor visually for color change into bluish color during the next 20 to 30 min of room temperature incubation of aliquots

   Continue if the aliquots stay reddish after the NaCl test

4. Add to each of the tubes

   200 μl 1% polyethylene glycol in aqua dest. for further stabilization of complexes, be sure to use PEG 20 (MW 20 000, for instance from Merck, Darmstadt), filter this solution through a 0.2 μm sterile filter; only freshly prepared solutions should be used (Horrisberger 1979)

5. Fill to the centrifuge tubes

   20 ml   phosphate-buffered saline (PBS)

   Check pH and incubate the solution for 30 min at 4 °C. Centrifuge to sediment the coated particles, e.g., 45 min at 21 000 g and 4 °C

   The gold now forms a black to deep red pellet and the supernatant may contain excess glycated BSA

6. Discard the supernatant, resuspend the pellet and fill tubes again with PBS plus polyethylene glycol as above. Centrifuge a second time

7. We prepare a stock solution by resuspending the pellet from four tube in 1 ml PBS (no PEG needed) and 10 μl pencillin-streptomycin solution as used in cell culture (for instance from Gibco Life Technologies, Eggenstein, FRG)

24.2.2 Coating Other Sizes of Gold Particles

Both the amount of protein needed for complex stabilization as well as the centrifugation conditions change with particle sizes. The smaller the particle size, the higher the overall surface area per ml of sol and the more protein is needed per ml for successful coating.

1. For 20 ml Au_{50} sol add 15 to 20 μl of glycated BSA (5 mg/ml) centrifuge for 30 min at 6000 g for sedimentation. Final stock solution may sediment slowly during storage.   **Procedure**
2. For 20 ml Au_5 sol add 50 to 60 μl of the glycated BSA solution (5 mg/ml). Centrifuge for 45 min at 100 000 g, 4 °C for sedimentation of coated particles; a red to black pellet is formed.
3. For 5 ml Au_2 sol add 50 μl of the glycated BSA solution (5 mg/ml). Centrifuge for 30 min at 150 000 g, 4 °C for sedimentation of particles, a brownish loose pellet results.

24.3 Application Example of Neoglycoprotein-Gold Complexes

An ever-increasing list of applications for these gold complexes can be made (Kolb-Bachofen in *Methods in Enzymology*, 1989). We will list briefly some of the methods where these gold particles are advantageous.

24.3.1 Electron Microscopy

This was the original application. Here the electron density of the metal is exploited for imaging binding sites, ligand endocytosis, etc. This has been described in great detail in a number of good reviews and books (Roth in *Techniques in Immunocytochemistry*, 1983b; Handley and Chien 1987).

24.3.2 Light Microscopy

Due to the red color, the gold complexes have also been used in light microscopy, where, for instance, accumulation of ligand within macrophages leads to red to black vacuoles easily recognized under the light microscope (Roth 1983a,b). The use in light microscopy, however, was restricted prior to the introduction of the silver enhancement technique (Holgate et al. 1983; De Waele et al. 1986). With this addition, labeling intensity and thus sensitivity is enlarged, so that in certain conditions it appears to be the most sensitive labeling technique available. Silver enhancers are commercially available as kits (for instance Aurion R-gent, Bio Trend, Köln, FRG) consisting of two solutions, enhancer and developer. Silver enhancement is the last step performed prior to embedding for visualization. The gold labeling protocols are well reviewed, we give the recipe for enhancement:

Procedure
1. Wash gold-labeled sections or cells and fix the label for 10 min at room temperature with 0.5% glutardialdehyde (Servea, Heidelberg, FRG) in phosphate- or Tris-buffered saline.
2. Wash thoroughly with aqua dest, to remove all Cl^- ions.
3. Mix the two silver enhancement solutions 1:1 and use immediately to cover the slide. A brownish color will develop. You may briefly control the staining intensity under the microscope and rinse with aqua dest after the desired result is obtained. Staining is usually complete after 2 to 8 min.

Slides are then processed for visualization following routine protocols. Some embedding resins may lead to fading of the silver signal (dark brown to black). We use Merckoglas (Merck, Darmstadt, FRG).

24.3.3 Western or Dot Blots (Ligand Blotting)

Glycated BSA-gold complexes are also successfully used to detect carbohydrate-specific binding proteins in nitrocellulose sheets as obtained by blotting techniques. Here again, silver enhancement gives a very good detection sensitivity (Moeremans et al. 1984) and is performed exactly as described above.

24.3.4 Other Applications

Since coating onto colloidal gold particles is an easy and cheap technique to immobilize proteins, rendering them centrifugeable, we have used these complexes in a number of additional applications.

Agglutination Assay. Glycated BSA complexes with larger gold particles (Au_{17}, Au_{50}) can be used analogous to latex particles. The presence of lectins in solutions (i.e., during purification) is determined rapidly by monitoring agglutination of NeoBSA-gold particles under the light microscope.

Absorption of Binding Activities from a Mixture. Carbohydrate-specific binding activities can be adsorbed from protein solutions by adding the respective neoBSA-gold particles under binding conditions (ion requirement) and then centrifuging to remove binding protein together with gold particles.

Purification of Binding Proteins. The same procedure as described for adsorption can also be exploited for further purification of binding proteins, provided binding is reversible. Reversibility may be achieved, for instance, by using specific ion requirements for binding. Then impure protein preparations are incubated with neoBSA-gold and centrifuged as above, the pellet resuspended under conditions where the specific ion is absent, thus allowing for detachment of the binding protein from the gold complex. A second centrifugation will remove neoBSA-gold from the suspension, yielding an easy purification step. We have recently used such a procedure to purify a receptor protein from macrophage plasma membranes (Alsdorff and Kolb-Bachofen 1992).

References

Alsdorff K, Kolb-Bachofen V (1992) Isolation and purification of a receptor for C-reactive protein (CRP) from rat macrophages. Eur J cell Biol 57 (Suppl 36): 5

Baschong W, Lucocq JM, Roth J (1985) Thiocyanate gold: small (2–3 nm) colloidal gold for affinity cytochemical labeling in electron microscopy. Histochemistry 83: 409–411

Brada D, Roth J (1984) "Golden Blot"-detection of polyclonal and monoclonal antibodies bound to antigens on nitrocellulose by protein A-gold complexes. Anal Biochem 142: 79–83

De Waele M, De Mey J, Renmans W, Labeur C, Reynaert PH, Van Camp B (1986) An immunogold-silver staining method for detection of cell-surface antigens in light microscopy. J Histochem Cytochem 34: 935–939

Faulk WP, Taylor GM (1971) An immunocolloid method for the electron microscope. Immunochemistry 8: 1081–1083

Frens G (1973) Controlled nucleation for the regulation of the particle size in monodisoperse gold solutions. Nature (London) Phys Sci 241: 20–22

Geoghegan WD, Ackerman GA (1977) Adsorption of horseradish peroxidase, ovomucoid and anti-immunoglobulin to colloidal gold for the indirect detection of concanavalin A, wheat germ agglutinin and goat anti-human immunoglobulin G on cell surfaces at the electron microscopic level: a new method, theory and application. J Histochem Cytochem 25: 1187–1200

Geoghegan WD, Scillian JJ, Ackerman GA (1978) The detection of human B lymphocytes by both light and electron microscopy utilizating colloidal gold labeled anti-immunoglobulin. Immunol Commun 7: 1–12

Handley DA, Chien S (1987) Colloidal gold labeling studies related to vascular and endothelial functions, hemostasis and receptor-mediated processing of plasma macromolecules. Eur J Cell Biol 43: 163–174

Holgate CS, Jackson R, Cowen PN, Bird CC (1983) Immunogold-silver staining: new method for immuno-staining with enhanced sensitivity. J Histochem Cytochem 31: 938–944

Horrisberger M (1979) Evaluation of colloidal gold as a cytochemical marker for transmission and scanning electron microscopy. Biol Cell 36: 253–258

Horrisberger M, von Lanthen M (1979) Multiple marking of cell surface receptors by gold granules: simultaneous localization of three lectin receptors on human erythrocytes. J Microscopy (Oxford) 115: 97–102

Hsu YH (1984) Immunogold for detection of antigen on nitrocellulose paper. Anal Biochem 142: 221–225

Kolb-Bachofen V (1989) Carbohydrate receptor binding using colloidal gold. In: Ginsburg V (ed) Methods in enzymology, vol 179. Academic Press, London New York, pp 111–121

Lee RT, Lee YC (1980) Preparation and some biochemical properties of neoglycoproteins produced by reductive amination thioglycosides containing an ω-aldehydoaglycon. Biochemistry 19: 156–163

Moeremans M, Daneels G, Van Dijck A, Langanger G, De Mey J (1984) Sensitive visualization of antigen–antibody reactions in dot and blot immune overlay assays with immunogold and immuno-gold/silver staining. J Immunol Methods 74: 353–360

Roth J (1983a) Applications of immunocolloids in light microscopy. II. Demonstration of lectin-binding sites in paraffin sections by the use of lectin-gold or glycoprotein-gold complexes. J Histochem Cytochem 31: 547–552

Roth J (1983b) The colloidal gold marker system for light and electron microscopic cytochemistry. In: Bulloch GR, Petrusz P (eds) Techniques in immunocytochemistry, vol 2. Academic Press, New York, pp 217–284

Springall DR, Hacker GW, Grimelius L, Polak JM (1984) The potential of the immunogold-silver staining method for paraffin sections. Histochemistry 81: 603–608

Teradeira R, Kolb-Bachofen V, Schlepper-Schäfer J, Kolb H (1983) Galactose-particle receptor on liver macrophages. Quantition of particle uptake. Biochem Biophys Acta 759: 306–310

Wolber RA, Beals TF (1989) Streptavidin-gold labeling for ultrastructural in situ nucleic acid hybridization. In: Hayat MA (ed) Colloidal gold: principles, methods, and applications, vol 2. Academic Press, London New York, pp 379–396

Zsigmondy R (1901) Z Anal Chem 40: 697–719

25 Lectins and Neoglycoproteins – Attractive Molecules to Study Glycoconjugates and Accessible Sugar-Binding Sites of Lower Vertebrates Histochemically

A. Danguy and H.-J. Gabius

Glycosylation of proteins and lipids is known to produce an individual pattern of the resulting modified products in various types of cells or at different stages of cellular development or disease. This chemical diversity of cellular glycoconjugates provides a potential for a broad range of functional capacity. Knowledge is growing concerning the biological importance of carbohydrate moieties (Rademacher et al. 1988; Slifkin and Doyle 1990; Gabius and Bardosi 1991; Vierbuchen 1991; Spicer and Schulte 1992). Therefore, their identification in situ constitutes an attractive chapter in histochemistry (Kiernan 1990; Lyon 1991). Due to the presence of tissue receptors for such carbohydrate sequences to constitute a production protein-carbohydrate recognition system lectins (Schrevel et al. 1981; Lis and Sharon 1986; Damjanov 1987; Alroy et al. 1988; Danguy et al. 1988; Walker 1988; Spicer and Schulte 1992) and neoglycoproteins (Stowell and Lee 1980; Schrevel et al. 1981; Kataoka and Tavassoli 1984; Monsigny et al. 1988; Danguy et al. 1991; Gabius and Bardosi 1991; Gabius and Gabius 1991) as histochemical reagents have received more and more attention during the past years within the study of both sides of such an interaction.

Our purpose is to present experimental procedures and useful comments using a panel of biotinylated lectins and neoglycoproteins to reveal the presence of specific carbohydrate sequences (by means of lectins) and histochemically accessible carbohydrate-binding structures (by means of neoglycoproteins) in vertebrate tissue sections.

Lectins commonly used as histochemical probes, their source, acronym and binding preference are listed in Table 1. Labeled neoglycoproteins used to evaluate the pattern of cell-associated, endogenous sugar receptors can be classified according to the type of carbohydrate moiety that is attached to the protein backbone.

25.1 Equipment and Reagents

Apart from the normal equipment indispensable to cut tissues and to stain tissue sections, the only special pieces of equipment needed are moist boxes or petri dishes which may contain microscopic slides and maintain them humid. The equipment needed and the methods used to prepare blocks of tissue and to cut sections are well described in standard books of reference (e.g., see Kiernan 1990; Lyon 1991).

Equipment, Solutions

Some attention must be paid to the slides. Indeed gelatin-coated slides are suitable in order to prevent lack of adhesiveness of the sections. Solution used for coating slides:

1. Melt 15 g of gelatin in 500 ml of warm water
2. Dissolve 1 g of chromealun in 220 ml of distilled water
3. Mix the two and add 70 ml of glacial acetic acid and 300 ml of 95% ethanol

Table 1. Biotinylated Lectins Commonly Used and Their Binding Specificities

Lectin: Latin name (common name)	Acronym	Sugar residues or sequences/ oligosaccharide structures binding preference
Dolichos biflorus (horse gram)	DBA	Terminal FP[a] > GalNAcα1, 3GalNAc > GalNAcα1, 3Gal
Glycine maximus (soybean)	SBA	Terminal α,βGalNAc > α,βGal
Helix pomatia (snail)	HPA	Terminal GalNAcα1, 3GalNAc = GalNAcα1,3Galβ1,3GlcNAc
Vicia villosa (hairy vetch)	VVA	Terminal GalNAcα1,3Gal > GalNAcα1,6Gal = GalNAc-serine
Archis hypogaea (peanut)	PNA	Terminal Galβ1,3GalNAc
Griffonia (= *Bandeiraea*) *simplicifolia*	GSA-I	Terminal αGal
Ricinus communis (castor bean)	RCA-I	Galβ1,4GlcNAc > β-Gal > αGal
Sophora japonica (pagoda tree)	SJA	Terminal Galβ1,3GalNAc > Galβ1,3GlcNAc > α,βGalNAc > α,βGal
Maclura pomifera (hedge apple tree)	MPA	Terminal Galβ1,3GalNAc > GalNAcα1, 6Gal
Artocarpus integrifolia (Jackfruit)	JAC	As MPA. Unlike PNA, JAC will bind this structure even in a mono- or disialylated form
Erythrina cristagalli (coral tree)	ECL	Tetraantennary and triantennary oligosaccharides containing four or three N-acetylactosamine branches respectively
Datura stramomium (thorn apple)	DSL	Terminal or internal GlcNAc(β1,4GlcNAc)$_{1-3}$ = Galβ1, 4GlcNAc
Phaseolus vulgaris leukoagglutinin (kidney bean)	PHA-L	GlcNAcβ1,2 Man triantennary complex oligosaccharides
Phaseolus vulgaris erythroagglutinin (kidney bean)	PHA-E	Bisected complex oligosaccharides
Triticum vulgare (wheat germ)	WGA	GlcNAc(β1,4GlcNAc)$_{1-2}$ > β1,4GlcNAc > NeuAc
Succinyl WGA (wheat germ)	S-WGA	GlcNAc(β1,4GlcNAc)$_{1-2}$
Solanum tuberosum (potato)	STL	GlcNAc(β1,4GlcNAc)$_2$β1,4GlcNAc
Griffonia (= *Bandeiraea*) *simplicifolia*	GSA-II	Terminal α,βGlcNAc, glycogen
Lycopersicon esculentum (tomato)	LEL	Oligosaccharides containing poly-N-acetylactosamine
Ulex europaeus (gorse seed)	UEA-I	L-Fucα1, 2Galβ1,4GlcNAcβ1,6
Lotus tetragonolobus (asparagus pea)	LTA	L-Fucα1, 2Galβ1,4 [L-Fucl,3] GlcNAcβ1,6
Aleuria aurantia (orange peel fungus)	OFA	L-Fuc > L-Fucl,3 = L-Fucα1,4
Canavalia eusiformis (jack bean)	Con-A	α-Man > α-Glc
Lens culinaris (lentil)	LCA	Fucosylated core region of bi- and triantennary N-glycosically linked oligosaccharides
Pisum sativum (pea)	PSA	
Galanthus nivalis (snow drop bulb)	GNA	Manα1,3 Man > Manα1,6Man > Manα1,2Man
Limulus polyphenus (horseshoe crab)	LPA	NeuAc
Limax flavus (slug)	LFA	NeuAc
Sambucus nigra (elderberry bark)	SNA	NeuAcα2,6Gal = NeuAcα2,6GalNAc
Maackia amurensis	MAA	NeuAcα2,3Galβ1,4GlcNAc

[a]FP: Forssman pentasaccharide GalNAcα1,3GalNAcα1,3Gall,4Galβ1,4GlcNAc; Man: mannose; Glc: glucose; Gal: galactose; GlcNAc: N-acetylglucosamine; GalNAc: N-acetylgalactosamine; Fuc: fucose; NeuAc: neuraminic acid

4. Store at room temperature
5. Immerse slides for 3–5 s and dry at room temperature
6. Store at room temperature and use as required

– Sugars (glucose, α-methyl-D-mannoside, N-acetyl-D-glycosamine, N-acetyl-D-galactosamine, L-fucose) were purchased from Janssen Chimica (Beerse, Belgium)
– Synthetic medium DPX: from BDH Chemicals, Poole, UK
– 3-3′-diaminobenzidine-tetrahydrochloride (DAB) was obtained from Janssen Chimica (Beerse, Belgium). DAB has been suspected as a carcinogen and should be handled with care
– Peroxidase Vectastain Elite ABC kit Standard PK 6100 was from Vector Laboratories, Inc (Burlingame, CA, USA)
– Avidin-biotin blocking kit SP 2001 (Vector Laboratories)
– Perhydrit tablets 1 g were from Merck (Darmstadt, FRG)
– Phosphate-buffered saline (PBS) is widely used for rinsing and as solvent for reagents in immunohistochemistry. Composition: 0.1 M phosphate buffer, pH 7.3 ± 0.1 1000 ml, sodium chloride 9 g

– Bouin's mixture

Saturated aqueous picric acid	750 ml
Formalin (36–40%)	250 ml
Glacial acetic acid	50 ml

This fixative is now more and more used in glycohistochemistry, because it preserves morphological features very well and does not perceptibly alter the binding pattern of most lectins and neoglycoproteins.

– Neutral buffered formalin

Sodium phosphate, monobasic ($NaH_2PO_4.H_2O$)	4.0 g
Sodium phosphate, dibasic (Na_2HPO_4)	6.5 g

Dissolve both salts in 750 ml of water, add 100 ml of formalin (36–40% HCHO), and make up to 1000 ml with water.

– Carnoy's fluid

Absolute ethanol	60 ml	mix just
Chloroform	30 ml	
Glacial acetic acid	10 ml	before using

Biotinylated Lectins and Neoglycoproteins

A panel of biotinylated lectins with a variety of sugar specificities were purchased from Vector Laboratories, Inc (Burlingame, CA, USA).

Biotinylated lectin kit I	cat. no. BK-1000
Biotinylated lectin kit II	cat. no. BK-2000
Biotinylated lectin kit III	cat. no. BK-3000

Biotinylated lectins from *Sambucus nigra* (SNA), *Maackia amurensis* (MAA), and *Galanthus nivalis* (GNA) were from Boehringer Mannheim (FRG).

A few lectins are exceedingly toxic, but most are harmless. Warnings about toxicity are given in catalogs and on the outsides of containers. The supplier will also provide information about the purity and specificity of each product.

The biotinylated neoglycoproteins were synthesized as outlined elsewhere in detail (Gabius and Bardosi 1991).

Enzyme Treatment. The occurrence of sialic acid was studied by digestion of the sections with neuraminidase from *Clostridium perfringens* (EC 3.2.1.18) (Boehringer Mannheim, FRG) before lectin staining. Sections were incubated for 30 min at 37 °C in a solution of acetate buffer pH: 5.5 containing 0.5 U neuraminidase per ml.

25.2 Experimental Procedures

25.2.1 Tissue Preparation

Preparation   Vertebrate tissues (skin, kidney, salivary glands, stomach, intestine, reproductive tract, etc.) are quickly removed and small fragments are fixed for 24 h at room temperature in Bouin's solution. The tissue specimens are then dehydrated in a graded ethanol series, cleaned in xylene, and embedded in paraffin wax. Sections are cut at a thickness of 5 µm and transferred to gelatin-coated slides. They are deparaffinized in xylene (or toluene) and hydrated through an alcohol series of descending concentration. After deparaffination, the tissue sections are incubated in methanol–0.3% H_2O_2 for 30 min at room temperature to block endogenous peroxidase activity. The sections are washed in distilled water for 5 min, followed by a wash in phosphate-buffered saline (PBS), pH 7.3 $\pm$ 0.1 for 5 min.

25.2.2 Lectin Cytolabeling of High Sensitivity

Procedure   1. The sections are removed from the PBS bath, and the excess solution is wiped from the glass slide with a paper tissue. All the following steps are performed in a moist box with a suitable airtight lid at room temperature. The air inside the box must be saturated with water vapor to prevent drying of the sections

2. In order to block nonspecific binding of biotin/avidin system reagents, the sections are covered with avidin D blocking solution for 15 min. Rinse briefly with PBS, then incubate for 15 min in the biotin-blocking solution. Excess solution is shaken off, and the slide around the tissue is blotted

3. Carefully apply the smallest possible drop of biotinylated lectin-containing solution (10 µg/ml PBS) to cover the sections to be labeled. Wait for 10 min. The optimal concentration of biotinylated lectin may vary between compounds purchased from different companies and those from different production series

4. The sections are rinsed twice in PBS, 5 min each

5. The sections are covered with Vectastain Elite ABC kit (Avidin DH and biotinylated horseradish peroxidase complex H) solution and incubated for 30 min

6. Rinse the slides in two changes of PBS, 5 min each

7. The sections are developed in PBS containing 0.1% (1 mg/ml) 3-3'-diamino-benzidine 4-HCl (DBA) mixed with an equal volume of 0.02% H_2O_2 for 10 min under microscope control to visualize the activity of peroxidase
8. The slides are rinsed with tap water, lightly couterstained with Groat's hematoxylin, cleared in toluene, mounted in the synthetic medium DPX, and coverslipped
9. The sections are examined with a light microscope; the staining intensity of the different structures is graded semi-quantitatively in five categories: 0: no reaction; 0/1: uncertain; 1: weak; 2: strong; 3: very intense

Controls include : (1) omission of the respective lectin; (2) omission of the ABC kit reagents; (3) incubation of the section with lectin solutions to which 0.2–0.3 M of the specific sugar had previously been added

25.2.3 Glycohistochemical Protocol

The various steps within the standardized staining protocol are outlined as follows:

1. As in step (1)
2. After rinsing in phosphate-buffered saline (PBS, pH 7.3 ± 0.1), the sections are preincubated with 0.1% carbohydrate-free bovine serum albumin (BSA)-0.1 M PBS solution for 15 min to block unspecific binding of BSA-biotin derivatives used in the following steps
3. Excess solution is blotted from the slides and the sections are incubated with 25–100 µg/ml of the different biotinylated carbohydrate-BSA conjugates (neo-glycoproteins) dissolved in PBS containing 0.1% carbohydrate-free BSA, for 45 min
4. As in steps (4)–(9). The staining intensity of the different structures was graded, as reported above. Controls to ascertain specificity of neoglycoprotein binding included:

Procedure

(a) Omission of any biotinylated probe to be able to exclude any binding of the kit reagents, e.g., the glycoproteins avidin and peroxidase
(b) Performance of the complete protocol with nonglycosylated, but biotinylated BSA at the same concentration as the labeled neoglycoprotein to be able to exclude any binding by protein-protein interaction
(c) Competitive inhibition experiments using the homologous unlabeled neoglyco-proteins, that were preincubated on the sections to saturate specific binding sites and then applied in a mixture with the labeled neoglycoprotein. Heterologous inhibition with a neoglycoprotein, carrying a carbohydrate residue with no or only a small affinity to the detected binding sites, complements this experimental series

25.3 Results and Comments

Owing to the fact that the type of tissue preparation may affect glycohistochemical binding patterns (Allison 1987; Alroy et al. 1988; Mason and Matthews 1988; Walker 1988; Schumacher et al. 1991), analysis of both paraffin and frozen sections should be performed. Advantages of cutting cryosections are speed and preservation of some lipid constituents, which are extracted during the course of dehydration and embedding in paraffin. Disadvantages are that sections are commonly too thick (e.g., 15–30 µm) for the resolution of structural detail within cells and the histological structures are very poorly preserved. In our hands no clear differences were observed in lectin and neoglycoprotein binding patterns between frozen and Bouin's solution-fixed paraffin-embedded specimens. Therefore, the main emphasis was placed upon fixed material in our studies. Unfortunately, there is not a generally perfect fixative for immuno- and cytochemistry. Some receptors and glycosubstances are well preserved with one but not with another fixative; it is thus imperative to exercise caution.

Optimal results for light microscopy have been achieved on tissues fixed 24 h with Bouin's solution before dehydration and embedment in paraffin. Carnoy fixation served better in infrequent cell types and nearly as well in a number of sites for preserving the reactivity to glycoproteins, but was found to be less effective for preserving lectin–binding capacity, particularly in membranes. Formalin, perhaps the most popular fixative, is apparently not the optimal choice for glycoconjugate histochemistry.

Protocols for staining with lectins and neoglycoproteins have been adapted from those which have proven satisfactory for immunohistochemistry (Sternberger 1979). Accordingly, the choice of marker lies between a fluorescent compound (fluorescein or rhodamine) or an enzyme, such as horseradish peroxidase. Colloidal gold has been used as a third type of label (Skutelsky et al. 1987). However, this last methodology is mainly employed as an ultrastructural marker (Beesley 1989). The advantage of using fluorescent dyes as label is their simple technical aspects, but a major drawback is background fluorescence, which interferes with interpretation of specific staining. Moreover, an additional disadvantage includes the impermanence of such labeled preparations. Our laboratory prefers 3,3'-diaminobenzidine-tetrahydrochloride (DAB) as the chromogen, because the brown granules of DAB polymers are insoluble in organic solvents and are stable, providing a permanent record, when slides are stored. The glycohistochemical method preferred in our laboratory

---

Fig. 1. a Lectin histochemistry for GSA-I binding sites in the epidermis of *Brachydanio rerio* (zebra fish). Mucous cells exhibit strong positive reaction (arrows). **b** Unstained photomicrograph from control section. This section was incubated with a solution containing GSA-I and 0.2 M of the inhibitory sugar (D-galactose). Bouin's solution-fixed paraffin-embedded specimens. *Bars* 10 µm

Fig. 2. a Tissue specimen (5 µm thick) of zebra fish epidermis fixed in Bouin's solution and embedded in paraffin. The section is stained with Con-A. The superficial cells are strongly labeled, the mucous cells are unstained (*arrows*). The other epithelial cells display a moderate labeling. **b** Frozen section of zebra fish epidermis. The binding pattern disclosed with Con-A is similar to that exhibited by the tissue processed in **a**. However, the labeling intensity is heavier and the resolution of structural detail poor due to the section thickness (10 µm). *Arrow* artificial space; *S* fragment of a scale. *Bars* 10 µm

Fig. 3. a Differential interference contrast (Nomarski) micrograph illustrating glycohistochemical localization of N-acetylglucosamine (GlcNAc)-specific binding sites in the kidney of *Pseudemys scripta elegans* (red-eared tortoise). One segment of the nephrotic tubules is strongly reactive (*large arrow*; others are unstained. Glomeruli (*GL*) remain unlabeled. **b** Kidney of control section to prove the specificity of the protein-carbohydrate interaction, the homologous unlabeled neoglycoprotein was employed in a two-step procedure. Section was preincubated with homologous GlcNAc-BSA conjugate to mask sugar-specific receptors. Subsequently, incubation was performed first with a mixture of biotinylated GlcNAc-BSA and unlabeled GlcNAc-BSA conjugate and then with ABC reagents and with chromogenic substrat. Nomarski optics. Tissue specimens were fixed in Bouin's fluid, embedded in paraffin and cut at a thickness of 5 µm. *Bars* 20 µm. All the sections are lightly couterstained with hematoxylin

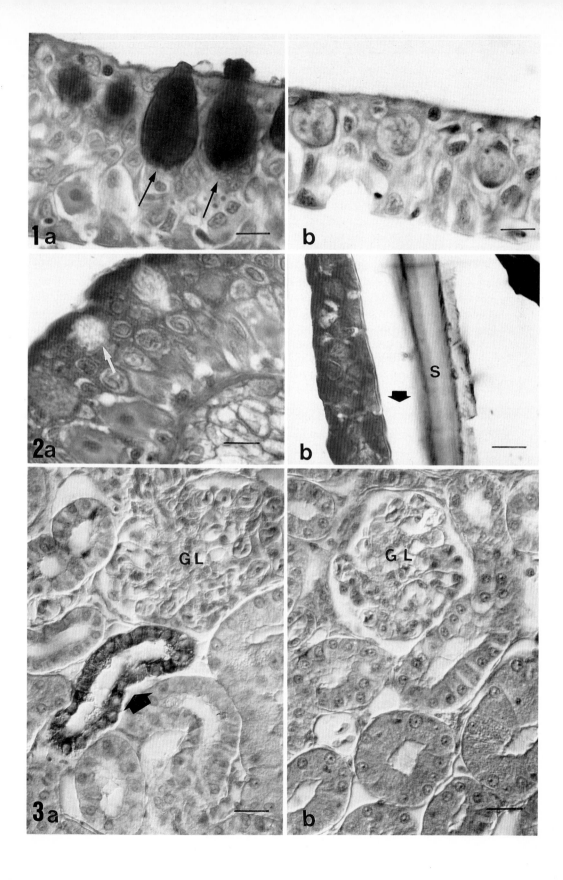

employs both biotinylated lectins and neoglycoproteins as probes and the avidin-biotin-peroxidase complex (ABC). This choice was based largely on experience in light microscopy, as exemplarily documented in Figures 1–3.

Avidin (MW 70 000) is a glycoprotein of egg-white. It has a very high affinity for the small molecule biotin (MW 244) (Green 1975) that can be covalently linked to several types of amino acid side chains. Since a broad spectrum of molecular probes, including lectins, neoglycoproteins, antibodies, and enzymes can be efficiently biotinylated, avidin-based revealing systems represent a widely employed technique for the detection and the visualization of various binding reactions (Hsu et al. 1981; Danguy and Genten 1990; Kiernan 1990). It is notable that the binding of the glycoproteins avidin as well as of horseradish peroxidase via endogenous sugar receptors may occur, namely in the presence of abundant mannose-binding sites (Kuchler et al. 1990).

In contrast to egg-white avidin, streptavidin, isolated from *Streptomyces avidinii*, has an isoelectric point close to neutral pH and contains no carbohydrate residues. Consequently, it has been reported that detection methods based on the biotin-streptavidin system do not suffer from the problems of nonspecific binding associated with egg-white avidin (Chaiet and Wolf 1964; Jones et al. 1987). The pitfall of endogenous biotin in the tissue can be circumvented by pre-exposure of the section to unlabeled avidin followed by biotin (Vector blocking kit) before addition of the biotinylated probe. In other respects, as reported by Jones et al. (1987), the eventual background obtained without the biotinylated probe can be strongly diminished, when the ionic strength of the buffer was raised.

In our laboratory, anyhow, the use of the Vectastain elite ABC kit and the blocking kit gave excellent specific and reproducible results.

Very recently, digoxigenin (DIG) has been introduced as a new hapten labeling system, and applied for glycoconjugate histochemistry (Sata et al. 1990). This steroid has been used to circumvent the necessity of pretreatment of the tissue sections to block the endogenous biotin (Wood and Warncke 1981). DIG occurs naturally only in plants of the digitalis family. Sata and colleagues reported patterns similar to those obtained with other methodologies. We used some DIG-conjugated neoglycoproteins in conjunction with horseradish peroxidase-conjugated anti-DIG antibodies, kindly provided by Dr. Haselbeek, Dr. Hösel, Boehringer Mannheim GmbH, Tutzing, FRG. This procedure has proven comparable in sensitivity and selectivity to the technique using the ABC complex.

The binding of a lectin molecule to a carbohydrate does not involve the formation of covalent bonds. It is similar in nature to the attachment of an antigen to its specific antibody. As is similarly the case with carbohydrate-specific antibodies, the affinity of a lectin is often influenced by the neighboring one to three units in the chain (Goldstein and Poretz 1986); consequently, a particular glycoconjugate may be able to bind different amounts of different lectins of the same group. This statement can also explain why in some cases inhibition of lectin binding may not be completely achievable even by using 0.3 M of the hapten sugar, especially seen with lectins which recognize complex oligosaccharide sequences such as PHA-E and PHA-L.

The reader may probably be surprised by the incubation time with biotinylated lectins, reported here. Our experience has proven that incubating tissue sections for 10 min with the biotinylated lectin purchased from Vector Lab. and Boehringer Mannheim gave the same results, all other parameters being equal, as a longer incubation time (30 min-1 h). Notably, low concentrations of biotinylated lectins, as employed in our laboratory (10 µg/ml), give excellent staining with the ABC system; this allows one to economize in the use of biotinylated reagents and to save money.

Sialyl residues occupy a terminal position in both N- and O-linked oligosaccharides, masking the penultimate β-Gal, Gal-β-1,3-GalNAc or α-D-GalNAc. Unmasking of these penultimate moieties requires cleaving the terminal sialyl residues by neuraminidase (sialidase) digestion. After removal of terminal neuraminic acid residues, the penultimate disaccharides are stainable with PNA or DBA. In situ localization of terminal trisaccharides, such as NeuAc-α-2,6-D-Gal-β-1,3-D-GalNAc, can be performed histochemically with the sialidase pretreatment-PNA sequence.

25.4 Troubleshooting

– Be sure tissue section is kept moist during all steps in the protocols
– Endogenous peroxidase may be developing the chromogenic product: block endogenous peroxidase
– ABC reagent may be binding to tissues for three major reasons

   a) Endogenous protein-bound biotin: use avidin-biotin blocking kit
   b) Endogenous mannose-specific lectin: add 0.2 M α-methylmannoside to the ABC diluent
   c) Ionic interactions: make up ABC reagent in buffer containing 0.5 M NaCl

– For lectins and ABC kit, the supplier's instructions should be followed, exactly
– It is advisable to aliquot biotinylated lectins and neoglycoproteins into suitable quantities and to store the aliquots frozen. Once an aliquot is thawed, it should not be re-frozen, as thawing and freezing may denature them, affecting the performance of this aliquot.

References

Allison RT (1987) The effects of various fixatives on subsequent lectin-binding to tissue sections. Histochem J 19: 65–74

Alroy J, Ucci A, Pereira MEA (1988) Lectin histochemistry: an update. In: DeLellis RA (ed) Advances in Immunohistochemistry, Raven Press, New York, pp 93–131

Beesley JE (1989) Colloidal gold: a new perspective for cytochemical marking. Oxford University Press, Oxford (Royal Microscopical Society Handbook Series, vol 17)

Chaiet L, Wolf FJ (1964) The properties of streptavidin, a biotin binding protein produced by Streptomyces Arch Biochem Biophys 106: 1–5

Damjanov I (1987) Lectin cytochemistry and histochemistry. Lab Invest 57: 5–20

Danguy A, Genten F (1990) Lectin histochemistry on glycoconjugates of the epidermis and dermal glands of *Xenopus laevis* (Daudin 1802). Acta Zool 71: 17–24

Danguy A, Kiss R, Pasteels JL (1988) Lectins in histochemistry: a survey. Biol Struct Morphol 1: 93–106

Danguy A, Genten F, Gabius HJ (1991) Histochemical evaluation of application of biotinylated neoglycoproteins for the detection of endogenous sugar receptors in fish skin. Eur J Bas Appl Histochem 35: 341–357

Gabius HJ, Bardosi A (1991) Neoglycoproteins as tools in glycohistochemistry. Progr Histochem Cytochem 22(3): 66 pp

Gabius HJ, Gabius S (1991) Tumor lectinology: a glycobiological approach to tumor diagnosis and therapy. In: Kilpatrick DC, Van Driessche E, Bog-Hansen TC (eds) Lectin Reviews, vol 1, Sigma Chemical Company, St-Louis, Missouri, pp 91–102

Goldstein IJ, Poretz (1986) Isolation, physiochemical characterization and carbohydrate-binding specificity of lectins. In: Liener IE, Sharon N, Goldstein IJ (eds) The lectins : properties, functions and applications in biology and medicine, Academic Press Inc, New York, pp 33–247

Green NM (1975) Avidin. Adv Prot Chem 29: 85–133

Hsu SM, Raine L, Fanger H (1981) Use of avidin-biotin-peroxidase complex (ABC) in immunoperoxidase techniques : a comparison between ABC and unlabelled antibody (PAP) procedures. J Histochem Cytochem 29: 577–580

Jones CJP, Mosley SM, Jeffrey IJM, Stoddart RW (1987) Elimination of the non-specific binding of avidin to tissue sections. Histochem J 19: 264–268

Kataoka M, Tavassoli M (1984) Synthetic neoglycoproteins. J Histochem Cytochem 32: 1091–1098

Kiernan JA (1990) Histological and histochemical methods. Theory and Practice, 2nd edn. Pergamon Press, Oxford

Kuchler S, Zanetta JP, Vincedon G, Gabius HJ (1990) Detection of binding sites for biotinylated neoglycoproteins and heparin (endogenous lectins) during cerebellar ontogenesis in the rat. Eur J Cell Biol 52: 87–97

Lis H, Sharon N (1986) Lectins as molecules and as tools. Ann Rev Biochem 55: 35–67

Lyon H (ed) (1991) Theory and strategy in histochemistry, Springer Verlag, Berlin

Mason GI, Matthews JB (1988) A comparison of lectin binding to frozen, formalin-fixed paraffin-embedded, and acid-decalcified paraffin-embedded tissues. J Histotechnol 11: 223–229

Monsigny M, Roche AC, Midoux P, Kieda C, Mayer R (1988) Endogenous lectins of myeloid and tumor cells. In: Gabius H.-J, Nagel GA (eds) Lectins and glycoconjugates in oncology, Springer Verlag, Heidelberg, pp 25–47

Rademacher TW, Parekh RB, Dwek RA (1988) Glycobiology. Ann Rev Biochem 57: 785–838

Sata T, Zuber C, Roth J (1990) Lectin-digoxigenin conjugates : A new hapten system for glycoconjugate cytochemistry. Histochemistry 94: 1–11

Schrevel J, Gros D, Monsigny M (1981) Cytochemistry of cell glycoconjugates. Progr Histochem Cytochem 14(2): 269 pp

Schumacher U, Brooks SA, Leathem AJ (1991) Lectins as tools in histochemical techniques : a review of methodological aspects. In: Kilpatrick DC, Van Driessche E, Bog-Hansen TC (eds) Lectin reviews, vol 1, Sigma Chemical Company, St-Louis, Missouri, pp 195–201

Skutelsky E, Goyal W, Alroy J (1987) The use of avidin-gold complex for light microscopic localization of lectin receptors. Histochemistry 86: 291–295

Slifkin M, Doyle RJ (1990) Lectins and their application to clinical microbiology. Clin Microbiol Rev 3: 197–218

Spicer SS, Schulte BA (1992) Diversity of cell glycoconjugates shown histochemically : a perspective. J Histochem Cytochem 40: 1–38

Sternberger LA (ed) (1979) Immunocytochemistry, 2nd edn. Wiley Medical Publ, New York

Stowell CP, Lee YC (1980) Neoglycoproteins. The preparation and application of synthetic neoglycoproteins. Adv Carbohydr Chem Biochem 37: 225–281

Vierbuchen M (1991) Lectin receptors. In: Seifert G (ed) Cell receptors. Morphological characterization and pathological aspects, Springer Verlag, Berlin, pp 271–361

Walker RA (1988) The use of lectins in histology and histopathology. A review. In: Bog-Hansen TC, Freed DLJ (eds) Lectins. Biology, biochemistry, clinical biochemistry, vol 6, Sigma Chemical Company, St Louis, Missouri, pp 591–600

Wood GS, Warncke R (1981) Suppression of endogenous avidin-binding activity in tissues and its relevance to biotin-avidin detection systems. J Histochem Cytochem 29: 1196–1204

26 In Situ Microdensitometry of Neoglycoprotein Staining: Detection of Tissue-Site Specific Patterns of Endogenous Lectin Expression

F. Vidal-Vanaclocha and D. Glaves

In this chapter we introduce a microdensitometric and image analysis procedure for the in situ quantification of endogenous lectin expression by normal and altered cells in different tissues and even at specific sites within a single tissue. The large family of sugar-binding molecules can be detected by biotinylated neoglycoproteins (see Danguy an Gabius, Chap. 25, and Sinowatz et al., Chap. 27, this vol.), using a peroxidase reaction associated to the avidin-biotin interaction. The chromophore concentration precipitated at defined tissue sites and components is proportional to the number of sugar binding sites, and quantification can be done by microdensitometry, i.e., by registrating the optical density (light intensity in negative images) at specific tissue areas (pixels) of digitalized images.

The procedure detailed below describes the basic steps required for this purpose. Illustrations are derived from our application of this quantitative approach to the study of site-associated differences in endogenous lectin expression by mouse colon and lung carcinoma cells (Glaves et al. 1989; Vidal-Vanclocha et al. 1990, 1991).

26.1 Equipment and Materials

Measurements can be made with an Olympus Vanox light microscope or equivalent fitted with a computed-regulated light source. Transmitted light is collected in a videcon (silicon) detector (SIT 66 TV camera, Dage-MTI, Inc., Michigan), with a sensitivity of 0.01 lx. Output from the detector is directed to a microcomputer-integrated, automatic image analysis system (Southern Microcomputer Instruments Inc., Atlanta, GA), and also displayed on a TV monitor. **Equipment**

Normal or neoplastic (Lewis lung carcinoma, colon 26 carcinoma) tissue samples from adult male C57BL/6 and Balb/c mice. **Materials**

Biotinylated neoglycoproteins produced by Dr. H.J. Gabius (Marburg): lactose-BSA, mannose-BSA, fucose-BSA, maltose-BSA, N-acetylglucosamine-BSA, N-acetylgalactosamine, rhamnose-BSA, α-D-glucose-BSA, melibiose-BSA, sialic acid-BSA, heparin, fucoidan, asialotransferrin, asialofetuin) (Gabius et al. 1988).

3-3'-diaminobenzidine (Sigma Chem. Co., St. Louis, MO). Hydrogen peroxide; methanol; ethanol; bovine serum albumin. Vectastain ABC kit (avidin DH and biotinylated horseradish peroxidase complex H; vector Burlingame, CA); hematoxylin and eosin.

26.2 Experimental Procedure

Preparation of Tissue Samples and Lectin Staining

Tissue samples are fixed in 95% ethanol at 4 °C and paraffin wax-embedded at low temperature. Sections are cut at 5 μm and section thickness homogeneity should be **Samples, Lectin Staining** evaluated by analytical interference microscopy (POL optics, Zeiss, Oberkochen, Germany). Following dewaxation and rehydration, endogenous lectins are identified by a modification of the ABC peroxidase method, using carbohydrates which are either directly biotinylated or coupled to biotinylated bovine serum albumin (BSA) neoglycoproteins. The panel of neoglycoprotein probes and their carbohydrate specificities have been described previously (Gabius et al. 1988) and used in some recent works (Vidal-Vanaclocha et al. 1990, 1991).

The staining procedure should be as follows: tissue sections are treated with 0.3% hydrogen peroxide in methanol to inhibit endogenous peroxidase activity, followed by treatment with 0.5% BSA to inhibit nonspecific background staining and then reacted with biotinylated neoglycoproteins at concentration of 50 and 200 µg/ml for 60 min. After washing in 0.05 M Tris-saline (pH 7.4), sections are treated with ABC reagent. Specific sites of reaction of endogenous lectins with these probes can be visualized in tissue sections using the chromogenic substrate, 3-3′-diaminobenzidine, in a peroxidase reaction. Control of nonspecific reaction should involve incubation with biotinylated but nonglycosylated BSA. Matching serial sections should be stained with standard hematoxylin and eosin (H & E) procedures (Fig. 1).

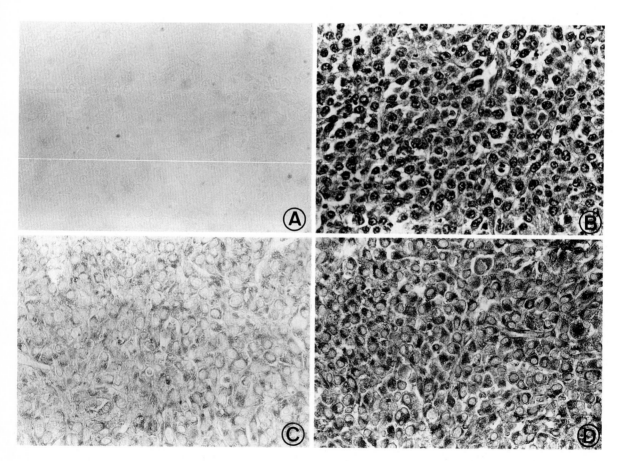

Fig. 1A–D. Histochemical localization of endogenous lectins with binding capacity for heparin (**C**) and fucoidan (**D**) in tissue samples obtained from a subcutaneously growing colon 26 carcinoma. Control sections were stained with biotinylated nonglycosylated BSA (**A**) and hematoxylin and eosin (**B**)

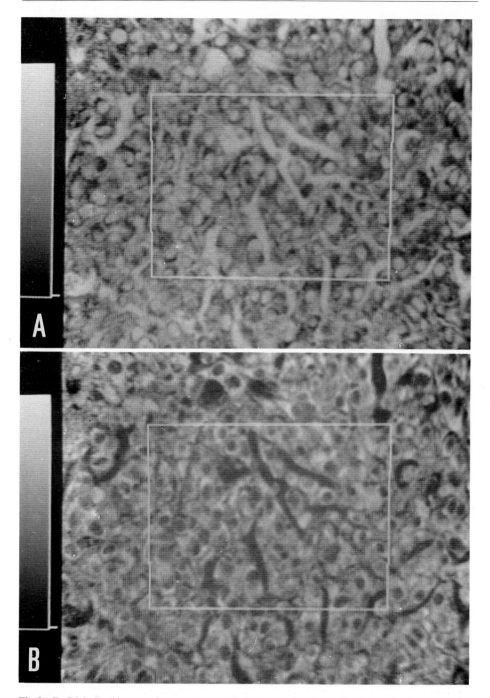

Fig. 2A,B. Digitalized images of a peroxidase-stained tissue section from a Lewis lung carcinoma growing in the kidney. The window used in image analysis corresponds to an approximate field of 3000 μm^2. The *left band* show the black-to-white scale in which all the pixel intensities are distributed. **A** Positive image in 255 gray levels. **B** The same image, after reversal, showing a negative image with 255 intensity levels, ranging from nonstained pixels (*black*) to highly stained pixels (*white*)

Microdensitometrs

Microden-
sitometry

Image processing and quantification of chromophore concentration at different tissues, or at specific sites within the studied tissue is carried out on digitalized images selected from tissue sections. On each section examined, under ×600 magnification, a significant number of morphologically intact tissue fields must be analyzed (at least ten). Each individual field should measure approximately 3000 μm² and contain 9421 pixels (3.14 pixels/μm²). This would allow the simultaneous image analysis of 40 cells as much (in compact tissues). The number of pixels on each digitalized image corresponds to the area of tissue examined, being one single pixel 0.32 μm² of tissue approximately (Fig. 2A).

Following electronic image reversal, a new negative image is obtained, in which the pixel intensity (optical density of the tissue) corresponds to the peroxidase staining intensity, i.e., chromophore concentration (Fig. 2B). Thus, the greater pixel intensity would correspond directly to the tissue structure with highest chromophore concentration. Light intensity at each pixel is recorded on a scale from 0 to 255 arbitrary intensity units. Once densitometry of a selected field has been done, all the recorded light intensity values are represented as a monoparametric pixel distribution by intensity units (Fig. 3). This histogram generates two parameters: integrated intensity (II), which represents the average pixel intensity related to the measured area and is expressed in arbitrary units; and the specific area occupied by each intensity value, which is expressed as percentage of the total area of the histogram. Moreover, the II of a defined intensity interval can be obtained by manually selecting its lower and upper boundaries in the intensity scale.

Before starting any quantitative analysis, it should be necessary that all the intensity values from the studied tissue be in the measurement scale. Thus, the same incident illumination level should be selected and maintained in the microscope for all the studied tissue sections. This would guarantee that even the highest chromophore concentration sites are in the measurement range.

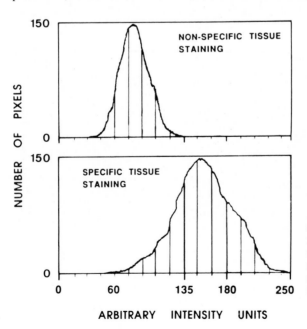

Fig. 3. Distribution of pixels by staining intensity in tissue sections incubated with biotinylated nonglycosylated-BSA (nonspecific tissue staining) and with biotinylated neoglycoproteins (specific tissue staining)

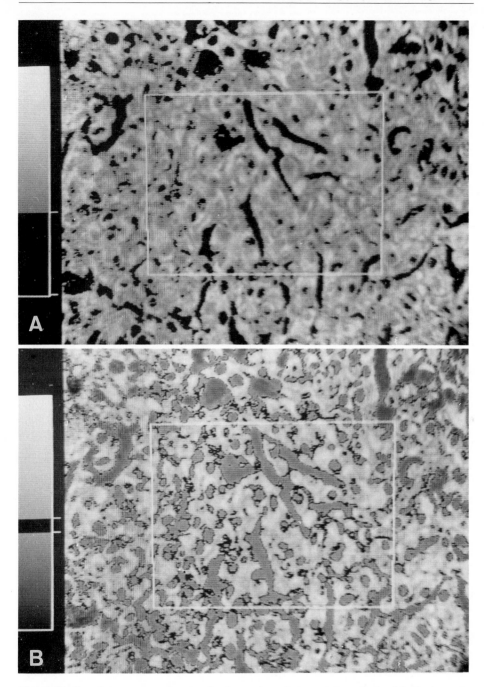

Fig. 4A–D. Complete sequence of a microdensitometric analysis carried out in the digitalized negative image shown in Fig. 2B. **A** The area corresponding to section gaps (black pixels) is defined in the *lower part* of the intensity scale. **B** The upper portion of the nonstaining interval corresponds to the small percent area occupied by extracellular matrix (*black pixels*). **C** The low staining interval corresponds to black pixels located in a significant percent area of cells, and in **D** in the upper part of the intensity scale, the high staining interval (*black pixels*) defines certain scattered cells of the selected field

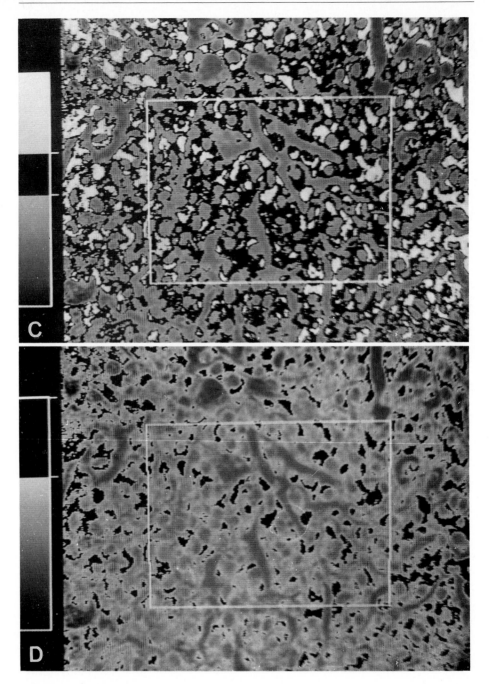

Fig. 4. (continued)

The nonspecific intensity threshold should also be determined for each analyzed tissue. This would allow one to define, in any pixel distribution profile, the percent area corresponding to nonstaining cells, extracellular matrix, and tissue gaps. The intensities of nonspecific reactions should be determined in different tissue sites by incubating tissue sections with nonbiotinylated BSA, in place of biotinylated neoglycoproteins probes. In the distributions of pixels obtained from these control sections (Fig. 3A), the highest values in the intensity distribution of pixels define the nonspecific threshold in the established intensity scale. This low intensity background should be subtracted from all intensity readings (normally this can be done automatically by interactive image analysis).

Below the nonspecific threshold, the noncellular areas corresponding to section gaps and unstained extracellular matrix can be determined by image analysis of H & E-stained sections, using "erosion" techniques (i.e., the manual selection of lower intensity pixels in the scale producing a progressive "erosion" in the analyzed image, with permanence of pixels included in the selected intensity interval) (Fig. 4A). Section gaps can largely vary from one tissue field to another, being normally below 15%. Unstained extracellular matrix tends to have the same proportion in different fields of a defined tissue (Fig. 4B).

Next, on the basis of prior visual inspection, and using the same procedure of intensity interval selection, two consecutive specific tissue staining intensity levels can be defined in the intensity histograms: low-staining interval and high-staining interval (see example shown in Fig. 5). These can then be used in the subsequent, objective interval analysis of integrated intensity distributions. Three classes are then identified on the basis of staining intensity, by microdensitometry, and the area

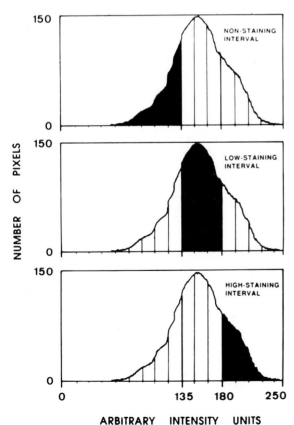

Fig. 5. Microdensitometric distribution of pixels by intensity in the three consecutive selected staining intervals. *Black area* of each distribution corresponds to the selected interval whose integrated intensity and percent area have been measured

occupied by these three classes can be determined by image analysis. The intensity classes are defined in the example of Fig. 5, as follows: (1) tissue structures and cells with specific high intensities (those lying between 165 and 250 AIU); (2) tissue structures and cells with specific low intensities (those lying between 135 and 164 AIU); and (3) tissue structures and cells with nonspecific staining, i.e., with very low to zero-staining intensities (those lying between 0 and 134 AIU).

Once section gaps and unstained extracellular matrix (ECM) have been defined, the specific areas they represent can be subtracted from the calculated nonspecific and specific intensity areas, to give the area occupied by cells with no detectable staining (i.e., "nonstaining cell area"):

Nonstaining cell area = 100 − (gaps area + ECM area)
  − (high intensity cell area + low intensity cell area).

Nonspecific extracellular matrix background and tissue gaps should be eliminated by interactive image analysis. Thus, in each field, the relative areas occupied by three classes of cells (high and low intensity, and nonstaining cells) could be determined. These three intervals in the integrated intensity histograms can be displayed on the negative digitalized images. As shown in Fig. 4C, white pixels correspond to high intensity; black pixels correspond to low intensity and gray pixels to nonspecific intensity. This format summarizes the three staining levels and permits the separate measurement of both integrated intensity classes and the areas occupied by them.

26.3 Concluding Remarks

This analytical approach permits, once tissue gaps and nonstained extracellular matrix have been subtracted, detection and individual quantitation of defined cell groups within a particular tissue, with specific endogenous lectin patterns. Image analysis of cell groups make it possible to assess precisely their detectable level of binding sites for neoglycoproteins (integrated staining intensity) and how much they are representative of the whole tissue (percent area occupied by pixels of a defined staining intensity). This information could go undetected in subjective microscopy, and in assays in which only total expression for whole tissues has been measured, as in many standard biochemical procedures.

Recent application of this quantitative analysis to the study of endogenous lectin expression by cancer cells, at different steps of their malignant progression, and at different sites of development as primary or metastatic tumors (Vidal-Vanaclocha et al. 1990, 1991) has shown significant time- and organ site-induced variations.

In recent years, histology has included the use of markers and molecular biologic probes which are often visualized through peroxidase reactions. Thus, this in situ quantitation of glycohistochemical tools validated here offers the potential for any other measurement of peroxidase reactions associated to molecular probes, where changes apparently can be subtle.

References

Gabius H-J, Bodanowitz S, Schauer A (1988) Endogenous sugar-binding proteins in human breast tissue and benign and malignant breast lesions. Cancer 61: 1125–1131

Glaves D, Gabius H-J, Weiss L (1989) Site-associated expression of endogenous tumor lectins. Int J Cancer 44: 506–511

Vidal-Vanaclocha F, Barberá-Guillem E, Weiss L, Glaves D, Gabius HJ (1990) Quantitation of endogenous lectin expression in 3LL tumors, growing subcutaneously and in the kidneys of mice. Int J Cancer 46: 908–912

Vidal-Vanaclocha F, Glaves D, Barberá-Guillem E, Weiss L (1991) Quantitative microscopy of mouse colon 26 cells growing in different metastatic sites. Br J Cancer 63: 748–752

27 Lectin and Neoglycoprotein Binding to Cells

F. Sinowatz, J. Plendl, H.J. Gabius, Ch. Neumüller, and W. Amselgruber

Lectins and neoglycoproteins have been used extensively in histochemistry, electron microscopy, and flow cytometry. Many techniques exist for each application, utilizing a variety of labels and instrumentation. For example, the use of lectins for light microscopic staining of cells can involve fluorochromes, enzymes, or gold labels and can be accomplished by direct, indirect, sandwich, or avidin-biotin methods (McCoy 1987). Comprehensive surveys of histochemical and cytochemical staining procedures are found in several excellent reviews (Schrevel et al. 1981; Roth 1983; Leathem 1986) and in this book (Danguy and Gabius; Egea; Kolb-Bachofen).

In order to detect lectin binding sites, advantage has also been taken of glycosylated enzymes, such as horseradish peroxidase, the carbohydrate portion of which is well characterized and whose activity leads to the formation of colored products (Straus 1981). Since only a limited number of such naturally occurring tools are available, synthetic markers have been made via chemical coupling of a neoglycoprotein or a carbohydrate derivative to a chosen enzyme under conditions which preserve activity (Gabius et al. 1987a, b, 1989; Naoi et al. 1987; Vorberg and Bundle 1990; Kim et al. 1992). Such neoglycoenzymes provide reasonable alternatives to fluorescent or radioactive probes. Radioiodination is outlined by Kojima (Chapter 31, this Volume).

This report focuses on several specific examples of lectin and neoglycoprotein techniques which can be used to label cells at the light microscopic and ultrastructural level. Information on the quantification of lectin and neoglycoprotein binding sites using flow cytometry is also presented.

27.1 Visualization of Membrane Lectins with Fluorescein-Labeled Neoglycoproteins

Synthesis of neoglycoproteins is described by Lee and Lee (Chap. 2, this volume). Chemically glycosylated BSA and naturally occurring glycoproteins that have been desialylated can be labeled with FITC (Sigma, Munich, FRG) as described by Kieda et al. (1979). The compounds are then separated from free label by gel filtration on Sephadex G-50 (Pharmacia, Freiburg, FRG), dialyzed extensively against water containing 0.9% NaCl and adjusted to a final concentration of 1 mg/ml of the resulting glycosylated BSA-fluorescein thiocarbamyl (FTC) derivatives.

Cell labeling Cell smears on glass slides are allowed to dry at RT for 30 min prior to fixation in a solution containing 6% $HgCl_2$, 1% sodium acetate and 0.1% glutaraldehyde (Schulte and Spicer 1983). Additional smears of unfixed cells are also prepared.

After repeated washing in 0.01 M PBS at pH 7.4, the slides are incubated with 50 µg/ml solutions of the FTC-labeled neoglycoproteins for 30 min at 4° or 37 °C.

The smears are subsequently washed in 0.01 M PBS (pH 7.4), dried, and examined with an epifluorescence microscope.

Controls to verify the specificity of protein-carbohydrate interaction include incu- **Controls**
bation of the slides in the presence of: (a) the corresponding nonfluorescent neoglycoprotein (1 mg/ml) and (b) the related free sugar (0.3 M). Further smears are incubated with FTC-BSA alone.

Results and Remarks

Direct visualization of membrane lectins in situ has been performed in several types of free cells, such as lymphocytes (Kieda et al. 1979) and spermatozoa. Distinct distribution patterns were especially evident in highly polarized cells, such as spermatozoa (Sinowatz et al. 1988). In bovine spermatozoa, pronounced fluorescence was seen in the postacrosomal area after incubation with mannose-BSA-FTC, cellobiose-BSA-FTC, mannose-6-phosphate-BSA-FTC, lactose-BSA-FTC, maltose-BSA-FTC and two asialoglycoproteins, namely desialylated lactoferrin-FTC and transferrin-FTC. The pattern of labeling seen for fluorescent neoglycoproteins is strongly influenced by the mode of fixation (Sinowatz et al. 1988). Best results are obtained using either unfixed cells, or fixation in a mixture containing mercury and glutaraldehyde, as described by Schulte and Spicer (1983).

27.2 Neoglycoenzymes

β-galactosidase from *E. coli*, *p*-aminophenyl glycosides, 1-ethyl-3-(3-dimethyl- **Materials**
aminopropyl)carbodiimide, 20 mM phosphate buffered saline (PBS, pH 7.4), Sepharose 4B or any gel filtration medium to separate glycosylated enzyme from reagents, Hank's balanced salt solution with 20 mM Hepes (pH 7.5) and 0.1% carbohydrate-free bovine serum albumin (Buffer 1), 100 mM Hepes (pH 7.0) containing 0.5% Triton X-100, 150 mM NaCl, 2 mM $MgCl_2$, 0.1% NaN_3, 0.1% carbohydrate-free bovine serum albumin and 1.5 mM chlorophenolred-β-D-galactopyranoside (Buffer 2), 0.2 M glycine (pH 10.5) and an ELISA-microtiter plate reader equipped with a 590 nm filter.

Synthesis of Neoglycoenzymes. Carbodiimide-mediated coupling of *p*-aminophenyl **Synthesis**
glycosides provides a nondestructive means of glycosylating *E. coli* β-galactosidase. After dissolving 8 mg of enzyme in 1 ml PBS at 4 °C, 16 µmol of *p*-aminophenyl glycoside and 32 µmol of 1-ethyl-3-(3-dimethylaminopropyl)carbodiimide are added. The solution is kept at 4 °C for 16 h under gentle stirring, then dialyzed extensively against PBS. Residual reagents are removed by gel filtration on a Sepharose 4 column (1.8 × 60 cm). The combined enzyme-containing fractions are dialyzed against PBS + 60% glycerol and stored at − 20 °C.

Cell Binding of Neoglycoenzymes. Cultured or isolated cells are carefully washed **Binding**
with Buffer 1 to remove any glycoproteins. Incubation of the cells with the neoglycoenzyme (nM quantities) is performed in microtiter plate wells for adhesive cells or in Eppendorf tubes for suspended cells at 4 °C to reduce uptake in 200–400 µl

Buffer 1 for 5–120, when plateau binding has been obtained in preceding experiments. Following this incubation period, the cells are quickly and thoroughly washed to remove unbound neoglycoenzyme and 200 µl of freshly prepared Buffer 2 are added. After 30–60 min of incubation in the dark at 37 °C, the enzymatic reaction is stopped by the addition of 200 µl of 0.2 M glycine (pH 10.5) and quantitated by measurement in a microtiter plate reader at 590 nm.The activity is calculated in fmol enzyme using activity graphs drawn for each individual enzyme preparation. Total binding of the probe to cells is reduced by the extent of nonspecific binding present, as determined using unmodified enzyme on aliquots from the same cell batch in parallel experiments.

Results and Remarks

Neoglycoenzymes facilitate quantification of cell surface receptors for exposed carbohydrate moieties that are located in spatial proximity (Fig. 1). Reduction in enzyme size from tetramer to monomer (β-galactosidase from *Aspergillus oryzae*) shows that a single enzyme can bind to more than one receptor, generating binding of high affinity (Gabius et al. 1990). Thus, the calculated extent of binding (B_{max}) sets the lower limit for actual receptor quantity. Any cell type can easily be monitored, including parasites, whose surface receptors can participate in infection (Schottelius and Gabius 1992). In comparison to fluorescence-activated cell sorting, it should be noted that this method will give an overall assessment without distinguishing subpopulations.

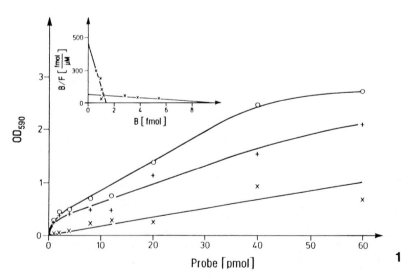

Fig. 1. Determination of specific binding (+) of N-acetyl-D-galactosaminylated *E. coli* β-galactosidase and Scatchard analysis of the binding data, plotted as *insert*, for cells of the human pre-B cell line BLIN-1 (4.3×10^4 cells per assay; $K_D^I = 3$ nM, $K_D^{II} = 142$ nM; $B_{Imax} = 1.9 \times 10^4$ bound enzymes per cell; $B_{IImax} = 1.3 \times 10^5$ bound enzymes per cell). Total binding (○) has to be reduced by the extent to unspecific binding (×)

27.3 Ultrastructural Localization of Lectin-Binding Sites

Tetrachloroauric acid (Merck, Darmstadt, Germany), trisodium citrate; polyethyl-   **Materials**
eneglycol; Carbowax-20 M; lectins; K_2CO_3; acetic acid; cacodylate buffer; osmium
tetroxide; bovine serum albumin; Epon; Araldite.

Preparation of Colloidal Gold. The general conditions under which colloidal gold   **Colloidal Gold**
can be formed and several methods for its preparation are extensively reviewed by
Roth (1983). The most popular reactions used involve reduction, and many of the
colloids prepared in this way have found application as cytochemical markers. In
our laboratory, reduction with trisodium citrate (Frens 1973) gives consistently good
results. The following procedure describes the preparation of colloidal gold with a
particle diameter of approximately 15 nm (Frens 1973). Gold particles of alternate
sizes (from 1 to 40 nm) are available commercially (e.g., BioCell Research Laboratories, Cardiff, UK).
   100 ml of a 0.01% aqueous solution of tetrachloroauric acid are heated in a
siliconized Erlenmeyer flask. Once the solution starts to boil, 4 ml of a 1% aqueous
solution of trisodium citrate are added rapidly. Five additional minutes of gentle
boiling completes the reduction process, as indicated by the appearance of a reddish-
orange color. Removal of free, uncomplexed protein and incompletely stabilized
gold particles is achieved by ultracentrifugation in a fixed angle rotor. Optimal
centrifugation conditions depend upon particle size. In the case of 15 nm gold
particles, $30\,000g$ for 60 min have been recommended (Roth 1983). This leads to the
formation of a loosely packed sediment of intense red color. Above this sediment, a
dark area of densely packed material is formed on the walls of the tube. This material
must be discarded, as it consists of precipitated, aggregated complexes or incompletely stabilized particles. The colorless supernatant contains free protein and must
therefore be aspirated and discarded as well. The protein-gold complexes are
contained in the sediment, which must be resuspended in an appropriate buffer
containing 0.2 mg/ml of polyethylene glycol (MW = 20 000) or Carbowax-20 M. A
second centrifugation may be performed to ensure the complete removal of free
protein.

27.3.1 One-Step Method of Lectin Labeling for Ultrastructural Studies

Preparation of Lectin-Gold Complexes. The lectins are dissolved in distilled water   **Lectin-Gold**
immediately before use and adsorbed onto 15 nm gold particles. According to Roth   **Complexes**
(1983), the following amounts of protein are needed to stabilize 10 ml of colloidal
gold: *Ricinus communis* I and II lectins, 130 µg; peanut lectin, 130 µg; *Helix pomatia*
lectin, 65 µg; soybean lectin, 130 µg; *Ulex europeus* I lectin, 250 µg; and *Bandeiraea
simplicifolia* isolectin B_4, 250 µg. In order for lectin-gold complexes to be formed,
optimal pH conditions for each lectin must be determined (Roth 1983). Table 1
presents data for pH adjustment of colloidal gold. Small quantities of 0.2 M K_2CO_3
or 1 M acetic acid are used for this purpose.

Labeling of Cells with Lectin-Gold Complexes. Either prefixed (glutaraldehyde   **Labeling**
0.5–1%) or unfixed cells in suspension may be labeled. Unfixed cells are washed in

Table 1. Data for adjustment of colloidal gold for some lectins. (Roth 1983)

Lectins	pH
Ricinus communis lectin I	8.0
Ricinus communis lectin II	8.0
Peanut lectin	6.3
Helix pomatia lectin	7.4
Soybean lectin	6.1
Lens culinaris lectin	6.9
Lotus tetragonolobus lectin	6.3
Ulex europeus lectin I	6.3
Bandeiraeae simplicifolia lectin	6.2

0.1 M cacodylate buffer, pH 7.4 (washing buffer) and incubated for either 20–30 min (room temperature or 37 °C) or 30–60 min (4 °C) with occasional slight agitation. The incubation mixture is diluted with washing buffer and cells are centrifuged at approximately $300g$ for 5 min. After removing the supernatant, the pellet is resuspended in washing buffer and recentrifuged twice. Fixation is then performed by adding 0.1% glutaraldehyde in 0.1 M cacodylate buffer, pH 7.4. The lectin-labeled cells are subsequently osmicated for 1 h in 1% OsO_4 in 0.1 M cacodylate buffer (pH 7.4), followed by several washes in the same buffer. Small quantities of these cells are then mixed with a drop of 20% bovine serum albumin (BSA), which is hardened by adding a drop of 25% glutaraldehyde. The specimens are subsequently dehydrated and embedded in Epon/Araldite. Ultrathin sections can be examined without counterstaining.

27.3.2 Ultrastructural Demonstration of Lectin and Neoglycoprotein Binding Sites with a Two-Step Method

The primary reagent (lectin; neoglycoprotein) in the two-step method is unlabeled. Native lectin bound to the cell is detected by incubation with an appropriate glycoprotein-gold complex. In the following, a procedure is described originally developed by Geoghegan and Ackerman (1977) for Con A and WGA-binding sites and successfully modified by Töpfer-Petersen et al. (1985) in order to demonstrate fucose-binding sites in the sperm plasma membrane.

Labeling   *Labeling of Colloidal Gold with Horseradish-Peroxidase (HRP) and with Ovomucoid Can Be Done According to Geoghegan and Ackerman (1977).* Colloidal gold (15 nm), prepared via the reduction of tetrachloroauric acid with 1% trisodium citrate (Frens 1973) as described above, is adjusted to pH 7.2–8.0 (for HRP) or pH 4.8–5.5 (for ovomucoid). 10 ml of either solution can be stabilized by adding 150 µg of HRP or ovomucoid, respectively. Successful absorption of protein onto the gold granules can be checked by negative staining with phosphotungstic acid.

Incubation   *Incubation of Cells.* Unfixed or glutaraldehyde-fixed cells (0.5–1% glutaraldehyde in 0.1 M sodium cacodylate buffer) are incubated with WGA (Serva, 0.25 mg/ml) or Con A (1 mg/ml) for 30 min at 4 °C under slight agitation. The cells are then washed three times with PBS, pH 7.4. An excess of (1–2 ml) gold-labeled glycoprotein (HRP-gold for the demonstration of Con A-binding sites and ovomucoid-gold for WGA-binding sites) is added to the cell suspension. Incubation is carried out at 4 °C for 30 min. The cells are then washed twice in PBS and centrifuged at 1500 rpm. After the first centrifugation, a dark red sediment and comparatively unstained supernatant are seen. This demonstrates that gold-labeled glycoprotein was present in excess. The further steps of cell fixation, embedding, and sectioning for EM observation are the same as those described above.

Results and Remarks

Labeled lectins have been used extensively to demonstrate carbohydrate groups in the membranes of many cell types (Figs. 2–4). Gold granules are particularly useful as probes, for their electron opacity makes them readily identifiable in TEM. This is exemplified by the distribution of Con A-binding sites on bovine spermatozoa using a two-step method according to Geoghegan and Ackerman (1977). Further advantages of colloidal gold are reviewed by Horisberger (1979) and Roth (1983). The most important of these include:

– only small amounts of the relatively expensive lectins are required;
– no reduction in the biological activity of the lectins occurs;
– gold markers show little nonspecific adsorption.

In the one-step method, steric hindrance affects the binding of lectin-labeled gold granules to the cell surface (Horisberger 1979; Roth 1983). This problem is particularly important in SEM studies where large colloidal gold granules are used, but has some significance for TEM studies as well. Absence of marking does not necessarily imply the absence of receptors, since the accessibility of gold markers to receptors is also dependent upon the size of the probe. Gold methods are therefore not considered suitable for the quantification of receptors on the cell surface (Horisberger 1979).

27.4 Fracture-labeling with Lectin-Colloidal Gold

Glutaraldehyde; paraformaldehyde; cacodylate buffer; glycyl-glycine buffer; bovine serum albumin; Freon 22; liquid nitrogen; glycerol; lectin-gold; concanavalin A; wheat germ lectin; horseradish peroxidase; ovomucoid; N-acetylglucosamine; methyl-α-D-mannopyranoside; osmium tetroxide; Epon; Araldite; uranyl acetate; lead citrate.   **Materials**

Freeze-Fracture of Cells. Cells are fixed in diluted Karnovsky fixative (1% glutaraldehyde, 1.25% paraformaldehyde in 0.1 M cacodylate buffer, pH 7.4) for 20 min and then washed in the same buffer containing 0.2 M glycine for the purpose of blocking free aldehyde groups. The cells are subsequently suspended in 30% bovine serum albumin (BSA) and centrifuged in an Eppendorf tube. The supernatant is discarded and the remaining BSA-cell suspension cross-linked with a drop of 25% glutaraldehyde. The progression of cross-linking can be monitored on the basis of a color change from yellow to amber.   **Freeze-Fracture**

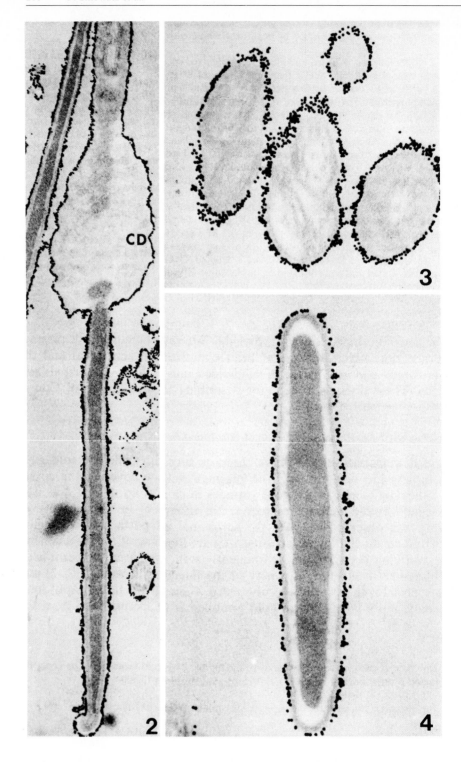

The resulting gels are sliced and washed several times in the glycine-containing buffer and subsequently impregnated in 10, 20, and 30% glycerol in PBS for 1 h each. Impregnated gels are then frozen in liquid Freon 22 in a brass container which has been precooled with liquid nitrogen (Sinowatz et al. 1989). Once the gels are frozen they are transferred into size S-5 glass homogenizer tubes, which are then both filled with and immersed in liquid nitrogen. Following the addition of a few drops of 30% glycerol/1% glutaraldehyde solution, the gels are sedimented to the base of the tube. They are then fractured repeatedly with a precooled pestle until they have been pulverized into fine fragments. The homogenizer tube should not be removed from the liquid nitrogen until the internal level of liquid nitrogen has been reduced to one-tenth of the original level. Two ml of glycerol/glutaraldehyde solution are then added. After thawing the glycerol, the tube should be immersed briefly in 30 °C water and then immediately transferred to an ice bucket for 15 min. The fragments are gradually deglycerinated via dropwise addition of 1 mM glycyl-glycine buffer at pH 7.4, or by overnight dialysis against the same buffer (osmolality = 320 mOsm).

Cytochemical Labeling of Freeze-fractured Cells with Lectin-Colloidal Gold. The fractured BSA-cell gels can be incubated directly in appropriate lectin-gold solutions. For some lectins, slight modifications of the techniques were also used. Binding sites for Con A and WGA can be demonstrated successfully by a two-step technique. Gels are incubated with 0.25 mg/ml of Con A for the electron-microscopic visualization of Con A-binding sites. The gel fragments are then washed several times in Sörensen phosphate buffer, followed by labeling with a colloidal gold-horseradish peroxidase complex (prepared according to the method of Geoghegan and Ackerman 1977). Incubation with unlabeled WGA (0.25 mg/ml) is followed by incubation with ovomucoid-coated colloidal gold (Geoghegan and Ackerman 1977).   **Labeling**

The samples are then osmicated (1% OsO_4 in 0.1 M sodium cacodylate buffer, pH 7.4) and embedded in an Epon/Araldite mixture. Ultrathin sections are collected on copper grids and stained routinely with uranyl and lead salts.

Controls are performed by pre-incubation in lectin-competitive sugars (e.g. 0.4 M methyl-α-D-mannopyranoside for experiments involving Con A, or 0.4 M N-α-D-glucosamine for those involving WGA) for 15 min at room temperature, followed by incubation with the lectin in the presence of the respective sugar.   **Controls**

Results and Remarks

Fracture labeling techniques, introduced by Pinto da Silva et al. (1981), represent a combined application of freeze fracture and cytochemical labeling. As in freeze

◄——————————————————————————————————

Fig. 2. Bovine spermatozoa from the caput epididymidis; Con A-HRP-gold technique. The plasma membrane covering sperm head and tail is heavily labeled with gold granules. *CD* cytoplasmic droplet. × 11 700

Fig. 3. Transverse section through the tails of bovine spermatozoa from the caput epididymidis; Con A-HRP-gold technique. × 32 400

Fig. 4. Transverse section through the head of a bovine spermatozoon from the caput epididymidis; Con A-HRP-gold technique. × 32 400

fracture techniques, fracture labeling involves the splitting of biomembranes along their bilayered continuum, thereby allowing the sidedness of membrane components to be determined. In some cases, transmembrane proteins can also be identified, as can their localization within the plane of the membrane. Fracture labeling can therefore be used to address questions concerning the expression of specific membrane components. Under favorable circumstances, the identification of transmembrane proteins is also possible.

Pinto da Silva et al. (1981) applied this technique to the initial characterization of Con A-binding site distribution on the fracture face of leucocytes. They later extended their study to the analysis of several other cell types, including sperm cells (Aguas and Pinto da Silva 1983). Using this technique, these authors were also able to demonstrate that the cytoplasmic surface of the inner acrosomal membrane is devoid of binding sites for both Con A and WGA (Aguas and Pinto da Silva 1985). Fracture labeling techniques may also be used to demonstrate endogenous lectins. Incubation of fractured spermatozoa with fucosyl-horseradish peroxidase-gold resulted in labeling of the acrosome at different fracture faces (Friess et al. 1987). Fucose binding sites are especially concentrated in the acrosomal matrix. Application of horseradish peroxidase-gold that had not been modified by the attachment of fucosyl residues ruled out the participation of carbohydrate moieties intrinsic to horseradish peroxidase (primary D-mannose).

27.5 Lectin and Neoglycoprotein Labeling of Cells for Analysis in Flow Cytometry

Materials   Buffer: phosphate buffered saline solution (PBS), pH 7.4 containing 0.1% bovine serum albumin and 0.1% sodium azide; FITC-lectins; biotin-lectins; biotin-neoglycoproteins; avidin-phycoerythrin (avidin-PE); avidin-FITC; 2% paraformaldehyde

Procedure   In order to keep cells alive and avoid clumping, all procedures must be performed on ice, and cold buffer and a refrigerated centrifuge should be used.

Single Lectin Labeling (FITC-Lectin). Cells are washed, centrifuged (5 min at 1000 rpm for most cells), and suspended in PBS. Cell density is adjusted to 10^6 cells/ml and FITC-Lectin is added to a final concentration of 2.5 µg/ml. After incubating the cells on ice for 30–45 min, they are washed twice and resuspended in PBS for subsequent analysis on a flow cytometer.

Double Lectin Labeling (FITC-Lectin and Biotin-Lectin/Avidin-PE). Cells are washed, centrifuged, and suspended (10^6 cells/ml) in PBS. Both FITC-lectin and biotin-lectin are added, each at a final concentration of 2.5 µg/ml. The cells are incubated on ice for 30–45 min, subsequently washed, and avidin-PE is added. They are then incubated on ice for another 30 min, washed twice, and resuspended in PBS for analysis in a flow cytometer.

Single Neoglycoprotein Labeling (Biotin-Neoglycoprotein/Avidin-FITC). Cells are washed, centrifuged and suspended (5×10^5 cells/ml) in PBS. Biotinylated neoglycoprotein is added at a final concentration of 25 µg/ml. The cells are incubated on ice for 1 h and subsequently washed in PBS. Then avidin-FITC is added and the cells are incubated on ice for another 30 min. Thereafter, they are washed twice and resuspended in PBS for analysis.

Fixation. Fixation should only be carried out if it is not possible to examine the cells immediately after the labeling procedure. The cells are washed and cold formaldehyde (2%) is added. Fixed cells can be stored at 4 °C in the dark for up to 1 week.

Results and Remarks

Flow cytometers are instruments in which multiple parameters (forward scatter, side scatter, fluorescence) of the cells in a single cell suspension are analyzed by a light

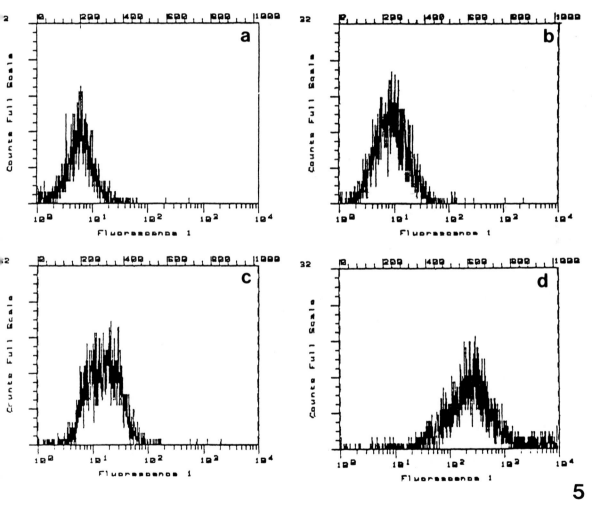

Fig. 5a–d. Flow cytometric analysis (FACSCAN) of lectin (*Dolichos biflorus* agglutinin) binding to cerebral endothelium. Cells were removed from culture dish by EDTA treatment, labeled with FITC-conjugated lectin and analyzed on a FACSCAN instrument using a 4-decade log amplification of fluorescence. **a,b** FACSCAN analysis of brain-derived endothelium, adult mouse. **a** Unlabeled cells. **b** DBA-FITC labeled cells. **c,d** FACSCAN analysis of brain-derived endothelium, embryonal mouse. **c** Unlabeled cells. **d** DBA-FITC labeled cells

source, usually a laser. Flow cytometry can be used to examine cell surface binding of fluorescein-conjugated lectins and neoglycoproteins. The advantage of this approach is that it allows quantification of the amount of lectin or neoglycoprotein bound (Shapiro1988; Ormerod 1990). The eye is only capable of detecting major differences in intensity and can be misled by variations in cell size or label distribution. For quantification of binding, direct staining is preferable to indirect staining. Since lectins can agglutinate cells, it is necessary to use subagglutinating concentrations.

Advances in flow cytometric instrumentation have made it possible to design experiments that make use of more than one label in a single sample (two-, three-, four-color fluorescence; Hoffman 1988). Fluorescein isothiocyanate (FITC) is by far the most popular fluorescent tag. Lectins conjugated to tetramethyl rhodamine isothiocyanate (TRITC) to phycobiliproteins such as phycoerythrin (PE) or to avidin are available commercially. Neoglycoproteins are coupled to a variety of labels, such as FITC or biotin.

Sources of cells for examination include blood cells and nonadherent culture cells, as well as those from adherent cultures that have to be brought into suspension by mild enzymatic and/or EDTA treatment.

Quantification of lectin and neoglycoprotein binding by flow cytometry has allowed the identification: (a) of cells, such as hemopoietic stem cells (Visser and Vries 1990), (b) of organ specificity in endothelial cells (Gumkowski et al. 1987), (c) of changes in cellular glycoproteins during aging (Auerbach et al. 1992; Plendl et al. 1993), and (d) of the role of membrane lectins in human colon carcinoma (Gabius et al. 1987). We have experience with flow cytometric analysis of different organ-specific endothelial cultures (see Fig. 5), several tumor cell lines, and embryonic cell cultures, to name but a few.

In general, flow cytometry is of less use when dissociation of cells is difficult, since dissociation procedures may lead to perturbations of the cell membrane and cell surface components.

Fixation should be avoided if possible, because it can change emission and wavelength characteristics of flurochromes, as well as the forward and side scatter of cells.

References

Aguas AP, Pinto da Silva P (1983) Regionalization of transmembrane glycoproteins in the plasma membrane of boar sperm heads is revealed by fracture label. J Cell Biol 97: 1356–1364

Aguas AP, Pinto da Silva P (1985) The acrosomal membrane of boar sperm: a Golgi-derived membrane poor in glycoconjugates. J Cell Biol 100: 528–534

Auerbach R, Plendl J, Kusha B (1992) Endothelial cell heterogeneity and differentiation. In: Maragoudakis ME (ed) Angiogenesis in health and disease. Plenum Press, New York, pp 55–62

Frens G (1973) Controlled nucleation for the regulation of the particle size in monodisperse gold solutions. Nature Phys Sci 241: 20–22

Friess AE, Toepfer-Petersen E, Schill WB (1987) Fracture labeling of boar spermatozoa for the fucose-binding protein (FBP). Histochemistry 87: 181–183

Gabius HJ, Engelhardt R, Hellmann T, Midoux P, Monsigny M, Nagel GA, Vehmeyer K (1987a) Characterization of membrane lectins in human colon carcinoma by flow cytometry, drug targeting and affinity chromatography. Anticancer Res 7: 109–112

Gabius HJ, Engelhardt R, Hellmann KP, Hellmann T, Ochsenfarth A (1987b) Preparation of neo-glycoprotein-enzyme conjugate using a heterobifunctional reagent and its use in solid-phase assays and histochemistry. Anal Biochem 165: 349–355

Gabius HJ, Bodanowitz S, Schauer A (1988) Endogenous sugar-binding proteins in human breast tissue and benign and malignant breast lesions. Cancer 61: 1125–1131

Gabius S, Hellmann KP, Hellmann T, Brinck U, Gabius HJ (1989) Neoglycoenzymes: a versatile tool for lectin detection in solid-phase assays and glycohistochemistry. Anal Biochem 182: 447–451

Gabius S, Schirrmacher V, Franz H, Joshi SS, Gabius HJ (1990) Analysis of cell surface sugar receptor expression by neoglycoenzyme binding and adhesion to plastic-immobilized neoglycoprotein for related weakly and strongly metastatic cell lines of murine model systems. Int J Cancer 46: 500–507

Geoghegan WD, Ackerman GA (1977) Adsorption of horseradish peroxidase, ovomucoid and anti-immunoglobulin to colloidal gold for the indirect detection of concanavalin A, wheat germ agglutinin and goat antihuman immunoglobulin G on cell surfaces at the electron microscopic level: a new method, theory and application. J Histochem Cytochem 25: 1187–1200

Gumkowski F, Kaminska G, Kaminski M, Morrissey LW, Auerbach R (1987) Heterogeneity of mouse vascular endothelium. In vitro studies of lymphatic, large blood vessel and microvascular endothelial cells. Blood Vessels 24: 11–23

Hoffman RA (1988) Approaches to multicolor immunofluorescence measurements. Cytometry Suppl. 3: 18–22

Horisberger M (1979) Evaluation of colloidal gold as a cytochemical marker for transmission and scanning electron microscopy. Biol Cellulaire 36: 253–258

Kieda C, Roche AD, Delmotte F, Monsigny M (1979) Lymphocyte membrane lectins. Direct visualization by the use of fluoresceinyl-glycosylated cytochemical markers. FEBS Lett 99: 329–331

Kim B, Behbahani I, Meyerhoff ME (1992) Lectin-based homogenous enzyme-lined binding assay for estimating the type and relative amount of carbohydrate within intact glycoproteins. Anal Biochem 202: 166–171

Leathem A (1986) Lectin histochemistry. In: Polak A et al. (eds) Immunocytochemistry. Wright, Bristol. pp 167–188

McCoy JP (1987) The application of lectins to the characterization and isolation of mammalian cell populations. Cancer Metast Rev 6: 595–613

Naoi M, Iwashita T, Nagatsu T (1987) Binding of specific glycoconjugates to human brain synaptosomes: studies using glucosylated β-galactosidase. Neurosci Lett 79: 331–336

Ormerod MG (1990) Flow cytometry. A practical approach. Oxford university press, Oxford.

Pinto da Silva P, Parkinson C, Dwyer N (1981) Freeze-fracture cytochemistry. Thin sections of cells and tissues after labeling on fracture faces. J Histochem Cytochem 29: 917–298

Plendl J, Hartwell L, Auerbach R (1993) Organ-specific change in *Dolichos biflorus* lectin binding by myocardial endothelial cells during in vitro cultivation. In vitro cell dev biol, 29A: 25–31

Roth J (1983) The colloidal gold marker system for light and electron microscopic cytochemistry. In: Bullock G, Petrusz P (eds) Techniques in immunocytochemistry, Vol. 2. Academic Press, London

Schottelius J, Gabius HJ (1992) Detection and quantitation of cell surface sugar receptor(s) of *Leishmania donovani* by application of neoglycoenzymes. J. Parasitol in press

Schrével J, Gros D, Monsigny M (1981) Cytochemistry of cell glycoconjugates. Progr Histochem Cytochem 14 (2): 1–269. G Fischer, Stuttgart, New York

Schulte BA, Spicer SS (1983) Light microscopic detection of sugar residues in glycoconjugates of salivary glands and the pancreas with lectin peroxidase conjugates. Histochem J 15: 1217–1238

Shapiro HM (1988) Practical flow cytometry. AL Liss, New York

Sinowatz F, Gabius HJ, Amselgruber W (1988) Surface sugar binding components of bovine spermatozoa as evidenced by fluorescent neoglycoproteins. Histochemistry 88: 395–399

Sinowatz F, Voglmayer JK, Gabius HJ, Friess AE (1989) Cytochemical analysis of mammalian sperm membranes. Progr Histochem Cytochem 19(4): 1–74. G Fischer, Stuttgart, New York

Straus W (1981) Cytochemical detection of mannose-specific receptors for glycoproteins with horseradish peroxidase as a ligand. Histochemistry 73: 39–47

Toepfer-Petersen E, Friess AE, Nguyen H, Schill WB (1985) Evidence for a fucose-binding protein in boar spermatozoa. Histochemistry 83: 139–145

Visser JWM, P de Vries (1990) Identification and purification of murine hematopoietic stem cells by flow cytometry. In: Darzynkiewicz S (ed) Methods in cell biology, vol. 33, Flow cytometry, Academic Press, New York

Vorberg E, Bundle DR (1990) Carbohydrate-enzyme conjugates for competitive EIA. J Immunol Meth 132: 81–89

28 Use of Lectins for the Phenotypic Characterization of Cultured Cell Monolayers

H.G. AUGUSTIN-VOSS

Lectins have been employed extensively to analyze glycoconjugates in various tissues using light and electron microscopic histochemical techniques (Damjanov 1987). They are also extremely useful for the phenotypic characterization of cultured cells, particularly if other markers, such as monoclonal antibodies, are not available. However, in contrast to the selective expression of a monoclonal antibody specific epitope, carbohydrates are expressed rather ubiquitously, for which reason lectin binding – with few exeptions – does not provide discrete cell-specific markers. While this limits the use of lectins as diagnostic instrument, semiquantitative cytochemistry, and quantitative lectin techniques (ELLA) can be used as powerful research tools to compare the phenotypic properties of different cell populations, as well as the different functional or metabolic states of a specific cell type.

The protocols detailed below summarize standard light and ultrastructural lectin cytochemical techniques and outline an enzyme-linked lectin assay that can be used to quantitatively compare relative lectin-binding intensities.

Materials   Conjugated and unconjugated lectins are available from a number of suppliers. Suppliers with a broad line of lectin products are EY Laboratories, Sigma, Vector Laboratories, and Accurate Chemical. Other immunological reagents including secondary antibodies and streptavidin-horseradish-peroxidase can be obtained from many different companies. Immunological gold reagent can be obtained from Amersham, EY Laboratories, and Vector Laboratories. Flexiperm silicon templates used to compartmentalize tissue culture dishes are produced by Heraeus, Germany.

28.1 Light Microscopic Lectin Cytochemistry of Cultured Cell Monolayers

Lectin cytochemical analysis of cultured cells can give a good assessment of their overall lectin-binding profile. It also allows the semiquantitative comparison of lectin-binding intensities of different cell populations as well as similar cell populations cultured under different conditions, provided that the technical variation between specimen can be eliminated. The basic protocol is summarized in Fig. 1. Cells can be grown on glass coverslips or in regular plastic tissue culture dishes (35 mm or larger to allow microscopic examination with standard laboratory microscope). We routinely use 100-mm petri dishes that are subdivided with self-adherent silicon templates into different compartments (Augustin-Voss and Pauli 1992). The silicon templates are cut from commercially available eight-well silicon inserts. This provides a conveniently sized dish for microscopic examination and, more importantly, identical cytochemical treatment of cells that were cultured in different silicon templates but stained identically once the templates are removed.

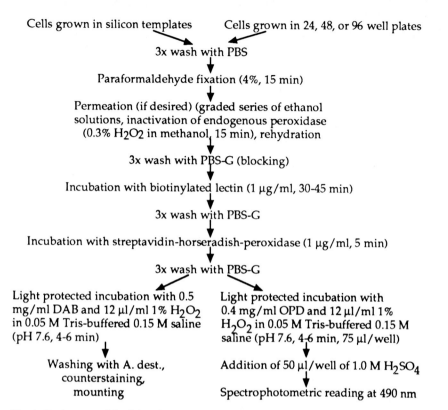

CYTOCHEMISTRY   **ELLA**

Cells grown in silicon templates   Cells grown in 24, 48, or 96 well plates

3x wash with PBS

Paraformaldehyde fixation (4%, 15 min)

Permeation (if desired) (graded series of ethanol
solutions, inactivation of endogenous peroxidase
(0.3% H_2O_2 in methanol, 15 min), rehydration

3x wash with PBS-G (blocking)

Incubation with biotinylated lectin (1 µg/ml, 30-45 min)

3x wash with PBS-G

Incubation with streptavidin-horseradish-peroxidase (1 µg/ml, 5 min)

3x wash with PBS-G

Light protected incubation with 0.5   Light protected incubation with
mg/ml DAB and 12 µl/ml 1% H_2O_2 0.4 mg/ml OPD and 12 µl/ml 1%
in 0.05 M Tris-buffered 0.15 M saline H_2O_2 in 0.05 M Tris-buffered 0.15 M
(pH 7.6, 4-6 min)   saline (pH 7.6, 4-6 min, 75 µl/well)

Washing with A. dest.,   Addition of 50 µl/well of 1.0 M H_2SO_4
counterstaining,
mounting   Spectrophotometric reading at 490 nm

Fig. 1. Basic protocol for light microscopic lectin cytochemistry and ELLA

The templates can also be used to rationalize the staining procedure if different
lectins or different lectin concentrations are to be tested on the same cell population:
the cells can be probed with different reagents prior to removal of the templates. The
subsequent steps (detection) can then be done after removal of the templates, which
allows the simultaneous staining of up to ten different cultures.

Cells to be used for lectin cytochemistry are washed three times with phosphate   **Procedure**
buffered saline containing 1.0 mM $CaCl_2$ and 1.0 mM $MgCl_2$ (PBS), after which they
are fixed with 4% buffered paraformaldehyde for 15 min (all procedures are
performed at room temperature unless stated otherwise). Fixed cells are washed
three times with PBS containing 0.2% gelatin (PBS-G) to block unspecific binding
sites for at least 15 min. Other blocking agents are equally efficient (0.4–10% BSA or
commercially available blockers (e.g., Megga-Block II, Onasco Biotechnologies,
Houston, TX), but gelatin treatment is usually sufficient to block unspecific binding.
After blocking, cells are incubated with 1 µg/ml of biotinylated lectin for 30 to
45 min. These are average incubation conditions which have been tested for several
cell types (endothelial cells, smooth muscle cells, fibroblasts) and various lectins.
Special applications might, however, require individual adjustments with respect to
concentrations and times of incubation. After lectin incubation, cells are thoroughly
washed with PBS-G (3x, 4 min each), then incubated with 1 µg/ml of streptavidin-

horseradish-peroxidase (5 min). Cells are again washed three times with PBS-G (4 min each) and the bound lectin complex is visualized by light-protected incubation with 0.5 mg/ml 3,3′-diaminobenzidine tetrahydrochloride and 12 µl/ml 1% hydrogen peroxide in 0.05 M Tris-buffered 0.15 M saline at pH 7.6 for 4–6 min, after which time the cells are washed with distilled water. For black and white documentation of staining results, cells are preferably counterstained with Methyl Green (1% in H_2O, 3 min). For color documentation, dishes are counterstained with Gill's #2 hematoxylin (3 min). For color documentation, some investigators might prefer to develop the bound lectin complex with 3-amino-9-ethylcarbazole (AEC), which provides a good color contrast with orange staining on blue background compared to the brown on blue background staining of DAB with hematoxylin counterstain (4 mg AEC in 1 ml N-N-dimethylformamide, 14 ml 0.1 M acetate buffer, pH 5.2, 0.15 ml 3% H_2O_2 (available as kit from Zymed). AEC staining will, however, fade after several weeks compared to the more permanent DAB staining, and is alcohol-soluble, for which reason alcohol-based solutions have to be avoided (e.g., Harris' hematoxylin). Finally, the stained cells are mounted with glass coverslips using a water-soluble mounting medium for cells grown on plastic or a xylene-soluble mounting medium for cells grown on glass.

This basic protocol will selectively stain cell surface glycoconjugates, since formaldehyde-fixed monolayers are relatively impermeable to large molecules such as antibodies and lectins, which can be demonstrated by the absence of immunostaining of cytoplasmic antigens. To stain total cellular glycoconjugates (i.e., particularly intracellular sites of storage such as Golgi apparatus and endoplasmic reticulum), fixed monolayers have to be permeated prior to lectin staining. This can be achieved by mild detergent treatment (0.01% Triton X-100 in PBS, 10 min) or by washing fixed monolayers with graded ethanol solutions, followed by a 15-min treatment with 0.3% H_2O_2 in methanol to inactivate endogenous peroxidase. After rehydration, cells are lectin-stained as described above.

28.2 Enzyme-Linked Lectin Assay (ELLA)

As a quantitative technique to analyze lectin-binding intensities, the enzyme-linked lectin assay is complementing lectin cytochemistry and provides an easily accessible tool to screen large numbers of samples of cultured cells (Fig. 1) (Gaveriaux and Loor 1987; Augustin-Voss et al. 1991). Cells are grown in 24, 48, or 96-well plates and are allowed to grow to the desired configuration (i.e., confluent, subconfluent). Care must be taken to tightly control the cells' growth configuration, as lectin-binding intensities vary considerably with respect to seeding density and confluency of the monolayer.

Procedure   The ELLA follows the same basic principles as outlined under lectin cytochemistry of cultured cell monolayers with respect to fixation, blocking, cell membrane permeation, lectin incubation, and streptavidin-horseradish-peroxidase incubation. Particularly when working with multiwell plates and multipipeters, and depending on the cell type's adhesiveness to the tissue culture plastic, care must be taken not to wash off cells during the different washings. In contrast to the cytochemical techniques, the bound lectin complex is developed with a soluble substrate [light protected incubation with 75 µl/well (96-well plate) of 0.4 mg/ml of o-phenylenediamine and 12 μl/ml of 1% H_2O_2 in 50 mM Tris-buffered 0.15 M saline (pH 7.6, 5 min)]. The reaction is stopped by addition of 50 μl/well (96 well plate) of 1.0 M H_2SO_4 and plates read immediately at 490 nm using a microplate spectrophotometer. Specific binding intensity is quantitated by measuring absorbance at 490 nm using the corresponding negative controls as blanks. The negative controls are also checked by measuring them against distilled water (shouldn't be higher than 0.05).

28.3 Ultrastructural Lectin Cytochemistry of Cultured Cell Monolayers

For ultrastructural analysis of lectin binding or determination of lectin receptor distribution, cell surface lectin binding sites can be visualized by electron microscopy (Augustin-Voss et al. 1991; Horisberger 1985). Whenever possible, preembedding cytochemical techniques should be employed to identify lectin receptors, i.e., the complete lectin cytochemical technique (lectin binding and detection) should be performed prior to plastic embedding of the cells. However, intracellular detection of

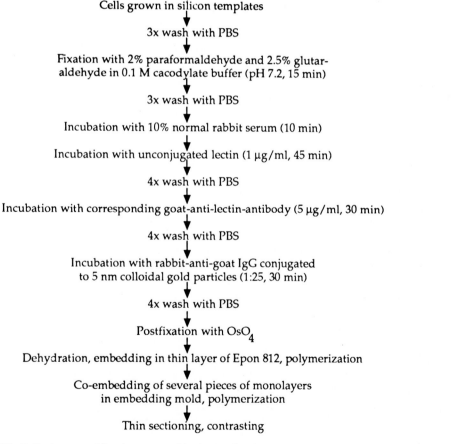

Outline of procedure

Cells grown in silicon templates
↓
3x wash with PBS
↓
Fixation with 2% paraformaldehyde and 2.5% glutaraldehyde in 0.1 M cacodylate buffer (pH 7.2, 15 min)
↓
3x wash with PBS
↓
Incubation with 10% normal rabbit serum (10 min)
↓
Incubation with unconjugated lectin (1 μg/ml, 45 min)
↓
4x wash with PBS
↓
Incubation with corresponding goat-anti-lectin-antibody (5 μg/ml, 30 min)
↓
4x wash with PBS
↓
Incubation with rabbit-anti-goat IgG conjugated to 5 nm colloidal gold particles (1:25, 30 min)
↓
4x wash with PBS
↓
Postfixation with OsO_4
↓
Dehydration, embedding in thin layer of Epon 812, polymerization
↓
Co-embedding of several pieces of monolayers in embedding mold, polymerization
↓
Thin sectioning, contrasting

Fig. 2. Basic protocol for ultrastructural lectin cytochemistry

glycoconjugates will usually require postembedding techniques (Roth 1983; Bendayan et al. 1987).

Procedure   Several procedures can be used for the ultrastructural analysis of cell surface lectin receptors. Gold-labeled lectins can be used for the direct electron microscopic detection of lectin binding sites. We prefer to use an indirect preembedding immunogold technique which appears to have a higher sensitivity (Fig. 2). Cells are grown in 60- or 100-mm culture dishes subdivided into several compartments with Flexiperm inserts as described above. They are fixed with 2% paraformaldehyde and 2.5% glutaraldehyde in 0.1 M cacodylate buffer (pH 7.2) for 15 min and washed three times with PBS. Monolayers are incubated with 10% normal rabbit serum for 10 min and exposed to unconjugated lectin for 45 min. After four washings with PBS, lectin-labeled cells are incubated with corresponding goat-anti-lectin-antibody at a concentration of 5.0 μg/ml (30 min). Thoroughly washed cells (four times with PBS) are then incubated with the secondary antibody for 30 min (1:25 dilution of rabbit anti-goat IgG conjugated to 5 nm colloidal gold). After removal of the Flexiperm insert, washed cells are postfixed with OsO_4, dehydrated in graded ethanol solutions, and embedded in a thin layer of Epon 812. Monolayer segments from two to four different areas of one culture are mounted together and co-embedded in Epon 812. Co-embedding of several pieces of plastic embedded cells helps to rationalize the labor-intensive electron microscopic specimen preparation and allows one to simultaneously examine several pieces from one monolayer, which is necessary to assess the heterogeneity of the monolayer. Thin sections of monolayer cross-sections are stained with uranyl acetate and lead citrate and examined by transmission electron microscopy. In contrast to other electron microscopic detection systems, gold labeling techniques permit simple quantitation of binding in-

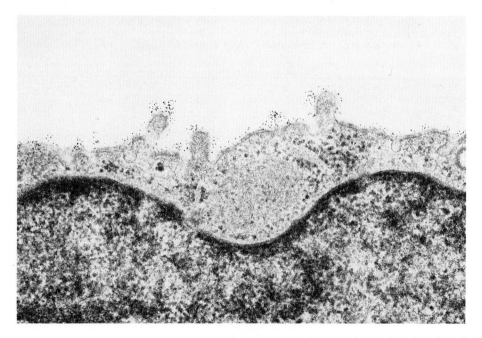

Fig. 3. Ultrastructural demonstration of WGA-binding sites on cultured bovine aortic endothelial cells

tensities by just counting the number of bound gold particles per examination unit (e.g., per µm cell surface) (Fig. 3). It can generally be recommended to use smaller gold particles (5 µm) over larger gold particles sizes, because the larger particles allow less sensitivity and will more easily interfere sterically with the specific binding reaction.

To evaluate the specificity of the lectin binding, monolayers to be used for cyto-chemical, ELLA, or ultrastructural analysis of cellular glycoconjugates are stained with lectins that were preincubated with the specific monosaccharide (0.2 M) for 20 to 30 min prior to lectin staining. This will completely eliminate binding of most lectins. Some lectins which preferentially bind more complex carbohydrate struc-tures may only partially be blocked by the corresponding monosaccharide.

Negative Controls

References

Augustin-Voss HG, Pauli BU (1992) Quantitative analysis of autocrine-regulated, matrix-induced, and tumor cell-stimulated endothelial cell migration using a silicon template compartmentalization tech-nique. Exp Cell Res 198: 221–227

Augustin-Voss HG, Johnson RC, Pauli BU (1991) Modulation of endothelial cell surface glycoconjugate expression by organ-derived biomatrices. Exp Cell Res 192: 345–351

Bendayan M, Nanci A, Kan FWK (1987) Effect of tissue processing on colloidal gold cytochemistry. J Histochem Cytochem 35: 983–996

Damjanov I (1987) Lectin cytochemistry and histochemistry. Lab Invest 57: 5–20

Gaveriaux C, Loor F (1987) An enzyme-linked lectin-binding assay on cells (CELLBA) for the comparison of lectin receptor expression on cell surfaces. J Immunol Methods 104: 173–182

Horisberger M (1985) The gold method as applied to lectin cytochemistry in transmission and scanning electron microscopy. Techniques in Immunocytochemistry, Academic Press, London, Vol 3, pp 155–178

Roth J (1983) Application of lectin-gold complexes for electron microscopic localization of glycocon-jugates on thin sectins. J Histochem Cytochem 31: 987–999

29 The Use of Lectins in the Purification of Cellular Subpopulations

I. LOPEZ-COLBERG, R.O. KELLEY, M.P. VAIDYA, and P.L. MANN

The embryonic cardiac chick tissue culture model has been widely used as a major experimental tool in the study of developmental processes for over six decades (Johnstone 1925). One limitation of this model is the cellular heterogeneity produced from the disaggregation of 14-day-old chick hearts. The two broad morphological types of cells are fibroblasts and myocytes. These cells exhibit very different proliferative and differentiative characteristics in vitro, and without intervention, the fibroblasts quickly overgrow the myocytes, preventing even short-term tissue culture studies. Several techniques have been used to purify the myocyte population, with varying degrees of success (Kaneko and Goshima 1982).

29.1 Objectives, Limitations, and the Outline of the Procedure

The methodology described herein identifies and utilizes specific lectin binding epitopes on myocytes as the basis for their separation from a heterogenous single cell suspension of 14-day-old chick embryo hearts. Cardiac myocytes have a transient, highly specific interaction with Wheat Germ Agglutinin (WGA) and Concanavalin A (Con A) lectins. Thus, when the specific lectin is conjugated to a solid phase (immobile) surface and the cells are brought into close contact with the surface, a highly selective interaction takes place. The myocytes are selectively bound to the solid surface and the contaminating fibroblast populations are easily rinsed off. The myocytes are then released from the surface by the addition of sufficient specific carbohydrate ligand to compete for the lectin binding site(s). This technique results in a seven fold increase in the myocyte concentration. These selected, purified myocytes have been shown to have subsequent utility for long-term studies.

Special attention is drawn to: the selectivity of the lectin interaction; the retention of the functional myocyte phenotype after selection; and possible subpopulations within the myocyte population itself.

29.2 Equipment and Materials

Equipment
- 300 capacity egg incubator (Lehay Mfg. Higginsville, Mo)
- Sterilized egg holders
- 100×15 mm bacteriological petri dishes (Falcon, Lincoln Park, NJ)
- Carlson 10-spot slides (Carlson Inc. Peotone, IL)
- Humidifed CO_2 incubator
- 0.2 µm Swinnex filters (Swinnex, Beddford, MA)
- 0.8 µm Gelman filters (Fisher)

– Wheaton serum bottles (Fisher)
– Leitz photo microscope equipped with epi-fluorescence
– (Quantitation can be achieved by an optional MPV photometer)
– Laboratory centrifuge, laminar hood, bolting cloth, and dry oven.

– Three dozen utility eggs   **Materials**
– Ethanol 70% (v/v)
– Tissue culture media (refer to media preparation below)
– Immu-Mount (Shandon, Pittsburgh, PA)

Lectins and competitive carbohydrates used in these studies include: wheat germ agglutinin (WGA, Calbiochem, San Diego, CA) and N-, NN'-, NN'N''-, and NN'N''N'''-oligomers of N-acetyl-D-glucosamine (GlcNAc, E-Y Labs, San Mateo, CA); concanavalin A (Con A, Pharmacia, Piscataway, NJ), and alpha-methyl-mannopyranoside (α-MMP); FITC conjugated lectins (Vector Labs, Burlingame, CA): concanavalin A (Con A), *Ulex europeus* agglutinin (UEA I), soybean agglutinin (SBA), wheat germ agglutinin (WGA), *Ricinus communis* agglutinin (RCA-120), peanut agglutinin (PNA), *Dolichos biflorus* agglutinin (DBA).

– Collagenase 4177 Cls II, and DNAase (Worthington, Freehold, NJ)
– Phosphate buffered saline (PBS 0.137 M NaCl, 0.003 M KCL, 0.009 M PO_4)
– Glutaraldehyde and formaldehyde (type II, Sigma, St. Louis, MO)
– 0.1 M Na_2CO_3 pH 9.0, 0.05 M $NaHCO_3$ pH 9.6
– PBS/BSA (1 mg/ml Fraction V, Sigma)
– Poly-L-lysine · HBr (Sigma, 100 kDa)
– D-glucose (Sigma), potassium free Ham's F-12 K, and TC 199 powdered media (Gibco, Gaithersberg, MD), L-glutamine (Gibco), fetal calf serum (FCS, Flow Labs, McLean, VA), horse serum (HS, Gibco), sodium salt of Penicillin G 10^6 u/bottle (Sigma).

29.3 Preparations

– A special medium (DeHann and Foggard 1975) is used throughout the experi-   **Media**
   ment.
   To 3.5 g of potassium free Ham's F-12 K powder
   add 1.38 g of TC 199 powdered medium
   in 462 ml double distilled water,
   then add 0.15 g of L-glutamine,
   pH to 6.9.
   Sterilize through a 0.2 μm filter,
   then add aseptically 20 ml FCS, 10 ml HS, and 2.5×10^5 u penicillin.

– DNAase at 10^5 Dornase units/ml in PBS, 0.2 μm filtered (stock)   **Dissociation**
– To 0.3 g Collagenase (4177 Cls II)   **Medium**
   add 150 ml of PBS
   and DNAase (0.75 ml), pH adjusted to 7.4
– Sterile filtered through 0.8 μm filter (47 mm) then a 0.2 μm filter
– Store at − 20 °C.

Cells
- Three dozen utility eggs (14 d) are processed at a time
- Determine embryo maturity according to Hamburger (Hamburger and Hamilton 1952)
- Individual eggs are dipped in 70% ethanol and transferred to a previously sterilized egg holder under a laminar flow hood. After approximately 20 min, the eggs are broken individually and the embryo transferred to a sterile petri dish. After decapitation, the chest is opened and the heart removed. The hearts are then transferred to another sterile petri dish with special medium at 37 °C. The hearts are split, rinsed to remove excess blood, and placed in dissociation medium. The tissue is then minced with fine scissors. A total volume of 2–3 ml of heart/enzyme suspension is rotated at 30 rpm at 37 °C in 30 ml Wheaton serum bottles for approximately 15 min. The large aggregates are allowed to settle and the single cell rich supernatant is removed for further processing. This step is repeated four to six times until the hearts are completely disaggregated. The supernatants are mixed with an equal volume of special medium and centrifuged at 500 rpm (approx. 75g) for 5 min. The cells are washed several times in PBS, counted and the viability (by Trypan Blue exclusion) determined.

Affinity Substrate
- All volumes are for 100×15 mm petri dishes
- 14 ml of 0.1% glutaraldehyde in 0.1 M Na_2CO_3 for 60 min at 20 °C
- Three washes with double distilled water
- Add poly-L-lysine at 20 μg/ml in 0.05 M $NaHCO_3$ at 4 °C
- Incubate for 16 h
- Three washes with 0.05 M $NaHCO_3$
- Add lectin at 1 mg/ml in 1% glutaraldehyde (0.05 M $NaHCO_3$)
- Heat at 40 °C for 16 h
- Three washes in PBS
- Block at 37 °C with PBS/BSA for 1 h.

Characterization

Fluorescence Characterization: Cell suspensions, affinity separated or unseparated, at 1×10^5/ml in complete medium are pipetted onto Carlson ten-spot slides (50 μl/ spot) and incubated at 37 °C for 16 h in a humidified CO_2 incubator. The slide cultures are washed in PBS (3×5 min each), fixed in 2.5% formaldehyde plus 5% D-glucose in PBS at pH 7.0 for 20 min at room temperature. After three PBS washes and 2–16 h block in PBS/BSA, the individual cultures are stained with the following FITC-lectins (Vector Lab) at 5 μg/ml in PBS/BSA: Concanavalin-A (Con A); *Ulex europeus* agglutinin (UEA I); soybean agglutinin (SBA); wheat germ agglutinin (WGA); *Ricinus Communis* agglutinin (RCA-120); peanut agglutinin (PNA); and *Dolichos biflorus* agglutinin (DBA). The staining is carried out at 37 °C for 90 min in humidifier boxes. All procedures with FITC lectins are carried out in subdued light to minimize photo bleaching. After three further washes in PBS, the slides are treated with Immu-Mount and covered with a coverslip. These preparations are viewed and photographed with Leitz photomicroscope equipped with epi-fluorescence. The quantitation of the lectin binding is carried out on the same microscope equipped with the MPV photometric analysis system. This system measures the photon output of the sample over a linear but arbitrary range of 0–100. The collimator on the MPV can be adjusted to view a 79 μ_m^2 area of a single cell. For this particular series of experiments, the collimator is adjusted to view a field which is just large enough for a single cell.

29.4 Procedure

The heterogeneous cell suspension is incubated in a polypropylene test-tube (in complete medium) for 60 min at 37 °C to allow regeneration of membrane receptorrs etc. This suspension is then filtered through sterile bolting silk to remove aggregates. The number of single cells is counted and viability determined. Approximately 1.6 $\times 10^7$ cells are plated into a 100×15 mm sterile petri dish coated with WGA or Con A. The cells are incubated at 37 °C for 60–90 min. The medium is gently agitated to resuspend loosely attached cells and the supernate is aspirated. The plates are washed gently with PBS and subsequently incubated at 37 °C for 60 min with 5 ml of pre-warmed competitor carbohydrate in PBS (5.8 mM GlcNAc for WGA and 2.6 mM α-MMP for Con A) is added and incubated for 60 min. The cells are resuspended with a Pasteur pipet and then transferred to a centrifuge tube, centrifuged, and counted. These cells are then used for subsequent studies.

Procedure

29.5 Results

The individual cell populations are identified morphologically, functionally (the beating of the myocytes) and structurally (electron microscopic evidence of myo-fibrillar depositions). Figure 1 shows the results of a screen of various lectins, which represent the major carbohydrate specificities for simple binding to cardiac myocytes and fibroblasts. The use of the MPV photometer aids in quantitation of the relative binding of the FITC lectins to the cell populations; however, simple qualitative estimates by eye with any fluorescence microscope provides the same basic information. Figure 1 shows that both Con A and WGA have a significantly higher staining ratio on the myocytes than on the fibroblasts. This same screening

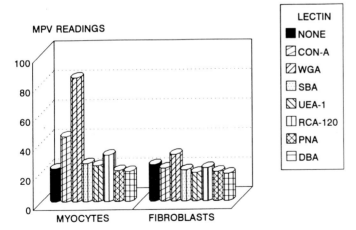

Fig. 1. Chick cardiac (14 day) myocytes and fibroblasts are identified morphologically and functionally, and their staining patterns with the listed lectins is assessed. Quantitation of FITC-lectin uptake is assessed by use of the MPV attachment on a Leitz photomicroscope equipped with epi-fluorescence. This is a typical results from one of nine experiments

protocol has been used to identify uniquely staining cell subpopulations in fresh and fixed whole tissue as well as free cell suspensions. These simple binding patterns have also been confirmed by FACS studies. The apparent background labeling of the fibroblasts (15–20 on the relative intensity scale) can be reduced to true background noise levels by using freshly prepared FITC conjugated lectins which have been purified on polyacrylamide gel electrophoresis or similar techniques. The results given here are routinely obtained with commercial lectin preparations.

Figure 2 shows the effect of the incubation time on the WGA affinity substrate as it influences the ultimate percentage of the enriched myocyte populations which adhered to the lectin coated plates. Incubation times between 60 and 90 min appear to provide maximal separation capacity with the highest purification ratio for the cell populations. The increase in fibroblast attachment to the WGA after 120 min of culture is probably more due to the fibroblasts natural adherence phenotype than to any increase in specificity for the lectin substrate.

Figure 3 shows the effect of specific competitor ligand on the recovery of myocytes from a WGA affinity substrate. Here four concentrations of GlcNAc are tested. All specific carbohydrate competitors show a dose-dependency, whereas nonspecific sugars do not. In some cases nonspecific sugars have been shown to induce the desorption of cells from the WGA substrate but the concentrations required were 10- to 100-fold higher than those observed with the specific carbohydrates. Care must be taken to test both specific and nonspecific carbohydrates in these studies. It is also important to test a number of specific carbohydrates with different nominal affinities for the specific interaction. By constructing a series of competitors, one can minimize the time of exposure of the cells to the purified carbohydrates, and perhaps more importantly we have observed indications that it is possible to desorb different functional myocyte populations from the WGA substrate with carbohydrates of different affinities.

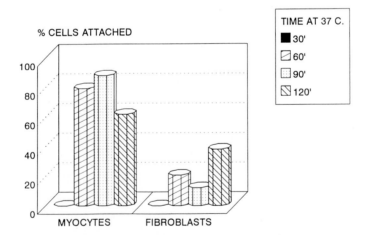

Fig. 2. The relative attachment of chick cardiac (14 day) myocytes and fibroblasts to a WGA affinity substrate. The substrate is prepared as described in the text. A heterogeneous mixture of cardiac cells is incubated with the substrate and the number of cells (myocytes and fibroblasts identified functionally and morphologically) are determined at various time points. Similar results are found with the Con A affinity substrate, and no specificity is observed in a limited number of experiments with an inappropriate substrate

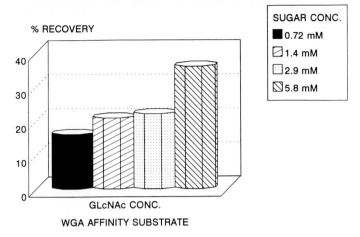

Fig. 3. The specificity of the interaction of the myocytes with the WGA substrate. Various concentrations of N-acetyl-glucosamine (GlcNAc) are incubated with the cell/substrate for 60 min and the percentage recovery is estimated. Controls included inappropriate affinity substrate and inappropriate competitors

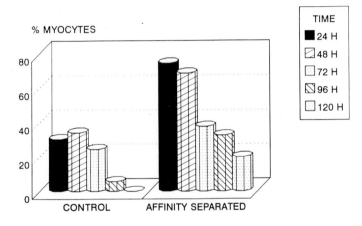

Fig. 4. The relative abundance of myocytes in affinity separated and control cultures as function of time in culture after separation. The myocytes are identified morphologically and functionally

Figure 3 shows that there is an approximately 40% recovery of attached myocytes when 5.8 mM GlcNAc competitor ligand solution is used. Virtually quantitative yields of myocytes can be obtained by using either higher concentrations of GlcNAc or lower concentrations of higher affinity sugars such as NN′N″N‴-tetra-acetyl chitotetrose (a tetramer of GlcNAc linked through β 1-3 linkages). These complex sugars are expensive, but may be useful if groups of myocytes are being prepared by serial desorption with increasing competitor carbohydrate concentrations.

Figure 4 shows data related to the effect of myocyte separation on subsequent performance in tissue culture. In these experiments the incidence of myocytes is tested over a 5-day culture period in both affinity substrate separated cultures and unseparated controls. The affinity substrate separated myocyte cultures remain viable and functional in culture for the 5 days. This extension in experimental time for these cultures is crucial for large numbers of experimental techniques.

29.6 Troubleshooting

Both PLL- and glutaraldehyde-treated surfaces in the preparation of the affinity substrates are used in the final protocol because they provide higher levels of reproducibility and stability than either PLL or glutaraldehyde by themselves. Whenever using glutaraldehyde, great care must be taken to remove all unconjugated aldehyde and to block any unreacted sites on the glutaraldehyde that is fixed to the surface. If any toxicity is observed after affinity separation, then more extensive washing of the substrates is required. If quantitative (better than 90%) yields of cells cannot be obtained (there is a variation in yield which is dependent on the affinity of the competitor carbohydrate used, see below) then more extensive blocking of the affinity substrate with BSA (try 1%, or add 10 mM glycine to the mixture) should be attempted.

The affinity substrate, once properly prepared and blocked, is stable for long periods of time. The substrate can be rinsed with water and dried, or left wet and wrapped in plastic shrink wrap. The advantages of preparing a number of affinity substrate dishes is mainly reproducibility of separation. These substrates can be re-used if care is taken to remove all cellular components. This can be achieved with the use of very high carbohydrate concentration (100 mM) for extended periods of time, followed by extensive water washing. The reproducibility of the separation procedure is severely reduced when the substrates are reused.

The initial experimental protocol involves extensive fluorescent lectin binding studies. These experiments establish the basic phenomenology, and are crucial for the selection of the affinity substrate for cellular subpopulation preparation. Many artifacts can express themselves at this stage of experimental development. Nonspecific fluorescence can result from high lectin concentrations or precipitation of extracellular matrix material, which is usually soluble and the presence of physically damaged cells in the preparation. All of these artifacts lead to the deposition of intense nonspecific fluorescence and should be avoided. One simple method of determining nonspecific fluorescence is that it is not directly concentration-dependent. Thus the initial quality control experiments should include lectin titration curve development (20–0.01 µg/ml), so that the concentration which provides the maximal fluorescence at the minimum concentration can be selected. The background fluorescence of the cells without any lectins should also be tested and minimized. High fluorescence at this point is usually due to the fixation technique or to desiccation of the cell populations at some point during preparation.

Specificity of the affinity substrate procedure should be tightly controlled. We use two controls for the substrate interaction itself. A nonspecific protein substrate such as BSA absorbed onto the PLL derivitized petri dish provides an assessment of the nonspecific attachment of the cells to an altered surface. An affinity substrate constructed of a lectin which has very low specificity for either cell population also provides a useful assessment of nonspecific substrate interactions. The specificity of the desorption process is also documented very carefully. A concentration gradient of the specific carbohydrate is useful in determining the relative affinity of the interaction. The use of nonappropriate carbohydrates provides an assessment of the specificity of the process.

Although this observation has not been explored by us, the control studies discussed above reveal that it is possible to remove three of four separate populations of myocytes from the affinity substrate by the propitious use of specific concentrations of various carbohydrate competitors (for example, 0.01 mM NN'N'' triacetyl chitotriose desorbs 88% of the myocytes, whereas 5.8 mM of GlcNAc is required to desorb 36% of these cells).

It is also very important when designing these experiments to use the minimal reagent concentrations required for reproducible results. Any manipulation of cells results in some level of functional impairment. This can be minimized by choosing the optimal conditions for the separation without exposing the cells to unnecessarily high doses of either lectin or carbohydrate. Myocytes prepared by these methods are functionally and morphologically indistinguishable from their control populations.

29.7 Summary

This study demonstrates the utility of lectins in the separation, and characterization and purification of specific functional cellular subpopulations from heterogeneous mixtures. The use of FITC-conjugated lectins to initially characterize the cellular population of interest in the original tissue is important. In the case of the myocytes, the functional definition of the cellular subpopulation of interest is relatively easy, because of the unique "beating" phenotype. In other cases the topological location of specifically labeling cells in the original tissue may be very important.

The preparation of the affinity substrate for this application required the use of both glutaraldehyde and poly-charged surfaces (PLL). Once prepared and characterized appropriately, these substrates perform in a stable reproducible manner. The major indications that the substrate has not been prepared correctly are: overt toxicity of the cells within 24 h of contact with the substrate, and the inability to desorb the attached cells quantitatively from the substrate.

Quality control measures which must be used include: assessment of the lectin specificity of the separation (before and after separation), assessment of the carbohydrate specificity (use of at least three specific sugars of differing affinities, and at least one nonspecific control), and the assessment of subsequent behavior of the separated cell populations in culture.

Using the procedures developed here, we have routinely achieved a sevenfold increase in the concentration of myocytes relative to the initial heterogeneous mixtures. This improvement in myocyte concentration permits a 5- to 7-day culture of these cells. Without the affinity separation technique, the fibroblasts outgrow the myocytes within 3 days of culture. The application of the WGA separation followed by Con A affinity separation results in even higher overall purity and therefore even more extended culture times, but the overall yields of cells lead us to the conclusion that there are other selection processes occurring which might affect the in vitro data. In other experiment arrangements this sequential separation by affinity substrate interaction could be useful. For instance, in experiments where specific phenotypically identical cellular subpopulations interacted as part of a differentiation sequence, these highly purified subpopulations could be useful in delineating the processes involved.

References

DeHann RL, Foggard HA (1975) Membrane response to current pulses in spheroidal aggregates of embryonic heart cells. J Gen Physiol 65:207–215

Hamburger V, Hamilton HL (1952) Normal stages of the chick. In: Lillie's development of the chick (H.L. Hamilton, ed) Holt, Rinehart and Winston, p74

Johnstone PN (1925) Studies on the physiological anatomy of the embryonic heart. Bull Johns Hopkins Hosp 36:299–305

Kaneko H, Goshima K (1982) Selective killing of fibroblast like cells in cultures of mouse heart cells by treatment with a calcium ionophore A23187. Exp Cell Res 142:407–411

30 Lectin-Binding Properties of Extracellular Matrix in the Chick Oral Membrane

R.E. WATERMAN, G.L. KRANNIG, M.P. VAIDYA, and P.L. MANN

The oral (buccopharyngeal, pharyngeal, oropharyngeal) membrane forms a temporary barrier between the stomodeal cavity and the lumen of the foregut during early development of vertebrate embryos. This barrier consists of a region of close approximation between the basal surfaces of the endodermal floor of the foregut and the overlying ectoderm of the stomodeum. The oral membrane (OM) gradually thins, perforates, and eventually disappears to allow direct continuity between the external environment and the lumen of the foregut. In higher vertebrates, the ectoderm and the endoderm are separated by a cell-free extracellular space during the early stages in the formation and perforation of the OM. There is circumstantial evidence to support the hypothesis that there is some functional difference between the extracellular matrix (ECM) within the OM and the matrix associated with the areas of the cranium in the immediate vicinity (Waterman and Schoenwolf 1980). Fluorescently conjugated lectins (FITC lectins) are used here to determine whether differences in lectin binding patterns between the OM and adjacent areas might be temporally and topographically correlated with the formation and perforation of the OM.

30.1 Objectives, Limitations, and the Outline of the Procedure

The objectives of this chapter include the development of procedures and techniques for the preparation of chick embryonic tissues suitable for FITC-lectin staining. These special techniques are required for the specific staining of the ECM, the target tissue of this study. Lectins are used as analytical probes to identify specific epitopes on the matrix material in the OM in stage 8 to 16 chick embryos (Hamburger and Hamilton 1951). The staining pattern of the OM has its own internal control which is used for comparison; namely, similar matrix material found in the same sections as the OM, just lateral to the OM in the cranial region.

Staining matrix with lectins presents special technical problems. The matrix itself tends to be a diffuse molecular "mat" of polymerized strands, which represents a relatively low density epitope array for lectins. Therefore, special attention is paid to reducing background nonspecific fluorescence and nonspecific lectin binding in order to maximize the limited staining potential of the matrix. Methodologies are developed which address the major areas of nonspecific interactions and background fluorescence; tissue fixation, stabilization of the ECM target; blocking of nonspecific sites; use of appropriate lectin concentrations; specificity of binding; affinity of binding; and quantitation of the binding.

In this study, chick embryos of stage 8 to 16 are lightly fixed with formaldehyde supplemented with cetylpyridinium chloride (CPC), embedded in paraffin, sectioned at 6 μm, transferred to slides, deparaffinized, hydrated, blocked, and stained with various lectins. Visual inspection and quantitative photomicrography are use to compare the specific staining patterns of the OM regions with those of the lateral cranial regions.

30.2 Equipment and Materials

Equipment
– 300 capacity egg incubator (Lehay Mfg. Higginsville, MO)
– Laboratory incubator (forced draft)
– Standard laboratory microtome
– Laboratory water bath
– Albuminized glass slides
– Light microscope equipped with epifluorescence (Nikon/Olympus)
– High speed ektachrome or tri-x film (Kodak, Rochester, NY)

Recommendations
a) Staging the embryos is extremely important for these studies and requires practice
b) Albumin on the slides aids in securing the sections, these can be prepared in advance
c) Standardize the photographic parameters, exposure time, film type, and printing exposures

Materials
– Fertilized white Leghorn chicken eggs (SPAFA, Norwich, CT)
– Immu-Mount (Shandon, Pittsburgh, PA)
– Saline solution (0.75%, w/v)
– Formaldehyde (2.5%) + cetylpyridinium chloride (CPC, 0.5%)
– Ethanol, xylene, paraffin [MP 58–60 °C], gelatin (Tousimis, Rockville, MD)

Various FITC conjugated lectins were tested in this study, these included; Concanavalin A (Con A), *Ulex Europeus* agglutinin (UEA I), soybean agglutinin (SBA), wheat germ agglutinin (WGA), *Ricinus communis* agglutinin (RCA-120), peanut agglutinin (PNA), and *Dolichos biflorus* agglutinin (DBA) (Vector Labs, Burlingame, CA). Optimal lectin concentrations for the lectins exhibiting specific labeling differences for this study are; PNA, 30 µg/ml; RCA, 10 µg/ml; and WGA, 2.5 µg/ml. Specific carbohydrate competitors for these lectins are: N-, NN′-, NN′N″-, and NN′N″N‴-oligomers of N-acetyl-D-glucosamine (GlcNAc, E-Y Labs, San Mateo, CA) for the WGA specificity; para-nitrophenyl β-D-galactose, and β-D-galactose for the RCA specificity; and the ortho nitrophenyl derivative of 2-aceto-2-deoxy-3-O-β-D-galactopyranosyl-D-galactose, and 1-O-methyl-β-D-galactopyranoside for the PNA specificity.

– Phosphate buffered saline (PBS, 0.137 M NaCl, 0.003 M KCl, 0.009 M PO_4)
– Bovine serum albumin supplemented PBS (BSA/PBS, 0.1% in PBS)
– Neuraminidase (*Clostridium perf*, 0.1 mg/ml, Worthington, Freehold, NJ)

Recommendations
a) Although it is not necessary for the solutions to be sterile, they should be kept free of contamination
b) The lectins are stored either lyophilized or at 10 mg/ml at − 20 °C until use
c) When the optimal lectin concentration is determined, 200 µl samples are frozen for subsequent use

d) The use of the neuraminidase must be optimized. The usual range is 0.01–0.1 mg/ml for 30 min at 37 °C
e) Both specific and nonspecific sugars should be tested at various concentrations

30.3 Detailed Procedure

– Embryos at stages 8–16 are used
– Eggs are opened into a finger bowl containing 0.75% saline
– The blastoderms are removed from the yolk
– They are fixed for 3–4 h at room temperature in formaldehyde/CPC
– Embroys are then dehydrated in ethanol and embedded in paraffin
– Sections are cut at 6 µm and floated onto warm water with gelatin
– Sections are placed on albuminized glass slides
– They are allowed to dry overnight at 45 °C
– Sections are deparaffinized in xylene
– They are re-hydrated
– Finally the sections are blocked in BSA/PBS

Preparation of OM sections

a) The OM develops and perforates during stage 8–16
b) The CPC is used to stabilize the ECM (0.5–1.0%)
c) The gelatin allows the sections to "spread" on the water
d) The albuminized slides give the sections a firm attachment to the slides
e) Once deparaffinized, the sections should be blocked and kept wet

Recommendations

– Determine the optimal lectin concentrations (50–1.0 µg/ml)
– Sections are flooded with 200 µl of lectin solution
– Slides are placed in humidified chambers and incubated for 60 min at 37 °C
– Slides are rinsed (2 × 5 min) in PBS
– Mount in Immu-Mount, examine and photograph

Lectin Labeling

a) Determine the minimal concentration of lectin which provides best label
b) Do not let any part of the sections dry
c) Use a grease pencil circle around section to stop fluid spread
d) Standardize the photographic parameters; exposure time, film type, and printing exposures.

Recommendations

Carbohydrate Competition:

– At least two specific and nonspecific carbohydrates are tested for each lectin
– Multiple concentrations to compete for at least 80% of binding
– Mix the sugar and lectin immediately prior to incubation
– Incubate for 90 min at 37 °C in a humidifier box to prevent drying
– Rinse, mount, and inspect as above
– Establish photographic comparison parameters with the control sections
– Use the same parameters for the sections with competitors
– Quantitate the data from photos

Quality Control

Recommendations
a) Choose specific/nonspecific competitors which are as appropriate as possible
b) Test specific competitor concentrations with 0–80% inhibition
c) Match nonspecific concentrations to specific for quantitation
d) Standardize the photographic parameters; exposure time, film type, and printing exposures

Biochemistry:

- Incubate sections in neuraminidase for 30 min at 37 °C
- Rinse sections with PBS (2 × 5 min) at room temperature
- Incubate treated sections with lectin for 60 min at 37 °C
- Process as described above

Recommendations
a) It is very important to use highly purified neuraminidase
b) Enzyme containing nonspecific pronases increases the nonspecific binding.
c) Use as little enzyme for as short a time as possible
d) The controls for this protocol include detailed lectin screens and multiple competitors

30.4 Results

Background fluorescence of the 6 μm control sections show no discernable anatomical landmarks under fluorescence illumination. This is one of the major control factors for the experimental protocol. Preliminary experiments indicated that the specific lectin labeling of the OM matrix with PNA was of low intensity in comparison to the intense staining of the neutral tube by WGA. One reason for this low intensity pattern is the nature of the target; matrix material is relatively low-density and tends to be space-filling. These attributes lead to difficulties in stabilizing the material and, once stabilized, even more difficulties in quantitative assessment of such a low density staining pattern. Thus, the first step in this procedure is to minimize the background noise as much as possible.

Figure 1 shows the anatomical landmarks of the sections studied in this experimental protocol. This figure was electronically abstracted for clarity from an actual photomicrograph. The photograph of the section was scanned and digitized into a computer (Mackintosh II x, on loan from Los Alamos National Labs, Los Alamos, NM), and then traced to show the major anatomical landmarks. The result is a simplified representation of the section which contains all the significant anatomical landmarks of a Stage 11 chick embryo at the level of the OM. As can be seen, the OM is a very small confined area, defined by a cell-free matrix with distinct cellular layers on the top and bottom. It is, however, the site of highly significant differentiation processes, involving the communication, phenotypic response, fusion, and finally the perforation of ectoderm and foregut endoderm. Visual tools are absolutely essential to dissect out the temporal and positional components of the sequence of events which leads to the development of the oral opening.

Table 1 shows the three lectins which showed prominent, consistent staining of the sections, and their nominal carbohydrate specificities. A panel of eight lectins was screened for specific lectin-staining patterns. These three were chosen from this panel.

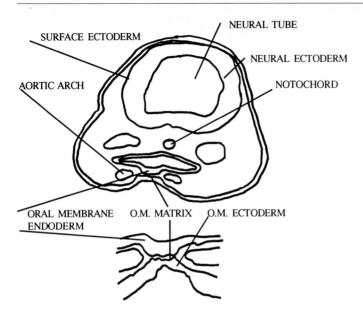

SURFACE ECTODERM

NEURAL TUBE

NEURAL ECTODERM

AORTIC ARCH

NOTOCHORD

ORAL MEMBRANE   O.M. MATRIX   O.M. ECTODERM
ENDODERM

ENLARGEMENT OF ORAL MEMBRANE

Fig. 1. Anatomical landmarks of the stage 11 chick embryo. This 6-µm section shows the OM region, and is prepared as described in the text. The section is photographed and printed using a standardized protocol to ensure accurate comparison between control and test sections. This print is then scanned at 300 DPI into a Mac IIx and processed with MacImage and then traced to show an outline of the important anatomical features of the section. The labels show these features. The OM region is magnified and presented in the *inset*

Table 1. The three major lectins which showed specific labeling patterns on the sections, and examples of their specific and nonspecific carbohydrate competitors

Specific lectin	Competitor	Noncompetitor
WGA	GlcNAc	GalNAc
RCA	PNP-Gal	PNP-Mann
PNA	(1-3)Gal/GalNAc	GlcNAc

Specificity of binding in these experiments is confirmed by the use of specific and nonspecific carbohydrate ligand competitors. Examples of both specific and non-specific sugars are given in Table 1.

WGA is a general stain for most of the major anatomical features in these sections. RCA, in general, stains similar structures but with less intensity. PNA is much more specific in its pattern and especially notable is its staining of the OM matrix.

Table 2 shows the major quantitative results of this protocol. As noted above, the WGA produces a very strong generalized staining pattern, not unusual for a tissue based high density epitope pattern. This suggests that glycosylation is extensive and that β1-3 linked GlcNAc residues are ubiquitous at this stage of development. Ricin staining (β-D-Gal) is also ubiquitous but noticeably less intense. PNA (β1-3 Gal-GalNAc), on the other hand, shows a high degree of anatomical specificity.

Table 2. An abstraction of the photographic quantitation of the lectin experiments. The scale used is a standard 0 to +4 intensity rating. Because all the photographic studies of the fluorescence were performed under standard conditions, the photographs could be rated by several individuals on a blind basis. The +4 rating is established within an experiment and all other patterns are rated according to this standard. Excellent correlation and reproducibility can be obtained using this method

Structure	WGA	RCA	PNA
Notochord	+ 4	+ 2	+ 4
Surface ectoderm	+ 4	+ 2	0
Neural ectoderm	+ 2	+ 1	+ 4
Aortic arch	+ 2	+ 1	0
OM endoderm	+ 2	+ 1	0
OM ectoderm	+ 2	+ 1	0
OM matrix	+ 1	+ 1	+ 4

The PNA binding is specific to the axial structures of the section. Specifically, the luminal and basal surfaces of the neuroepithelium, perinotochordal region, and the matrix tissue which lies within the OM. A high magnification analysis of the OM region reveals that the staining is diffuse and concentrated specifically between the two cellular layers which define the region. No cells are visible in this diffuse matrix-rich area.

The specific matrix-dependent PNA staining is present from stage 8 through perforation (stages 14–15). PNA staining is also observed in the ectoderm of Rathke's pouch. This specific staining, unlike the matrix staining, appears only at the later stages of OM development (stages 14–15). At stages 14–15, WGA and RCA have intense binding patterns to the matrix between the ectoderm and endoderm of the closing plate. PNA does not stain these structures.

Neuraminidase treatment of the sections results in the appearance of generalized PNA staining through out the sections, indicating that the specific PNA staining seen in the OM is probably due to the unique display of penultimate β1-3 Gal-GalNAc residues, in an otherwise general glycosylation process.

This unique PNA staining pattern is the first evidence of a compositional difference between matrix in the oral region and the matrix observed more laterally in the cranial region. This type of study in which positional, temporal, and biochemical data can be correlated in situ is very important in the ultimate elucidation of the molecular mechanisms responsible for the definition, preparation, and final development of significant anatomical features. In turn, these studies will be the first steps in the early diagnosis and management of aberrant development of these regions.

30.5 Troubleshooting

Tracing differentiation mechanisms in embryonic tissues is complicated by the extremely small numbers of cells and cellular products involved in initiating subsequent growth and differentiation reactions. Thus, high sensitivity is required for these lectin staining protocols. Sensitivity of detection is as much a matter of reducing background noise levels as of increasing the signal intensity. In this particular case, where specific signal is so limiting, greater attention must be paid to noise reduction. This is especially crucial when working with lectins. Bi- or multi-functional lectins at high concentrations (usually $> 20\,\mu g/ml$ for tissue displayed epitopes and $> 50\,\mu g/ml$ for more diffuse epitopes) lead to nonspecific micro-crystalline aggregates of the lectins themselves. Although these produce intense coloration and therefore, are sometimes mistaken for specific lectin staining, these micro-crystallites are artifacts. They tend to spread well beyond the original epitope boundaries and result in much higher nonspecific deposition of lectin.

While fixation of embryonic tissue with glutaraldehyde results in better structural preservation, 2.5% formaldehyde is used since the bi-functional nature of glutaralde-hyde tends to increase the nonspecific background binding of the lectins.

The CPC additive is used to stabilize the normally partially soluble matrix material. Alcian Blue and other molecular chelators have been used for the same purpose. When assessing these materials, care must be taken not to introduce reagents with inherent fluorescent properties or materials which will lead to increased nonspecific background fluorescence values. CPC is an excellent choice for the stabilization of extra cellular matrix, but it appears to be very harsh on the cells themselves. When the structure of the matrix is examined at the EM level it is observed that the cellular morphology is poor when CPC is used.

Fluorescence decay is an exponential process. Therefore, the maximal signal is obtained immediately after irradiation of the specimen with UV light. The few seconds that it normally takes to focus and prepare to obtain photographic evidence can result in an much as a 50% reduction in the total signal output. Again, this is especially crucial in this study where the OM matrix is diffuse low density staining material. Typically, the sections are viewed, focused, and prepared for photography under minimal white light illumination (with a blue filter), then in complete darkness the camera shutter is opened and the UV illumination initiated for a timed exposure. This simple procedure provides maximal light capture and a method for obtaining quantitative photographic evidence of the experiment. When this method is stand-ardized with independent quantitative measurements of light output, excellent correlation over a wide dynamic range is found. More modern automatic cameras are often difficult to use under these circumstances.

Blocking of unwanted chemistry between the surface of the section and the lectin must also be approached with caution. If blocking is omitted completely, nonspecific absorption of the lectin reagents is unacceptably high. If, on the other hand, the blocking procedure is too intense (high protein concentrations, very extended periods of time), then much low intensity specific staining will be obliterated (stearic hindrance). We commonly use 0.1% BSA in PBS as a starting point for the blocking procedure. If nonspecific staining is still a problem, 10 mM glycine is added to the mixture, followed by increasing the BSA concentration to 1% if necessary.

One clear indication that the fixation, sectioning, and blocking procedures are adequate is given by the conformational specificity of the lectin staining itself. For instance, in preliminary studies, strong RCA staining is observed but only back-ground levels of *Bandeiraea simplicifidia* staining. Since the primary difference between these lectins is the α and β conformation of the Gal residues, one can be reasonably satisfied that there is an adequate balance between the background and specificity of the lectin interactions.

Another quantitative indication that the specific binding event is primarily epitopic in nature, without excessive amounts of surface-specific co-operativity (leading to irreversible binding), is the ability to construct dose-dependent carbohydrate ligand competition curves. For instance, the binding of RCA can be completely inhibited with β D-Gal. Para-nitrophenyl β D-Gal is four to ten fold better on a concentration basis in inhibiting the interaction. Typically, two to four separate ligand competitors are used to establish the specificity of interaction. These competitor experiments also provide some quantitative information about the interaction. If simple visual quantitation of binding is used, then enough concentrations of each sugars (6–12) must be tested to provide for the assessment of the 50% inhibition point of the original fluorescence (use a $+4$ to $+2$ reduction criteria). The sugar concentrations required for some standard reduction in fluorescence intensity (50%) can then be plotted against the nominal published IC_{50} doses for these same sugars (Goldstein and Hayes, 1978). A linear relationship indicates that the major binding entity of the lectin is directed towards the sugar epitope itself. If, however, only a limited proportion of the total binding can be competed for in this way, then the residual binding is probably due to cooperativity, and should be traced back to fixation or blocking problems.

30.6 Summary

– A protocol is established which minimizes background tissue fluorescence
– A protocol is established for specific lectin staining of embryonic structures
– A unique β 1-3-Gal-GalNAc-dependent epitope is identified in the OM matrix
– Stage-specific, differentiation-dependent changes in the OM matrix are identified

This study forms the basis for a detailed examination of the mechanism(s) involved in OM initiation, growth, differentiation, and final rupture to form the oral opening. The data presented here suggest that there is a unique processing of the matrix associated with the OM. This matrix appears phenotypically distinct from matrix in the lateral cranial region, a distance of only a few tenths of a micron displaced from the OM. This level of spacial discrimination is an indication of how useful lectin reagents are in these types of mechanistic studies. Future investigations could involve the study of endogenous lectins in this tissue region and their correlation with the observed carbohydrate ligand studies. Studies have shown that these endogenous lectins form another component of the control mechanism for the expression of carbohydrate-dependent control over cellular/tissue differentiation mechanisms (Bardosi et al. 1989).

References

Bardosi A, Dimtri T, Wosgien B, Gabius HJ (1989) Expression of endogenous receptors of neo-glycoproteins, especially lectins, that allow fiber typing on formaldehyde-fixed muscle biopsy specimens. J Histochem Cytochem 37: 989–998
Goldstein IJ, Hayes CE (1978) The lectins: carbohydrate-binding proteins of plant and animals. Adv Carbohydr Chem Biochem 35: 127–340
Hamburger V, Hamilton HL (1951) A series of normal stages in the development of the chick embryo. J. Morph. 88: 49–92
Waterman RE, Schoenwolf GC (1980) The ultrastructure of OM formation and rupture in the chick embryo. Anat Rec 197: 441–470.

31 Biodistribution of Radioactive Lectins

S. Kojima

It has been reported that certain lectins exhibit tumor-cell specificity (Inbar et al. 1960; Sela et al. 1970; Louis and Wyllie 1981; Louis et al. 1981), which makes them invaluable as tools for elucidating changes in cell surface architecture upon malignant transformation, or for studies of surface plasticity during normal developing processes. Judging by these characters of lectins, they appear to be promising powerful tools for cell biology, drug-carriers for therapy, and diagnosis of various diseases such as tumor, through carbohydrates.

We had already reported that *Pisum sativum* agglutinin (PSA) specifically bound with Ehrlich ascites tumor cells (EAT) in vitro and accumulated in Ehrlich solid tumor (EST) in vivo, and suggested the possible application of this lectin to tumor imaging radiopharmaceuticals (Kojima and Jay 1986). In this chapter, biological activity, biodistribution, and one of the application of radiolabeled lectins are described.

31.1 Radiolabeling of Lectins

Lectins (Affinity purified, E.Y. Laboratory Inc., San Meteo, USA); *Phaseolus vulgaris* **Materials** (PHA), *Pisum sativum* (PSA), *Bauhinia purpurea* (BPA), *Maclura promifera* (MPA), wheat germ (WGA), *Griffonia simplicifolia* [I] (GS-[I]), *Griffonia simplicifolia* [II] (GS-[II]), *Griffonia simplicifolia* [I-B4] (GS-[I]-B4), *Concanavalia ensiformis* (Con A), *Ulex europeaus* [I] (UEA-[I]), *Arachis hypogaea* (PNA), *Glycine max* (SBA), *Dolichos biflorus* (DBA), *Riccinus communis* [I] (RCA-[I]), *Riccinus communis* [II] (RCA-[II]), *Rana japonica* (RN)

^{125}I-NaI (17.4 Ci/mg, NEN, Boston, Mass., USA), ^{67}GaCl$_3$ (carrier-free, Nihon Mediphysics, Japan)

Diethylene triamine penta acetic acid (DTPA, Dojin Chemical Lab., Japan)

Deferoxamine (DFO) methylate (Chiba Geigy, Tokyo, Japan)

1,3,4,6-Tetrachloro-3α-6α-diphenylglycoluril (Iodogen, PIERCE, Rockford, Illinois, USA)

31.1.1 Labeling of Lectins with Different Radioactive Nucleides

The lectins are easily lebeled with ^{125}I-NaI, without loss of the biological activity, by the Iodogen method. Iodinated compounds, however, were often subject to problems such as dehalogenation, loss of biological activity, etc. In order to overcome these problems, labeling methods for obtaining an *in vivo* stable and highly biological activite radiolabeled lectin should be established. In this chapter, the labeling methods of lectins with ^{125}I and radioactive metals, e.g., ^{67}Ga or ^{111}In, using bifunctional chelating agents, are described. In addition, the effect of labeling on the *in vitro* biological activities and the biodistribution of lectins in tumor bearing mice are examined.

31.1.2 Conjugation of DTPA and DFO to Lectins

DTPA and DFO were conjugated with lectins basically by the method of Krejcarek and Tucker (1977) and Yokoyama et al. (1982), respectively, with a slight modification as follows:

Preparation of DTPA Lectins

Conjugation
1. DTPA (10 mg, 0.025 mmol) and triethylamine (12.5 mg, 0.125 mmol) are dissolved in 200 μl of H_2O with gentle heating, and the solution was freeze dried
2. Pentatriethylammonium produced by the reaction described above was dissolved in 200 μl of acetonitrile with gentle heating
3. The reaction mixture was then cooled at 4 °C and isobutylchloro-formate (3.5 mg, 0.025 mmol) was added. The mixture was stirred for 0.5 h in a cold room
4. The resulting mixed carboxycarbonic anhydride of DTPA was then added to a solution of the lectin (10 mg) in 10 ml of 0.1 M $NaHCO_3$ The reaction mixture was kept in a cold room overnight and subsequently dialyzed against 0.01 M acetate buffer (pH 5.0).
5. The DTPA-lectin conjugate was isolated in void fractions by Sephadex G-25 (or G-50) column chromatography, eluted with 0.01 M acetate buffer (pH 5.0). Fractions corresponding to DTPA lectin were monitored by measuring the UV absorption at 280 nm, and dialyzed against 0.1 M glycine-HCl buffer (pH 3.5)

Preparation of DFO Lectins

1. Ten μl of 1% (w/v) glutaraldehyde are added to 1 ml of 60 mM DFO mesylate dissolved in phosphate buffered saline (PBS, pH 7.4), and the mixture was stirred for 5 min at room temperature
2. An aliquote of the reaction mixture (e.g., 0.1 ml) was added to 10 ml of 10 mM lectin solution dissolved in PBS buffer, and the mixture was stirred for 45 min at 4 °C
3. After 0.1 mg of sodium borohydrate had been added, the reaction mixture was stirred for another 15 min
4. The DFO lectin conjugate was isolated in void fractions by Sephadex G-25 (or G-50) column chromatography, eluted with PBS

31.1.3 Radiolabeling of Lectin with ^{125}I, and of DTPA Lectins and DFO Lectins with [67]Ga

Radioiodination of lectins was performed by the Iodogen method (Fraker and Speck 1978) with a slight modification as follows:

Radiolabeling of Lectins with [125]I

Radiolabeling
1. One hundred μl of 1.0 mM Iodogen dissolved in methylene chloride was applied to a glass test tube (14 × 75 mm) and thoroughly dried under a stream of dry nitrogen gas at room temperature

2. A solution of 1 mg of lectin dissolved in 2 ml of PBS and 1.85 MBq of ^{125}I-NaI were added to the test tube coated with Iodogen

3. After being kept at 4 °C for 30 min, the reaction mixture was applied on Sephadex G-25 (or G-50) column chromatography, eluted with PBS

Radiolabeling of DTPA-Lectins and DFO-Lectins with ^{67}Ga

1. An aliquot of DTPA lectin dissolved in 0.1 M glycine-HCl buffer (pH 3.5) was incubated with 370 KBq of ^{67}GaCl$_3$ at room temperature for 30 min

2. DFO lectin was also labeled with ^{67}Ga in similar way.

3. Each radiolabeled lectin was separated by Sephadex G-25 (or G-50) gel column chromatography, eluted with PBS

31.2 Assay Methods of Biological Activities of Radiolabeled Lectins

31.2.1 Agglutinating Activity of EAT Cells

1. The cells are harvested from abdominal cavity of EAT-bearing mice and washed three times with ice-cold physiological saline   **Agglutinating**

2. Solutions of the unmodified-, ^{125}I-, ^{67}Ga-DTPA-, and ^{67}Ga-DFO lectins were each serially two-fold diluted in microtest tubes containing 0.1 ml of 0.9% NaCl

3. To each test tube, 0.1 ml of EAT cell suspension (1.0×10^7 cells/ml in 0.9% NaCl) is added

4. The mixture is incubated at room temperature for 40 min, and the degree of cell clumping is determined according to the criteria of Vesely et al. (1972)

31.2.2 In Vitro Binding Activity of Radiolabeled Lectins to EAT Cells

1. The EAT cells are adjusted to a final concentration of 4×10^7 cells/ml in PBS   **Binding**

2. 0.1 ml of EAT cells are mixed with 1 KBq (1.25 µg) of radiolabeled lectins and incubated at 0 °C for 1 h.

3. The reaction is stopped by adding 2 ml of ice-cold PBS and washed twice with the same buffer by centrifugation at 3000 rpm for 5 min

4. The bound radioactivity of the cell pellets is determined with a γ-counter

5. The extents of radiolabeled lectins are calculated as the percentage of the retained radioactivity in the precipitated cells to the added total dose

31.3 Biodistribution of Radiolabeled Lectins

Differences of the biodistribution among ^{125}I-, ^{67}Ga-DTPA-I, and ^{67}Ga-DFO lectins are described in this chapter. A protocol using *Pisum sativum* agglutinin (PSA) as a lectin is shown, as follows:

1. Each radiolabeled PSA, i.e., ^{125}I-PSA, ^{67}Ga-DTPA-PSA, or ^{67}Ga-DFO-PSA, is   **Biodistribution** injected into the tail vein of Ehrlich solid tumor (EST)-bearing mice nine days

after the s.c. implantation of EAT cells (2×10^5 cells/head). Dose of each radiolabel is about 37 KBq/5 µg/mouse

2. At 24 h postinjection, they are sacrificed, and organs and tumor removed, weighed, and assayed for radioactivity with a γ-counter
   Blood sample is withdrawn from the abdominal aorta

3. From these data, the percentage injected dose per g of wet tissue (%ID/g tissue) is calculated

4. The ratio (%ID/g tissue) of each radiolabeled-PSA is compared

31.4 Results

31.4.1 Biological Activity and Biodistribution of Radiolabeled Lectins

Gel chromatography patterns of ^{125}I-PSA, ^{67}Ga-DTPA-PSA and ^{67}Ga-DFO-PSA are shown in Fig. 1. When PSA is radioiodinated with ^{125}I-NaI by the Iodogen method, labeling efficiency is usually about 50%. ^{67}Ga-DTPA-PSA and ^{67}Ga-DFO-PSA are both easily prepared by simple incubation, at low pH, of each conjugate with ^{67}GaCl$_3$. A labeling efficiency of 95% ~ 100% is obtained in both cases. A loss of radioactivity for these radiolabels, being kept at 4 °C for one week, is negligible.

Comparison of agglutinating activities of the intact PSA, ^{67}Ga-DTPA-PSA, and ^{67}Ga-DFO-PSA is shown in Table 1. EAT cells are agglutinated at 16 µg/ml of the intact PSA. ^{125}I-PSA labeled by Iodogen method show the same activity as the parent lectin. When PSA is conjugated with DTPA, the activity is almost completely

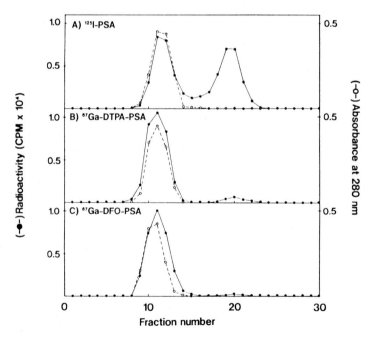

Fig. 1. Gel Chromatography of ^{125}I, ^{67}Ga-DTPA- and Ga-DFO-PSA on Sephadex G-25

lost, whereas the activity is well retained in DFO-PSA. The following radiolabeling with metal nucleide, ^{67}Ga, has no effect on the agglutination activity. The in vitro binding activity of ^{125}I-PSA, ^{67}Ga-DTPA-PSA and ^{67}Ga-DFO-PSA with EAT cells is shown in Table 2. The ratio of ^{125}I-PSA is 36.2% binding to the cells, ^{67}Ga-DTPA-PSA shows substantially lower binding (5.6%), but reasonably high binding (26.7%) in ^{67}Ga-DFO-PSA. ^{67}GaCl$_3$, ^{67}Ga-DTPA and ^{67}Ga-DFO themselves represent low binding, (6.0%, < 0.2%, and < 0.3%, respectively). The 24 h biodistribution of ^{125}I-PSA, ^{67}Ga-DTPA-PSA, and ^{67}Ga DFO-PSA in EST-bearing mice is summarized in Table 3. The biodistributions of ^{125}I-PSA and ^{67}Ga chelating agent conjugated PSAs are significantly different. The radioactivity of ^{125}I-PSA accumulated in tumor tissue is 2.73% ID̂/g tissue. The value of ^{67}Ga-DTPA-PSA is markedly lower (1.4% ID/g tissue). Furthermore, the radioactivity remaining in the blood after IV injection of ^{67}Ga-DTPA-PSA is much higher than that in the case of ^{125}I-PSA. On the other hand, ^{67}Ga-DFO-PSA gives a much higher accumulation in the tumor tissue, approximately twice that of ^{125}I-PSA. The value (%ID/g tissue) of ^{67}Ga-DFO-PSA in the liver, however, is almost the same as that of ^{67}Ga-DTPA-PSA, with much higher excretion ratio into kidney. A marked difference between ^{125}I-PSA and ^{67}Ga-chelating agent conjugated PSA is that the radioactivity concentrates into the liver after IV injection of ^{67}Ga-DTPA- and ^{67}Ga-DFO-PSA. It has been well known that antibodies labeled with radioiodine are unstable in vivo and show decreased immunological activity (Pettit et al. 1982). Another disadvantage of this labeling method is loss of the label from the antibodies by

Table 1. Agglutinating activity of PSA to Ehrlich ascites tumor cells[a]. (Kojima and Jay 1987)

PSA	Concentration (µg/ml)					
	4	8	16	32	62	128
PSA	−	±	+	+ +	+ +	+ + +
^{125}I-PSA	−	±	+	+ +	+ +	+ + +
DTPA-PSA	−	−	−	−	±	+
DFO-PSA	−	±	+	+ +	+ +	+ + +
^{67}Ga-DFO-PSA	−	±	+	+ +	+ +	+ + +

[a]One plus sign (+) designates only small macroscopically observable clumps resisting gentle shaking of the test tube; two pluses (+ +), clumps at least twice big as one plus clumps; three plus (+ + +), very large clumps only. The plus minus sign covers a range of microscopical clumps mixed with numerous free single cells.

Table 2. In vitro binding of radiolabeled PSA with Ehrlich ascites tumor cells[a]. (Kojima and Jay 1987)

PSA	Binding ratio (%)
^{125}I-PSA	36.2
^{67}Ga-DTPA-PSA	5.6
^{67}Ga-DFO-PSA	26.7
^{67}GaCl$_3$	9.0
^{67}Ga-DTPA	< 0.2
^{67}Ga-DFO	< 0.3

[a]Each value is the mean of four independent assays and represents the binding (%) with 4×10^6 cells.

Table 3. The 24-h biodistributions of ^{125}I-PSA, ^{67}GA-DTPA-PSA, and ^{67}Ga-DFO-PSA in Ehrlich solid tumor-bearing mice[a]. (Kojima and Jay 1987)

Tissue	^{125}I-PSA	^{67}GA-DTPA-PSA	^{67}Ga-DFO-PSA	^{67}Ga-DTPA	^{67}Ga-DFO
Blood	0.50 ± 0.10	1.11 ± 0.13	0.79 ± 0.06	1.22 ± 0.19	0.01 ± 0.00
Liver	4.66 ± 1.43	17.00 ± 0.55	17.90 ± 1.86	2.56 ± 0.20	0.06 ± 0.01
Kidneys	10.92 ± 2.25	8.80 ± 1.56	27.80 ± 1.25	3.43 ± 0.08	0.73 ± 0.01
Spleen	9.00 ± 2.44	11.21 ± 2.13	14.90 ± 1.22	1.86 ± 0.33	0.02 ± 0.00
Heart	2.56 ± 0.82	1.08 ± 0.13	2.52 ± 0.09	0.63 ± 0.06	0.01 ± 0.00
Lungs	5.91 ± 1.02	1.73 ± 0.13	10.80 ± 1.14	1.19 ± 0.17	0.03 ± 0.00
Muscle	0.52 ± 0.19	0.24 ± 0.05	0.92 ± 0.20	0.25 ± 0.02	0.01 ± 0.00
Tumor	2.73 ± 0.50	1.27 ± 0.21	4.69 ± 0.78	2.15 ± 0.44	0.03 ± 0.00

[a]Each value represents % injected dose/g tissue (%ID/g tissue), and the mean $\pm$ SD for five mice.

dehalogenation (Halpern et al. 1983; Hagan et al. 1985). On the other hand, the use of a metallic radionucleide such as ^{67}Ga or ^{111}In as a label in conjugation with a bifunctional chelating agent may provide higher labeling efficiency, greater stability, and better retention of biological activity. In this chapter, the labeling efficiency, biological activity, and biodistribution of radiolabeled PSAs labeled with different radioactive nuclides, i.e., ^{125}I and ^{67}Ga, are compared. ^{67}Ga-chelating agent conjugated lectins could be prepared with higher labeling efficiency than ^{125}I-lectins. Agglutinating activity towards various tumor cells is one of the biological activities of lectins (Vesely et al. 1972). The activity of ^{67}Ga-DFO-PSA is almost the same as the ^{125}I-PSA, but significantly reduced when the lectin is conjugated with DTPA. DTPA and DFO are both generally used as bifunctional chelating agents when a protein is to be labeled with ^{67}Ga or ^{111}In. DFO is well known as a specific chelating agent for iron. This agent appears to form a stable one to one chelate with Ga(III) or In(III) ions by coordination bond through the three hydroxamate groups, and contains an amino group of comparatively high reactivity, which is available as a functional group for coupling reactions with proteins. On the other hand, in the case of DTPA, a carboxyl group is available for coupling to form DTPA-protein conjugates. Differences of agglutinablity between DTPA lectins and DFO lectins may be due to this fact. In addition, DFO forms an unchanged complex with protein, and might therefore have relatively little influence on the activity of the protein. Difference of the in vitro binding activity to EAT cells between ^{67}Ga-DTPA-PSA and ^{67}Ga-DFO-PSA accords well with that of agglutinating activity. Namely, the binding activity of ^{67}Ga-DFO-PSA is well kept together with the agglutinating activity. The in vivo behavior of radiolabeled proteins is influenced by many factors and is not always in agreement with the in vitro reactivities. As shown in Table 3, it is noteworthy that a larger amount of ^{67}Ga-DFO-PSA, which showed little change of in vitro biological activity in comparison with the parent lectin, was taken up by EST tissue. This may be due to the greater in vivo stability of ^{67}Ga-DFO-lectins, ^{125}I-lectins may be deiodinated in vivo. One of the big problems, however, is that high nonspecific concentration of radioactivity in other organs, especially liver and spleen are observed when ^{67}Ga (or ^{111}In)-chelating agent-lectins are used. Even if there are a few problems to be solved, radiolabeled lectins with a high in vivo stability and without loss of biological activity are seriously required for biodistribution studies.

31.4.2 Application of Lectins for Tumor Imaging Radiopharmaceuticals

The application of radiolabeled monoclonal antibodies to tumor imaging radio-pharmaceuticals in nuclear medicine has been extensively studied. Although radiolabeled antibodies provided adequate tumor images, there are still some problems in this approach. One of the greatest problems is that background subtraction by computer techniques is necessary to obtain clear images within an early postinjection time. This is due to the slow clearance of radiolabeled antibodies from the blood circulation. The remaining radioactivity in patients not any exposes them to additional radiation, but also adds several potential sources of error. It is already known that lectins react with corresponding receptor sites on the cells surface in the same way as antibodies.

As certain lectins show tumor cell specificity, meaning presence of the specific receptor on the cell surfaces, as described in introduction, we can expect an application of lectins to radiopharmaceuticals for tumor imaging. So we tried to find out which lectins might concentrate in tumor lesions and examined the possibility of using radio-labeled lectins as tumor imaging agents. As shown in Table 4, PHA, Con A, and PSA demonstrated high affinity for EAT cells in an in vitro binding assay. Although these lectins showed a relatively high accumulation in the tumor in comparison with other lectins, their in vitro binding specificity to EAT cells was not always in agreement with the in vivo accumulation ratio in tumor tissues. This fact indicates that the in vivo biodistribution is determined by various factors other than an interaction between lectin and tumor cells. Namely, PHA never concentrate in the EST tissue in spite of the in vitro high affinity to the EAT cells, as shown in Table 5. On the other hand, PSA with a moderate affinity shows significantly high concentration in the tumor tissue. It is reported that PSA, similarly to Con A, hemagglutinates all types of red cells and binds specifically to α-D-mannosyl and α-D-glucosyl residues (Van Wauwe et al. 1975; Kornfield et al. 1981). We observed that the in vitro binding of PSA to EAT cells was also, inhibited by mannose and glucose.

Table 4. In vitro binding of ^{125}I-lectins with Ehrlich ascites tumor cells[a]. (Kojima and Jay 1986)

Lectins	Abbreviation	Binding ratio (%)
Phaseolus vulgaris	PHA	85.2
Concanavalia ensiformis	Con A	57.7
Pisum sativum	PSA	41.6
Bauhinia purpurea	BPA	19.1
Maclura promifera	MPA	17.1
Wheat germ	WGA	16.5
Griffonia simplicifolia [I]	GS [I]	14.4
Riccinus communis [II]	RCA[II]	12.2
Riccinus communis [I]	RCA [I]	6.2
Ulex europeaus [I]	UEA	5.5
Arachis hypogaea	PNA	3.1
Glycine max	SBA	2.5
Griffonia simplicifolia [I-B4]	GS [I-B4]	1.8
Dolichos biflorus	DBA	0.9
Griffonia simplicifolia [II]	GS [II]	0.6

[a]Each value is the mean of four independent assays and represents the binding with 4×10^6 cells.

Table 5. The 24 h biodistribution of ^{125}I-lectins in Ehrlich solid tumor-bearing mice[a]. (Kojima and Jay 1986)

Tissue	^{125}I-PSA	^{125}I-Con A	^{125}I-PSA	^{125}I-BPA	^{125}I-WGA	^{125}I-PNA
Blood	0.60 ± 0.04	0.77 ± 0.18	0.36 ± 0.02	0.08 ± 0.03	0.88 ± 0.02	0.16 ± 0.04
Liver	2.42 ± 0.17	2.44 ± 0.33	2.22 ± 2.22	3.56 ± 0.34	2.43 ± 0.14	0.59 ± 0.03
Kidneys	3.02 ± 0.51	4.26 ± 0.25	7.56 ± 0.60	0.53 ± 0.06	9.26 ± 0.69	2.94 ± 0.38
Spleen	12.20 ± 4.30	4.34 ± 0.24	7.52 ± 0.79	1.64 ± 0.18	4.18 ± 0.54	0.41 ± 0.07
Heart	1.65 ± 0.29	0.92 ± 0.13	1.42 ± 0.09	0.04 ± 0.01	0.55 ± 0.14	0.13 ± 0.02
Lungs	6.99 ± 1.74	6.39 ± 0.95	6.77 ± 0.28	0.09 ± 0.02	1.44 ± 0.19	0.58 ± 0.10
Muscle	0.28 ± 0.05	0.25 ± 0.06	0.36 ± 0.02	0.03 ± 0.00	0.29 ± 0.04	0.05 ± 0.01
Tumor	0.88 ± 0.02	1.00 ± 0.18	2.35 ± 0.23	0.12 ± 0.02	0.68 ± 0.08	0.28 ± 0.10

[a]Each value represents % injected dose/g tissue (%ID/g tissue), and the mean ± SD for five mice.

Table 6. Effect of time on tissue distribution of ^{125}I-PSA in Ehrlich solid tumor-bearing mice[a]. (Kojima and Jay 1986)

Tissue	3 h	6 h	12 h	24 h	48 h	72 h
Blood	3.59 ± 0.42	1.78 ± 0.21	0.71 ± 0.05	0.36 ± 0.02	0.18 ± 0.02	0.15 ± 0.03
Liver	14.90 ± 0.69	10.77 ± 0.57	5.26 ± 0.07	2.22 ± 0.22	0.76 ± 0.13	0.36 ± 0.03
Kidneys	29.00 ± 2.43	27.89 ± 3.64	16.10 ± 1.82	7.56 ± 0.60	2.40 ± 0.14	1.14 ± 0.17
Spleen	16.30 ± 1.84	15.40 ± 4.44	11.08 ± 0.32	7.52 ± 0.79	4.17 ± 0.57	2.52 ± 0.26
Heart	3.03 ± 0.14	3.10 ± 0.07	2.52 ± 0.23	1.42 ± 0.09	0.58 ± 0.11	0.32 ± 0.03
Lungs	13.48 ± 1.65	14.23 ± 0.89	10.34 ± 0.78	6.77 ± 0.28	3.57 ± 0.58	2.50 ± 0.43
Muscle	0.58 ± 0.03	0.51 ± 0.01	0.47 ± 0.02	0.36 ± 0.02	0.21 ± 0.02	0.14 ± 0.02
Tumor	3.72 ± 0.44	4.71 ± 0.40	3.63 ± 0.61	2.35 ± 0.23	1.47 ± 0.19	1.02 ± 0.16

[a]Each value represents % injected dose/g tissue (%ID/g tissue), and the mean ± SD for five mice.

These facts suggest that PSA also interacts with EAT cells through α-D-mannosyl and α-D-glucosyl residues present on the cell surface. In a atudy of the biodistribution of ^{125}I-PSA as a function of time, high levels of radioactivity were observed in kidney, spleen, liver, lung, and tumor (Table 6). Tumor uptake was greatest 6 h postinjection, and clearance of the radioactivity from the blood pool was fairly rapid. The tumor to blood (T/B) ratio, one of the important factors for imaging of tumors, is 2.6 at 6 h postinjection. This ratio of ^{125}I-PSA promises to give a clear tumor image within an early postinjection time. On the basis of our preliminary examination radiolabeled lectins appear to be promising radiopharmaceuticals for tumor imaging.

References

Fraker PJ, Speck JC (1978) Protein and cell membrane iodination with a sparingly soluble chloroamide. 1,3,4,6-tetra-chloro-3α, 6α-diphenylglycoluril. Biochem Biophys Res Commun 80: 849–857

Hagan PL, Halpern SE, Chen A, Krishnan L, Frinke J, Bartholomew RM, Daivid GS, Carlo D (1985) In vivo kinetics of radiolabeled monoclonal anti-CEA antibodies in animal model. J Nucl Med 26: 1418–1423

Halpern SE, Hagan PL, Garver PR, Koziol J, Chen AWN, Frincke JM, Bartholomew RM, David GS, Adams TH (1983) Stability, characterization, and kinetics of ^{111}In-labeled monoclonal anti-tumor antibodies in normal animals and nude mouse-human tumor models. Cancer Res 43: 5347–5355

Inbar M, Rabinowitz, Sachs (1969) The formation of variants with a reversion of properties of transformed cells: III Reversion of the structure of the cell surface membrane. Int J Cancer 4: 690–696

Kojima S, Jay M (1986) Application of lectins to tumor imaging radiopharmaceuticals. Eur J Nucl Med 12:836–840

Kojima S, Jay M (1987) Comparisons of labeling efficiency, biological activity and biodistribution among ^{125}I- , ^{67}Ga-DTPA- and ^{67}Ga-DFO-lectins. Eur J Nucl Med 13: 366–370

Kornfeld K, Reitman ML, Kornfeld R (1981) The carbohydrate binding specificity of pea and lentil lectins. J Biol Chem 256: 6633–6640

Krejcarek GE, Tucker KL (1977) Covalent attachment of chelating groups to macromolecules. Biochem Biophys Res Commun 77: 581–585

Louis CJ, Wyllie RG (1981) Fluorescein-Concanavalin A conjurates distinguish between normal and maglignant human cells: a preliminary report. Experientia 37: 508–509

Louis CJ, Whllie RG, Chou ST, Sztynda T (1981) Lectin-binding affinities of human epidermal tumors and related conditions. Am J Clin Pathol 75: 642–647

Pettit WA, Bennet SJ, Deland FH, Goldenberg DM (1982) Iodonation and acceptance testing of antibodies. In: Tumor imaging. Burchiel SW, Rhodes BA, Friedman MA (eds) Masson USA, New York, pp99–109

Sela BA, Lis H, Sharon N, Sachs L (1970) Different locations of carbohydrate-containing sites in the surface membrane of normal and transformed mammalian cells. J Membrane Biol 3: 267–279

Simpson DL, Thorne DR, Loh HH (1978) Lectins: endogeneous carbohydrate-binding protein from vertebrate tissues: functional role in recognition processes. Life Science 22: 727–748

Van Wauwe JP, Loontiens FG, De-Briyne CK (1975) Carbohydrate binding specificity of the lectins from the pea (*Pisum sativum*). Biochem Biophys Acta 379: 456–461

Vesely P, Entlicher G, Kocourek J (1972) Pea phytohemagglutination of tumor cells. Experietia 28: 1085–1086

Yokoyama A, Ohmomo Y, Horiuchi K, Saji H, Tanaka H, Yamamoto K, Ishii Y, Torizuka K (1982) Deferoxamine, a promising bifunctional chelating agent for labeling proteins with gallium: Ga-67 DF-HSA: Concise Communication. J Nucl Med 23: 909–914

32 Lectin Interactions with Intestinal Epithelial Cells and Their Potential for Drug Delivery

B. NAISBETT, C.-M. LEHR, and H.E. JUNGINGER

Prolongation of the residence time of drug delivery systems in the intestinal tract is a major challenge to improve drug absorption and to reduce dosing frequency. Most approaches using bio (muco) adhesive polymers which are generally able to stick nonspecifically to the mucous linings of the gut wall have failed to fulfill these aims. Some of the reasons are the rapid turnover of the mucus and the de-activation of the mucoadhesive properties of the systems by soluble mucins – amply available in the gut – before reaching the gut wall. Suitable lectins, however, may show real improvements in achieving these goals, as they are able to interact specifically with mucosal membrane sugars and show marked prolongation of binding to the gut wall. Lectin-coated drug delivery systems may therefore be useful to overcome the disadvantages of nonspecifically binding systems. It has also been demonstrated that lectins are absorbed by the gut and may show some potential as carriers for improved drug absorption. Finally, lectin-stimulated endocytosis by means of receptor-mediated invagination processes may be an approach to design sophisticated drug carrier systems especially for oral absorption of peptides and proteins.

This chapter will cover the experimental techniques for studying lectin interactions with intestinal epithelial cells.

32.1 Binding to Intestinal Rings

In selecting lectins which may have bioadhesive properties in the intestine, it is necessary to have a quick, simple screening protocol to determine the nature and extent of binding of an individual lectin to the intestinal mucosa. In this procedure, radiolabeled lectins are incubated for a short period with rings of intestinal tissue, in a tissue culture medium. After incubation, intestinal rings are dissolved and counted for radioactivity. The amount of lectin binding can then be calculated. Incubation of the lectins with different competing sugars prior to incubation with the gut rings will determine whether the lectin binds to the intestinal tissue via specific sugars, or whether the binding mechanism is simply by nonspecific adhesion.

As well as being quick and simple to perform, other advantages of this method include the ability to select defined regions of the intestine for study, (duodenum, jejunum, ileum, colon), and the fact that the tissue preparation method allows much of the contaminating mucus to be removed prior to incubation. Using gut rings, however, does not allow any determination of uptake mechanisms to be made.

Materials

Equipment
Thermostated oscillating water bath
50 ml Erlenmeyer flasks with silicone stoppers
Notched glass rod for everting the intestine
Ethicon braided silk sutures (Mersilk 1.5, Ethicon, Edinburgh, Scotland)

5 ml volumetric flasks
Medical oxygen O_2/CO_2 (95%/5%)
Glass dish for everting the intestine, set up to have oxygen bubbling through it
Pipets (gravity feed)
Gamma counter

Chemicals
Tissue culture medium 199 (ICN Flowlabs, England)
Radiolabeled lectin solution – labeled with ^{125}I
Competing sugars
Ice-cold saline (0.85% w/v, 1 M NaOH)

Tissue Preparation   **Procedure**
Adult male Wistar rats (250–300 g) are starved for 24 h and killed by cervical dislocation. The small intestine is excised and immediately placed in warm (37 °C), oxygenated (O_2/CO_2; 95/5%) tissue culture medium 199 (TC199). The intestine should be washed through gently with the same medium to remove luminal mucus and any remaining food debris and placed in fresh TC199. The intestine may be cut in half and one length everted on a notched glass rod.

To evert the intestine, a glass rod of diameter 3–4 mm, length 30–35 cm, is used, with a smooth groove or notch, 4–5 cm from one end. The length of intestine is carefully pushed over the glass rod from the shortest end until it reaches just past the groove. The intestine is held in place on the rod by tying a knot around the intestinal tissue at the position of the groove with a suture. The knot should be tight enough to hold the intestine firmly but not cut into the tissue. The bulk of the tissue can then be everted by pushing it gently over the rod.

The tissue is removed from the rod by cutting it near the groove, and is immediately placed in warm, oxygenated TC199. Whilst still bathed in TC199, the damaged ends of the intestine are removed with a sharp scalpel and everted and noneverted tissue is then cut into rings of approximately 0.5 cm in width.

Incubation
Nine ml of TC199 is placed into individual Erlenmeyer flasks in a thermostated (37 °C) water bath. Medical oxygen is then bubbled through each flask and the flasks are stoppered. Three gut rings are placed carefully in each flask and are left to equilibrate for 5 min. Medical oxygen is again bubbled carefully into the flasks and 1.0 ml of radiolabeled lectin substrate at an appropriate concentration is added at time zero. Substrate concentrations used are generally around 2 μg/ml and are made up in TC199. Flasks are re-stoppered and the rings are incubated, with shaking (70 oscillations/min) for 15 min.

At the end of the incubation, rings are removed from the flasks, blotted dry and washed four times in ice-cold saline, to remove any non-bound radiolabeled lectin. Rings are then placed individually in 5 ml volumetric flasks and dissolved in 1 M NaOH (overnight at room temperature or 1–2 h in a 37 °C water bath). One ml samples of dissolved gut tissue are then taken for radioactive counting (γ-counter), and 0.1 ml samples are analyzed for protein content.

Protein Assay
A sample of dissolved gut ring solution (0.1 ml) is made up to 1.0 ml with distilled water and analyzed for protein according to the method of Lowry et al. (1951), as modified by Peterson (1983). BSA is used as a standard. Samples and standards are run in triplicate.

Calculations

Binding of the radiolabeled lectins to gut rings is expressed as nanograms of lectin bound per milligram of gut ring protein. This is calculated from the amount of radioactivity measured in the dissolved tissue and from knowing the specific activity of the radiolabeled lectin (μCi/mg lectin)

Competing Sugars

To determine whether the binding of the lectins to the tissue is specific binding, mediated through intestinal carbohydrates, it is necessary to perform the incubations in the presence of competing sugars. A range of sugars can be chosen, but it is usual to include the monosaccharides to which many lectins are known to bind, for example, mannose, galactose, N-acetylgalactosamine, N-acetylglucosamine, and fucose. Some lectins, such as tomato lectin, are also known to have specificity for di-, tri-, and tetrasaccharides, and these may also be tested.

The radiolabeled lectin is preincubated in TC199, (15 min, 37 °C) with a range of competing sugars, all present in tenfold (molar) excess of the lectin prior to incubation with the gut rings. After this preincubation, gut rings are prepared as before and incubated, (15 min, 37 °C) in TC199 with the radiolabeled lectin-competing sugar solutions. Binding of the lectins to the gut rings is calculated as before, and the percentage inhibition of binding for each of the sugars tested can be determined.

Results   Typical results for these experiments are demonstrated by the incubation of ^{125}I-tomato lectin with everted and noneverted small intestinal rings, shown in Fig. 1. ^{125}I-Bovine serum albumin and ^{125}I-PVP have been used as macromolecular controls. As is shown in Fig. 1, everting the intestine in this case had no effect on the extent of lectin (or control macromolecules) binding to the rings. Although this may seem anomalous, it is simply an artifact of the tissue preparation process, as rings of

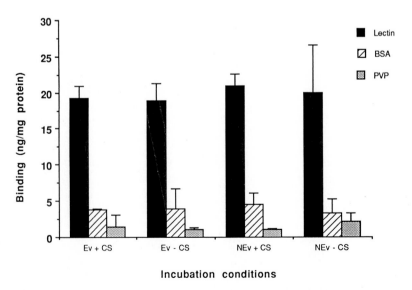

Fig. 1. Binding of macromolecules to intestinal rings. Incubation conditions: *Ev* everted intestine; *NEv* noneverted intestine; *+ CS* with fetal calf serum; *− CS* minus fetal calf serum; *BSA* bovine serum albumin; *PVP* polyvinylpyrrolidone. Data are means $\pm$ SE $n = 16$ (Naisbett and Woodley 1990)

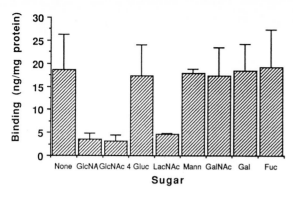

Fig. 2. Effect of competing sugars on the binding of tomato lectin to intestinal rings. Rings are noneverted, minus fetal calf serum. *None* lectin only; *Gluc* glucose; *GlcNAc* N-acetylglucosamine (monomer); *GlcNAc 4* N-acetylglucosamine tetramer; *LacNAc* N-acetyllactosamine; *Mann* mannose; *GalNAc* N-acetygalactosamine; *Gal* galactose; *Fuc* fucose. Data are means $\pm$ SE n = 16 (Naisbett and Woodley 1990)

noneverted tissue "curl up" outwards, thus presenting a large area of mucosal surface on the external face of the rings, available for binding. When everted tissue is cut into rings however, it does not curl up, therefore still presenting a full mucosal surface available for binding. In experimental situations, this results in no statistically significant differences being found in the binding of molecules to everted and noneverted tissue.

Figure 2 demonstrates the effect of pre-incubating tomato lectin with a range of competing sugars prior to incubation with gut rings. These data indicate that the binding of tomato lectin to small intestinal rings is specific and it can be concluded that the lectin may be binding to the gut via specific sugar residues present in the carbohydrates, glycoproteins and/or glycolipids at the enterocyte surface. In vivo, the lectin could also interact with sugars present in the mucus glycoproteins which are known to contain N-acetylglucosamine (GlcNAc) residues. With the in vitro gut ring system, however, very little mucus remains on the intestinal surface.

32.2 Binding to Isolated Fixed Enterocytes

The following protocol provides an alternative method to study lectin adhesion to the intestinal epithelium. Instead of rings of whole intestinal tissue, isolated enterocytes are used. After fixation, the cell preparation can be stored for several weeks and used for comparative binding experiments. As these cells are no longer alive, binding may only occur to surface bound sugars and is not followed by cellular internalization. We found this approach in particular useful in performing quantitative lectin binding studies as well as in studying the pH dependence and cross-reactivity with mucus glycoproteins.

Equipment   **Materials**
Fume hood, lead protection, gamma counter, and gamma monitor
Refrigerator (0–5 °C) and freezer (− 20 °C)
Centrifuge (5000 g or less)
15-ml centrifuge tubes
Disposable test tubes (1 × 4 cm) with stoppers for gamma counting
Econopack 10 DG disposable desalting columns (Biorad)

1000 NMWL PLGC Ultrafree-MC filter units (Millipore)
Light microscope and counting chamber (Improved Neubauer Hemocytometer)

Chemicals
50 mM Phosphate buffer (Pi), pH = 7

$NaH_2PO_4 \cdot H_2O$ (fw 137.99, 50 mM) 6.9 g/l
bring pH to 7 with 1 N NaOH

Phosphate buffered saline, pH = 7.4, refrigerated:

NaCl	(fw 58.45, 139 mM)	8.125 g/l
KCl	(fw 74.55, 2.5 mM)	0.186 g/l
$Na_2HPO_4 \cdot 7H_2O$	(fw 268.07, 8 mM)	2.145 g/l
KH_2PO_4	(fw 136.09, 1.5 mM)	0.204 g/l

10 mM EDTA-buffer, pH = 6.8 (Knutton et al. 1985):

NaCl	(fw 58.45, 96 mM)	5.611 g/l
KCl	(fw 74.55, 1.5 mM)	0.112 g/l
$Na_2HPO_4 \cdot 7H_2O$	(fw 268.07, 5.6 mM)	2.145 g/l
KH_2PO_4	(fw 136.09, 1.5 mM)	0.204 g/l
EDTA	(fw 292.24, 10 mM)	2.922 g/l

1, 3, 4, 6, -Tetrachloro-3α,6α-diphenylglucoril (Iodogen, Sigma or Pierce)
Dichloro-methane (CH_2Cl_2, ca 5 ml)
Purified lectin (e.g., from Sigma, Vector, or EY-Labs)
0.5 mCi $Na^{125}I$, in 5 μl NaOH solution (IMS-30, Amersham)
Glutaraldehyde (20% solution, Sigma)
Sodium azide (NaN_3, Sigma, prepare 20% stock solution for safe and easy handling)
Bovine serum albumin (fraction V powder, Sigma)

Procedure   *Radio-labeling of Lectins*
(**Important:** Take adequate measures for working with free ^{125}I. Work in fume hood behind 3 mm lead (or thicker acryl) shield, wear goggles and double plastic gloves. Use gamma monitor to detect nonshielded sources of irradiation and spills!)

Prepare reaction vessels by coating the bottom of test tubes with Iodogen by dispensing 100 μl aliquots of an Iodogen-solution in CH_2Cl_2 (10 mg/ml) and drying on air (fume hood!).

Dissolve 0.1–1 mg lectin in 100 μl 50 mM Pi and transfer to Iodogen coated tube. Add 5 μl (= 0.5 mCi) $Na^{125}I$. Close tube immediately. Let reaction proceed for 5–15 min under occasional shaking.

Stop reaction by adding 1 ml PBS and transfer the contents of the tube to desalting column. Elute by adding PBS in 6–10 fractions of 1 ml and collect separately in test tubes. Count 10 μl samples of each fraction. The iodinated lectin peak usually appears in fractions 4 and 5. Pool these fractions and store in lead container at − 20 °C.

As the iodinated lectin slowly decomposes during storage (ca. 10%/week), free or low MW-associated ^{125}I should be removed immediately prior to its use by repeated ultrafiltration. (Detailed protocol available from Millipore).

Note. Although the Iodogen method is one of the mildest available, the lectin may be damaged. It is therefore advisable to check the biological activity of the iodinated lectin by other methods (e.g., hemagglutination test). Smaller amounts of lectin in the reaction mixture may require shorter reaction times (e.g., 5 min for 0.1 mg)

Preparation of Isolated Pig Enterocytes

Porcine small intestinal tissue (ca. 10 cm from jejunum or ileum) should be used within a few hours after slaughtering. The segment is opened longitudinally and carefully washed in cold PBS to remove mucus and other debris, and then incubated for 5 min in ca. 50 ml EDTA-buffer. After this treatment, the enterocytes detach easily and can be separated by mild shearing with a wide-bore pipet.

Dispense the enterocyte suspension in centrifuge vials and spin down for 2 min at 100 g. Resuspend the pellet several times in PBS to remove mucus and smaller contaminating cells. Enterocytes are easily identified by their shape and size under a light microscope.

After the last washing step, resuspend the enterocytes in PBS containing 2% glutaraldehyde and fix for 1 h at 4 °C. Wash cells again to remove the fixative and resuspend in PBS containing 0.2% NaN_3. Store at 4 °C.

Enterocyte Radio-Binding Assay (ERBA)

Dilute enterocytes in PBS to a final concentration of ca. 10^6 cells/ml. (Neubauer counting chamber). Add 0.5% (w/v) BSA to suppress nonspecific binding.

Dispense 50 µl-aliquots of enterocytes in coded test tubes. Add 50 µl lectin solution of various concentrations in PBS (0.5% BSA), spiked with a known amount of radiolabeled lectin and incubate for 1 h at room temperature. Triplicates for each concentration are recommended. Nonspecific binding at each concentration is determined at each point in the presence of a lectin-specific inhibiting sugar.

The lectin solutions are conveniently prepared by serial dilution of a concentrated stock solution (e.g., 1 mg lectin/ml, spiked with 10^5 cpm/ml tracer). Count aliquots of each dilution to determine the specific radioactivity (cpm/µg lectin).

Results

Determination of Specific Affinity and Number of Binding Sites. The affinity constant k_d and the number of binding sites at the saturation maximum B_{max} were determined for the specific binding of tomato lectin (TL) to fixed pig enterocytes. Total binding was calculated as the mean of a triplicate determination per each TL concentration. Nonspecific binding in presence of 1 mM Tetraacetyl-chitotetraose [$(GluNAc)_4$, Sigma] was calculated by linear regression of the single data points. Specific binding was obtained as the difference between average total and nonspecific binding. Under the assumption of a one ligand-one receptor model, the data were fitted to the model equation by nonlinear regression as shown in Fig. 3 for one representative experiment.

pH-Dependence of Lectin Binding and Cross-Reactivity with Mucus. For this purpose, ERBA was performed in various aqueous buffer solutions at a lectin concentration of 250 µg/ml (approx. $= K_d$). Buffers were prepared according to the monograph "standard buffer solutions" in the United States Pharmacopoeia (USP) Vol. XXII. For pH 2.0–5.8, a 0.2 M phthalate buffer was used, for pH 5.8–8.0 a 0.2 M phosphate buffer. All solutions were made isotonic by adding the necessary amount of glucose.

Figure 4 shows the percentage of total and specific binding in a pH range between 2 and 8 (mean ± SD, $N = 3$). Specific binding in PBS (pH = 7.4) was used as a reference value and assumed to be 100%. In an acidic milieu, binding was con-

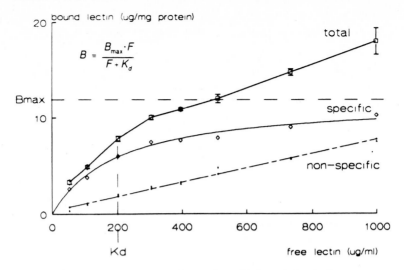

Fig. 3. Total, nonspecific and specific binding of TL to isolated pig enterocytes as found in one representative experiment (Lehr et al. 1992). Data for specific binding were fitted according to the model equation shown in the *upper left corner*. *F* concentration of free lectin (µg/ml); *B* amount of lectin bound per mass unit cellular protein (µg/ml) at saturation maximum. K_d is an apparent affinity constant, indicating the lectin concentration at which 50% of the binding places are occupied

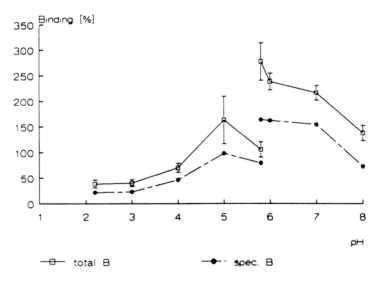

Fig. 4. pH dependence of TL binding to isolated pig enterocytes in two different buffer systems (see text); 100% was defined as specific binding measured in PBS (pH 7.4), set equal to 100%. (Lehr et al. 1992)

siderably decreased. The discrepancy in binding at the overlapping pH = 5.8 indicates an additional effect of the qualitative composition of the buffer system.

Inhibition of Lectin Binding by Mucus Glycoproteins. To study the inhibition of TL binding to enterocytes, ERBA was performed in the presence of various amounts of crude pig gastric mucin (Sigma). The cross-reactivity with this glycoprotein can be

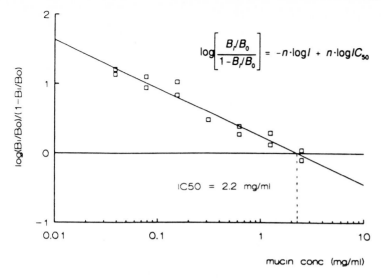

$$\log\left[\frac{B_i/B_0}{1-B_i/B_0}\right] = -n \cdot \log I + n \cdot \log IC_{50}$$

Fig. 5. Indirect Hill ('logit-log') plot of specific TL binding to pig enterocytes inhibited by various concentrations of pig crude gastric mucin. B_i is the amount of lectin specifically bound in presence of mucin at the concentration I; B_0 is the amount of lectin specifically bound without mucin. $IC_{50\%}$ is the mucin concentration which inhibits specific binding by 50%, and was found as the x-intercept at approx. 2.2 mg/ml. (Pooled data from two experiments: Lehr et al. 1992)

read from an indirect (logit-log) Hill plot of the binding data (Weiland and Molinoff 1981), as shown in Fig. 5.

32.3 Cellular Uptake and Transport

32.3.1 Improved Everted Intestinal Sac

Increased adherence of lectins to the intestinal mucosa as demonstrated earlier may enhance a lectin's uptake into enterocytes by endocytosis. This is now beginning to be particularly important in drug delivery, as the endocytic capture of macromolecules has been used to try to achieve selective delivery of macromolecular drugs and prodrugs (Duncan 1987), and lectins are now being investigated as drug carriers in these systems.

The method described here demonstrates how an improved in vitro everted gut sac culture system may be used to investigate the interaction and endocytosis of a lectin, (transcellular route) with the small intestinal surface. The methods enable us to determine the kinetics of the interaction and to elucidate the mechanism of uptake.

Radiolabeled lectin is incubated in tissue culture medium with sacs of everted intestine for periods up to 2 h. Lectin is measured within the tissue and in the serosal fluid inside the sacs. Metabolic inhibitors and varying substrate concentrations are used to determine if the process of uptake is energy- and concentration-dependent. These data are used to establish the mechanism of the lectin uptake.

Materials   *Equipment*
As for gut ring experiments but –
25 ml volumetric flasks
Three small artery clamps

Chemicals
As for gut ring experiments, but –
Glucose test kit (GOD-PERID™, Boehringer Mannheim)
Metabolic inhibitors

Procedure   *Tissue Preparation and Incubation*
Everted intestinal sacs were prepared using the method of Bridges (1980). The small intestine of an adult male rat was excised and everted as before and placed in fresh, warm, oxygenated TC199. One end of the intestine was secured with an artery clamp, medium was introduced into the intestine through a gravity-feed pipet and the intestine was then sealed with a second clamp. This large gut sac was divided into 12×1.5 cm sacs using braided silk sutures. Each sac was then separated and preincubated individually in oxygenated TC199 (37 °C). These tasks should take approximately 15 min to perform.

Sacs were incubated as for the gut rings and individual sacs were removed at intervals up to 2 h and blotted dry. The sacs were opened and the serosal fluid drained into a small flask. The volume of serosal fluid in each sac was measured using a syringe, and any sacs that had leaked were discarded. Gut sac tissue was washed and digested as for gut rings, and the protein content measured. Samples of incubation medium, gut tissue, and serosal fluid were taken for radioactive counting. Uptake of radiolabeled substrate was calculated as nanograms of substrate bound to gut tissue or present in the serosal fluid per milligram of gut sac protein.

Gut Sac "viability"
Viability or "leakiness" of the sacs during the incubation may be monitored by measuring glucose uptake and utilization by the intestinal tissue. The tissue culture medium contains 1 mg/ml glucose, and if the gut tissue is "alive" during incubation, it should be able to transport glucose against a concentration gradient and accumulate glucose within the gut sacs. Glucose may be measured in the incubation medium and the serosal fluid using the GOD-PERID test kit, following the manufacturer's instructions. Briefly, 0.1 ml of gut sac incubation medium, serosal fluid and glucose standard (91 µg/ml) were incubated with 5.0 ml of ABTS [diammonium 2,2'-azino-bis(3-ethylbenzothiazoline-6-sulphonate)] reagent at 37 °C for 15 min. The absorbance of the standard, samples, and a blank containing 0.1 ml H_2O were measured at 436 nm. The concentration of glucose was calculated from the absorbance of the glucose standard.

Although an accumulation of glucose should be seen within the sacs, the overall amount of glucose within the system is likely to fall, due to the utilization of glucose by the intestinal tissue. If a higher concentration of glucose is not found within the sacs however, results from those sacs should be discarded as the sacs have leaked or the tissue is no longer viable.

Mechanism of Uptake
During pinocytosis, substrates may be captured in two possible ways, either in solution (fluid-phase pinocytosis) or attached to the invaginating plasma membrane (adsorptive pinocytosis), or by a combination of both fluid and adsorptive modes.

During both these modes of uptake, an increase in concentration of the extracellular medium results in uptake (expressed in nanograms) increasing linearly with concentration. For a substrate entering in the absorptive mode, however, the sites to which the substrate binds will become saturated and uptake will level off.

Because the binding of a substrate to the cell surface does not require metabolic energy or the cytoskeletal system, metabolic and cytoskeletal inhibitors should have little or no effect on the binding component of substrate association with the enterocytes. However, these inhibitors should block uptake by fluid-phase pinocytosis as energy is needed for the movement and fusion of membranes and vesicles. Low temperature also has the same effect as metabolic inhibitors.

It is possible therefore to determine the nature of the mechanism of lectin uptake by incubating everted gut sacs with various metabolic inhibitors (sodium azide and sodium fluoride at 10^{-3} M), cytoskeletal inhibitors (colchicine at 5×10^{-5} M), low temperature and different substrate concentrations, (see Rowland and Woodley 1981).

Results

The binding and uptake of ^{125}I-labeled tomato lectin, (at a concentration of 2 μg/ml), with everted gut sacs are shown in Fig. 6. From the graph it can be seen that association of the lectin with gut tissue increases linearly with time (rate = 26 ng/h/mg protein). Uptake of radioactivity into the serosal fluid was also linear, but much slower, the rate being 1.7 ng/h/mg protein.

The binding and uptake of ^{125}I-tomato lectin with everted gut sacs at 4 °C is also shown in Fig. 6. At 4 °C, tomato lectin interaction with the gut mucosa was reduced by 78% but not totally negated. This shows that the majority of the lectin activity found in the tissue at 37 °C represents internalized lectin, with a binding component of approximately 20%. These results are indicative of adsorptive pinocytosis.

The effect of concentration on uptake is shown in Fig. 7a. Sacs have been incubated for 1 h. The rate of uptake by the tissue and the rate of appearance of the macromolecule in the serosal fluid both increased with substrate concentration up to (substrate) = 10–15 μg/ml. At concentrations greater than 15 μg/ml, saturation occurred. This provides further evidence of adsorptive endocytosis being the mechanism of uptake.

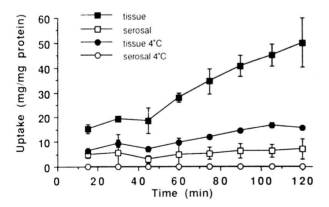

Fig. 6. Uptake of tomato lectin at 37 and 4°C by everted intestinal sacs. Final lectin concentration = 2 μg/ml. Data are means ± SE n = 12. (Naisbett and Woodley 1989)

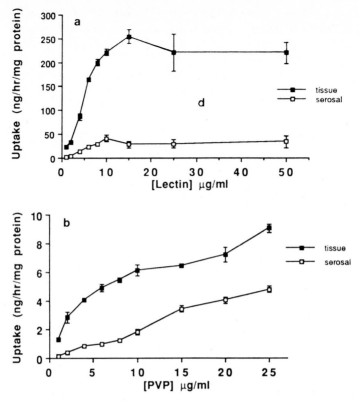

Fig. 7a,b. Effect of substrate concentration on tomato lectin and PVP uptake by everted intestinal sacs at 37 °C. Sacs were incubated for 1 h. Data are means $\pm$ SE $n = 12$ (Naisbett and Woodley 1989)

It is interesting to compare these data obtained for a control macromolecule, polyvinylpyrrolidone (PVP). The effect of ^{125}I-PVP concentration on uptake is shown in Fig. 7b. Uptake of PVP by the mucosa and appearance in the serosal fluid increases linearly with substrate and no saturation of uptake can be obtained at substrate concentrations up to 25 μg/ml. This result is compatible with a fluid-phase mode of endocytic uptake. Together with data showing neglible uptake at 4 °C, (not shown), this indicates that PVP is endocytosed in a fluid-phase mode, and supports the use of PVP as a fluid-phase marker molecule in in vitro intestinal cell preparations.

32.3.2 Caco-2 Cell Monolayers

The human colon carcinoma cell line Caco-2 is a unique model to study transport processes across in vitro cultured intestinal epithelium. After seeding on microporous filters, these cells spontaneously differentiate into polarized, columnar cells, and form confluent, tight monolayers which both structurally and functionally resemble the small intestinal epithelium. In contrast to native intestinal tissue, Caco-2 monolayers do not contain goblet cells which produce mucus that might interfere with the lectin binding. In addition, this model allows the study of binding and transport processes in both apical to basolateral and opposite direction in order to identify potential vectorial transport.

Equipment

General equipment for cell culture (laminar flow hood, incubator, inverted microscope, etc.)

Equipment for protein iodination and gamma counting (see previous methods in this chapter)

Six-well plates with 24 mm-Transwell-COL cell culture chamber inserts (Costar)

Epithelial Volt-Ohm meter (EVOM, World Precision Instruments) with one pair of chop stick electrodes

Caco-2 cells (available, e.g., from the American Type Culture Collection, ATCC, 12301 Parklawn Drive, Rockville, Maryland 20852, USA, Catalog #: ATCC HTB 37)

Chemicals

Dulbecco's Modified Eagle's Medium (MEM), substituted with 10% fetal calf serum, 1% nonessential amino acids benzylpencillin (100 U/ml) and streptomycin (100 µg/ml) (available separately from Gibco or Sigma)

Glutathione-bicarbonate Ringer's solution (GBR, pH = 7.4, Schoenwald and Huang 1983)), possibly, supplemented with 0.5% BSA:

Stock-solution A:		*Stock solution B:*	
NaCl	26.08 g	$CaCl_2 \cdot 2H_2O$	0.612 g
KCl	1.412 g	$MgCl_2 \cdot 6H_2O$	0.636 g
$NaH_2PO_4 \cdot H_2O$	0.412 g	D-Glucose	3.600 g
$NaHCO_3$	9.816 g	Glutathione, reduced	0.368 g
Water ad	2.0 l	Water ad	2.0 l

Mix equal amounts of A and B, adjust tonicity to 300 mOsm (more A to increase, more B to decrease) and check pH after perfusion with O_2/CO_2 (95 + 5).

Ice-cold PBS (ph = 7.4), (see Sect. 32.2.2)
0.3 N NaOH, containing 1% Triton X-100

A comprehensive description of elementary cell culture techniques can be found, e.g., in Freshney (1987). Caco-2 cells are seeded on collagen coated Transwells at a density of 60 000 cells/cm² and are incubated at 37 °C in a humidified atmosphere (5% CO_2). The medium (1.5 ml apical, 2.5 ml basolateral) is changed every other day. Tight monolayers are formed within 6–9 days. Full differentiation was reported to be reached by day 15, while some enzymes continued to increase through day 20 in culture (Hidalgo et al. 1989). We used the monolayers after 21 ± 2 days.

Prior to the experiment, measure the transepithelial electrical resistance (TEER). At 37 °C, 24 mm-Transwell-COL should give a reading of 250–300 Ohm (monolayer + membrane).

Note. By comparing the readings obtained with transwells of different sizes, we found a nonlinear relationship between surface area and resistance, not obeying Ohm's law. The TEER can therefore not be easily expressed as Ohm × cm². Relative comparison of direct readings (Ohm), however, still allows the detection of damaged monolayers.

Remove culture medium and wash cells two times with warm, oxygenated GBR. Fill the compartments with GBR containing 0.5% BSA (1.5 ml top, 2.5 ml bottom) to block nonspecific adsorption of proteins. Incubate for 1 h at the desired temperature for the experiment. Measure TEER again.

Dissolve the lectin and radioactive tracer (ca. 10^{6-7} cpm/ml, less than 2% free ^{125}I) in GBR (+ 0.5%BSA) at a concentration ten times higher than the final concentration. Count $3 \times 10 \, \mu$l to determine the specific radioactivity. At $t = 0$, remove either 150 μl GBR (+ 0.5%BSA) from the top (apical binding, A→B transport) or 250 μl from the bottom (basolateral binding, B→A transport) of the transwells and replace by the same volume of lectin solution.

During 5 h incubation, take a 1.0 ml sample every hour from the downstream (receiver) compartment and count the radioactivity. Replace by 1.0 ml fresh GBR (+ 0.5%BSA). (Keep in incubator along with the transwells, to be assured of the right temperature and pH.) At the end of the experiment, check TEER again and retain a 100 μl sample from the upstream (donor) compartment for counting. Remove all media and dispose as radioactive waste.

Place transwell plate on ice. Make a small (ca. 5 mm) incision into each monolayer using a sharp lancet blade to disrupt the barrier between apical and basolateral compartment. Wash three times with 2 ml ice cold PBS and remove immediately. To determine the cell-associated radioactivity, add 1 ml 0.3 N NaOH/1% Triton X-100 and incubate for 15 min at 37 °C. Transfer the lyzed cells quantitatively into a counting tube by rinsing with 2×1 ml water.

Results   Lectin binding/uptake by and transport across Caco-2 cell monolayers was studied using different plant lectins: tomato lectin (TL) and two *Phaseolus vulgaris* isolectins,

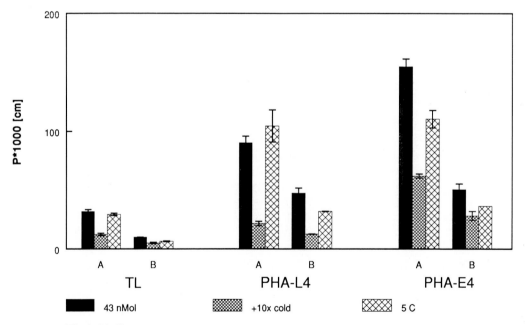

Fig. 8. Binding/uptake of different plant lectins by Caco-2 cell monolayers after 5 h incubation (mean ± SD, $n = 2$–3). (Lehr and Lee 1992)

PHA-E4 and PHA-L4 (Lehr and Lee 1992). To account for the different volumes of the apical and basolateral compartments of the transwell system, the results are expressed as a penetration index (P), which is given by Eq. (1) and has the unit of cm.

$$P = \frac{cpm_{transported\ or\ bound}}{cpm/cm^3_{donor} \times cm^2_{area}}. \tag{1}$$

As Fig. 8 shows, binding of all three lectins to Caco-2 cells was inhibited by adding the tenfold amount of cold lectin, the apical side of the monolayers in general binding more lectin than the basolateral side. Tomato lectin appeared to be less bioadhesive than the two PHA isolectins, of which E_4 binds stronger than L_4. At 4 °C, binding was not reduced for TL and PHA-L_4 when applied apically, indicating merely adsorption rather than endocytosis. By contrast, membrane invagination processes appear to contribute to the binding of PHA-E_4 to both the apical and basolateral cell membrane, as well as to the basolateral binding of the two other lectins.

Transport of lectins across the membranes (Fig. 9) is ranked similarly to binding/uptake, but at least one order of magnitude smaller and was less than 1% of the dose during 5 h. Transport at 4 °C was reduced by 80% or more, suggesting that membrane vesiculation processes are involved. In contrast to binding, transport was not reduced in some cases by the tenfold amount of cold lectin, which is indicative of fluid phase rather than adsorptive transcytosis.

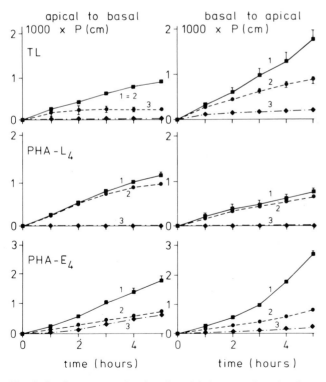

Fig. 9. Lectin transport (total radioactivity) across Caco-2 cell monolayers (mean $\pm$ SD $n = 2$–3) *1* 43 nM, 37 °C; *2* 430 nM, 37 °C; *3* 43 nM, 4 °C. (Lehr and Lee 1992)

Acknowledgements. B. Naisbett would like to thank Dr. John Woodley and Dr. Janine Bridges (University of Keele, Staffordshire, UK) for their help and guidance during her PhD study, in course of which part of these results were obtained. C.-M. Lehr would like to thank Prof. Vincent H.L. Lee (University of Southern California, Los Angeles, USA) for the invitation to spend one year in his department where part of this work was done. NATO and DAAD (Bonn, Germany) are thanked for financial support.

References

Bridges JF (1980) PhD Thesis, University of Keele, Staffordshire, England

Duncan R (1987) In Controlled drug delivery; fundamentals and applications Eds. Robinson JR & Lee VHL, Marcel Dekker, Inc. New York

Freshney RI (1987) Culture of animal cells, 2nd edn., Alan, R. Liss, New York

Hidalgo IJ, Raub TJ, Borchardt RT (1989) Characterization of the human colon carcinoma cell line (Caco-2) as a model system for intestinal epithelial permeability. Gastroenterology 66: 736–49

Knutton S, Loyd DL, Candy DCA, McNeish AS (1985) Adhesion of enterotoxigenic *Escherichia coli* to human small intestinal enterocytes. Infect Immun 48: 824–31

Lehr C-M, Lee VHL (1992) Binding and transport of bioadhesive lectins in the Caco-2 cell model. Proceed. 19th Int. Symp. Control Rel Bioact Mater, July 26–31, Orlando, Florida, USA

Lehr C-M, Bouwstra JA, Kok W, Noach ABJ, de Boer AG, Junginger HE (1992) Bioadhesion by means of specifically binding tomato lectin. Pharm Res 9: 447–53

Lowry OH, Rosebrough NJ, Farr AL, Randell RJ (1951) Protein measurement with the folin phenol reagent. J Biol Chem 193: 262–275

Naisbett B, Woodley JF (1989) Uptake of tomato lectin by the adult rat small intestine in vitro. Biochem Soc Trans 17: 883

Naisbett B, Woodley JF (1990) The binding of tomato lectin to the intestinal mucosa and its potential for oral drug delivery. Biochem Soc Tr0ans 18: 879

Peterson GL (1983) Determination of total protein. Methods in Enzymology 91: 95–119

Rowland RW, Woodley JF (1981) Uptake of free and liposome-entrapped ^{125}I-labelled PVP by rat intestinal sacs in vitro – evidence for endocytosis. Bioscience Reports 1: 399–406

Schoenwald RD, Huang HS (1983) Corneal penetration behaviour of β-blocking agents: I. Physicochemical factors. J Pharm Sci 72: 1266–72

Weiland GA, Molinoff PB (1981) Quantitative analysis of drug-receptor interactions: I. Determination of kinetic and equilibrium properties. Life Sci 29: 313–330

33 Neoglycoprotein-Liposome Conjugates for Studies of Membrane Lectins

N. Yamazaki, S. Gabius, S. Kojima, and H.-J. Gabius

The easiest method to study involvement of individual classes of sugar receptors such as membrane lectins and their carbohydrate ligands in cellular interactions in a model is established by deliberately reducing the intriguing complexity of surface receptors or ligands, resulting in surrogate systems. Guided by this framework, liposomes are prepared whose surface can be aimingly modified by chemical conjugation of potential receptors or ligands to be tested. Their interaction with respective ligand- and receptor-exposing surfaces, liposomes or cells helps to infer the generated strength of this type of receptor-ligand recognition. Moreover, these devices may be helpful for cell type-specific targeting. This chapter outlines the preparation, characterization, and selected application of neoglycoprotein-liposome conjugates which are instrumental in analysis of cell surface lectins.

33.1 Outline of Procedure

The preparation procedure for neoglycoprotein-liposome conjugates, mainly described in this chapter, is divided into two parts which are further subdivided as follows:

1. Formation of liposomes by a flow-through cholate dialysis method
   a) Preparation of lipid/detergent mixed micelles
   b) Flow-through dialysis of the mixed micelles
2. Covalent coupling of neoglycoproteins to liposomes by a two-step reaction method.
   a) Periodate oxidation of gangliosides in the liposome membrane
   b) Coupling of neoglycoproteins to oxidized liposomes by reductive amination

An outline of the reaction scheme for neoglycoprotein/liposome conjugation is given in Fig. 1.

33.2 Equipment and Materials

Equipment

- Automatic adjustable pipets with disposable tips
- 10-ml round-bottomed flask for rotary evaporator with ground-glass joint and cap
- Magnetic stirrer with Teflon-coated stirring bars (5 mm)
- N_2-gas cylinder with pressure regulator
- Rotary evaporator with an inlet connected to N_2-gas source
- Desiccator equipped with a high vacuum pump
- Water bath with temperature regulator and circulator

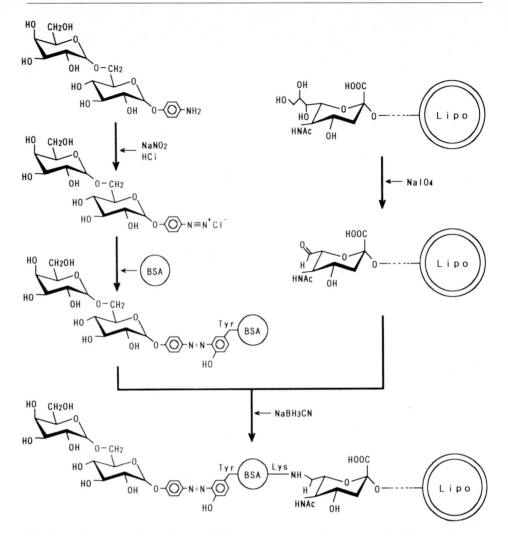

Fig. 1. Reaction scheme for neoglycoprotein/liposome conjugation. Preparation of melibiosylated bovine serum albumin (BSA)-coupled liposomes (*Lipo*) was chosen as an example

- Ultrasonicator (bath type) with time regulator
- Ultrafiltration cell, 10 ml (Amicon Model 8010, equipped with Amicon PM 10 membrane or Nuclepore 0.03 μm polycarbonate membrane and a concentration/dialysis selector Amicon CDS 10 connected to a 800-ml reservoir)
- 0.45-μm Sterile filters (Millipore Millex-HV) with disposable sterile syringes
- Refrigerator with precise control of temperature
- Assorted glass tubes with Teflon-lined screw cap

Materials
- L-α-Dipalmitoylphosphatidylcholine(DPPC) (Sigma)
- Cholesterol (Chol) (Sigma)
- Dicetylphosphate (DCP) (Sigma)
- Gangliosides (Type III from bovine brain, Sigma)

- Sodium cholate (Sigma)
- Salts and solvents are all analytical grade or HPLC grade from either Merck, Sigma or Wako.
- Neoglycoproteins, e.g., prepared as follows: sugar-free bovine serum albumin (BSA from Biomol, Hamburg, Germany) as a carrier protein was coupled with the diazo derivatives of *p*-aminophenyl glycosides, which resulted in yields of 10 ± 2 carbohydrate moieties per carrier molecule, as described in detail elsewhere (McBroom et al. 1972 or see Lee and Lee Chap.2, this Vol.)

- TAPS-buffered saline (pH 8.4); 10 mM TAPS buffer (pH 8.4) containing 150 mM NaCl
- 0.2 M Sodium periodate in TAPS-buffered saline (pH 8.4)
- Phosphate-buffered saline (pH 8.0); 10 mM phosphate buffer (pH 8.0) containing 150 mM NaCl
- 2 M Sodium cyanoborohydride in phosphate-buffered saline (pH 8.0)
- Phosphate-buffered saline (pH 7.2); 10 mM phosphate buffer (pH 7.2) containing 150 mM NaCl

33.3 Experimental Procedure

33.3.1 Formation of Liposomes by a Flow-through Cholate Dialysis Method

Preparation of Lipid/Detergent Mixed Micelles

(1) In a 10-ml round-bottomed flask introduce a mixture of lipids, composed of **Protocol** DPPC (17 mg), Chol (12 mg), DCP (2 mg) and gangliosides (15 mg) at a molar ratio of 35:45:5:15, and 48 mg of sodium cholate, to give a lipid to detergent molar ratio of 0.6.
(2) Dissolve the solids in 3 ml of chloroform/methanol (1:1, v/v), evaporate the solvent under a stream of nitrogen on a rotary evaporator at about 30 °C in a water bath, pump under vacuum in a desiccator for about 1 h, and store the flask overnight in vacuo to ensure complete removal of organic solvents. At this stage, the solids should dry onto the wall of the flask as a film.
(3) Add 3 ml of TAPS-buffered saline (pH 8.4) in the flask, replace the air in the flask by N_2, seal the flask with a cap, and dissolve the dried lipid/detergent film by gently stirring for 1 h to make a homogeneous suspension.
(4) Sonicate the lipid/detergent suspension in a bath-type sonicator for 1 h, with an intermission of 3 s every 2 min, to form a homogeneous mixed micelles suspension. For the preparation of homogeneous unilamellar liposomes, the suspension of lipid/detergent mixed micelles obtained should be as transparent as possible before starting the following dialysis.

A Flow-Through Dialysis of the Mixed Micelles. Prepare a 10-ml ultrafiltration cell fitted with a PM 10 membrane, introduce the mixed micelle suspension in the cell and diafiltrate with approximately 100 ml of TAPS-buffered saline (pH 8.4) for 24 h at room temperature, by applying about 2 atm N_2 pressure. (The volume of the prepared liposome suspension is increased after diafiltration.) Filtrate the suspension

(5–7 ml) through a 0.45 μm Millex-HV membrane into a glass tube, and store the liposome suspension in the tube in a refrigerator at 4 °C.

33.3.2 Preparation of Neoglycoprotein-Liposome Conjugates: Covalent Coupling of Neoglycoproteins to Liposomes by a Two-Step Reaction Method

Periodate Oxidation of Gangliosides in the Liposome Membrane

Protocol
1. In a 10-ml glass tube introduce 5–7 ml of the liposome suspension, containing approximately 46 mg of total lipid in TAPS-buffered saline (pH 8.4), and add 0.5–0.7 ml of 0.2 M sodium periodate to make the periodate concentration 20 mM.
2. Incubate by gently stirring the mixture at room temperature in the dark for 2 h.
3. Prepare a 10-ml ultrafiltration cell fitted with a 0.03 μm polycarbonate membrane, introduce the reaction mixture in the cell, and diafiltrate overnight with approximately 100 ml of phosphate-buffered saline (pH 8.0) at about 7 °C by applying 1 atm N_2-pressure to separate residual periodate from the liposomes and to change the buffer. Store the oxidized liposomes under a nitrogen atmosphere in a tightly closed tube at 4 °C

Coupling of Neoglycoproteins to Oxidized Liposomes by Reductive Amination

Protocol
1. In a 2-ml glass tube introduce 1–2 mg of solid neoglycoprotein and 0.5–0.7 ml of oxidized liposome suspension, containing approximately 4.5 mg of the total lipid in phosphate-buffered saline (pH 8.0), and incubate for 2 h at room temperature with stirring.
2. Add 10 μl of 2 M sodium cyanoborohydride per ml reaction mixture and incubate overnight by gently stirring at about 10 °C.
3. Prepare a 10-ml ultrafiltration cell fitted with a 0.03 μm polycarbonate membrane, introduce the reaction mixture in the cell, and diafiltrate overnight with approximately 100 ml of phosphate-buffered saline (pH 7.2) at about 7 °C by applying about 1 atm N_2 pressure to separate unbound neoglycoproteins from the liposomes and to change the buffer and pH environment. Filtrate the dialyzed liposome suspension through a 0.45μm Millex-HV membrane, and store the suspension of neoglycoprotein-liposome conjugates under a nitrogen atmosphere in a tightly closed tube at 4 °C.

33.3.3 Characterization of Neoglycoprotein-Liposome Conjugates

Determination of Protein and Lipid Composition. The amount of neoglycoproteins coupled to the membrane surface of liposomes is determined by the modified Lowry method in the presence of sodium dodecyl sulfate (Markwell et al. 1978). The lipid concentration can be determined as follows: the cholesterol content of the individual liposome suspensions is first analyzed by using a commercial kit for total-cholesterol determination, e.g., Determiner TC555 (Kyowa Medex, Tokyo, Japan), and then the amount of lipids in the liposome preparations can be calculated by using the data and the molar ratio of lipids.

Analysis of Purity and Stability. The purity and the size distribution of liposomes or neoglycoprotein-conjugated liposomes can be estimated by gel-permeation chromatography on a Sephacryl S1000 superfine column (10 mm × 100 cm) equilibrated and eluted at 0.1 ml/min with PBS (pH 7.2) at 25 °C. Liposome content in the effluent can be monitored by 90° light scattering at 633 nm using a fluorimeter. For calibration of the column with polystyrene latex particles, the column is pre-equilibrated and then eluted with PBS (pH 7.2) containing 0.5% Triton X-100. The use of Triton X-100 prevents aggregation of the latex particles. Size estimates obtained by this technique have been shown, for the present liposome preparations, to correlate closely with dynamic laser light scattering measurement. For analyzing the stability of neoglycoprotein-coupled liposomes, liposome suspension is loaded on the column, after keeping at 4 °C for several months and incubated at 37 °C for 24 h.

Results and Remarks

The present experimental procedure for preparation of liposomes and for coupling of neoglycoproteins to liposomes is a modified version of the published method, described elsewhere (Yamazaki 1987, 1989; Yamazaki et al. 1992), which is an adaptation of the controlled detergent dialysis method for liposome preparation by Zumbuehl and Weder (1981) and the two-step method for protein coupling by Heath et al. (1981). The prepared liposomes composed of DPPC, Chol, DCP and ganglio-sides at a molar ratio of 35:45:5:15 are homogeneous in size as illustrated in Fig. 2. The size distribution of neoglycoprotein-coupled liposomes is identical with that of noncoupled liposomes, and the neoglycoprotein–liposome conjugates do not contain free neoglycoproteins (Fig. 2). Coupled protein/total lipid ratios in the final

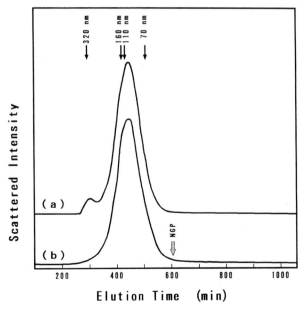

Fig. 2. Gel-permeation chromatography on Sephacryl S1000 superfine column. Elution profiles of *a* non-coupled liposomes, and *b* mannosylated BSA-coupled liposomes. The column was calibrated, as depicted by *arrows* above the profile on *a*, with four sizes of polystyrene latex particles: 320, 160, 110 and 70 nm. The *arrow* above the profile on *b* denotes the elution position of free neoglycoproteins (NGP)

products are usually between 0.07 and 0.25 g/g. These preparations are stable and do not change their property after keeping at 4 °C for several months. The present procedure has been optimized and simplified, however, it could be further developed by a process of trial and error.

33.3.4 Interaction of Neoglycoprotein-Liposome Conjugates with Cells and Tissues

Inhibition Assay of Neoglycoenzyme Binding to Cells. This assay is based on the principle that neoglycoprotein-coupled liposomes or neoglycoproteins, as inhibitors being tested, compete with neoglycoenzymes, as probes detecting sugar receptors, in binding specific cell-surface sugar receptors on cultured cells (Gabius et al. 1989; Yamazaki et al. 1992).

Cells (human colon adenocarcinoma cell line COLO 205) are grown for 24 h after trypsinization in 96-well polystyrene microtiter plates and adhesive cells are washed with Hank's balanced salt solution. Neoglycoenzyme solution is added to each well containing cells in the presence or absence of inhibitors, and the plate is incubated at 4 °C for 30 min. After three rinses of the cells with Hank's balanced salt solution, the enzymatic reaction of the cell-associated enzymes is initiated by adding the substrate of chlorophenolred-β-D-galactopyranoside. The reaction is stopped after incubation for 1 h at 37 °C in the dark by addition of 0.5 M glycine and development of the colored reaction product is quantitatively assessed at 590 nm in a microtiter plate reader.

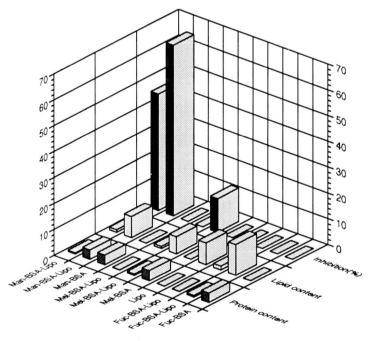

Fig. 3. 3-D diagram of the inhibitory effect of three types of neoglycoprotein-coupled liposomes and free neoglycoproteins, together with a noncoupled liposome, on the binding of neoglycoenzyme probes to human colon adenocarcinoma cells. (After Yamazaki et. al. 1992)

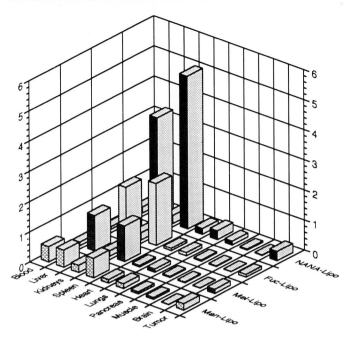

Fig. 4. 3-D diagram of the biodistribution of five types of [^{125}I]neoglycoprotein-coupled liposomes in Ehrlich solid tumor-bearing mice after 24 h. (After Yamazaki et al. 1992)

Radioiodination and Biodistribution of Neoglycoprotein-coupled Liposomes. Radio-iodination is performed by chloramine-T method (Kojima and Gabius 1988; Yamazaki et al. 1992). After injection of liposomes into the tail vein of Ehrlich solid tumor-bearing mice, biodistribution is determined. [^{125}I]Neoglycoprotein-coupled liposome injections are given 7 days after Ehrlich ascites tumor cells have been inoculated into the left rear leg of male ddY mice.

Results and Remarks

Two examples of interaction of neoglycoprotein-coupled liposomes with cells and tissues are illustrated in Figs. 3 and 4. Both in vitro and in vivo assays show individual responses among different types of neoglycoprotein-conjugated liposomes. These neoglycoprotein-liposome conjugates will find a wide field of application.

References

Gabius S, Hellmann K-P, Hellmann T, Brinck U, Gabius H-J (1989) Neoglycoenzymes: a versatile tool for lectin detection in solid-phase assays and glycohistochemistry. Anal Biochem 182: 447–451

Heath TD, Macher BA, Papahadjopoulos D (1981) Covalent attachment of immunoglobulins to liposomes via glycosphingolipids. Biochim Biophys Acta 640: 66–81

Kojima S, Gabius H-J (1988) Biodistribution of neoglycoproteins in mice bearing solid Ehrlich tumor. J Cancer Res Clin Oncol 114: 468–472

Markwell MAK, Haas SH, Bieber LL, Tolbert NE (1978) A modification of the Lowry procedure to simplify protein determination on membrane and lipoprotein samples. Anal Biochem 87: 206–210

McBroom CR, Samanen CH, Goldstein IJ (1972) Carbohydrate antigens: coupling of carbohydrates to proteins by diazonium and phenylisothiocyanate. Meth Enzymol 28: 212–219

Yamazaki N (1987) Multivalent interaction between polysaccharide fragments and lectin-conjugated lipid vesicles: an affinity system to probe the structural features of carbohydrates. In: Zlatkis A (ed) Advances in chromatography 1986, Part II. Elsevier, Amsterdam, pp 371–380

Yamazaki N (1989) Analysis of the carbohydrate-binding specificity of lectin-conjugated lipid vesicles which interact with polysaccharide fragments. J Membr Sci 41: 249–267

Yamazaki N, Kojima S, Gabius S, Gabius H-J (1992) Studies on carbohydrate-binding proteins using liposome-based systems. I. Preparation of neoglycoprotein-conjugated liposomes and the feasibility of their use as drug-targeting devices. Int J Biochem 24:99–104

Zumbuehl O, Weder HG (1981) Liposomes of controllable size in the range of 40 to 180 nm by defined dialysis of lipid/detergent mixed micelles. Biochim Biophys Acta 640:252–262

Lectins and Biosignaling

34 Effects of Lectins on Adenylylation and Phosphorylation of Plasma Membrane Proteins

E. San José, A. Benguría, H.-J. Gabius, and A. Villalobo

One of the most interesting approaches in the study of lectins is the recognition of defined biochemical pathways and specific cellular responses modulated by these glycoconjugate-binding proteins. In this context, the transduction of signals from lectin receptors, located in the plasma membrane, to specific intracellular target systems is an integral part of these mechanisms and deserves special attention. Biosignaling is, certainly, crucial to understand the mode of action of lectins, a particularly important subject of study in normal and tumor cells when lectins from plant and animal origins are used (Olden and Parent 1987; Gabius 1991).

The earlier steps to understand lectin-mediated biosignaling processes could logically be the identification of specific lectin receptors located in the plasma membrane, and the recognition of changes in regulatory posttranslational modifications of plasma membrane proteins, such as phosphorylation and adenylylation (San José et al. 1990). To attain this end, it is necessary to have available suitable experimental model systems. We therefore describe in this chapter the preparation of plasma membranes from normal rat hepatic tissue and from a rat hepatocarcinoma cell line, where the concomitant adenylylation and phosphorylation of a set of polypeptides, of 130, 120, 110 and 100 kDa in normal tissue, occur (Church et al. 1988; San José et al. 1990). Furthermore, some effects of β-galactose-specific lectins from plant and mammalian origin have been observed on both posttranslational modifications of this set of plasma membrane proteins from normal liver (San José et al. 1993).

These membrane preparations could be useful to identify as well as to isolate lectin-binding polypeptides, putative plasma membrane lectin receptors, in normal and neoplastic cells.

34.1 Materials

The equipment described in the protocols is that currently used in our laboratories, and it is only mentioned as an orientation to perform the experiments. Other instruments with analogous characteristics are equally suitable to carry out the experiments.

Equipment

Centrifuges and Rotors: For the preparation of plasma membrane fractions and the preparation of supernatants from solubilized membrane fractions, a high speed refrigerated centrifuge Beckman J2-21M/E model provided with a JA-14 rotor, and a refrigerated ultracentrifuge Beckman L8-70 or L8-70M model provided with swinging buckets SW-28, SW-41Ti, and SW-60Ti rotors, and a fix-angle 60Ti rotor are required. We have also used an ultracentrifuge Kontron Centrikon T-2070 model, compatible with the same set of rotors. To perform the washes of the ascites tumor cells, a top table centrifuge (Kubota 2010 model) capable of performing stable low speed centrifugations at $90g_{max}$ provided with a swinging bucket rotor with holders for 50 ml tubes is required. To pellet the proteins precipitated with trichloroacetic acid, we used a top table centrifuge (WIFUG) provided with a rotor for 10 ml tubes. It is also required a small top table centrifuge for Eppendorf tubes for the adenylylation and phosphorylation experiments. For this purpose we have used a Biofuge A (Heraeus Sepatech). All centrifugation times described in this work are measured from the moment in which the rotor reaches the maximum selected speed, and do not include the acceleration and the breaking times.

Homogenators: For the homogenization of liver it is required a 65 ml glass homgenator provided with a tight Teflon piston (Braun-Biotech) mounted on a firmly hold variable-speed driller (Bosch). It is useful to have the driller mounted in a cold room (4 °C).

For uniform resuspension of the isolated membranes a small 15 ml emery glass homogenator with an emery glass piston is required.

To resuspend trichloroacetic acid-precipitated proteins a thin (2 mm) glass rod with round tip is useful. This rod can be conveniently prepared by sealing the tip of a short Pasteur pipet on a gas flame burner.

For the homogenization of the plasma membrane preparations and homogenization of the tumor cells, a Polytron (SEV-geprúft) provided with PTA 10S and PTA 7K heads is required.

Electrophoresis and Autoradiograph Equipment: A standard refrigerated vertical slab gel electrophoresis apparatus to run simultaneously two (14 cm wide per 16 cm tall) gels connected to a regulable (0–200 mA) constant-current power supplier (FisherBiotech FB 600) was used. We performed the electrophoresis at room temperature when low current (10–12 mA) and long processing times (overnight) were used. When higher current (30–50 mA) is employed, we recommend to use the refrigeration accessory. We have also used custom-made metacrylate electrophoresis cuvets provided with platinum wires (for 11.5 cm wide per 15.5 cm tall gels) with excellent results. Care should be taken to avoid electric shocks when noncovered custom-made electrophoresis cuvets are used.

We used a custom-made two-cylindrical compartments (45 ml each) metacrylate gradient mixer and a magnetic stirrer for the preparation of linear gradient polyacrylamide slab gels.

A gel dryer with regulated temperature and operation time settings (Ephortec) is connected to a nonoiled electric vacuum pump (General Electric, 5KH36KM419T model). It is very important to connect a 1 l trap, covered with dry ice, and containing 200 g of solid calcium carbonate between the gel dryer and the vacuum pump to avoid the entrance of methanol and acetic acid fumes into the pump.

A set of 6–12 casettes for 20.3 cm per 25.4 cm radiography films (Cronex) with intensifier screens (Quanta Fast-detail).

An automatic X-ray film developer (AGFA Curix 60).

A transmittance/reflectance scanning photodensitometer and plotter with integration capability (Hoefer Scientific Instruments, model GS 300) for the quantitative determination of the intensities of the radiolabeled bands from the X-ray films.

Surgical Instruments: The following basic instrumentation is required:

One pair of small and one medium-sized sharp scissors with pointed tips
Two medium-sized tweezers with round tips
A set of scalpel blades (number 15) with handle
A set of disposable 5 ml syringes with needles (18 G $\times$ 1$\frac{1}{2}''$)

Spectrophotometers: We used a double-beam/dual-wavelength UV/Vis SLM-Aminco DW-2C model spectrophotometer, and a single wavelength, split-beam UV/Vis spectrophotometer (Perkin-Elmer 320 model) for measuring enzymatic activities. In addition, we used a Spectronic 20D colorimeter for protein and inorganic phosphate assays.

Radioprotection Equipment: A conditioned fume hood provided with a 45 $\times$ 90 cm trial covered with one-side impermeabilized absorbent paper and a 1.5 cm thick metacrylate shield (40 cm wide $\times$ 50 cm high) for protection against β-radiations during labeling assays
   Two shielded metacrylate racks for 20 Eppendorf test tubes
   The stock radiolabeled ATP solutions should be maintained in lead canisters during the labeling assays
   A Geiger counter (Technical Associates PUG1 model)
   Liquid radioactive residues should be disposed in large-opening glass bottles and solid radiocative residues in 18 $\times$ 18 $\times$ 30 cm metacrylate boxes

Other Equipment: A fraction collector for 80 tubes with automatic drop-counter device (Bio-Rad 2110 mode).

An Abbe's refractometer (Bausch & Lomb)
A water pump connected to a tap
An animal guillotine

Chemicals

Radiolabeled [γ-^{32}P]ATP (triethylammonium salt) (3000 Ci/mmol) and radiolabeled [α-^{32}P]ATP [tetra(triethylammonium) satl] (3000 Ci/mmol) from New England Nuclear. X-OMAT™ AR X-ray blue-sensitive films (20.3 $\times$ 25.4 cm) from Eastman. Carbohydrate-free bovine serum albumin from Biomol. Sepharose 4B from Pharmacia. Standard proteins for molecular weight determination, Coomassie brilliant blue R-250, TEMED, and bis-acrylamide from Bio-Rad. Tris from Serva Hepes, acrylamide, sucrose, and divinyl sulfone from Merck. Concanavalin A-agarose, concanavalin A, CNBr-Sepharose 4B, wheat germ lectin, *p*-nitrophenyl glycosides, bovine serum albumin, *p*-nitrophenylphosphate, *p*-nitrophyenyl-5′-thymidylate, glucose-6-phosphate (disodium salt), cytochrome c (horse heart), succinate (sodium salt), pyruvate (sodium salt), deoxycholate (sodium salt), dithiothreitol, rotenone, *β*-NADH (disodium salt), *β*-NADPH (tetrasodium salt), *α*-lactose, mannose, ouabain, histone (calf thymus), EDTA, EGTA, Triton X-100, AMP (sodium salt), and ATP (sodium salt) from Sigma.

All other chemicals should be of analytical grade

Solutions   *Sucrose and other Solutions*[1]: To avoid a common error in the preparation of sucrose solutions, we would like to indicate in this section that all the sucrose solutions required for the preparation of the different discontinuous sucrose gradient centrifugations for the isolation of plasma membrane fractions described in this work, are expressed in % (weight/weight), rather than the most common % (weight/volume) expression. The all important densities resulting from the preparation of these solutions expressed in % (w/w) or % (w/v) are understandably different. A good practice is to check at a standard temperature (5 or 20 °C) the densities of the different sucrose solutions, measuring the refraction indexes with an Abbe's refractometer and cross-check the results with appropriated tables (Dawson et al. 1986). Also should be noted not to adjust the pH of the sucrose solutions once the sugar has been dissolved, but rather use a 5 mM Tris-HCl buffer with a present pH 8 for the preparation of all sucrose solutions. This is done to avoid the erratic pH-electrode response that occurs in a highly concentrated sucrose solution.

The following stock sucrose solutions are required for the preparation of crude or further purified plasma membrane fractions: 32% (w/w), 34% (w/w), 35% (w/w), 39% (w/w), 40% (w/w), 41% (w/w), and 60% (w/w).

For Enzymatic Assays

50 mM KCN
(do not store, prepare fresh)

15 mg/ml cytochrome c
(store at − 20 °C)

1% (w/v) bovine serum albumin
(store at − 20 °C)

1 mM dithiothreithol
(do not store, prepare fresh)

9 mM NADH
(do not store, prepare fresh)

20 mM NADPH
(do not store, prepare fresh)

100 mM glucose-6-phosphate
(store at − 20 °C)

3 mM 2,6-DPIP
(store at − 20 °C)

10 mM benzylamine
(do not store, prepare fresh)

15 mM ouabain
(store at − 20 °C)

1 M NaCl
(store at room temperature)

200 mM KCl
(store at room temperature)

[1]All solutions except rotenone should be prepared in distilled water.

100 mM MgCl$_2$
(store at room temperature)

20% (w/v) SDS
(store at room temperature)

50 mM p-nitrophenylphosphate
(store at − 20 °C)

10 mM HNa$_2$PO$_4$
(store at 4 °C)

10% (w/v) Triton X-100
(store at 4 °C)

50% (w/v) Triton X-100
(store at 4 °C)

0.01 N NaOH
(store in plastic bottle
at room temperature)

0.1 N NaOH
(Store in plastic bottle
at room temperature)

1 N NaOH
(store in plastic bottle
at room temperature)

0.01 N HCl
(store at room temperature)

0.1 N HCl
(store at room temperature)

1 N HCl
(store at room temperature)

0.6 mM rotenone. This solution should be prepared in absolute ethanol (store in a screw-cap glass test tube in the dark at − 20 °C)

The pH of the following solutions should be adjusted with NaOH:

10 mM EDTA (pH 7.4)
(store at 4 °C)

40 mM ATP (pH 7)
(store at − 20 °C)

40 mM AMP (pH 7.4)
(store at − 20 °C)

100 mM succinate (pH 7)
(store at − 20 °C)

20 mM p-nitrophenyl-5′-thymidylate
(pH 7.4)
(store at − 20 °C)

25 mM pyruvate (pH 7.4)
(store at − 20 °C)

For Electrophoresis

Stock acrylamide:

30% (w/v) acrylamide
0.8% (w/v) bis-acrylamide

This stock solution iş used to make the solutions containing 5 and 20% acrylamide required for the preparation of the gel gradient, and is described to indicate the adequate acrylamide/bis-acrylamide ratio. The extent of cross-linkage of the gel can be modified. The pore size of the gel depends on the extent of this cross-linkage and will therefore become smaller or larger by increasing or decreasing, respectively, the concentration of bis-acrylamide with respect to the concentration of acrylamide.

1.5% (w/v) ammonium persulfate
This solution keeps only for a short time and has to be prepared fresh each time

Sample buffer:

60 mM Tris-HCl (pH 6.8)
20% (w/v) glycerol
5% (w/v) β-mercaptoethanol
10% (w/v) SDS
0.02% (w/v) bromophenol blue

5% gel solution: 375 mM Tris-HCl (pH 8.8)
                            5%   (w/v) acrylamide
                            1.5% (w/v) glycerol
                            0.1% (w/v) SDS

The polymerization was initiated upon addition of 0.03% (w/v) ammonium persulf-ate plus 0.1% (w/v) TEMED

20% gel solution: 375 mM Tris-HCl (pH 8.8)
                            20% (w/v) acrylamide
                            7%   (w/v) glycerol
                            0.1% (w/v) SDS

The polymerization was initiated upon addition of 0.02% (w/v) ammonium persulf-ate plus 0.1˙ (w/v) TEMED

Stacking gel solution: 350 mM Tris-HCl (pH 6.8)
                            4%   (w/v) acrylamide
                            0.1% (w/v) SDS

The polymerization was initiated upon addition of 0.04% (w/v) ammonium persulf-ate plus 0.15% (w/v) TEMED

Gel staining solution: 50%   (v/v) methanol
                            10%   (v/v) acetic acid
                             0.25% (w/v) Commassie brilliant blue R-250

Gel destaining solution: 50% (v/v) methanol
                            10% (v/v) acetic acid

Gel swelling solution: 10% (v/v) acetic acid

For Adenylylation and Phosphorylation Assays

If the amount of radiolabeled compound purchased is large enough, we recommed making small aliquots and store them at $-20\,°C$. Freeze-thaw cycles should be avoided. The radioactive ATP stock solutions are freshly prepared for each experiment adding 2 µl of the concentrated solutions of $[\alpha\text{-}^{32}P]ATP$ (3000 Ci/mmol) or $[\gamma\text{-}^{32}P]ATP$ (3000 Ci/mmol) to 50 µl of 200 µM of nonradiolabeled ATP, and the volume is completed with distilled water to obtain 100 µl at a final concentration of 100 µM ATP and a specific activity of 2 Ci/mmol

10 mM EGTA (pH 7.4)   1 mg/ml histone
The pH was adjusted with NaOH   (store at $-20\,°C$)
(store at $4\,°C$)

1 % (w/v) Triton X-100   60 mM $MgCl_2$
(store at $4\,°C$)   (store at room temperature)

For Preparation of Immobilized Lectins

0.5 M ethanolamine. The solution pH should be adjusted to 8.2 with HCl

For Protein Determination

Solution A: 2% (w/v) Na_2CO_3 in 0.1 N NaOH
                            (store at room temperature)

Solution B:	2% (w/v) Na,K-tartrate (store at 4 °C)
Solution C:	1% (w/v) CuSO$_4$ (store at room temperature)
Solution D:	Folin-Ciocalteu phenol reagent diluted onefold with distilled water (the concentrated reagent should be stored at 4 °C, discard when it turns greenish)
BSA standard:	1 mg/ml bovine serum albumin (store at − 20 °C in small aliquots)

For Phosphate Determination

Reagent A:	150 mM sodium acetate 3 mM ammonium molybdate 1 N SO$_4$H$_2$ (store at room temperature)
Reagent B:	10% (w/v) ascorbic acid (store at 4 °C)
Standard phosphate:	10 mM HNa$_2$PO$_4$ (store at 4 °C)

All buffers should be stored at 4 °C

Buffers

For Plasma Membrane Preparations

Buffer A:	5 mM Tris-HCL (pH 8) 150 mM KCl
Buffer B:	5 mM Tris-HCl (pH8)
Buffer C:	25 mM Na-Hepes (pH 7.4). This buffer is prepared using Hepes (free acid) and the pH adjusted with NaOH
Buffer NKT:	20 mM Tris-HCl (pH 7.4) 150 mM NaCl 5 mM KCl

For Enzymatic Assays

100 mM phosphate (pH 7)   200 mM phosphate (pH 7.4)
1 M glycine (pH 8.5)   1 M glycine (pH 9.6)
100 mM glycyl-glycine-imidazol (pH 7.5) 1 M Tris-HCl (pH 10.5)
400 mM Na-citrate (pH 5)   100 mM Na-maleate (pH 6.5)

For Lectin Immobilization

For concanavalin A immobilization in a CNBr-Sepharose matrix are required:

100 mM H$_3$BO$_3$ and 400 mM NaCl (pH 8.3)
25 mM Na-Hepes and 150 mM NaCl (pH 7.4)

For lectin immobilization in a DVS-activated Sepharose matrix is required:

0.2 M Na_2HPO_4/KH_2PO_4, 0.9% (w/w) NaCl (pH 8.6)

For Lectin Affinity Chromatography

For concanavalin A affinity chromatography are required:

Solubilization and equilibration buffer: 25 mM Na-Hepes (pH 7.4)
5% (w/v) glycerol
1% (w/v) Triton X-100

Elution buffer: 25 mM Na-Hepes (pH 7.4)
5% (w/v) glycerol
1% (w/v) Triton X-100
50 mM mannose

For mistletoe lectin l and bovine heart 14 kDa lectin affinity chromatographies are required:

Solubilization buffer: 20 mM Tris-HCl (pH 7.8)
0.5 M NaCl
4 mM β-mercaptoethanol
2 mM EDTA
100 mM lactose
0.5% (w/v) Triton X-100
0.1% (w/v) Na-deoxycholate
0.1 mM PMSF
5 μg/ml leupeptine
5 μg/ml antipapain
5 μg/ml chymostatin

Equilibration buffer: 20 mM Tris-HCl (pH 7.8)
150 mM NaCl
2 mM β-mercaptoethanol
0.02% (w/v) Triton X-100

Wash buffer: 20 mM Tris-HCl (pH 7.8)
150 mM NaCl
1 mM dithiothreitol
0.02% (w/v) Triton X-100

Elution buffer: 20 mM Tris-HCl (pH 7.8)
150 mM NaCl
1 mM dithiothreitol
0.02% (w/v) Triton X-100
0.3 M lactose

Dialysis buffer I: 20 mM Tris-HCl (pH 7.8)
150 mM NaCl
2 mM β-mercaptoethanol
0.02% (w/v) Triton X-100

Dialysis buffer II: 20 mM Tris-HCl (pH 7.8)
2 mM EDTA
4 mM β-mercaptoethanol
0.05% (w/v) Triton X-100

Dialysis buffer III: 7.5 mM Tris-HCl (PH 7.8)
                                        2 mM EDTA
                                        4 mM β-mercaptoethanol
                                        0.05% (w/v) Triton X-100

For Adenylylation and Phosphorylation Assays

100 mM Na-Hepes (pH 7.4)

For Electrophoresis

Stacking buffer: 1.5 M Tris-HCl (pH 6.8)
This stock buffer is used to prepare the stacking gels

Separating buffer: 1.5 M Tris-HCl (pH 8.8)
This stock buffer is used to prepare the running gels

Electrophoresis buffer × 10:   250 mM Tris (base)
                                        1.92 mM glycine
                                        1% (w/v) SDS

This stock electrophoresis buffer is kept without pH adjustment and diluted with distilled water tenfold before use. This pH is set to 8.3 with HCl if necessary after dilution. The diluted buffer is used for electrophoresis development.

34.2 Preparation of Plasma Membrane Fractions

The isolation of rat liver plasma membrane fractions essentially follows the protocol described by Brown et al. (1976) with some slight modifications done by us (Church et al. 1988; San José et al. 1990). Among these modifications it is of interest to indicate that we have found more reproducible results when the required alkali pH of the different sucrose solutions were maintained with a diluted buffer (5 mM Tris-HCl, pH 8), rather than with the addition of a few drops of diluted NaOH, as described in the original protocol (Brown et al. 1976). The whole procedure is carried out at 4 °C. It is possible to prepare by this method a crude plasma membrane fraction, and a further purified plasma membrane fractions. We have used this procedure for the preparation of plasma membrane fractions from normal adult rat liver, neonatal rat liver, regenerating rat liver after partial hepatectomy, and from the rat ascites hepatocarcinoma cell line AS-30D (Church et al. 1988; San José et al. 1990).

Preparation

34.2.1 Crude Plasma Membranes from Normal Liver

We currently use three Sprague–Dawley male albino rats (250–300 g) in each preparation, which are sacrificed by decapitation using a guillotine. Alternatively, we have also used Wistar male albino rats of the same weight (Church et al. 1988). The peritoneal cavity of the animals is opened immediately and the liver removed and placed in a 250 ml beaker containing 100 ml of ice-cold buffer A (150 mM KCl, 5 mM Tris-HCl at pH 8). The main external blood vessels and biliary ducts are removed, and the liver is finely minced with scissors and distributed in four 100 ml

Crude Plasma Membranes

beakers containing buffer A to a total volume of 50 ml. The contents of each beaker are independently homogenized in the 65 ml glass-Teflon homogenizer with ten firm up-down strokes using a low speed setting for the driller. Care should be taken to avoid heating the contents of the homogenizer by performing the whole operation in a cold-room (4 °C) and putting the beakers on ice.

The homogenized material is taken to a final total volume of 300 ml with buffer A and centrifuged in two polypropylene bottles of 250 ml in the JA-14 rotor at $2000g_{max}$ for 15 min at 4 °C. The supernatant is carefully removed by pouring, and the pellet is placed in a preweighed and precooled 250 ml beaker to determine its weight (from 25 to 30 g) on a precision balance. Thereafter, three times its weight of a 60% (w/w) sucrose solution is added, and the mixture is homogenized in the 65 ml glass Teflon homogenizer ten times, as indicated above. This results in the preparation of a suspension of subcellular material, containing plasma membranes, in 45% (w/w) sucrose. Alternatively, we have prepared this 45% (w/w) sucrose-subcellular fraction suspension by adding solid sucrose in steps with continuous stirring at 4 °C.

Aliquots of the above mentioned 45% (w/w) sucrose containing a suspension of the particulated subcellular cell fraction are distributed (15 ml each) in the bottom of six 38 ml Ultra-Clear™ (25 × 86 mm) centrifuge tubes for the SW-28 rotor, and successive layers of 41% (w/w) sucrose (8 ml), 39% (w/w) sucrose (8 ml), and 35% (w/w) sucrose (7 ml) are added, as indicated in Fig. 1. The centrifugation at $150\,000g_{max}$ is performed for 1 h. The membrane fraction migrating between the 35–39% (w/w) sucrose interface constitutes the crude plasma membrane fraction and is collected with a Pasteur pipet.

If the preparation is ended at this point, the isolated membrane fraction is diluted with 10 volumes of buffer C (25 mM Hepes-Na at pH 7.4) and centrifugated in 38 ml polycarbonate tubes in the fix-angle 60Ti rotor at $150\,000g_{max}$ for 30 min. The supernatant is carefully discarded with a water pump, and the pellet resuspended in 1–1.5 ml of buffer C, thoroughly homogenized in a small (15 ml) glass-glass homogenator, aliquoted in 100–200 μl fractions, and deep-frozen in 1 ml screw-cap criovials (Nunc) in liquid nitrogen. Care should be taken to freeze also smaller volume aliquots for the determination of protein and enzymatic markers. The samples are stored at − 70 °C and maintain their activity for at least 1 month, although we usually use the membranes within 1 week. Freeze and thaw cycles have to be avoided. A typical preparation of crude plasma membrane fractions yields 25–35 mg of protein.

34.2.2 Further Purified Plasma Membranes from Normal Liver

Purified Plasma Membranes

In order to obtain further purified plasma membranes, the material collected in the 35–39% (w/w) sucrose interface, as described in the preceding section, is diluted with 10 volumes of buffer B (5 mM Tris-HCl at pH 8) and centrifugated as above. After eliminating the supernatant, the pellet is resuspended in 12 ml of buffer B and homogenized with a Politron provided with the PTA 10S head for 15 s at setting 7. The homogenized material (2 ml each) is distributed on the top of six discontinuous sucrose gradients in 11 ml Ultra-Clear (14 × 89 mm) tubes for the SW-41Ti rotor. This second sucrose gradient is prepared as follows: 40% (w/w) sucrose (3 ml), 34% (w/w) sucrose (3 ml), and 32% (w/w) sucrose (3 ml). The centrifugation takes place

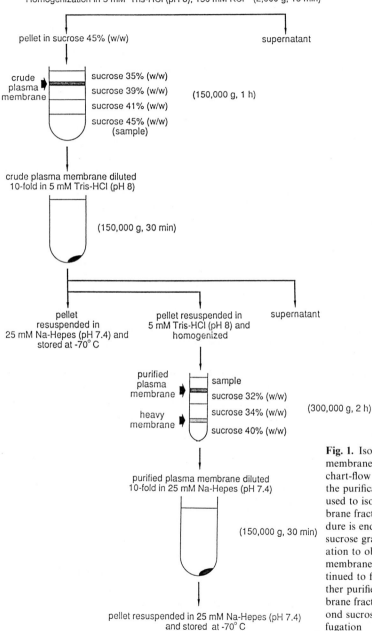

Fig. 1. Isolation of plasma membrane fractions. The chart-flow scheme represents the purification procedure used to isolate plasma membrane fractions. The procedure is ended after the first sucrose gradient centrifugation to obtain crude plasma membrane fractions, or continued to finally obtain further purified plasma membrane fractions after the second sucrose gradient centrifugation

for 2 h at $300\,000g_{max}$. Two distinct membrane fractions result from this centrifugation (see Fig. 1). A heavy fraction migrating to the 34–40% (w/w) sucrose interface, and a light fraction migrating to the 32% (w/w) sucrose-Tris buffer interface are obtained. The light fraction that constitutes the further purified plasma membranes is collected with a Pasteur pipet, diluted with 10 volumes of buffer C and placed in 38 ml polycarbonate tubes and centrifuged at $150\,000g_{max}$ for 30 min in the fix-angle 60Ti rotor. The pellet is resuspended in 1–1.5 ml of buffer C, aliquoted in 100 µl fractions, deep frozen in 1 ml screw-cap criovials (Nunc) in liquid nitrogen, and stored at $-70\,°C$. Care should be taken to freeze also smaller volume aliquots for the determination of protein and enzymatic markers. The viability and use of the further purified plasma membrane fractions is identical to that of the crude fraction already described. A typical preparation of further purified plasma membranes yields 12–15 mg of protein.

34.2.3 Plasma Membranes from the Ascites rat Hepatoma AS-30D Cells

Plasma Membranes

We have also employed with excellent results the plasma membrane isolation procedure described above for the preparation of plasma membrane fractions from an ascites rat hepatoma cell line. The purity of the isolated membrane fractions from the tumor cells has been described (Church et al. 1988). Nevertheless, the yield of membrane obtained is approximately 50% of that in the case of normal liver.

Origin of the Tumor Cell Line. The transplantable ascites variant of the rat hepatoma denoted AS-30D was isolated by Smith et al. (1970). The hepatoma was induced with 3-methyl-4-dimethylaminoazobenzene, and was classified as a hepatocarcinoma (Smith et al. 1970). We obtained the cell line as a generous gift of Dr. Peter L. Pedersen from the Department of Physiological Chemistry, The Johns Hopkins University School of Medicine, Baltimore MD, USA.

Maintenance of the Tumor Cell Line. The tumor cell line was serially maintained in the peritoneal cavity of Wistar albino male rats (Charles River) weighing 250–300 g by the injection (using a thick needle) of 1 ml of undiluted ascitic fluid from a donor animal to six or eight receptor animals. The normal time of growth of a sufficient number of tumor cells in the peritoneal cavity is 7 days. It is not desirable to prolong growth periods. We have frequently observed that some animals are refractory to the inoculum. Therefore, in order to avoid the loss of the cell line and/or to obtain low cell yield, it is wise to inoculate a larger number of animals than will normally be used. The tumor cells collected from three to six animals are sufficient to make one plasma membrane preparation.

Procedure

Seven days after the inoculation, the animals are sacrificed by decapitation and the ascitic fluid collected by performing an abdominal incision. The ascitic fluid is filtered through a double layer of gauze set on a funnel to remove blood clots and other debris, and diluted to 250 ml with NKT buffer (150 mM NaCl, 5 mM KCl, 20 mM Tris-HCl at pH 7.4) precooled at $4\,°C$. The cells are washed four to six times in 50 ml centrifuge tubes using a total volume of 1.5–2 l of NKT buffer. The cells are recovered by very low speed centrifugation at $90g_{max}$ for 4 min in a swinging bucket rotor. The tumor cells prepared in this way were virtually depleted to red blood cells. It is essential to perform this centrifugation at the indicated low speed to prevent cell breaking and to achieve an efficient separation from the erythrocytes that are lighter than the tumor cells and therefore remain in the supernatant.

The tumor cell suspenion was homogenized in a total volume of 100 ml of buffer A (150 mM KCl, 5 mM Tris-HCl at pH 8) using two 30 s bursts in a Polytron homogenizer operating at setting 8. The larger PTA 7K head is recommended.

The homogenized material was centrifugated in 250 ml polypropylene bottle for the JA-14 rotor at $2000g_{max}$ for 15 min. The pellet was resuspended in a minimum volume of 5 mM Tris-HCl (pH 8), and was added solid sucrose in steps to attain a 45% (w/w) sucrose-subcellular material suspension. This fraction was later processed in a manner identical to that described above to obtain the crude plasma membrane fraction, or the preparation is continued as described above to obtain the further purified plasma membrane fraction (see Fig. 1).

34.2.4 Enzymatic Markers for Quality Control

The purity of the different isolated plasma membrane fractions can be controlled by measuring, in the crude tissue homogenate and the isolated membrane fractions, the specific activities of a set of enzymes originally located in different subcellular compartments and/or subcellular membranes. We have performed the following determinations:

Plasma membrane markers: Na^+, K^+-ATPase (ouabain sensitive), 5'-nucleotidase, 5'-phosphodiesterase, and alkaline phosphatase. Endoplasmic reticulum markers: NADPH-cytochrome c reductase (rotenone insensitive), and glucose-6-phosphatase. Lysosome marker: acid phosphatase. Inner mitochondrial membrane markers: succinate dehydrogenase and succinate-cytochrome c reductase. Outer mitochondrial membrane marker: monoamine oxidase. Cytoplasmic marker: lactate dehydrogenase.

The relative enrichments of both crude and further purified plasma membrane fractions with respect to the crude tissue homogenate from normal adult rat liver using a series of enzymatic markers are presented in Table 1. Additional data for the enrichment of the further purified plasma membrane fraction (light fraction), and the heavy membrane fraction obtained in the second sucrose gradient centrifugation in preparations from adult, neonatal, and regenerating normal rat liver, and the AS-30D rat hepatoma can be found in Church et al. (1988).

Na^+, K^+-ATPase

Enzymatic Assays

The ouabain-sensitive Na^+, K^+-ATPase activity was assayed according to the method of Post and Sen (1967) with some modifications as follows: The standard assays (from 50–100 µg crude homogenate or 5–15 µg plasma membrane proteins) were carried out in duplicates at 37 °C for 15 min in a total volume of 1 ml of a medium containing 100 mM NaCl, 20 mM KCl (when added), 1.5 mM ouabain (when added), 20 mM glycyl-glycine-imidazol (pH 7.5), 2 ml $MgCl_2$ 0.5 mM EDTA, and 4 mM ATP. The reaction was initiated upon addition of ATP and arrested upon addition of 1 ml of 20% (w/v) SDS.

For each determination, two sets of three 10 ml glass test tubes should be set up. One is blank, in which the ATP is added only after the addition of SDS to prevent any enzymatic reaction, a second tube to determine the total ATPase activity, containing NaCl and KCl, and in the absence of ouabain as indicated above, and a

Table 1. Activities of different enzymatic markers in the crude and further purified plasma membrane fractions

Subcellular location	Enzymatic activity	Specific activity (nmol/min/mg prot.)			Fold enrichment	
		Crude homogenate	Crude membrane	Purified membrane	Crude membrane	Purified membrane
Plasma membrane	Na^+,K^+-ATPase	42 ± 16 (8)	202 ± 53 (8)	822 ± 102 (4)	4.8	19.6
	5'-Nucleotidase	69 ± 4 (11)	855 ± 98 (13)	2775 ± 475 (6)	12.4	40.2
	Alkaline phosphatase	6 ± 0.5 (5)	104 ± 15 (5)	325 ± 123 (3)	17.3	54.2
	5'-Phosphodiesterase	164 ± 13 (5)	3952 ± 417 (5)	ND	24.1	ND
Endoplasmic reticulum	NADPH-cytochrome c reductase	8 ± 1 (4)	16 ± 5 (5)	19 ± 4 (5)	2.0	2.3
	Glucose-6-phosphatase	50 ± 9 (4)	54 ± 26 (2)	91 ± 6 (4)	1.1	1.8
Lysosomes	Acid phosphatase	88 ± 25 (2)	146 ± 14 (2)	ND	1.6	ND
Mitochondria inner membrane	Succinate dehydrogenase	5 ± 0.2 (2)	ND	< 0.05 (4)	ND	< 0.01
	Succinate cytochrome c reductase	22 ± 6 (2)	38 ± 9 (2)	ND	1.7	ND
Mitrochondria outer membrane	Monoamine oxidase	3 ± 0.1 (2)	ND	4 ± 1 (4)	ND	1.3
Cytoplasm	Lactate dehydrogenase	938 ± 149 (2)	ND	122 ± 13 (4)	ND	0.1

The results are given as mean values ± SEM. The numbers in parenthesis indicate the number of preparations tested. ND indicates not determined.

third tube in the absence of KCl and in the presence of ouabain to determine the ouabain-resistant activity. It is important to withdraw the KCl at the same time that ouabain is added, since potassium ion partially prevents the inhibitory action of ouabain. The ouabain-sensitive ATPase activity was determined by subtracting the activity in the absence minus the presence of ouabain.

The inorganic phosphate liberated to the medium from ATP was determined by the method described in Heldt and Klingenberg (1967) with small modifications. After arresting the reaction 2 ml of reagent A (150 mM sodium acetate, 3 mM ammonium molybdate, and 1 N sulfuric acid) and 1 ml of reagent B [10% (w/v) ascorbic acid] were added. Color development is allowed to proceed for 30 min in each tube, and read in a spectrophotometer at 660 nm. Care should be taken to read the absorbance in each tube at identical time, since the molybdate reagent slowly hydrolyzes ATP. Therefore, it is important to stagger the start and the stop of the reaction, as well as the reading of the color developed, in each tube for a period of time (30 s or 1 min) to allow comfortable reading.

A standard phosphate calibration plot should be prepared for each determination adding 0, 50, 100, 200, 500 and 1000 nmol of HNa_2PO_4 in 1 ml of distilled water. The standards are processed as above for color development. The calibration of phosphate gives a linear plot up to a concentration of 1 mM.

5'-Nucleotidase

The 5'-nucleotidase activity was assayed following the hydrolysis of AMP according to the method of Heppel and Hilmoe (1955) with some modifications as follows: The standard assays (from 100–150 μg crude homogenate or 10–20 μg plasma membrane proteins) were performed in 10 ml test tubes at 37 °C for 30 min in 1 ml of a medium containing 100 mM glycine (pH 8.5), 10 mM $MgCl_2$, and 4 mM AMP. The reaction was started upon addition of AMP and arrested upon addition of 1 ml of 20% (w/v) SDS. The determination of the inorganic phosphate liberated from the AMP was measured, as indicated above.

5'-Phosphodiesterase

The 5'-phosphodiesterase assay was based on following the hydrolysis of p-nitrophenyl-5'-thymidylate spectrophotomectrically as described in Brown et al. (1976) and summarized as follows: the standard assays (from 100–200 μg crude homogenate or 20–50 μg plasma membrane proteins) were performed at 37 °C in a 3 ml glass spectrophotometer cuvette containing 100 mM glycine (pH 9.6), 0.1% (w/v) Triton X-100, and 2 mM p-nitrophenyl-5'-thymidylate. The reaction was initiated upon addition of the crude homogenate or plasma membrane fractions, and the initial rate of liberation of p-nitrophenol was measured at 400 nm. An extinction coefficient ($E_{cm,mM}$) of 14 was used in the calculations.

Alkaline Phosphatase

The alkaline phosphatase was assayed by measuring the formation of p-nitrophenol from p-nitrophenylphosphate at 400 nm following the method of Pekarthy et al. (1972). An extinction coefficient ($E_{cm,mM}$) of 14 was used for the p-nitrophenol. About 500 μg crude homogenate or 10–50 μg plasma membrane proteins were incubated at 37 °C in a 3 ml glass spectrophotometer cuvet containing 200 mM Tris-HCl (pH 10.5), 5 mM $MgCl_2$, 5% (w/v) Triton X-100, and 1 mM p-nitrophenylphosphate.

The reaction was initiated upon addition of the crude homogenate or plasma membrane fractions.

Acid Phosphatase

The acid phosphatase was assayed by determination the formation of p-nitrophenol from p-nitrophenylphosphate at 400 nm following the method described in Ostrowski and Tsugita (1961). Crude homogenate (100 μg) or plasma membrane (50 μg) proteins were incubated at 37 °C for 10 min in 0.2 ml of 200 mM Na-citrate (pH 5), and 10 nM p-nitrophenylphosphate. The reaction was initiated upon addition of the p-nitrophenylphosphate and arrested upon addition of 2.8 ml of 100 mN NaOH. The samples were read in a spectrophotometer at the indicated wavelength. The calibration is prepared with p-nitrophenol standard solutions containing 0, 20, 50, 100, 200 and 500 nmol in 3 ml of the same reaction medium.

NADPH-Cytochrome c Reductase

Rotenone-insensitive NADPH-cyctochrome c reductase activity was measured by assessing the reduction of cytochrome c in a dual-wavelength spectrophotometer at 550–555.6 nm or in a split-beam single-wavelength spectrophotometer at 550 nm. An extinction coefficient ($E_{cm, mM}$) of 19.6 or 28 was used for the dual-wavelength or single-wavelength determinations, respectively. The assays follow the method of Parker and Thompson (1970): 400–500 μg crude homogenate or 150 μg plasma membrane proteins were incubated in a 3 ml glass spectrophotometer cuvet containing 85 mM phosphate buffer (pH 7), 0.5 mM KCN, 2 μM rotenone, 133 μM NADPH, and 0.025% (w/v) cytochrome c (oxidized form). The reaction was initiated upon addition of NADPH.

A concentrated stock solution of rotenone in ethanol is prepared to add no more than 1 or 2% (v/v) of ethanol in the final assay. Rotenone is added to inhibit the mitochondrial complex l, and therefore to prevent the oxidation of the pyridine nucleotide via the transdehydrogenase, and KCN is added to prevent the reoxidation of cytochrome c by the mitochondrial cytochrome c oxidase.

Glucose-6-Phosphatase

Glucose-6-phosphatase activity was assayed by measuring the liberation of inorganic phosphate from glucose-6-phosphate according to Swanson (1955) as follows: 100–200 μg crude homogenate or 30–100 μg plasma membrane proteins were assayed at 37 °C for 30 min in 0.5 ml of 60 mM Na-maleate (pH 6.5) and 20 mM glucose-6-phosphate. The reaction was initiated upon addition of the protein fractions and arrested by adding 0.5 ml of 20% (w/v) SDS. The colorimetric determination of free phosphate, liberated to the medium, was done as described in Section Na^+,K^+ATPase, although the assays and standards were scaled down to half the volumes used previously.

Succinate Dehydrogenase and Succinate-Cytochrome c Reductase Assays

Succinate dehydrogenase activity was measured according to the method of Matlib and O'Brien (1975) following the reduction of 2,6-DPIP at 600 nm. Alternatively, we assayed the succinate-cytochrome c reductase activity following the reduction of cytochrome c at 550 nm. Extinction coefficients ($E_{cm,mM}$) of 21 for 2,6-DPIP and of 28 for cytochrome c (reduced form) were used. The assays were performed with

400 µg crude homogenate or 150 µg plasma membrane proteins in both cases. It is important to carefully adjust the pH of the stock solution of succinate to 7.

The succinate dehydrogenase (500 µg crude homogenate or 100 µg plasma membrane proteins) was assayed at room temperature in a 3 ml glass spectrophotometer cuvet containing 10 mM phosphate buffer (pH 7.4), 10 mM succinate, and 0.1 mM 2,6-DIPI. The reaction was initiated upon addition of succinate.

The succinate-cytochrome c reductase (500 µg crude homgenate or 100 µg plasma membrane proteins) was assayed at 37 °C in a 3 ml glass spectrophotometer cuvet containing 85 mM phosphate buffer (pH 7), 0.5 mM KCN, 0.025% (w/v) cytochrome c (oxidized form), and 7 mM succinate. The reaction was initiated upon addition of succinate.

Monoamine Oxidase

Monoamine oxidase activity was measured monitoring the formation of benzalde-hyde as described by Tabor et al. (1954), but using a dual-wavelength spectro-photometer at 250–300 nm. A millimolar extinction coefficient ($E_{cm, mM}$) of 11.2 was used for benzaldehyde. The assays were carried out as follows: 100–200 µg crude homogenate or plasma membrane proteins were assayed at room temperature in a 3 ml quartz spectrophotometer cuvet containing 10 mM phosphate buffer (pH 7.4) and 1 mM benzylamine. The reaction was initiated upon addition of the crude extract or the plasma membrane fractions.

Lactate Dehydrogenase

Lactate dehydrogenase was measured spectrophotometrically following the oxida-tion of NADH at 340 nm as described by Peters (1976). An extinction coefficient ($E_{cm, mM}$) of 6.22 was used. They assays were carried out as follows: 2–5 µg crude extract or 50–100 µg plasma membrane proteins were incubated at room temper-ature in 1 ml quartz spectrophotometer cuvet containing 10 mM phosphate buffer (pH 7.4), 0.1% (w/v) bovine serum albumin, 0.1 mM dithiothreitol, 2.5 mM pyruvate, and 0.3 mM NADH. An initial measurement of the rate of oxidation of NADH in the absence of pyruvate was done to determine the activity of NADH-oxidase that has to be subtracted from the rate of oxidation of NADH in the presence of pyruvate.

34.2.5 Determination of Protein Concentrations

For the determination of protein concentration in the isolated membrane fractions we regularly use the method of Lowry et al. (1951). However, we perform an initial protein precipitation step to eliminate any soluble interfering compound as follows: From 5 to 10 µl of isolated membrane fractions are diluted in 1 ml of distilled water and 1 ml of 20% (w/v) trichloroacetic acid is added. The tubes are kept on ice for a few minutes and later centrifuged in a bench top centrifuge at 4500 rpm for 15 min. The supernatant is discarded and the test tubes should be inverted, resting on top of a thick layer of filter papers until most of the remaining trichloroacetic acid has gone into the paper. Thereafter, 0.5 ml of distilled water is added, mixed, and the protein determined with a standard protocol adding 2.5 ml of a mixture of reagents A, B, and C (at a 50:0.5:0.5 ratio), 10 min later adding 0.5 ml of the Folin-Ciocalteu phenol reagent, previously diluted onefold. The color that had developed after 30 min was read in a spectrophotometer at 660 nm. A calibration plot is prepared using 0, 5, 10, 25, 50 and 80 µg bovine serum albumin in 0.5 ml distilled water.

Protein Concentration

For the determination of protein concentrations eluted from the lectin chromatography columns and the determination of the efficiency of coupling of covalently bound lectins to immobilized resins, however, we used the dye-binding assay of Redinbaugh and Campbell (1985).

34.3 Preparation of Neoglycoproteins

The preparation of neoglycoproteins has already been addressed by Lee and Lee (Chap. 2). Determination of covalently bound carbohydrate residues for quality control was carried out spectrophotometrically (Monsigny et al. 1988).

34.4 Purification of Lectins

A survey of lectin purification has already been provided. In both cases, affinity chromatography on lactose-Sepharose 4B, obtained by divinyl sulfone activation, was performed.

34.5 Preparation of Immobilized Lectins

Immobilization   Coupling to divinyl sulfone-activated agarose has been found to produce affinity matrices with low leakage (Lihme et al. 1986). Thus, we employed this method for two lectins, while also using CNBr-activated resin alternatively for another lectin.

Mistletoe Lectin I (VAA)

Purified mistletoe lectin I (3 mg) was incubated with 10 ml of divinyl sulfone-activated Sepharose 4B dissolved in 8 ml of 0.9% (w/v) NaCl, 0.2 M Na_2HPO_4/KH_2PO_4 (pH 8.6), for 12 h at 4 °C. The solution was carefully withdrawn without sucking the resin to dryness and the matrix was thoroughly washed with the coupling buffer prior to deactivating residual active sites with β-mercaptoethanol for 2 h at room temperature. The capacity of the resin was checked by the binding of asialofetuin as previously described (Gabius et al. 1991).

Bovine Heart 14 kDa Lectin

The coupling of bovine heart 14 kDa lectin of divinyl sulfone-activated Sepharose 4B was done as described above for the mistletoe lectin l. However, the bovine heart lectin was previously stabilized by carboxamidomethylation with 0.1 M iodoacetamide in 50 mM Tris-HCl (pH 8.6) during elution with 0.3 M lactose according to the protocol of Powell and Whitney (1984).

Concanavalin A

The coupling of concanavalin A to CNBr-activated Sepharose was done as follows: We added 0.8 g of preswelled CNBr-activated Sepharose 4B to a 250 ml beaker containing 100 ml of 1 mN HCl, and shook gently for 10 min. The gel was retained in a Buchner funnel fitted with Whatman 3 MM Chr filter paper, and washed three times with a borate buffer (100 mM H_3BO_3 at pH 8.3, and 400 mM NaCl) at 4 °C. The pH of the filtered solution was checked with a pH-test paper to ascertain that HCl has been totally neutralized. The resin retained in the paper was taken with a spatula and placed in a 250 ml beaker containing 100 ml of the borate buffer and 50 mg of concanavalin A. The mixture was incubated overnight at 4 °C under gentle stirring. The use of a heavy magnetic fly should be avoided to prevent grinding of the gel beads. If possible, a suspended magnet should be used. Thereafter, the resin was again filtered in the Buchner funnel, retained in Whatman 3 MM Chr filter paper and transfered to a 250 ml beaker containing 100 ml of 0.5 M ethanolamine (pH 8.2). The incubation was continued overnight at 4 °C. The ethanolamine treatment blocks the still unreacted sites of the resin. The resin then was packed in a glass column (0.5 cm internal diameter × 5.4 cm height) and thereafter extensively washed with a buffer containing 25 mM Na-Hepes (pH 7.4) and 150 mM NaCl. We store the column at 4 °C and add 0.02% (w/v) sodium azide to preserve the column resin until use.

34.6 Chromatographic Procedures with Immobilized Lectins

Immobilized lectins can be used to isolate lectin-binding proteins, and we have used these chromatographic procedures to isolate a series of plasma membrane-bound proteins from rat liver using immobilized mistletoe lectin I and immobilized 14 kDa bovine heart lectin. We have also isolated with immobilized concanavalin A the adenylylable plasma membrane glycoproteins from the same origin (San José et al. 1993).

Mistletoe Lectin I and Bovine Heart 14 kDa Lectins

Purified plasma membranes (43 mg of protein) were solubilized in a medium containing 20 mM Tris-HCl (pH 7.8), 0.5 M NaCl, 4 mM β-mercaptoethanol, 2 mM EDTA, 100 mM lactose, 0.5% (w/v) Triton X-100, 0.1% (w/v) Na-deoxycholate, and a cocktail of protease inhibitors (0.1 mM PMSF, 5 µg/ml leupeptine, 5 µg/ml chymostatin, and 5 µg/ml antipapain). To remove the sugar, we dialyzed overnight at 4 °C against 2 l of 20 mM Tris-HCl (pH 7.8), 150 mM NaCl, 2 mM β-mercaptoethanol, and 0.02% (w/v) Triton X-100, changing the buffer four times. Half of the volume of the extract was passed through the mistletoe lectin l-Sepharose column, and the other half through the bovine heart 14 kDa lectin-Sepharose column. Each column contained 5 ml of bed volume and 280 µg of bound lectin/ml. The columns were preequilibrated with 20 mM Tris-HCl (pH 7.8), 150 mM NaCl, 1 mM dithiothreitol, and 0.02% (w/v) Triton X-100. After extensive washing with this buffer (50–55 bed volumes), the elution of the bound proteins was performed with the same buffer containing 0.3 M lactose. The eluted fractions were first dialyzed against 1 l of 20 mM Tris-HCl (pH 7.8), 2 mM EDTA, 4 mM β-mercaptoethanol, and 0.05% (w/v) Triton X-100 with five changes, and a second dialysis was performed against 1 l of the same buffer containing only 7.5 mM Tris-HCl (pH 7.8) with three changes. For the analysis of the isolated proteins the samples were subjected to a chloroform extraction to remove the Triton X-100, and the samples were lyophilized and thereafter stored at − 60 °C until used. The yield from the column of immobilized mistletoe lectin I was 65 µg protein, and from the column of immobilized bovine heart 14 kDa lectin was 105 µg of protein.

Chromatography

Concanavalin A

Purified plasma membranes (10–15 mg of protein) were solubilized for 10 min at 4 °C in 3 ml of a medium containing 25 mM Na-Hepes (pH 7.4), 5% (w/v) glycerol, and 1% (w/v) Triton X-100. Five ml of the same buffer were added and thereafter centrifuged in 2 ml Ultra-Clear™ tubes (11 × 60 mm) in a SW-60Ti rotor at $100\,000g_{max}$ for 15 min. The resulting supernatant was passed through a commercially available concanavalin A-agarose column (5 ml bed volume containing 14 mg of concanavalin A/ml), or the one prepared by us (1.3 ml bed volume), and equilibrated with the same buffer. After extensive washing with the above mentioned buffer, the bound protein was eluted with a buffer containing 25 mM Na-Hepes (pH 7.4), 5% (w/v) glycerol, 1% (w/v) Triton X-100, and 50 mM mannose. The eluted fractions (0.8 ml) were collected with a fraction collector and stored at − 70 °C until use. We normally used these fractions for the determination of the adenylylable polypeptides.

34.7 Labeling Assays

Labeling The standard adenylylation assays were carried out by labeling the polypeptides with [³²P]AMP using [α-³²P]ATP as substrate. Alternatively, we have used [2,5′,8-³H]ATP or [2,8-³H]ATP as substrate (Church et al. 1988). We have also detected guanylylation (covalent binding of GMP), using [α-³²P]GTP as substrate (Church et al. 1988). The standard phosphorylation assays were carried out by labeling the polypeptides with ³²P using [γ-³²P]ATP as substrate. Alternatively, we have used [γ-³²P]GTP as substrate (Church et al. 1988). The labeled polypeptides were detected after electrophoretic separation of the different polypeptides and autoradiography as described in Sections 34.8 and 34.9.

34.7.1 Adenylylation Assays

Adenylylation These assays were carried out in 1 ml Eppendorf capped test tubes in a total volume of 100 or 200 μl. An adequate amount of plasma membrane fractions (30–120 μg of protein) was incubated at 37 °C for 5 min in a medium containing 20 mM Na-Hepes (pH 7.4), 1 mM EDTA, and 10 μM [α-³²P]ATP (1.5–5 μCi). The addition of EDTA is optional, although its use is recommended, since it increases the amount of labeling from 20 to 40-fold (San José et al. 1993). The reaction was initiated upon addition of the radiolabeled ATP, and arrested upon addition of 100–200 μl (equal to the assay volume) of ice-cold 20% (w/v) trichloroacetic acid. The tubes are centrifuged in a bench top centrifuge at maximum speed for 15 min. The supernatant was carefully removed and the precipitated proteins processed for electrophoresis as described in Section 34.8.

In Fig. 2 (panel A), the patterns of polypeptides labeled in normal adult rat liver plasma membrane with [α-³²P]ATP (lane 1) and [2,5′,8-³H]ATP (lane 2), and in the rat hepatocarcinoma AS-30D plasma membrane labeled with [α-³²P]ATP (lane 3) and with [2,5′,8-³H]ATP (lane 4) are presented (see also Church et al. 1988; San José et al. 1990).

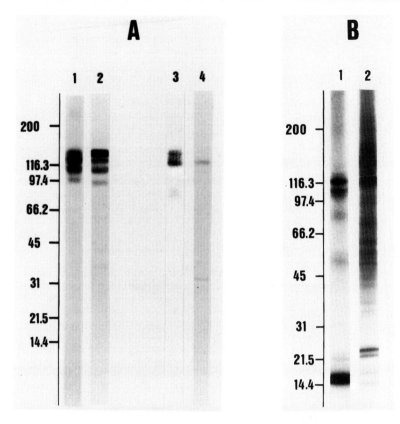

Fig, 2A,B. Adenylylation and phosphorylation of plasma membrane polypetides in normal rat liver and the AS-30D rat hepatocarcinoma. **A.** Autoradiographs of the adenylyĺated polypeptides. *Lane 1* normal rat liver plasma membrane polypeptides labeled with [α-^{32}P]ATP; *lane 2* normal rate liver plasma membrane polypeptides labeled with [2,5',8-^{3}H]ATP; *lane3* AS-30D rate hepatocarcinoma plasma membrane polypeptides labeled with [α-^{32}P]ATP; *lane 4* AS-30D rat hepatocarcinoma plasma membrane polypeptides labeled with [2,5',8-^{3}H]ATP; **B** Autoradiographs of the phosphorylated polypeptides. *lane 1* normal rat liver plasma membrane polypeptides labeled with [γ-^{32}P]ATP; *lane 2* AS-30D rat hepatocarcinoma plasma membrane polypeptides labeled with [γ-^{32}P]ATP. The phosphorylation experiments were performed in the presence of 6 mM MgCl$_2$. Molecular weight of standard proteins in kDa is indicated

Effects of Lectins: To test the effects of different lectins on the adenylylation process, we assayed the adenylylation of the plasma membrane proteins as described above in the presence of different concentrations of lectins (from 0 to 500 µg/ml) but adding to the assays 0.1% (w/v) Triton X-100 to permeabilize the plasma membrane vesicles and to allow the binding of the lectins to both sides of the membranes (if required). It is recommended to incubate the lectins with the membranes for 20 min at 4 °C before starting the reaction.

In Fig. 3, the inhibitory effects of wheat germ lectin (WG), concanavalin A (Con A), mistletoe lectin I (ML-I), and the 14 kDa bovine heart lectin (14K-bh), on the adenylylation of normal rat liver plasma membrane polypeptides, is presented (see also San José et al. 1993).

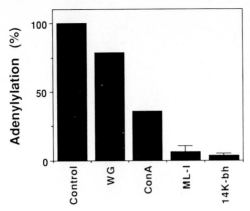

Fig. 3. Inhibitory effects of different lectins on the adenylylation of the plasma membrane polypeptides from normal rate liver. The plot presents the inhibitory effect of 500 µg/ml wheat germ lectin (*WG*), 500 µg/ml concanavalin A (*Con A*), 100 µg/ml mistletoe lectin l (*ML-l*), and 100 µg/ml bovine heart 14 kDa lectin (*14K-bh*) on the adenylylation (labeled with [α-^{32}P]ATP) of the 130, 120, 110, and 100 kDa polypeptides from normal rat liver plasma membrane. The *error bars* represent the standard deviation of two different experiments

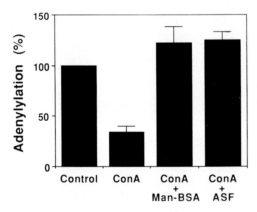

Fig. 4. Protective effects of a neoglycoprotein containing mannose residues and of asialofetuin on the concanavalin A-induced inhibition of adenylylation. The assays were carried out in the presence of 500 µg/ml concanavalin A (Con A), 1 mg/ml bovine serum albumin conjugated with mannose (Man-BSA), or 1 mg/ml asialofetuin (ASF) as indicated. The plot presents the adenylylation (labeled with [α^{32}P]ATP) of the 130, 120, 110, and 100 kDa polypeptides from normal rat liver plasma membrane. The *error bars* represent the standard deviation in the labeling of the four polypeptides

Effects of Neoglycoproteins on Lectin Action: To identify the glycosidic residues to which the different lectins bind and affect the adenylylation process, we have performed assays, as described in Section 1.7.1.1, but in the presence of 1 mg/ml of different neoglycoproteins. A prevention of the inhibitory effect of the tested lectin by a given neoglycoprotein identifies the glycosidic residues essential for the adenylylation process. We have used neoglycoproteins containing mannose, fucose, lactose, mannose-6-phosphate and sialic acid, and identified mannose residues as essential for the inhibitory action of concanavalin A on the adenylylation process (San José et al. 1993). Figure 4 presents a typical experiment in which the protective effects of the neoglycoprotein containing mannose residues (Man-BSA) and of asialofetuin (ASF) on the inhibitory action of concanavalin A (Con A) upon the adenylylation process is observed.

34.7.2 Phosphorylation Assays

Phosphorylation   These assays were performed in 1 ml Eppendorf capped test tubes in a total volume of 100 or 200 µl. An adequate amount of plasma membranes (from 30 to 100 µg of protein) was incubated at 37 °C for 1 min in a medium containing 20 mM Na-Hepes (pH 7.4), 6 mM MgCl$_2$ (when added), and 10 µM [γ^{-32}P]ATP (2–3 µCi). We have carried out experiments in the absence and in the presence of magnesium ion. The reaction was initiated by the addition of the radiolabeled ATP, and stopped upon addition of 100 or 200 µl (equal to the assay volume) of ice-cold 20% (w/v) trichloroacetic acid. The tubes were centrifuged in a bench top centrifuge at maximum speed for 15 min. The supernatant was carefully removed and the pellet processed for electrophoresis as indicated in Section 34.8. Figure 2 (panel B) presents the patterns of phosphorylated polypeptides from the normal adult rat liver (lane 1), and the AS-30D hepatocarcinoma (lane 2) plasma membranes (see also Church et al. 1988; San José et al. 1990).

Effects of Lectins: The effect of lectins (from 0 to 500 µg/ml) on the phosphorylation of plasma membrane proteins was assayed, as described above in Section 34.7.2, except that 0.1% (w/v) Triton X-100 was added to permeabilize the plasma membrane vesicles and to allow the binding of the different lectins to both sides of the membranes (if required). It is recommended to incubate the lectins with the membranes for 20 min at 4 °C before initiating the reaction. We have observed that the phosphorylation of the adenylylable polypeptides is less sensitive to inhibition by mistletoe lectin I and bovine heart 14 kDa lectin than the adenylylation process (San José et al. 1993).

34.8 Electrophoresis

The polyacrylamide slab gel electrophoresis method developed by Laemmli (1970) was modified to obtain linear gradient polyacrylamide gels from 5 to 20% in the presence of 0.1% (w/v) SDS.

Electrophoresis

Gel Casting

The gels were cast between two 19 × 19 cm or 14.5 × 15.5 cm glass plates pouring the decreasing acrylamide solutions developed from a metacrylate gradient mixer. The volume of the gel to be prepared should be measured before setting the stock solutions in the gradient mixer, since it varies depending on the thickness of the spacer used. This is done by measuring the volume of water that can be contained between the mounted glass plates. We normally prepare 1 mm thick gels.

Sample Preparation

The trichloroacetic acid-precipitated proteins were processed for electrophoresis grinding the pellet with a thin glass rod and progressively adding sample buffer up to a volume of 50–100 µl. We added 10 µl of 1.5 M Tris-HCl (pH 8.8) to readjust to an alkali pH. We continue grinding with the glass rod until total resuspension. The samples are then boiled for 5 min.

The Eppendorf test tubes used should be firmly closed, and we insert an injection needle in the lid to avoid accidental opening during the boiling process by the built vapor pressure. This method prevents significant sample evaporation. The samples should be well shaken before sample loading (30–50 µl).

Electrophoretic Separation

A standard electrophoretic separation takes place for 12–15 h at 12 mA. If two gels are simultaneously run with the same power supply, the applied current should be double amount as that indicated to obtain the same separation time.

Protein Staining

To stain the protein bands, the gels are incubated for 15 min in the Coomassie blue staining solution, and thereafter incubated in the destaining solution for several hours with multiple solution changes to obtain the desirable low background staining. Thereafter, the gels are first placed in the acetic acid solution for 15 min and afterward in distilled water for another 15 min before drying them. If the gels are not dried, they can be conserved in distilled water for a long time in appropriate containers.

Gel Drying

The gels are carefully placed on top of a Whatman 3 MM Chr filter paper, covered with an ultrathin transparent plastic film of polyethylene and set in the gel dryer under vaccum for 1 h at 70 °C.

34.9 Autoradiography

Autoradiography The dried gels are kept in the dark in close contact with a radiographic film inside a metal casette at − 20 ° or − 70 °C for an appropriate period of time (from 1 to 15 days). We normally use intensifier screens to enhance the film exposure. The X-ray films are developed in an automatic developer machine.

34.10 Isolation of Adenylylable Glycoproteins by Concanavalin A Affinity Chromatography

Using the chromatographic procedure with immobilized concanavalin A described above and upon elution with mannose, we have isolated a set of active adenylylable polypeptides from rat liver plasma membrane fractions. The chromatographic elution profile of adenylylable polypeptides obtained from the concanavalin A-agarose column upon elution with mannose has been described (San José et al. 1993).

Adenylylation Assays

The adenylylation of the polypeptides present in the mannose-eluted fractions from the concanavalin A-agarose column was assayed using 50–100 µl of the peak fractions in the same conditions as previously described.

In Fig. 5 (panel A) the patterns of adenylylated polypeptides from the rat liver plasma membrane in its membrane-bond form (track 1) and the polypeptides solubilized and isolated from the concanavalin A-agarose column (track 2) are presented.

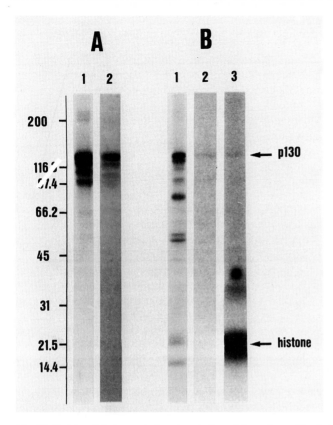

Fig. 5A,B. Adenylylation and phosphorylation of the polypeptides isolated by concanavalin A affinity chromatography. In **A** the patterns of adenylylated polypeptides from normal rat liver plasma membrane (labeled with [α-^{32}P]ATP) in its native membrane-bound form (*lane 1*) and after solubilization and concanavalin A affinity chromatography (*lane 2*) are presented. In **B** the patterns of phosphorylated polypeptides from normal rat liver plasma membrane (labeled with [γ-^{32}P]ATP and in the absence of added Mg^{2+}) in its native membrane-bound form (*lane 1*) and after solubilization and concanavalin A affinity chromatography (*lanes 2 and 3*) are presented. The assay in *lane 3* contains in addition 1 mg/ml histone. The *arrows* point to the 130 kDa phosphopolypeptide (p130), and phosphorylated histone. Molecular weight of standard proteins in kDa is indicated

Phosphorylation Assays

The phosphorylation of the polypeptides present in the mannose-eluted fractions from the concanavalin A-agarose column were assayed using 50–100 µl of the peak fractions in the same conditions as used in Section 34.7.2. Alternatively, we added to the assays 100 µg/ml histone to test for phosphorylation of exogenous substrate. We have described the phosphorylation of histone by protein kinase(s) present in the mannose-eluted fractions from the concanavalin A-agarose column (San José et al. 1993).

In Fig. 5 (panel B) the patterns of phosphorylated polypeptides from the rat liver plasma membranes (track 1) and from the mannose-eluted fractions from the concanavalin A-agarose column in the absence (track 2) and presence (track 3) of histone are presented. It is observed that the 130 kDa polypeptide is phosphorylated. Furthermore, histone is phosphorylated as well.

34.11 Summary

Our results show that the plasma membrane fractions prepared by the method described in this chapter are useful to study lectin-induced effects on two different posttranslational modifications of plasma membrane-bound proteins (i.e., adenylylation and phosphorylation). Moreover, affinity chromatographic procedures that utilize immobilized lectins allow us to isolate different lectin-binding plasma membrane proteins, among which we find the adenylylable polypeptides. The different sensitivities of the adenylylation and the phosphorylation of the polypeptides under study to the inhibitory actions of lectins indicate that they act at multiple sites. Furthermore, we have discriminated the two systems that perform the adenylylation and the phosphorylation of these polypeptides by the use of concanavalin A affinity chromatography, since some of the polypeptides (120, 110 and 100 kDa) under study, present in the mannose-eluted fractions, are adenylylated but are not phosphorylated, while other (the 130 kDa polypeptide) undergoes both posttranslational modifications. This contrasts with the concomitant adenylylation and phosphorylation of these polypeptides when in their native membrane-bound form (San José et al. 1990).

Acknowledgements. The work in the authors' laboratories was funded by the Dirección General de Investigación Científica y Técnica (grant PB89-0079), and the Consejería de Educación de la Comunidad de Madrid (grant C174-90), Spain (to A. V.), grants from the Dr. M. Scheel-Stiftung fü Krebsforschung and the BMFT-program Alternative Methoden der Krebsbekämpfung, Germany (to H.-J.G.), and the Acciones Integradas (42A) between Germany and Spain (to H.-J.G. and A.V.).

References

Brown AE, Lok MP, Elovson J (1976) Improved method for the isolation of rat liver plasma membrane. Biochim Biophys Acta 426: 418–432

Church JG, Ghosh S, Roufogalis BD, Villalobo A (1988) Endogenous hyperphosphorylation in plasma membrane from an ascites hepatocarcinoma cell line. Biochem Cell Biol 66: 1–12

Dawson RHC, Elliot DC, Elliot WH, Jones KM (Eds) (1986) in: Data for Biochemical Research 3rd Edition p. 545, Oxford University Press, New York

Gabius H-J (1991) Detection and functions of mammalian lectins – with emphasis on membrane lectins. Biochim Biophys Acta 1071: 1–18

Gabius H-J, Wosgien B, Hendrys M, Bardosi A (1991) Lectin localization in human nerve by biochemically defined lectin-binding glycoproteins neoglycoprotein and lectin-specific antibody. Histochemistry 95: 269–277

Heldt HW, Klingenberg M (1967) Assay of nucleotides and other phosphate-containing compounds by ultramicroscale ion-exchange chromatography. Methods Enzymol 10: 482–487

Heppel LA, Hilmoe RJ (1955) "5" Nucleotidases. Methods Enzymol 2: 546–549

Laemmli UK (1970) Cleavage of structural proteins during the assembly of the head of bacteriophage T4. Nature 227: 680–685

Lihme A, Schafer-Nielsen C, Larsen KP, Müller KG, Bog-Hansen TC (1986) Divinyl sulfone-activated agarose. Formation of stable and non-leaking affinity matrices by immobilization of immunoglobulins and other proteins. J Chromatogr 376: 299–305

Lowry OH, Rosebrough NJ, Farr AL, Randall RJ (1951) Protein measurement with the Folin phenol reagent. J Biol Chem 193: 265–275

Matlib MA, O'Brien PJ (1975) Compartmentation of enzymes in the rat liver mitochondrial matrix. Arch Biochem Biophys 167: 193–203

Monsigny M, Petit C, Roche AC (1988) Colorimetric determination of neutral sugars by a resorcinol sulfuric acid micromethod. Anal Biochem 175: 525–530

Olden K, Parent JB (Eds) (1987) Vertebrate lectins. Van Nostrand Reinhold Co New York

Ostrowski W, Tsugita A (1961) Purification of acid phosphomonoesterase from the human prostate gland. Arch Biochem Biophys 94: 68–78

Parker JG, Thompson W (1970) The composition of phospholipids in outer and inner mitochondrial membranes from guinea-pig. Biochim Biophys Acta 196: 162–169

Pekarthy JMI, Short J, Lansing AI, Lieberman 1 (1972) Function and control of liver alkaline phosphatase. J Biol Chcm 247: 1767–1774

Peters TJ (1976) Analytical subcellular fractionation of jejunal biopsy specimens: methodology and characterization of the organelles in normal tissue. Clin Sci Mol Med 51: 557–574

Post RL, Sen AK (1967) Sodium and postassium-stimulated ATPase. Methods Enzymol 10: 762–768

Powell JT, Whitney PL (1984) Endogenous ligands of rat lung β-galactoside-binding protein (galaptin) isolated by affinity chromatography on carboxy-amidomethylated-galaptin-Sepharose. Biochem J 223: 769–776

Redinbaugh MG, Campbell WH (1985) Adaptation of the dye-binding protein assay to microtiter plates. Anal Biochem 147: 144–147

San José E, Benguria A, Villalobo A (1990) A novel adenylylation process in liver plasma membrane-bound proteins. J Biol Chem 265: 20653–20661

San José E, Villalobo E, Gabius H-J, Villalobo A (1993) Inhibition of the adenylylation of liver plasma membrane-bound proteins by plant and mammalian lectins. Biol Chem Hoppe-Seyler 374: 133–141

Smith DF, Walborg EF Jr, Chang JP (1970) Establishment of a transplantable ascites variant of a rat hepatoma induced by 3'-methyl-4-dimethylaminoazobenzene. Cancer Res 30: 2306–2309

Swanson MA (1955) Glucose-6-phosphatase from liver. Methods Enzymol 2: 541–543

Tabor CW, Tabor H, Rosenthal SM (1954) Purification of amine oxidase from beef plasma. J Biol Chem 196: 645–661

35 Lectin-Induced Alterations in the Level of Phospholipids, Inositol Phosphates, and Phosphoproteins

H. WALZEL, H. BREMER, and H.-J. GABIUS

Lectins on cell surfaces are involved in cell-cell interaction and recognition phenomena by combining with complex carbohydrates of glycoconjugates of opposing cells (Sharon and Lis 1989). The interaction between lectins and complementary carbohydrates is quite stimulating, as sugars are candidates for the transmission of biological information through their structural diversity. Therefore, it is conceivable that soluble endogenous lectins, analogous to mitogenic lectins of plant origin (Grier and Mastro 1988; Danilov and Cohen 1989), may induce signal-transducing events after binding to cell surface glycoproteins with signaling potency (Kornberg et al. 1991). A possible candidate for such a signal molecule is phosphatidylinositol-4,5-bisphosphate (PIP_2), which is located in the cell membrane. PIP_2 breakdown by phospholipase C to 1,2-diacylglycerol and inositol-1,4,5-triphosphate (IP_3) has been shown to be involved in a number of transmembrane signaling events (O'Brian and Ward 1989; Rana and Hokin 1990). Diacylglycerol activates protein kinase C together with Ca^{2+}, which is mobilized from intracellular stores by IP_3.

The experiments detailed below describe the induction of changes in the level of phospholipids, inositol phosphates and in the metabolism of phosphoproteins of THP-1 cells induced by soluble β-galactoside-specific lectins.

35.1 Materials

Cell line The human monocytic cell line THP-1 was obtained from ATCC, No. TIB 202

Buffers Buffer A with LiCl, containing 125 mM NaCl, 5 mM $Na_2HPO_4 \cdot 2H_2O$, 5 mM glucose, 10 mM Hepes, 1 mM $CaCl_2 \cdot 6H_2O$, 1 mM $MgCl_2 \cdot 6H_2O$, 10 mM LiCl, 0.1% BSA, pH 7.3
Buffer A, containing 140 mM NaCl, 3 mM KCl, 1 mM $MgCl_2 \cdot 6H_2O$, 10 mM glucose, 1.8 mM $CaCl_2 \cdot 6H_2O$, 20 mM Hepes, pH 7.4
PBS, containing 145 mM NaCl, 2.1 mM $NaH_2PO_4 \cdot H_2O$, 7.8 mM $Na_2HPO_4 \cdot 2H_2O$, pH 7.4

Chemicals RPMI-1640 medium, Hepes, heat-inactivated FCS, and kanamycin were purchased from GIBCO BRL, Berlin, FRG. The radiochemicals myo-2-[^{3}H]inositol, carrier-free [^{32}P]orthophosphate and Hyperfilm-β max were from Amersham Buchler, Braunschweig, FRG and precoated, aluminum-backed silica gel thin layer chromatography (TLC) plates 60 F_{254} (20×20 cm) from Merck, Darmstadt, FRG

Tris, Dowex 1X8 200–400 mesh, PMSF, aprotinin, Servalyt pH range 3–10 isodalt grade, SDS and the other electrophoresis chemicals were obtained from SERVA, Heidelberg, FRG. The standard phospholipids PIP_2, PIP, PC, PI, PA and PE, Nonidet P40 as well as deoxyribonuclease and ribonuclease were purchased from Sigma Chemical Co. *Staphylococcus aureus* nuclease was from Boehringer,

Mannheim, FRG and SDS-PAGE low molecular weight standards from Pharmacia, Uppsala, Sweden

35.2 Experimental Procedure

Cell Culture. The human monocytic cell line THP-1 was maintained over many passages in RPMI-1640 medium supplemented with 10% (v/v) heat-inactivated FCS, 5×10^{-5} M 2-mercaptoethanol, and 0.1 mg/ml kanamycin at 37 °C under 5% CO_2 in a humidified incubator

β-Galactoside-Specific Lectins. The human placenta β-galactoside-binding lectin and the lectin I from mistletoe (MLI) were isolated, as described by Hirabayashi and Kasai (1984) or by Ziska et al. (1978), respectively. Before use, the lectins were purified by affinity chromatography on loctosyl-Sepharose 4B and analyzed by SDS-PAGE and silver staining

35.2.1 Analysis of Inositol Phosphates and Phosphatidylinositols of [³H] Inositol-Labeled Cells

THP-1 cells at 2×10^6/ml were cultured in inositol-free RPMI-1640 medium with 4 μCi myo-2-[³H]inositol for 24 h at 37 °C. The FCS applied to supplement the culture medium was dialyzed against 25 mM Hepes, pH 7.4. Then the cells were washed three times and resuspended at 5×10^6 cells/ml in buffer A containing 10 mM LiCl. The cells (5×10^6 cells/tube) were incubated for 15 min at 37 °C before the addition of either PBS (controls) or the β-galactoside-binding lectin from human placenta. After various periods of time, the reaction was terminated by addition 1 ml of ice-cold $CHCl_3$:CH_3OH 2:1 (v/v). After mixing for 1 min the tubes were immediately frozen in a dry ice/ethanol bath. Then the cells were extracted by vigorous mixing while thawing on ice. Separation of the two phases is then obtained by centrifugation for 5 min at 10000*g*. **Labeling and Stimulation**

For analysis of inositol phosphates, the aqueous upper phase is pipeted into a 10 ml reagent tube. To convert all inositol phosphates into the charged state, one volume of 50 mM Tris/formate, pH 8.0 is added followed by two volumes of double distilled water to decrease the ionic strength. The specimen are loaded onto a 2 ml prefilled, formiate-equilibrated Dowex 1X8 column. Free inositol was removed by washing with 20 ml H_2O and glyceroinositol phosphates with 10 ml 60 mM ammonium formiate. IP, IP_2, and IP_3 were then sequentially eluted with 10 ml each of 0.2, 0.4, and 1.0 M ammonium formiate in 0.1 M formic acid, respectively (Berridge et al. 1983) and collected in scintillation vials. After drying at 110 °C and resuspension of the residue in 0.1 ml of water radioactivity was determined by liquid szintillation counting. The kinetics of lectin-induced alterations in the level of inositol phosphates in THP-1 cells is summarized in Fig. 1. **Inositol Phosphates**

For analysis of phosphatidylinositols, the organic lower phase is pipeted into a 2 ml reagent tube. Total cellular lipids were isolated from the cell pellet by repeated extraction (all at 0–4 °C) with $CHCl_3$:CH_3OH:2.4 M HCl 1.5:1:1.5 (Downes and **Phosphatidyl-inositols**

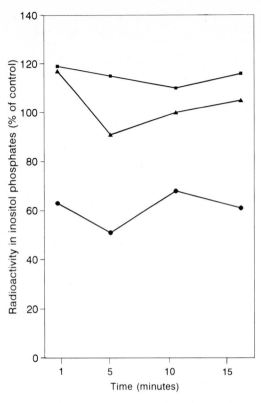

Fig. 1. Kinetics of lectin-induced release of inositol phosphates in THP-1 cells. [³H]inositol-labeled THP-1 cells at 5 ×10⁶ cells/ml received PBS either with or without (control) 5 µg lectin from human placenta at 37 °C. The treatment was terminated by addition of 1 ml ice-cold $CHCl_3:CH_3OH$ 2:1 after the periods of time indicated. The extracted inositol phosphates were separated by ion exchange chromatography on Dowex 1X8. The release of ■ IP, ● IP₂ and ▲ IP₃ is expressed as radioactivity in inositol phosphates in % relative to controls. *Points* are averages of triplicates. Standard deviations are less than 12%

Michel 1982). The lower $CHCl_3$-phase is pipeted into the 2 ml reagent tube. Extracted lipids were evaporated to dryness under nitrogen gas and resuspended in 50 µl of $CHCl_3$. The lipids were separated by a two-step one-dimensional thin layer chromatography (Medh and Weigel 1989). The TLC plates were soaked for 90 s in a 54 mM oxalic acid solution adjusted to pH 7.2 with KOH. After drying at room temperature the plates were activated at 110 °C for 15 min. The THP-1 lipid extracts (20 µl) were spotted on different lanes of the plate followed by spotting of 30 µg PIP₂ and 30 µg PIP dispersed in $CH_3OH:H_2O$ 1:1 and 60 µg PI in $CHCl_3$ as standard phospholipids. The separation was performed by ascending chromatography in two different solvent systems. In the first step, the separation was stopped when the acidic solvent system composed of $CHCl_3:CH_3OH:CH_3COOH:H_2O$ 55:43:3:4 reached the top of the plate. In the second step, the separation was finished when the front of the basic solvent system containing $CHCl_3:CH_3OH:NH_4OH:H_2O$ 40:70:10:20 reached a distance of 46 mm. The separated [³H]-labeled cellular PIP₂, PIP and PI were visualized by use of comigrating standard phosphatidylinositols by exposure the plates to iodine vapors. Spots were cut out for liquid scintillation counting. The time dependence of lectin-induced alterations in the level of cellular phosphatidylinositols is shown in Fig. 2.

35.2.2 Analysis of Phospholipids and Phosphoproteins of [³²PO₄]-Labeled Cells

Labeling and Stimulation   Cultured THP-1 cells were washed twice with buffer A and were then incubated for PO₄-depletion in the same medium at 2×10⁶ cells/ml for 1 h at 37 °C. After washing, the cells at 1×10⁷/2 ml buffer A were labeled with 0.5 mCi [³²P] orthophosphate for

2 h at 37 °C. Then the labeled cells were collected by centrifugation and washed twice with buffer A. The cells at a density of 5×10^6 cells/ml buffer A were stimulated with MLI at 37 °C. For analysis of phosphoproteins, the stimulation was stopped by placing the tubes on ice. The cell pellets obtained by centrifugation at 0 °C were immediately frozen in a dry ice/ethanol bath.

Total cellular lipids were isolated by repeated extraction and separated by ascending two-step one-dimensional thin layer chromatography as described under Section 2.2.3. The separated [^{32}P]-labeled phospholipids were visualized by use of the comigrating standard phospholipids 30 µg PIP_2 and 30 µg PIP dispersed in $CH_3OH : H_2O$ 1 : 1, 20 µg PA in $CHCl_3 : CH_3OH$ 1 : 1, 50 µg PC, 20 µg PE and 60 µg PI in $CHCl_3$ by exposure the TLC plates to iodine vapors. The spots were cut out for liquid scintillation counting. Lectin-induced alterations in the level of phospholipids of THP-1 cells are summarized in Table 1. **Phosphoproteins**

THP-1 cells were lyzed and treated with nuclease as described by Radke and Martin (1979) and Garrels (1979) with some modifications. The cells were lyzed in 0.5% (w/v) Nonidet P40, 50 µg of *Staphylococcus aureus* nuclease per ml, 2 mM $CaCl_2$, 1% (w/v) aprotinin, and 20 mM Tris/HCl, pH 8.8, by adding 50 µl buffer directly onto the thawing cell pellet. After treatment on ice for 10 min, the lysate was brought to 1 mM PMSF and then deoxyribonuclease was added to 100 µg/ml, and ribonuclease A to 50 µg/ml in 5 mM $MgCl_2$ and 50 mM Tris/HCl, pH 7.0. After treatment for additional 15 min on ice, the lysates were transferred to tubes containing enough urea to bring the concentration to 9.5 M. Then one volume of lysis buffer was added containing 9.5 M urea, 2% (v/v) Servalyt pH range 3–10, 2% (w/v) Nonidet P40 and 5% (v/v) 2-mercaptoethanol. **Phosphoproteins**

Cell lysates were separated by two-dimensional polyacrylamide gel electrophoresis (O'Farrell 1975; see also Cowles et al., Chap. 9, this Vol.). For the first dimension separation by isoelectric focusing cell lysates were subjected to cylindrical, 14-cm-long acrylamide gels containing 9.2 M urea and pH 3–10 Servalytes. After

Table 1. Effect of MLI on the level of phospholipids in THP-1 cells

Phospholipid	[^{32}P]-cpm/10^7 cells	
	Control	MLI treatment
PIP_2	6280 ± 202	$15\,964 \pm 472$
PIP	3386 ± 141	4516 ± 242
PI	7266 ± 746	6247 ± 884
PA	$36\,207 \pm 1716$	$37\,174 \pm 1996$
PC	$10\,386 \pm 472$	$12\,404 \pm 349$
PE	$11\,878 \pm 702$	$10\,417 \pm 576$

THP-1 cells at 4×10^7 cells/3 ml buffer A were labeled with 100 µCi of [$^{32}PO_4$] for 2 h at 37 °C. The cells then received PBS either with or without (control) 0.5 µg MLI and the incubation was continued for 15 min. The tubes were placed on ice and the cells were pelleted by centrifugation of 0 °C. After washing twice in ice-cold buffer A the cellular phospholipids were extracted with $CHCl_3 : CH_3OH : 2.4$ M HCl 1.5 : 1 : 1.5 and separated by two-step one-dimensional thin layer chromatography. Radioactivity in the spots was measured by liquid scintillation counting. Results are averages of triplicates $\pm$ standard deviation.

separation, the thin, porous, first dimension gels were stored for only 5 min without agitation in equilibration buffer with 10% (w/v) SDS. The increased SDS content allowed entry of enough SDS to bind the proteins and carry them into the second dimension, minimized the loss of protein, and increased the resolution by decreased diffusion. The thin focusing gels fit tightly between the glass plates and adhere to the stacking gel of the second dimension so that embedding in agarose is not required. After electrophoresis gels were silver stained (Blum et al. 1987) and dried. $[^{32}P]$-labeled polypeptides were located by autoradiography.

35.3 Discussion

Endogenous phosphatidylinositol-specific phospholipase C (PI-PLC) is the key enzyme of phosphatidylinositol turnover, generating two secondary messengers upon stimulation of cells. In mammalian cells, several calcium-dependent PI-PLCs have been isolated and characterized involved in signal transduction which are specific for PI, PIP and PIP_2 (Rhee et al. 1989). In this present study, we described in detail that β-galactoside-binding lectins induce alterations in the level of phospholipids and inositol phosphates. Interaction of the human placenta lectin with $[^3H]$ inositol-prelabeled THP-1 cells increased significantly the level of PIP and PIP_2 in a time dependent manner, however, did not influence the PI-level (Fig. 2). THP-1 cells responded to binding of this lectin with significantly decreased levels of IP_2 and poorly increased levels of IP and IP_3 after 1 min (Fig. 1). Similarly, addition of mistletoe lectin I to $[^{32}PO_4]$-prelabeled THP-1 cells increased only poorly the level of $[^{32}P]$-PC and $[^{32}P]$-PIP and that of $[^{32}P]$-PIP_2 2.5-fold (Table 1).

Receptor activation having tyrosine kinase by growth factors caused the stimulation of phosphatidylinositol turnover in cell culture (Hepler et al. 1987), and phospholipase C-γ was phosphorylated at tyrosine residues in growth factor-treated cells (Nishibe et al. 1989). Therefore, phospholipase C-γ is considered as a possible

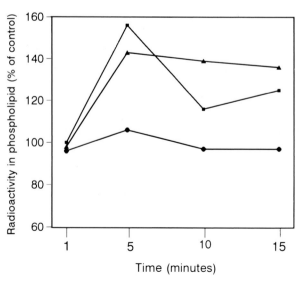

Fig. 2. Kinetics of lectin-induced alterations in the level of phosphatidylinositols of THP-1 cells. $[^3H]$inositol-labeled THP-1 cells at 5×10^6 cells/ml were treated with 5 μg lectin from human placenta at 37 °C. Cellular phosphatidylinositols were extracted with ice-cold $CHCl_3 : CH_3OH : 2.4$ M HCl 1.5 : 1 : 1.5 and separated by two-step one-dimensional thin layer chromatography. The time dependence of lectin-induced alterations in the level of ● PI, ■ PIP and ▲ PIP_2 is expressed as radioactivity in phospholipid in % relative to controls. *Points* are averages of triplicates. Standard deviations are less than 10%

substrate for tyrosine kinase. One of the roles of tyrosine kinase in signaling processes may be the activation of the phosphatidyl turnover. Therefore, we have analyzed the phosphorylation pattern of lectin-treated THP-1 cells. However, the pattern of lectin-treated cells and of controls were found to be qualitatively identical.

Our results indicate that β-galactoside-binding lectins induce alterations in the level of inositol phosphates and phosphatidylinositols of THP-1 cells in culture.

References

Berridge MJ, Dawson RMC, Downes CP, Heslop JP, Irvine RF (1983) Changes in the level of inositol phosphates after agonist-dependent hydrolysis of membrane phosphoinosides. Biochem J 212: 473–482

Blum H, Beier H, Gross HJ (1987) Improved silver staining of plant proteins, RNA and DNA in polyacrylamide gels. Electrophoresis 8: 93–99

Danilov YN, Cohen CM (1989) Wheat germ agglutinin but not concanavalin A modulates protein kinase C-mediated phosphorylation of red cell skeletal proteins. FEBS Lett 257: 431–434

Downes P, Michel RH (1982) Phosphatidylinositol-4-phosphate and phosphatidylinositol-4,5-bisphosphate: lipids in search of a function. Cell Calcium 3: 467–502

Garrels JI (1979) Two-dimensional gel electrophoresis and computer analysis of proteins synthesized by clonal cell lines. J Biol Chem 254: 7961–7977

Grier CE, Mastro AM (1988) Lectin-induced phosphatidylinositol metabolism in lymphocytes is potentiated by macrophages. J Immunol 141: 2585–2592

Hepler JR, Nakahata N, Lovenberg TW, DiGuiseppi J, Herman B, Earp HS, Harden TK (1987) Epidermal growth factor stimulates the rapid accumulation of inositol (1,4,5)-triphosphate and a rise in cytosolic calcium mobilized from intracellular stores in A431 cells. J Biol Chem 262: 2951–2956

Hirabayashi J, Kasai K-J (1984) Human placenta β-galactoside-binding lectin. Purification and some properties. Biochem Biophys Res Commun 122: 938–944

Kornberg LJ, Earp HS, Turner CE, Prockop C, Juliano RL (1991) Signal transduction by integrins: increased protein tyrosine phosphorylation caused by clustering of β_1 integrins. Proc Natl Acad Sci USA 88: 8392–8396

Medh JD, Weigel PH (1989) Separation of phosphatidylinositols and other phospholipids by two-step one-dimensional thin layer chromatography. J Lipid Res 30: 761–764

Nishibe S, Wahl MI, Rhee SG, Carpenter G (1989) Tyrosine phosphorylation of phospholipase C-II in vitro by the epidermal growth factor receptor. J Biol Chem 264: 10335–10338

O'Brian CA, Ward NE (1989) Biology of the protein kinase C family. Cancer and Metastasis Rev 8: 199–214

O'Farrell PH (1975) High resolution two-dimensional electrophoresis of proteins. J Biol Chem 250: 4007–4021

Radke K, Martin GS (1979) Transformation by Rous sarcoma virus: effects of src gene expression on the synthesis and phosphorylation of cellular polypeptides. Proc Natl Acad Sci USA 76: 5212–5216

Rana RS, Hokin LE (1990) Role of phosphoinositides in transmembrane signaling. Physiol Rev 70: 115–164

Rhee SG, Suh PG, Ryu SH, Lee SY (1989) Studies of inositol phospholipid-specific phospholipase C. Science 244: 546–550

Sharon N, Lis H (1989) Lectins as cell recognition molecules. Science 246: 227–233

Ziska P, Franz H, Kindt A (1978) The lectin from Viscum album L.: purification by biospecific affinity chromatography. Experientia 34: 123–124

36 Measurement of Intracellular Calcium Levels by Flow Cytometry Following Treatment of Murine Macrophage/Monocytes with Mistletoe Lectin

S.S. JOSHI and H.-J. GABIUS

The purified mistletoe lectin, ML-I, with galactoside specificity has a dramatic antitumor effect and also causes secretion of cytokines such as tumor necrosis factor-α, interleukin-1 and interleukin-6 (Hajto et al. 1990; Joshi et al. 1991; Gabius et al. 1992). The precise mechanism(s) of action of ML-I at the cellular level is not known. Some of the cellular regulatory factors can lead to rapid changes in a variety of cellular functions, including changes in intracellular second messenger systems affected by the binding of lectin ML-I. These intracellular messengers may further mediate the induction of nuclear activities essential for the production of immune regulating/stimulating cytokines and other factors. One of the main factors involved in intracellular second messenger pathways is changes in kinetics of intracellular calcium mobilization/concentration. The effects of lectins on the levels of intracellular calcium levels can be measured using appropriate fluorescent dyes such as Indo-1 or Flura-2 or Fluo-3 and flow cytometry techniques (June and Rabinovitch 1991). The flow cytometer can be used to measure various functional parameters that are of interest to lectinologists. The flow cytometry can be used to measure the concentration of intracellular free ions such as calcium in single living cells using appropriate fluorescent probes. The commonly used fluorescent probes are Indo-1 and Fluo-3. Indo-1 requires a flow cytometer which is capable of UV illumination, whereas in the case of Fluo-3 dye, the UV source of illumination is not required. Fluo-3 requires regular argon ion laser. This chapter focuses on the determination of intracellular calcium levels in a murine macrophage/monocyte cell line J774A.1 following treatment with purified mistletoe lectin ML-I using flow cytometry techniques using Fluo-3 as fluorescent probe. Fluo-3 is a fluorescein-based calcium probe originally developed by Minta et al. (1989).

36.1 Materials and Methods

Materials   Murine macrophage/monocyte J774A.1 cells
2 mg/ml Fluo-3 acetoxymethyl ester (Fluo-3 AM; Molecular Probes #F-1241)
1 mg/ml ionomycin (Calbiochem #407950)
Pluronic detergent F-127 (Molecular Probes)
Dimethyl sulfoxide (DMSO, Sigma Chem.)
Phosphate-buffered saline (PBS)
12×75 mm polypropylene tube
37 °C water bath
Fluorescence microscope
Flow cytometer with argon ion laser
Purified mistletoe lectin, ML-I

The murine macrophage/monocyte cell line J774A.1 was maintained in vitro in a monolayer culture. The cells were grown in Dulbecco's modified Eagle's medium supplemented with 10% fetal calf serum, antibiotics, and L-glutamine (2 mM). The cell suspensions were prepared by dislodging the cells, using a cell scraper. The dislodged cells were repeatedly but gently pipetted to disperse the cells into single cells. The cells were counted using an electronic cell counter and the cell concentration was adjusted to 5×10^6 cells/ml.

Preparation of Cells

Loading the cells with Fluo-3. The J774A.1 cells were suspended in DMEM medium. Five million cells in 1 ml volume was added with Fluo-3/AM ester at a concentration of 100 μg/5×10^6/ml. A nonionic surfactant, pluronic F-127, was used to aid solubilization of AM ester into aqueous medium. One microliter of a 25% (w/w) stock solution of pluronic F-127 in dry dimethyl sulfoxide was mixed with every 10 nmol of AM ester before thoroughly dispersing the mixture into aqueous solution. Loading the cells in the presence of a detergent such as pluronic F-127 significantly decreases the heterogeneity of Fluo-3 uptake by the cells (Vandenberghe and Ceuppens 1990). The mixture was incubated for 1 h with intermittent shaking at 37 °C and washed three times with DMEM medium. The cells were then resuspended gently in medium (do not vortex). For washing, the cells were centrifuged for 6 minutes at 950 rpm (180 g). The loaded cells were stored at 22 °C and protected from light until analysis. In order to confirm loading the cells with Fluo-3, the loaded cells were examined using a fluorescent microscope. The Fluo-3/AM loaded cells were then analyzed on a Coulter EPICS-5 flow cytometer.

Loading of Cells

The intensity of green fluorescence with and without ML-I treatment was determined. The Fluo-3/AM-loaded cells were excited with 488 nm wavelength of an argon ion laser and the fluorescence was collected at 525 nm using a linear amplification. The cells were maintained at 37 °C using a warm water circulating system. The cells were analyzed by gating on forward and right angle scatter. The dead cells were excluded by gating out cells without Fluo-3 fluorescence. The cells were analyzed as above with and without different concentrations of purified mistletoe lectin ML-I. The change in fluorescence intensity following treatment with ML-I was determined. As a control, the Fluo-3/AM-loaded cells were also added with ionomycin 1 μg/ml final concentration and analysis was repeated. The results were analyzed by determining the mean response versus time and the fraction of responding cells versus time.

Flow Cytometric Analysis

36.2 Results

Our results demonstrated that upon stimulation of J774A.1 murine macrophage/monocyte cells with purified mistletoe lectin ML-I, the intracellular Fluo-3 fluorescence intensity was increased markedly, as shown in Fig. 1. The unloaded J774A.1 cells did not show a significant antifluorescence. There was about 25% increase in fluorescence intensity of intracellular Fluo-3 and about four fold increase in the number of cells with intracellular Fluo-3 indicating the change in calcium flux following treatment with ML-I.

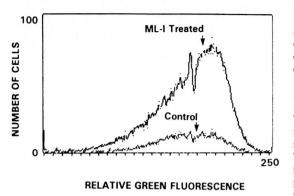

RELATIVE GREEN FLUORESCENCE

Fig. 1. Flow cytometric analysis of J774A.1 murine macrophage/monocyte cell line for the intracellular Ca^{2+} concentration following treatment of the cells with purified mistletoe lectin ML-I; 5×10^6 cells were loaded with Fluo-3/AM by incubating the cells for 1 h at 37 °C and washed with medium. The loaded cells were added with lectin ML-1 (10 ng/ml) and analyzed for intracellular Ca^{2+} content after 5 min. The histogram shows the intensity of fluorescence in each channel. *Control* peak 93 at channel 187; *ML-I treated* peak 114 at channel 187

36.3 Discussion

The effects of lectins on intracellular calcium ion concentration, which in turn might play an important role in cellular functions, can be measured using flow cytometry with appropriate fluorescent probes such as Fluo-3. The positive aspect of the Fluo-3 probe is its similarity to fluorescein probe since it is a derivative of fluorescein and does not require a UV laser for analysis. However, it is more difficult with Fluo-3 to determine the percent responding cells than when Indo-1 is used. There is a significant amount of heterogeneity in the distribution of Fluo-3 in cells which results in a relatively poor separation of responding cells and non-responding cells. Rijkers et al. (1990) have reported an improved resolution of responses by analysis of the ratio of Fluo-3 to Snarf-1 fluorescence. In summary, the lectin-mediated changes in the intracellular calcium can be measured using Flow-3 probe and flow cytometry. This fluorescein-related probe can be used in a simple flow cytometric technique. For the detailed analysis of lectin-mediated changes in intracellular calcium mobilization/concentration and kinetics of calcium flux, one should use Indo-1 as probe.

References

Gabius HJ, Walzel H, Joshi SS, Kruip J, Kojima S, Goke V, Kratziu H, Gabius S (1992) The immunomodulatory galactoside-specific lectin from mistletoe: partial sequence analysis, cell and tissue binding, and impact on intracellular biosignaling of monocytic leukemia cells. Anticancer Res 12: 669–676

Hajto T, Hostanska K, Frei K, Rordorf C, Gabius HJ (1990) Increased secretion of tumor necrosis factor-α, interleukin-1 and interleukin-6 by human mononuclear cells exposed to β-galactoside specific lectin from clinically applied mistletoe extract. Cancer Res 50: 3322–3326

Joshi SS, Komanduri KC, Gabius S, Gabius HJ (1991) Immunotherapeutic effects of purified mistletoe lectin, ML-I, on murine large cell lymphoma. In: Lectins and Cancer. HJ Gabius and S Gabius (Eds.), Springer-Verlag, Heidelberg, pp. 207–217

June CH, Rabinovitch PS (1991) In: Current protocols in immunology. JE Coligan, AM Kruisbeek, EM Shevaeh and W Strober (Eds.), John Wiley & Sons, New York, pp. 5.5.2

Minta A, Kao JPY, Tsien RY (1989) Fluorescent indicators for cytosolic calcium based on rhodamine and fluorescein chromophores. J Biol Chem 264: 8171–8178

Rijkers GT, Justement LB, Grifferen AW, Cambier JC (1990) Improved method for measuring intracellular Ca^{++} with Fluo-3. Cytometry 11: 923–927

Vandenberghe PA, Ceuppens JL (1990) Flow cytometric measurement of cytoplasmic free calcium in human peripheral blood T-lymphocytes with Fluo-3, a new fluorescent indicator. J Immunol Methods 127: 197–205

37 Measurement of Lectin-Induced Superoxide Release from Human Neutrophils

A.V. Timoshenko and H.-J. Gabius

Lectin-carbohydrate interaction can trigger diverse cellular responses. Plasma membranes of human neutrophils contain NADPH oxidase (Cross and Jones 1991), which can be activated by different stimuli including lectins (Cohen et al. 1982). The key component of this superoxide anion (O_2^-)- generating system, cytochrome b-245, was shown to be a glycoprotein with about 21% carbohydrate (Harper et al. 1985), even enabling direct interaction with some exogenous lectin. There are different methods to determine the activity of NADPH oxidase and respiratory burst of neutrophils: luminol-dependent chemiluminescence (Hallett and Campbell 1983), nitroblue tetrazolium dye reduction assay (Das et al. 1990), electron microscopic technique (Ohno et al. 1982), fluorometric assay (Hyslop and Sklar 1984) and reduction of cytochrome c (Markert et al. 1984). The last method is most useful due to its high specificity to O_2^-. The method can be applied in various modifications: continuous and discontinuous assay (Markert et al. 1984) and microassay (Mayo and Curnutte 1990). Employing cytochrome c, any lectin-dependent effect on the generation of this compound, involved in host defence, can easily be monitored.

37.1 Experimental Procedures

37.1.1 Isolation of Human Neutrophils

Materials

- 40 ml fresh human blood with 3.8% citrate
- plastic syringe for 50 and 10 ml
- plastic tubes for 10 ml (Greiner, Nürtingen, FRG)
- Ficoll-Paque (Pharmacia, Uppsala, Sweden)
- buffer A: phosphate-buffered saline (PBS), pH 7.3 containing 10 mM Na_2HPO_4/KH_2PO_4, 137 mM NaCl, 2.7 mM KCl
- buffer B: PBS containing 5.55 mM D-glucose, 0.9 mM $CaCl_2$ and 0.5 mM $MgCl_2 \cdot 6H_2O$ (PBSG)
- 6% solution of Dextran T70 (Pharmacia, Uppsala, Sweden) in 0.9% NaCl
- hypertonic solution of 0.6 M NaCl
- centrifuge Universal II (Hettich, Tuttlingen, FRG)

Procedure

1. Twenty ml of blood is drawn into each of two syringes containing 10 ml of 6% Dextran T70 and, after mixing, the syringe is allowed to stand vertically at room temperature for 1 h.
2. The neutrophil-rich upper layer is collected in four plastic tubes and centrifugation for 10 min at 300g at room temperature follows. The pellets are resuspended together in 10 ml PBS and centrifugation is repeated.
3. Three ml of ice-cold bidistilled water is added to final pellet for 20–30 s to lyze the erythrocytes. Then 1 ml of ice-cold 0.6 M NaCl is added to restore the isotonicity.

4. Centrifugation of the cell suspension for 4 min at $100g$ is carried out and hypotonic lysis is repeated with the cell pellet.
5. The cells are pelleted again and suspended in 8 ml PBS.
6. Equal volumes (4 ml) of the cell suspension are layered on two 6-ml Ficoll-Paque solutions and spun at 400 g for 40 min at room temperature.
7. The pellet containing neutrophils is suspended in 10 ml PBSG, washed once at $300g$ for 5 min and resuspended in 5 ml PBSG.

Before using the suspension of neutrophils is kept on ice. Cell viability, as assessed by trypan blue exclusion, in generally more as 95%.

37.1.2 Measurement of O_2^- Release

Materials
– Eppendorf tubes for 1.5 ml, multipet 4780 and centrifuge
– 5415 C (Eppendorf-Netheler-Hinz, Hamburg, FRG)
– Solution of PBSG (see above)
– Solution of cytochrome c (type III, Sigma, Munich): 16 mg/ml in PBSG
– Solutions of lectins: 1 mg/ml in PBSG
– Stock solution of superoxide dismutase (SOD, Sigma, Munich, FRG): 3 mg/ml in PBSG (can be stored in aliquots in 100 µl at $-20\,°C$ and thawed only once)
– Suspension of neutrophils: 8×10^6 cells/ml in PBSG

Procedure
1. The following reagents are added in consecutive order at room temperature into each Eppendorf tube: 510 µl PBSG, 40 µl lectin-containing solution, 200 µl neutrophil suspension, and 50 µl cytochrome c solution. Control tubes contain 200 unit/ml SOD. Final concentrations of the constituents of the assay are: neutrophils 2×10^6 cells/ml, lectin 10–100 µg/ml, cytochrome c 80 µM or 1 mg/ml.
2. The specimens are kept at $37\,°C$ for 15 min; the reaction is stopped by placing the tubes into ice and centrifugation at 10 000 rpm in an Eppendorf centrifuge follows.
3. The absorbance at 550 nm of the supernatants is determined, using a 80 µM cytochrome c solution as a reference (750 µl PBSG and 50 µl cytochrome c stock solution).
4. The release of O_2^- (in nmol) is calculated with the formula: O_2^- (nmol) $= 47.7 (A_1 - A_2)$, where A_1 is the absorbance at 550 nm of the specimen without SOD, A_2 is the absorbance at 550 nm of the specimen with SOD (Markert et al. 1984).

37.2 Results

Figure 1 shows changes in absorbance at 550 nm of cell supernatants of lectin-stimulated neutrophils in presence and absence of SOD. Obviously, the lectin-induced reduction of cytochrome c is almost completely connected with O_2^- release from neutrophils in the cases of the galactose-specific mistletoe lectin and the mannose-specific concanavalin A. The stimulation is abrogated in the presence of the inhibitory sugar. Notably, various lectins elicit a nonuniform response, as illustrated in Fig. 2.

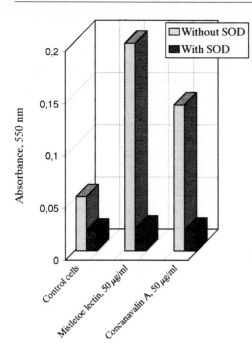

Fig. 1. Effect of superoxide dismutase (SOD, 200 U/ml) on the reduction of cytochrome c (80 μM) by lectin-treated neutrophils

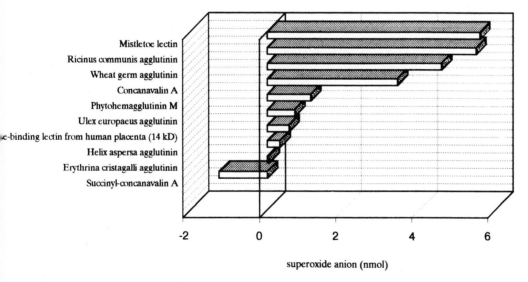

Fig. 2. Lectin-induced superoxide anion release from human neutrophils (the release of superoxide anion from lectin-treated cells is set into relation to control cells without lectin treatment)

37.3 Remarks

According to the described procedure, the average yield of neutrophils was $74 \pm 33 \times 10^6$ cells from 40 ml blood ($n = 27$). It is sufficient to perform 40–50 assays. It is very important to note that neutrophils can produce a certain amount of superoxide without stimulation. The average value of A_1 for unstimulated neutrophils was 0.049 ± 0.034 ($n = 26$). Hence, some lectins can show a negative effect relative to the untreated cells.

Acknowledgment. A.V. Timoshenko is a recipient of a scholarship of the Deutscher Akademischer Austauschdienst (DAAD).

References

Cohen HJ, Chovaniec ME, Wilson MK, Newburger PE (1982) Con A-stimulated superoxide production by granulocytes: reversible activation of NADPH oxidase. Blood 60: 1188–1194

Cross AR, Jones OTG (1991) Enzymic mechanisms of superoxide production. Biochim Biophys Acta 1057: 281–298

Das UN, Padma M, Sagaz PS, Ramesh G, Koratkar R (1990) Stimulation of free radical generation in human leukocytes by various agents including tumor necrosis factor is a calmodulin-dependent process. Biochem Biophys Res Commun 167: 1030–1036

Hallett MB, Campbell AK (1983) Two distinct mechanisms for stimulation of oxygen-radical production by polymorphonuclear leucocytes. Biochem J 216: 459–465

Harper AM, Chaplin MF, Segal AW (1985) Cytochrome b-245 from human neutrophils is a glycoprotein. Biochem J 227: 783–788

Hyslop PA, Sklar LA (1984) A quantitative fluorometric assay for the determination of oxidant production by polymorphonuclear leukocytes: its use in the simultaneous fluorometric assay of cellular activation processes. Anal Biochem 141: 280–286

Markert M, Andrews PC, Babior BM (1984) Measurement of O_2^- production by human neutrophils. The preparation and assay of NADPH oxidase-containing particles from human neutrophils. Meth Enzymol 105: 358–365

Mayo LA, Curnutte JT (1990) Kinetic microplate assay for superoxide production by neutrophils and other phagocytic cells. Meth Enzymol 186: 567–575

Ohno YI, Hirai KI, Kanoh T, Uchino H, Ogawa K (1982) Subcellular localization of hydrogen peroxide production in human polymorphonuclear leukocytes stimulated with lectins, phorbol myristate acetate, and digitonin: an electron microscopic study using $CeCl_3$. Blood 60: 1195–1202

38 Isolation of Lymphocytes and Accessory Cells for Assays of Mitogenicity

D.C. KILPATRICK

Many lectins have the ability to stimulate cells to grow and divide. Lymphocytes are the usual target cells for mitogenic assays and the study of lectin-lymphocyte interaction has made a substantial contribution to our understanding of lymphocyte activation and its control (Kilpatrick 1991). Crude and variable preparations of mononuclear cells are often used for assessment of mitogenicity, but it is well established that accessory cells have an influence on the response of T lymphocytes, and that some lectin-lymphocyte interactions are more accessory cell-dependent than others.

It is therefore desirable to isolate T lymphocytes when investigating mitogenic properties, so that the activity towards the isolated T (and/or B) cells can be compared to the original mix containing accessory cells, and indeed to reconstituted preparations to which individual cell components have been added back.

Cell and cell subset separation and isolation can be achieved with a high degree of purity by a number of means. Fluorescence activated cell sorting, exploiting cell or subset specific monoclonal antibodies coupled to one or more fluorescent dyes, is excellent in principle, but requires very expensive equipment and is not a suitable method for obtaining large numbers of cells quickly. The use of magnetic beads coupled to specific monoclonal antibodies constitutes a facile and rapid technology, but is relatively expensive for preparing cell populations in bulk. Separation of cells based on panning methods utilizing antibodies attached to a solid surface is also a feasible strategy.

Described below, however, are methods used regularly in this laboratory when investigating the mitogenic characteristics of lectins, which provide an adequate quantity of cells with the use of inexpensive materials and readily available equipment. These methods are applicable to human blood samples from normal, healthy individuals, but would not necessarily be equally suitable for blood from other species or for human blood obtained from patients with leukoproliferative disorders. For a comprehensive account of alternative methods of lymphocyte and accessory cell preparation, the reader is referred to the monograph by Klaus (1987).

38.1 Objectives and Limitations

It is assumed that usually the principal aim will be to obtain a good preparation of T lymphocytes and/or monocytes as accessory cells. By this perspective, other cells (erythrocytes, platelets, granulocytes) would be viewed as contaminants to be eliminated. However, it is appreciated that there may be occasions when the influence of these individual cell types is to be examined and therefore the isolation of the latter would be necessary. Therefore, both options will be considered.

Pure blood cannot be used as starting material, for it would soon clot. Therefore, the addition of anti-coagulants or alternative anti-clotting devices is necessary, and the optimal choice depends on the precise objectives of cell isolation. If platelet preparation were desired, EDTA would be the anti-coagulant of choice, but often it is desirable to eliminate platelets from the start. Defibrination is an excellent means of doing so, but this procedure may reduce the yield of monocytes and indeed other cells if not performed skilfully enough. The use of carbonyl iron is the next best option for platelet removal and this procedure should give good mononuclear cell yields and also an improved separation of granulocytes from lymphocytes. Carbonyl iron is best used in conjunction with heparinised blood. All materials should be sterilized (e.g., by autoclaving) before use, and all open operations should be carried out in a Class II microbiological safety cabinet.

38.2 Isolation or Removal of Platelets

Materials

EDTA Solution	(5% w/v Na_2 EDTA $\cdot$ $2H_2O$ [Sigma]).
EDTA Buffer	(9 mM Na_2 EDTA $\cdot$ $2H_2O$/26 mM Na_2 $HPO_4 \cdot 2H_2O$/0.1 M NaCl, adjusted to pH7 with HCl)
Dextran solution	(10% w/v dextran grade B, mw 150–200 000 (BDH), in saline)
Saline	(0.85% w/v Na Cl)
Heparin	(Porcine heparin, lithium salt, Grade IV, Sigma)
Carbonyl iron	(reduced pentacarbonyl iron, Sigma)

Principle   Platelets are best prepared from blood anticoagulated with EDTA, which prevents platelet clumping and so facilitates separation from leukocytes. If platelets are to be removed and discarded, however, this can be achieved by methods superior to differential centrifugation. Clotting can be induced by moving glass beads, creating a compact clot and leaving a large volume of usable defibrinated blood. Alternatively, platelets can be induced to adhere to iron particles in suspension after which they can be retained by a magnetic field.

Preparation of Platelets

Preparation
1. Add fresh blood to commercially purchased EDTA blood collection tubes or, if unavailable, add 1 part EDTA Solution to nine parts blood.
2. Mix anticoagulated blood with an equal volume of saline, then add dextran solution to a final concentration of 1.5%. Mix fully, and incubate at 37 °C for 15 min in a water bath. (30 ml Universal containers from Sterilin Ltd are suitable).
3. Remove the upper (leukocyte-rich) layer by aspiration with a Pasteur pipet, and centrifuge at 400g for 5 min.
4. Aspirate the supernatant and spin again as before. The platelet-rich supernatant is again aspirated. [The pellets (granulocytes, lymphocytes, etc.) may be used as a source of leukocytes, if so required].
5. Harvest the platelets by centrifugation at 1000g for 10 min.
6. Resuspend in EDTA Buffer and wash twice.

Removal of Platelets by Defibrination

Defibrination
1. Pour fresh blood into McCartney bottles containing around 25 glass beads (2 mm diam.) and gently shake by hand for 10 min. Motion in an arc is desirable, and it is essential to keep the beads in motion.
2. Aspirate platelet-free blood from the clot formed around the beads.

Removal of Platelets with Carbonyl Iron

1. Add blood (anti-coagulated with heparin) to McCartney bottles containing carbonyl iron (0.5*g* per 20 ml blood) and shake gently to suspend the particles.   **Carbonyl Iron**
2. Incubate in a waterbath at 37 °C for 20 min, then place on a roller shaker for a further 5 min.
3. Dilute with an equal volume of saline and add Dextran solution as previously described. Place the diluted blood on a strong magnet and leave for 15 min.
4. Aspirate the leukocyte-rich (upper) layer, which will be essentially free of platelets.

38.3 Separation of Red Blood Cells

Erythrocytes outnumber leukocytes by a ratio of 700:1, so their preparation from blood is not a problem, but their presence is a major obstacle in the preparation of leukocytes, since even a high degree of separation of red blood cells may still leave a significant absolute number as contaminants. If a high yield of leukocyte-free erythrocytes be desired, the lower layer obtained after dextran sedimentation of defibrinated or anticoagulated blood (see under Platelets) should be used as starting material. Often the contamining red blood cells in the upper leukocyte-rich layer will still provide a sufficient yield. In either case, a pure erythrocyte preparation is obtained by density gradient centrifugation after centrifugation through a double layer of Ficoll solutions, as described in detail below.

38.4 Separation of Granulocytes, Mononuclear Cells and Erythrocytes

– Ficoll-diatrizoate solution, S.G. 1.077 (Histopaque 1077, Sigma)   **Materials**
– Ficoll-diatrizoate solution, S.G. 1.119 (Histopaque 1119, Sigma)

The separation of blood cells by density gradient centrifugation over the polymer,   **Principle**
Ficoll, exploits the tendency of erythrocytes towards rouleaux formation and the relatively large density difference granulocytes and other leukocytes. By using two layers of Ficoll solution, one at the density of lymphocytes (B cells and monocytes are only marginally less dense than T cells) and the other at the density of granulocytes, mononuclear cells are trapped at the first interphase, granulocytes are trapped at the second interface and erythrocytes form a pellet at the bottom (English and Andersen 1974).

1. Use as starting material the leukocyte-rich 400*g* pellet obtained after dextran   **Procedure**
sedimentation of defibrinated or anti-coagulated blood. Wash twice and re-suspend in saline.
2. Layer 5 ml of the leukocyte suspension onto a double layer of 5 ml Histopaque 1119 (bottom) and 5 ml Histopaque 1077 (top) in a 15 ml centrifuge tube. (The gradient media and overlay suspension are best introduced by tilting the centrifuge at a 45 °C angle and allowing them to run down the side of the tube from a Pasteur pipet).

3. Centrifuge at 700g for 25 min at room temperature.
4. Aspirate mononuclear cells (upper interface), granulocytes (lower interface) and erythrocytes (pellet) and wash free of Ficoll by centrifugation at 400g for 8 min.
5. Apply each leukocyte fraction separately to a single layer of the appropriate density of Histopaque and centrifuge, and collect from the interface as before.
6. Wash each cell preparation in saline or culture medium as required.

38.5 Separation of Lymphocytes and Monocytes

Materials
- RPMI 1640 culture medium (Gibco)
- Phosphate Buffed Saline (Dulbecco A, Oxoid)
- Gelatin (from Swine skin, type II, Sigma)
- Autologous serum (from clotted blood incubated at 37 °C for 2 h)

Principle
Lymphocytes may be separated from monocytes on the basis of the adhesive properties of the latter. Various experimental conditions have been used to exploit this difference; the procedure described here follows that of Leb et al. (1983).

Procedure
1. Prepare in advance gelatin-coated plates as follows. Pour 0.1% (w/v) aqueous gelatin solution (11 ml) into 90 mm diameter plastic petri dishes (Sterilin) and leave for 2 h at 4 °C. Pour off the gelatin solution, and leave to dry at 37 °C for 2 days. Store the dry gelatin-coated dishes at room temperature until ready for use.
2. Use as starting material the mononuclear cell fraction prepared by density gradient centrifugation as described above, wash in RPMI-1640 culture medium then resuspend to a cell density of about 10^7 cells/ml in RPMI 1640 to which fresh autologous serum had been added to 10% (v/v).
3. Place 6.5 ml of the cell suspension into a gelatin coated petri dish and leave for 1 h at 37 °C in a 5%-CO_2 incubator.
4. Gently pour the cell suspension from the petri dish into another container, and wash the dish very gently three times with 8 ml of PBS prewarmed to 37 °C. This should remove all the nonadherent cells (lymphocytes), which can be collected by centrifugation (400g; 8 min).
5. Place the washed petri dishes on ice-water for 30 min
6. Recover the adherent cells (monocytes) by vigorous washing with ice-cold PBS. (Repeated squirting with a Pasteur pipet is an effective technique). Collect the monocytes by centrifugation (400g; 8 min).

38.6 Separation of T and B Lymphocytes

Materials
- Ovine blood
- Neuraminidase (Sigma)
- Fetal calf serum (Globefarm Ltd)
- Polybrene (Sigma)
- Nycodenz monocytes (Nyegaard UK Ltd)

T lymphocytes have the ability to bind to and form rosettes with ovine erythrocytes via E-receptors (CD2 molecules) on the T cell surface. This rosette formation works best if the sheep red blood cells are modified chemically or enzymatically with neuraminidase and is greatly enhanced with fetal calf serum. The rosettes can easily be separated from B cells by density gradient centrifugation, and the T cells recovered from the rosettes by osmotic lysis of the erythrocytes. Finally, passage through a hypertonic medium (Nycodenz monocytes) removes any remaining monocytes or red blood cell ghosts.

1. Prepare desialated sheep red blood cells (SRBC) as follows. Isolate the cells from diluted ovine blood by centrifugation over Ficoll gradients just as described before for human erythrocytes. The SRBC are then suspended in PBS at 6% (v/v) and incubated with neuraminidase (0.03 units/ml) at 37 °C for 30 min. Wash four times in PBS, then resuspend at 2.5% (v/v) in PBS containing EDTA (10 mM) and gelatin (0.1%).

2. Resuspend the lymphocyte preparation (after separation of adherent cells described above) at a cell density of $3-5 \times 10^6$/ml in RPMI 1640 culture medium containing 20% fetal calf serum. Mix with an equal volume of neuraminidase-treated SRBC suspension prepared as described. Mix and leave on a roller mixer for 5 min at room temperature. The cells are then harvested by light centrifugation ($250g$ for 5 min, or just enough to bring all the red blood cells down), and then very gently resuspended. Check rosette formation microscopically at this point, while the cells are being gently harvested again as before.

3. Resuspend the cells/rosettes and layer onto an equal volume of Histopaque 1077 and centrifuge at $600g$ for 25 min.

4. Collect the B cells from the interface, harvest and wash.

5. Aspirate and discard the remaining Histopaque, then wash the pellet three times in saline or PBS.

6. Add 2.5 ml of ice-cold 0.1% NaCl (saline diluted eightfold in distilled water) to each 0.5 ml of rosette-containing pellet, vortex for 5 s, then add 12 ml of physiological saline and mix immediately. Recover the T lymphocytes from the lyzed erythrocytes (ghosts) by three washes in saline.

7. Carefully layer this T cell preparation (3 ml) over 3 ml of Nycodenz monocytes and centrifuge at $800g$ for 15 min. The T lymphocyte pellet obtained is then washed in saline or culture medium.

38.7 Discussion

No mention has been made of the volume of blood to be used. This would depend on the number of particular cells required for a specific purpose. Typically, 20 ml of venous blood might provide an adequate number of mononuclear cells or T lymphocytes for mitogenic assays, but if monocyte effects are to be studied, perhaps 50–100 ml of blood may be required. A 20 ml blood sample contains approximately 120×10^6 leukocytes from which typically 30×10^6 mononuclear cells, 10×10^6 purified T lymphocytes, and 1.5×10^6 monocytes may be recovered. These figures can be used as a rough guide when planning experiments.

A complete scheme for recovering all cellular blood components is summarized graphically (Fig. 1). In practice, this should be modified according to the options

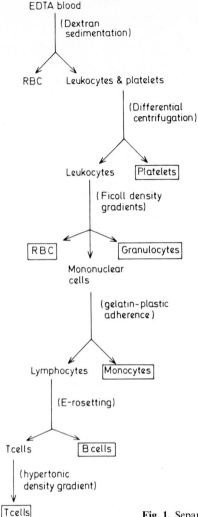

Fig. 1. Separation of different cell types from samples of human blood

outlined in the text, depending on the precise requirements of the experimental work. It is impossible to withdraw blood without altering the properties of platelets to some extent, so all platelet preparations are of dubious equivalence to the in vivo situation. Usually, platelets will be regarded as a nuisance. It is difficult to remove platelets completely from leukocytes by differential centrifugation alone, and impossible to do so without significantly decreasing the recovery of leukocytes. Therefore, unless a platelet preparation is specifically desired, it is best to defibrinate or remove platelets by carbonyl iron treatment.

The procedures described yield virtually pure preparations of platelets, erythrocytes, granulocytes and T lymphocytes. The adherent cell preparation is likely to be about 90% (or less) monocytes, and the B cells might be of similar purity. Further refinement of those preparations requires the use of one of the other techniques alluded to in the Introduction, of which magnetic immunobead technology would probably be the most appropriate (Klaus 1987).

References

English D, Andersen BR (1974) Single step separation of red blood cells, granulocytes and mononuclear leukocytes on discontinuous density gradient of Ficoll-hypaque. J Immunol Methods 5: 249–252

Kilpatrick DC (1991) Lectin interactions with human leukocytes: mitogenicity, cell separation, clinical applications In: Lectin reviews Vol 1, Sigma Chemical Company, St Louis, pp 69–80

Klaus GGB (1987) Lymphocytes – a practical approach. IRL Press, Oxford

Leb L, Crusberg T, Fortier N, Snyder LM (1983) Evaluation of methods using adherence to substrate and density gradient for the isolation of human monocytes. J Immunol Methods 58: 309–321

39 Determination of Lectin-Mediated Augmentation of Natural Killer Cell-Mediated Cytotoxicity

S.S. Joshi

Natural killer (NK) cells are the cytotoxic cells which manifest cytolytic activity without prior sensitization or stimulation by an antigen or activation by a lymphokine. These cells are believed to play a major role in the body in antitumor response, against virally infected cells and in regulation of hematopoiesis. Elimination of these effector cells using antibodies reacting with these cells in vivo increases the tumorigenicity and/or metastatic properties of certain tumor cells, indicating their importance in antitumor mechanisms. The other kind of antitumor cytotoxic cell population found in the body are lymphokine activated killer (LAK) cells which require prior activation with lymphokines such as interleukin-2 (IL-2). The LAK cells are capable of killing not only NK-sensitive tumor cells but also NK-resistant tumor cells. The origin of these naturally occurring cytotoxic cells is still being questioned and discussed by the experts in the field. The detailed discussion of their origin is beyond the scope of this chapter. Since the NK cells do play a role in antitumor immune response, it would be ideal to investigate certain compounds which might increase the number and cytolytic activity of NK cells in vivo. Among such compounds, certain lectins are known to be useful biological response modifiers which enhance immune response, including the activities of NK cells. Recently, we have reported the immunostimulatory effects of a galactoside specific lectin, ML-I, purified from mistletoe extract (Joshi et al. 1991; Gabius et al. 1992).

The purified mistletoe lectin, ML-I, is a galactoside specific lectin with an approximate molecular mass of 63 kDa. ML-I consists of three subunits with different functional domains, the two subunits (AB) with an apparent molecular weight of 29 and 27 kDa termed A1 and A2, and the carbohydrate-binding (galactoside specific) subunit with an apparent molecular weight of 34 kDa, termed B. At a very low dose, the purified ML-I has demonstrated in rabbits, mice and in humans an antitumor/immunostimulatory effect in vivo (Hajto et al. 1989; Gabius and Gabius 1990; Joshi et al. 1991). The mistletoe extract is known to enhance multiple arms of the immune system (Joshi et al. 1991; Gabius et al. 1992). In this report, the details of determination of immunostimulatory effects of a lectin, ML-I, with special reference to NK cell mediated cytotoxicity in mice will be discussed. With appropriate modifications, the techniques can be applied to other lectins or to similar compounds in other species.

39.1 Materials and Methods

Materials
- Cytotoxic effector cells from treated and control animals
- Complete RPMI medium containing 10% fetal calf serum
- Chromium-51 (carrier-free as sodium chromate)
- 1% (v/v) Triton-X 100
- 96-well round-bottomed microtiter plates
- Skatron's supernatant harvester

Purification of ML-I. Purified mistletoe lectin was prepared from crude extract obtained from mistletoe plant (Gabius 1990). The homogeneity of the purified lectin was confirmed by SDS-polyacrylamide gel electrophoresis.

Purification

In vivo Treatment of ML-I. A group of 12 normal Balb/c mice weighing about 20 g (6–8 weeks old) were injected intraperitoneally with 10 ng/kg purified mistletoe lectin, ML-I, in 0.1 ml of phosphate buffered saline. [Alternatively, the lectins can be injected intravenously via tail vein with a lower dose (1 ng/kg).] On alternate days, up to day 9, after initiation of treatment, the levels of NK cells and functions of NK cells in the spleens of those treated were determined using flow cytometric analysis and in vitro cytotoxicity assay. Spleen cells obtained from untreated normal mice served as controls in these studies.

Treatment

Preparation of Spleen Cells. The spleen from control and lectin-treated animals was removed after proper euthanization of the mice, and gently disrupted in the presence of medium RF10 (consisting of RPMI-1640, 10% fetal calf serum, antibiotics and L-glutamine), using a plastic tuberculin sterile syringe without a needle. The cell suspension was then transferred to a 15-ml conical centrifuge tube and allowed to settle on ice for 5 min. The suspended cells were transferred to another 15-ml conical centrifuge tube and centrifuged at 200g for 10 min, using a swinging bucket rotor. The spleen cell pellet was resuspended in 2 ml of Tris-buffered ammonium chloride (pH 7.2) for 5 min at room temperature to lyze the red blood cells, and washed with RF10 medium. The cells were resuspended in RF10 medium and the concentration of the cells was adjusted as needed for each of the following assays.

Preparation

Flow Cytometric Analysis. In order to determine the frequencies of NK cells in the spleens of control and lectin-treated mice, the flow cytometric technique was employed. One million spleen cells from control and ML-I lectin-treated mice were placed in 96-well microtiter plate. Rabbit polyclonal antibody to asialo GM-1 (10 µg in 0.1 ml) (Wako Chemicals, Dallas, TX) was added to these cells. Alternatively, appropriate amounts of other antibodies such as PK136, which specifically mark NK cells, can also be used. The cells were incubated on ice for 45 min and washed three times with phosphate buffered saline (PBS) containing 2% fetal calf serum (FCS). The cells were then stained with fluoresceinated anti-rabbit immunoglobulin by incubating the cells with antibody for 30 min on ice and followed by three washes with PBS with 2% FCS. Final cell suspensions for flow cytometric analysis were brought up in 2 ml of PBS with 2% FCS containing 3% formalin. The percent positive cells were quantitated using an Ortho 50H cytofluorograph equipped with a Cytomation Cicero Data Acquisition Controlled Analysis System (Denver, CO). The total percentage of positive cells was calculated by subtracting the 3% positive fluorescent background from the percentage of positive cells from each sample (Joshi et al. 1987).

Flow Cytometry

Cytotoxicity Assay. The antitumor cytotoxicity levels of spleen cells from control and lectin-treated animals were determined using in vitro cytotoxicity assay. In this assay the chromium-51-labeled murine YAC-1 lymphoma cells were used as target tumor cells and the spleen cells from lectin-treated and control animals were used as effector cells. The concentration of the spleen cells was adjusted to 2×10^7 cells/ml in RF10 medium. A serial dilution of the effector cells was prepared in RF10 and used for the cytotoxicity assay. Five million YAC-1 tumor target cells were incubated with 100 µCi of chromium-51 (as sodium chromate) for 45 min at 37 °C. The labeled cells

Cytotoxicity

were washed twice with RF10 medium by centrifugation of the cells at 800 rpm. The cell concentration of the target cells was adjusted to 2×10^5 cells/ml in RF10 medium. The cytotoxicity assay was performed using varying effector to target ratios (E : T), i.e. 100 : 1, 50 : 1, 25 : 1, and 12.5 : 1. Cells were incubated in a round-bottomed microtiter plate for 4 h at 37 °C in a total volume of 0.2 ml (ideally by mixing 0.1 ml effector cells and 0.1 ml of target cells). Spontaneous release was determined by incubating only target cells and total release was determined by adding 0.1 ml of 1% Triton-X 100 to same amount of target cells used in the assay. Triplicate samples were used for each experimental point. After the incubation period, supernatant was harvested using a Skatron supernatant harvester system. (Alternatively, the supernatant can be taken from each well following centrifugation of the plate in a plate holder. In this case, care should be taken not to collect the cells or cellular debris from the wells while collecting the supernatant.) Chromium-51 release was measured using an LKB gamma counter and the percentage of target cell lysed was calculated as follows:

$$\% \, \text{Cytotoxicity} = \frac{\text{experimental CPM} - \text{spontaneous CPM}}{\text{maximum CPM} - \text{spontaneous CPM}} \times 100,$$

where experimental CPM = effector plus target cells, spontaneous CPM = target cells plus medium, maximum CPM (total release) = target cells plus 1% Triton-X 100.

Statistics

Statistical Analysis. The statistical significance of the results obtained was determined using the Student's t-test.

39.2 Results

Frequencies of NK Cells. The percentage of NK cells in the lectin-treated mice were about 150% of the control mice as determined by flow cytometric analysis using antibodies to asialo GM1 as a marker for NK cells. The percentage of positive cells in control animals was 20.5 ± 0.2. The levels of NK cells in lectin-treated animals were significantly higher than control with $p < 0.05$.

Cytotoxicity Levels of Spleen Cells. The antitumor cytotoxicity levels of spleen cells from animals treated with mistletoe lectin ML-I was significantly higher than the control. In general, there was about 100% increase in percentage cytolysis when compared to spleen cell cytotoxicity from control untreated animals. This significant increase in cytotoxicity levels which resulted from lectin treatment might have a significant role in host defense mechanisms against tumorigenicity and metastatic properties of cancer cells in vivo.

39.3 Discussion

This report has focused on the methodology for the evaluation of NK cell frequencies and their functions in animals treated with lectins such as galactoside specific lectin ML-I. It has been reported by Hajto et al. (1990) that the purified ML-I also induces enhanced secretion of cytokines such as interleukin-1, interleukin-6, and tumor necrosis factor-α. These cytokines may also activate other cytokine production indirectly including IL-2 which might activate NK cells to function as LAK cells which are potent antitumor effector cells. In addition, the in vivo administration of biologic response modifiers such as purified lectin might also enhance the number of LAK cell precursors which give rise to functional antitumor effector cells. If one wants to determine the frequency of LAK precursor cells from the animals treated with lectins, the spleen cells isolated as above will be activated with high dose recombinant IL-2 (1000 U/ml) in a RPMI-based complete tissue culture medium containing 2-mercaptoethanol (10^{-4} M) for 3 to 5 days in vitro. Such activated cells will then be used as effector cells in a cytotoxicity assay described above using NK-sensitive and NK-resistant tumor target cells.

References

Gabius H-J (1990) Influence of type of linkage and spaces on the interaction of β-galactoside-binding proteins with immobilized affinity ligands. Anal Biochem 189: 91–94

Gabius H-J, Gabius S (1990) Tumor lektinologie: Status und Perspektiven Klinischer Anwenchung. Naturwissenschaften 77: 505–514

Gabius H-J, Walzel H, Joshi SS, Kruip J, Kojima S, Gerke V, Kratzin H, Gabius S (1992) The immunomodulatory β-galactoside specific lectin from mistletoe: partial sequence analysis, cell and tissue binding, and impact on intracellular biosignalling of monocytic leukemia cells. Anticancer Res 12: 669–676

Hajto T, Hostanska K, Gabius H-J (1989) Modulatory potency of the β-galactoside specific lectin from mistletoe extract (Iscador) on the host defense system in vivo in rabbits and patients. Cancer Res 49: 4803–4808

Hajto T, Hostanska K, Frei K, Rordorf C, Gabius H-J (1990) Increased secretion of tumor necrosis factor-α, interleukin-1 and interleukin-6 by human mononuclear cells exposed to β-galactoside-specific lectin from clinically applied mistletoe extract. Cancer Res 50: 3322–3327

Joshi SS, Tilden PA, Jackson JD, Sharp JG, Brunson KW (1987) Cell surface properties associated with malignancy of metastatic large cell lymphoma cells. Cancer Res 47: 3551–3557

Joshi SS, Komanduri KC, Gabius S, Gabius H-J (1991) Immunotherapeutic effects of purified mistletoe lectin ML-1 on murine large cell lymphoma. In: Lectins and cancer, Gabius and Gabius (Eds.), Springer-Verlag, Heidelberg, pp 207–217

40 Effects of *Pseudomonas aeruginosa* PA-I and PA-II Lectins on Tumoral Cells

N. Gilboa-Garber and D. Avichezer

Pseudomonas aeruginosa is an opportunistic pathogenic Gram-negative bacterium. It is unique in its ability to withstand the antimicrobial activity of the currently used antibiotics and other drugs. As a result, this bacterium, which had been formerly used to cure streptococcal infections, has gained the status of a serious medical problem (Sabath 1980; Cryz 1984). It endangers the life of hospitalized patients suffering from extensive burns, chronic lung (e.g., cystic fibrosis), kidney and other diseases, as well as patients exhibiting congenital, acquired (e.g., AIDS), or therapeutically induced (e.g., cancer patients following chemotherapy or radiotherapy and subjects treated for organ transplantation) immuno-deficiency (Sabath 1980; Cryz 1984). The high virulence of this bacterium is due to two groups of products: (a) sophisticated virulence factors and (b) lectins plus lectinoid adhesins (Gilboa-Garber and Garber 1992). Protection against *P. aeruginosa* infections has therefore been sought in vaccines (Sabath 1980; Cryz 1984) that stimulate specific immunity against the bacterium toxic products (for prevention of their effects) or against its lectins, or other adhesions, which enable the first step of infection via the adhesion of the bacterium to the host cells.

P. aeruginosa produces high levels (comparable to those found in legume seed extracts) of two lectins, PA-I and PA-II, that resemble the phytohemagglutinins in their specificity, properties and applications (Gilboa-Garber 1988; Gilboa-Garber and Garber 1992). These two lectins, the first galactophilic and the second specific for L-fucose and D-mannose, were purified using affinity chromatography (Gilboa-Garber 1982). The purified lectins have been used for stimulation of human peripheral lymphocytes (Gilboa-Garber 1988; Gilboa-Garber and Garber 1992) and murine splenocytes (Avichezer and Gilboa-Garber 1987) and for protection of mice (as a very effective vaccine) from otherwise lethal *P. aeruginosa* infections (Gilboa-Garber 1988; Gilboa-Garber and Garber 1992). Recently, the gene coding for PA-I lectin was isolated and sequenced and the amino acid sequence of the PA-I lectin subunit was deduced (Avichezer et al. 1992).

The PA-I and PA-II lectins were found by us to be of medical importance for cancer diagnosis when applied for the estimation of peripheral T lymphocyte mitogenic response (instead of the classical mitogenic lectin of *Phaseolus vulgaris*) using thymidine incorporation assay (Gilboa-Garber et al. 1986). They have also been used in the field of experimental cancer (AKR lymphoma and Lewis lung carcinoma) cell suppression (Avichezer and Gilboa-Garber 1991). The interaction of the *P. aeruginosa* lectins with the cancer cells was shown by agglutination tests, by suppression of the tumor cell viability and growth in vitro, and by reduction of its tumorigenicity (but preservation of immunogenicity) in vivo (Gilboa-Garber et al. 1986, Leibovici et al. 1987, Avichezer and Gilboa-Garber 1991). The most interesting and important were the observations on the PA-I and PA-II lectin diverse, but complementary effects on Lewis lung carcinoma cells in vitro (Avichezer and Gilboa-Garber 1991); while PA-I had a direct suppressive effect, PA-II effect was mediated by activation of cytotoxic immune cells (e.g., lymphocytes, macrophages, polymorphonuclear and natural killer cells). Deduction from the results in mice to human beings is still premature; however, such a possibility exists, since recent findings indicate that both PA-I and PA-II lectins interact with different human blood group antigens, including those regarded as associated with cancer such as P, T and Tn, Le^x, and Le^y, and the stage-specific embryonic or oncogenic antigens (SSEA) of different types (Levene et al. 1993). An appropriate combination of PA-I and PA-II lectins may function synergistically as antitumor therapy, leading to tumor cell destruction by acellular and cellular immune mechanisms, and provide, at the same time, protection against *P. aeruginosa* infections which endanger cancer patients.

The present chapter describes the direct effects of PA-I and PA-II lectins on AKR lymphoma and 3LL cells exhibited in cell agglutination, reduced survival and proliferation in vitro, as well as decreased tumorigenicity in vivo (with preserved immunogenicity). It also describes the in vitro, cell-mediated antitumoral effect of PA-II lectin on 3LL cells.

40.1 Murine Tumoral and Normal Cells

The murine tumoral cells used are AKR lymphoma cells, which induce tumors in inbred AKR/Cu mice, and Lewis lung carcinoma (3LL) cells, which induce tumors in inbred C57B1/6J mice. The normal murine cells used are splenocytes (both as a control for the lymphoma cells and as responder immune cells for the cell-mediated antitumoral effect on the 3LL cells) and lung cells (as a matching for the 3LL cells).

40.1.1 AKR Lymphoma Cells

The two variants of AKR lymphoma cells used, one exhibiting low malignancy (LM) and the other high malignancy (HM), were described by Gross (1957). The first produces large subcutaneous (s.c.) tumors and no (or small) metastases in the lymph nodes, and the latter is highly metastatic in AKR/Cu mice (Leibovici et al. 1987).

AKR lymphoma cell suspensions are prepared from subcutaneous (LM) and mesenteric lymphatic (HM) tumors produced in AKR/Cu mice. These tumors are removed aseptically, excised, and pressed through a sterile 50-mesh stainless steel screen (with the aid of a plunger from a 5-ml syringe) into 5 ml of Medium A (antibiotic-containing medium) composed of RPMI 1640 medium (Gibco) supplemented with 100 units penicillin G, 100 µg dihydrostreptomycin and 0.2 µg *p*-hydroxybenzoic acid butyl ester/ml. Cell clumps are removed by sedimentation. The unclumped cell suspensions are washed twice in Dulbecco's phosphate buffered saline (DPBS) [prepared as follows: (a) NaCl 8 g, KCl 0.2 g, Na$_2$HPO$_4$ 1.15 g, KH$_2$PO$_4$ 0.2 g in 800 ml water; (b) CaCl$_2$ 0.1 g in 100 ml water; (c) MgCl$_2 \cdot$6H$_2$O 0.1 g in 100 ml water. The (a), (b), and (c) solutions are autoclaved separately, cooled, and then mixed] by centrifugation (100g for 7 min in a clinical centrifuge) and resuspended in RPMI medium (when used for both routine cell maintenance and in vivo experimental assays) or in Medium A supplemented with 5% heat-inactivated (30 min at 56 °C) fetal calf serum (FCS), 5×10^{-5} M 2-mercaptoethanol and 2 mM L-glutamine (when used for in vitro experimental assays). The viable cell number is determined by counting in a hemocytometer (see below).

Preparation

Maintenance. AKR lymphoma cells are maintained in vivo by s.c. inoculation of 2×10^5 viable cells, in a volume of 0.2 ml, in the back of AKR/Cu (6–10-week-old) mice.

Viability Assay   The cells are counted in a hemocytometer, following addition of 0.01% trypan blue (Sigma) dissolved in saline (0.85% NaCl) solution. Viable cells exclude the dye while dead cells are stained by it.

40.1.2 Lewis Lung Carcinoma (3LL) Cells

The 3LL cells used were described by Sugiura and Stock (1955). They produce large subcutaneous tumors in inbred C57B1/6J mice.

Preparation   The 3LL cells may be prepared from either in vivo serially passaged subcutaneous tumors transplanted in the back of C57B1 mice (the tumor is minced, excised in RPMI medium, and filtered through several layers of gauze to remove debris), or in vitro cell (10^4 cell inoculum) cultures grown in plastic petri dishes (e.g., from Nunc, Denmark or Cel-Cult, England) in 5 ml of the Medium A supplemented with 10% FCS. Single-cell suspensions are prepared from these cell cultures by a brief incubation (2 min at 37 °C) in the presence of 0.25% trypsin (ICN Biochemicals, OH) and 0.05% EDTA in Puck's solution (NaCl 8 g, KCl 0.4 g, glucose 1 g and NaHCO$_3$ 0.35 g/l). The cells are washed (three times) by centrifugation (as described above) and resuspended either in RPMI medium or in medium A with 10% FCS for the in vivo or in vitro experiments, respectively.

Maintenance. The 3LL tumors are maintained in vivo as previously described using C57B1 mice. The cell cultures are maintained at 37 °C in a 5% CO$_2$ humidified atmosphere for 3–4 days or until their growth is confluent.

Remarks   After long-period serial passages of 3LL cells in animals their malignancy may increase. This well-known phenomenon should be taken in mind in keeping the number of serial passages of the tumors minimal. The tumor developmental stage may also be related to malignancy. It is preferable to use fully developed tumors.

Long-Term Preparation. For long-period conservation, the 3LL cells are suspended to a concentration of 1×10^6 cells per ml of the Medium A containing 10% of each FCS and dimethylsulfoxide, and stored at -70 °C until used. Frozen cells are gradually thawed at room temperature and cultured in Medium A supplemented with 20% FCS.

Quality control   The viability of the 3LL cell suspension is checked before both in vitro and in vivo experiments, using the trypan blue exclusion assay (Sect. 40.1.1). Only tumor cell suspensions exhibiting at least 80% viability are further used.

40.1.3 Normal Splenocytes

Preparation   Splenic cells are obtained from aseptically removed spleens of C57B1 mice (6–8 weeks old). Each spleen is pressed through a sterile steel screen (see above) and cell clumps are removed by sedimentation. The cells are washed (three times) with DPBS (see above), then subjected to "erythrocyte lyzing" buffer (0.15 M NaCl, 0.01 M $KHCO_3$) treatment for 30 s and washed twice with 5 volumes of DPBS. Incubation of the splenocytes with the "lyzing buffer" should not be carried out for more than 30 s because prolonged incubation causes leukocyte lysis.

Quality Control. The splenocyte viability assay is performed as previously described.

40.1.4 Normal Lung Cells

Murine lung cell suspensions are obtained by mincing the excised pulmonary (C57B1) tissue in Medium A and filtration through several layers of gauze. **Preparation**

The lung cell viability assay is performed as described above. **Quality control**

40.2 Lectins

40.2.1 Culturing of the Lectin-Producing *P. aeruginosa* Cells and Their Extraction

The lectin-producing *P. aeruginosa* (ATCC 33347) cells are grown at 28 °C with vigorous shaking for 3 days with supplementation of choline (0.2% w/v) daily to the culture medium (Gilboa-Garber 1982). **Culturing**

For predominance in PA-I production, the culture medium used is Grelet's medium modified according to Eagon (GE medium), which contains the following ingredients per liter: KNO_3 10 g, K_2SO_4 0.174 g, KH_2PO_4 6.8 g, $MgSO_4 \cdot 7H_2O$ 0.123 g, $CaCl_2 \cdot 4H_2O$ 0.183 g, $MnSO_4 \cdot 4H_2O$ 2.23 mg, $Fe_2(SO_4)_3$ 20 mg, $ZnSO_4 \cdot 7H_2O$ 14.4 mg, KOH 0.33 g and 4 g yeast extract (Difco), at pH 7. The first three components may be autoclaved together, but each of the remaining salt solutions must be autoclaved separately.

For predominance in PA-II production, the medium used for the bacterial growth is nutrient broth (Difco).

The 3-day-old *P. aeruginosa* cells are harvested by centrifugation (10 000g for 20 min). Since most of the lectin activity is cell-associated, the cells are disrupted by ultrasonic vibrations and their debris are removed by centrifugation at 30 000g for 20 min. The supernatant is used for the subsequent purification steps including heating (for the denaturation and removal of heat-sensitive proteins), precipitation by ammonium sulfate (70%) and affinity chromatography (Gilboa-Garber 1982).

40.2.2 PA-I Lectin Purification Procedure

PA-I lectin is purified from crude extracts prepared from bacteria grown in GE medium. These extracts exhibit lectin activity inhibitable mainly by D-galactose. It may also be purified from crude extracts exhibiting both PA-I and PA-II lectin activities, after separation of the PA-II lectin activity (see below). Foreign heat-labile proteins are removed from the crude extract by heating at 65 °C for 15 min and **PA-I Purification** centrifugation (30 000g for 30 min). The lectin is precipitated from the supernatant by 70% saturation of ammonium sulfate (7 volumes of saturated ammonium sulfate solution are added gradually by stirring at 0 °C, to 3 volumes of the supernatant). After 30 min, the precipitate is collected by centrifugation at 18 000g for 30 min and dissolved in distilled water to one-fifth of the original volume. The preparation obtained is loaded on a Sepharose 4B column which was previously equilibrated with a Tris-buffered saline (TBS, 0.01 M, pH 8.5). Unadsorbed proteins are removed by extensive washing of the column with TBS, and then the purified PA-I lectin is eluted with 0.3 M D-galactose. After dialysis against large volumes of phosphate buffered saline (0.01 M, pH 7), with at least two changes daily during 4 days, the samples are concentrated by lyophilization. The purified lectin preparations are stored in small aliquots at − 20 °C.

40.2.3 PA-II Lectin Purification Procedure

PA-II
Purification

PA-II lectin is purified from the extracts of the bacteria grown in nutrient broth (exhibiting high lectin activity inhibitable by the sugars D-mannose and L-fucose). The first two purification steps (heating and precipitation by ammonium sulfate) are as described in Section 40.2.2. For affinity chromatography, D-mannose-bearing Sepharose 4B column is used. This modified Sepharose is obtained by activation of 15 g of washed-suction-dried Sepharose 4B with 15 ml of 1,4-butanediol diglycidyl ether (Aldrich Chemical Co.) and 15 ml of 0.6 N NaOH containing 30 mg of NaBH$_4$ (toxic!). The mixture is rotated mechanically at room temperature for 10 h. The epoxide-activated Sepharose is subsequently collected on a sintered-glass funnel and washed with a large excess of distilled water. Four g of this Sepharose are mixed with 400 mg of D-mannose dissolved in 8 ml of 0.1 N NaOH. The sugar-containing mixture is rotated mechanically at 37 °C for 24 h, after which the affinity adsorbent is collected by filtration and washed with 500 ml of 0.1 M borate buffer (at pH 8), followed by a large volume of distilled water. Thirty ml of the lectin preparation are loaded onto the modified Sepharose column (4 × 20 cm) previously equilibrated with TBS. After adsorption of the PA-II lectin, the column is washed with TBS for the removal of the unadsorbed proteins and then with 0.3 M D-mannose for the elution of the purified PA-II lectin. The eluate is dialyzed against large volumes of PBS (0.01 M pH 7), concentrated and stored at − 20 °C (see Sect. 40.2.2).

Remarks

Recently, the PA-I lectin has been commercialized and may be purchased from Sigma. PA-II is not yet commercially available.

Quality control

The protein purity is analyzed by conventional sodium dodecyl sulfate poly-acrylamide (15%) gel electrophoresis and Coomassie brilliant blue staining.

The sugar-specific lectin activity is examined by the classical hapten inhibition technique using hemagglutination test with papain-treated human erythrocytes (Gilboa-Garber 1982). PA-I and PA-II hemagglutinating activities are inhibited by 0.15 M galactose and 0.15 M mannose solutions, respectively (Gilboa-Garber 1982).

40.2.4 Plant Lectins

The lectins from the plants *Phaseolus vulgaris* (PHA) and *Canavalia ensiformis* (Con A) are purchased from Difco and Sigma, respectively.

40.3 Direct Effects of the Lectins on AKR Lymphoma and 3LL Cells

The direct effects of the lectins on the cells were examined in vitro at several levels: (a) agglutination of the cells, (b) survival of the cells (shown by vital staining), and (c) proliferation of the cells (shown by labeled thymidine incorporation test), as well as in vivo, by checking the tumorigenicity of the cells and their immunogenicity.

40.3.1 Agglutination of the Cells by the Lectins

Agglutination of cells, due to lectin binding to target carbohydrate molecules on their surface, may be used for detection of tumor cell transformation associated with an altered surface carbohydrate composition. Generally, lectin binding to the surface of the transformed cells is much higher than that of the normal ones. In addition to their ability to differentiate between cancer and normal cells, certain lectins may also be used for separation between them and for discrimination between different tumor types, as well as for evaluation of tumor malignancy status and patient prognosis.

    The use of the *P. aeruginosa* PA-I and PA-II lectins in such tests may be of particular interest, since both exhibit special sugar specificities and detect blood cell antigens such as the stage specific embryonic or oncogenic antigens (Levene et al. 1993).

Background and Objectives

Phase contrast microscope
Thermostated incubator at 37 °C
Glass microscope slides
Pasteur pipets equipped with a rub
DPBS or PBS (0.01 M, pH 7.2)
D-galactose and D-mannose

Equipment and Chemicals

The freshly washed normal and tumor cells (Sect. 40.1) are suspended to a concentration of $1-2 \times 10^5$ in 0.2 ml PBS without or with the lectin (40–200 µg) either without or together with the lectin-specific sugar at 0.15 M final concentration (D-galactose for PA-I and D-mannose for PA-II). Following 2–4 h incubation at 37 °C, the intensity of the cell agglutination is examined by the naked eye and by phase contrast microscopy. The agglutination intensity is graded from + (a few cells clumped) to + + + + (all the cells are clumped).

Procedure

Results

Agglutination of AKR Lymphoma Cells by the Lectins. AKR lymphoma cell variants of both high (HM) and low malignancy (LM, showing a weak spontaneous aggregation in the absence of lectins) are strongly agglutinated by both the *P. aeruginosa* (PA-I and PA-II) and plant (PHA and Con A) lectins at concentrations of 0.2–1 mg/ml (Table 1, and Leibovici et al. 1987).

    The specificity of PA-I lectin-induced agglutination is shown by its inhibition with D-galactose (0.15 M) and those of PA-II lectin and Con A by 0.15 M D-mannose.

    Normal splenocytes of C57B1 mice, examined under the same conditions as the tumor cells, exhibit somewhat lower agglutinability with PA-I, Con A and PHA, but somewhat higher with PA-II (Table 1).

Table 1. Agglutination of murine cancer and normal cells by the lectins[a]

Lectins: Cell type	None	PA-I	PA-II	PHA	Con A
Tumor cells					
LM-AKR lymphoma	±	+ + +	+ +	+ + + +	+ + + +
HM-AKR lymphoma	–	+ + +	+ +	+ + + +	+ + +
Lewis lung carcinoma (3LL)	±	+ + +	+ + +	+ + + +	+ + + +
Normal cells					
Splenocytes	–	+ +	+ + +	+ + +	+ +
Lung cells	±	+ +	+ +	+ + +	+ +

[a]The data represent mean values of results obtained in nine experiments. The slight cell agglutination (±) observed in the absence of the lectins represents the spontaneous aggregation of the cells.

Agglutination of 3LL Cells by the Lectins. 3LL cells are found to be agglutinated by both the bacterial and plant lectins at concentrations of 0.2–1 mg/ml (Table 1). Normal lung cells from C57B1 mice, examined under the same conditions, are also remarkably agglutinated by these lectins, but to a lower extent (Table 1). The agglutination could be inhibited by the suitable sugars, as described above.

Remarks. Certain preparations of either tumor or normal cells, particularly those stored at 4 °C, exhibit some clumping even without the addition of lectin. This aggregation is not inhibited by the lectin-specific sugars. Therefore, fresh cell suspensions have to be used for these experiments, and spontaneous cell aggregation in absence of lectins should always be checked and used for comparison.

Quality Control. Agglutination by the lectins is regarded as specific only if the cells alone and with the lectin-specific sugar mixtures are not (or slightly) agglutinated while those exposed to the lectins (alone or with the nonspecific sugars) are strongly agglutinated.

40.3.2 Survival of the Lectin-Treated Cells

Background and Objectives

The cytotoxic/static effects of the lectins can be studied in vitro by following the ability of the 3LL cells to adhere to plastic dishes, using vital cell staining with crystal violet, as described by Drysdale et al. (1983) as the cytomorphological assay.
    There is a special interest in comparing the effects of the PA-I and PA-II lectins on cancer cells because both are derived from the same *P. aeruginosa* cells and exhibit similar physical properties and biological effects but they differ in sugar specificities. 3LL cell survival in the presence of the lectins was examined using this cytomorphological assay and was compared to that obtained either without any lectin or with lectins and their specific sugars.

Equipment and Chemicals

5% CO_2 humidified and thermostated incubator at 37 °C
Microplate photometer (MR-600 Dynatech, or the equivalent)
24 "U"-shaped well plates (e.g. from Nunc, Denmark or Cel-Cult, England)
96 "U"-shaped well plates (e.g. from Nunc, Denmark or Cel-Cult, England)
Pasteur pipets equipped with a rub.

Lectins (Sect. 40.2)
DPBS (Sect. 40.1.1)
0.2% Crystal violet solution in 2% ethanol
1% Sodium dodecyl sulfate solution

The tumor cells (3×10^5) suspended in 1 ml of Medium A supplemented with 10% **Procedure**
FCS are cultured in the 24-well plates without or with the lectins in the absence or
presence of the respective specific sugars, at 37 °C in a 5% CO_2 humidified
atmosphere. After 48 h incubation, the supernatant (containing the dead cells) is
discarded by a Pasteur pipet. The adherent viable cells are stained with 1 ml of the
crystal violet solution at 37 °C for 10 min. The plates are then gently rinsed three
times with DPBS and the stained cells are solubilized by the addition of 0.4 ml of a
1% sodium dodecyl sulfate solution to each well, at 37 °C for 10 min. The resulting
lysates are diluted with 5 volumes of distilled water. Aliquots of 0.2 ml (in quad-
ruplicate) from each test sample are transferred to a 96-well microplate and their
absorbance at 570 nm is measured with the aid of a microplate photometer.

The percentage of cytotoxicity is calculated according to the following formula:
$[(a - b) \times 100]/(a - c)$, where a, b and c are the absorbances of: tumor cells in
medium without lectins, tumor cells in medium with lectins, and medium alone (as
background), respectively.

Results

Examination of the in vitro cytotoxic activity of the PA-I and PA-II lectins on 3LL
cells, assayed by the crystal violet vital staining, has revealed that PA-I exerts a
significant dose-dependent inhibition of the 3LL cell viability (Avichezer and
Gilboa-Garber 1991). A 50% cytotoxic effect was obtained with 4 µg PA-I/ 0.2 ml,
whereas PA-II, under the same conditions, was almost ineffective (Avichezer and
Gilboa-Garber, 1991).

Remarks. Some cautions should be kept in mind: high lectin doses lead to an
extensive cell agglutination; prolonged incubations and excessive tumor cell inoculum result in a rapid cell confluency and death. The optimal experimental
conditions usually must not exceed: (a) cell inoculum of 3×10^5/well; (b) maximal
incubation time of 48 h; and (c) lectin doses giving only a slight or no cell
agglutination.

Quality Control. The results are regarded as specific when the control cultures
containing the lectins, together with their specific sugars, behave like the control
cultures without the lectins.

40.3.3 Proliferation of the Lectin-Treated Tumor Cells

The ^{3}H-thymidine incorporation test is used for the assay of tumor cell growth **Background**
in vitro and of toxic effects of lectins on it. Pulses of ^{3}H-thymidine are given at **and Objectives**
different time intervals after the addition of lectin with or without the specific sugar,
and its incorporation to the tumor cell DNA is determined.

Equipment
and Chemicals

Liquid scintillation counter
5% CO_2 humidified and thermostated incubator at 37 °C
Automatic cell harvester
96-"v"-shaped well microplates (e.g., from Nunc, Denmark or Cel-Cult, England)
Glass fiber filters (Whatman 934-AH)
Lectins (Sect. 40.2)
Methyl α-D-galactoside (Sigma)
L-fucose
Labeled [methyl-^{3}H] thymidine
Scintillation fluid

Procedure

The tumor cells (AKR lymphoma or 3LL cells) are suspended to a concentration of 1×10^6 and 1×10^5 cells, respectively, in 1 ml of the appropriate culture medium (See above). The lectins are added to them (20 and 100 μg/ml, respectively) with or without the specific sugar (0.12 M methyl-α-D-galactoside for PA-I and 0.05 M L-fucose for PA-II). Samples of the cells (0.2 ml) with or without the lectins are grown in quadruplicates in microtiter plates ("v"-shaped wells) in a 5% CO_2 humidified atmosphere at 37 °C for 24–72 h. They are labeled with 1 μCi ^{3}H-thymidine 20 h before the cells are harvested by filtration onto glass fiber filters, using an automatic cell harvester. The ^{3}H-thymidine incorporation is determined by liquid scintillation counting.

Results

The Lectin Effects on AKR Lymphoma Cells. The two AKR lymphoma LM and HM variants (grown in cultures for 72 h) have been shown to differ in their sensitivity (tested by ^{3}H-thymidine incorporation assay) to the lectins (20 μg/ml).

The LM cells exhibit a very high sensitivity to Con A, PA-I and PHA, but low sensitivity to PA-II (Table 2). The HM cells are also highly sensitive to Con A, somewhat less sensitive to PHA and PA-I, but significantly more sensitive to PA-II than the LM cells (Table 2, and Leibovici et al. 1987).

The Lectin Effects on 3LL Cells. Examination of the same lectin effects on the ^{3}H-thymidine incorporation into 3LL cells (cultured in the presence of 20 μg

Table 2. Cytotoxic effects of the lectins (20 μg/ml) on the LM and HM-AKR lymphoma cells, exhibited by ^{3}H-thymidine incorporation into cells grown at 37 °C for 72 h

	AKR Tumor type	None	PA-I	PA-II	PHA	Con A
				Lectin added		
^{3}H-thymidine incorporation (CPM)	LM	69 255	849	49 060	3540	327
	HM	34 088	13 446	12 605	6335	488
Cytotoxicity (%)	LM		98	29	95	99
	HM		60	63	81	98

The data represent mean values of the results obtained in 4–7 cultures. The *p* value of all of them was highly significant (0.0005) except that of PA-II with the LM cells, which was < 0.1.

Table 3. Cytotoxic effects of the lectins (20 µg/0.2 ml) on the 3LL cells, exhibited by ^{3}H-thymidine incorporation into cells grown at 37 °C for 24 h

		Lectin added			
	None	PA-I	PA-II	PHA	Con A
^{3}H-thymidine incorporation (CPM)	104 952	4905	92 577	95 246	47 647
Cytotoxicity (%)		95	12	9	55
p value		< 0.0005	< 0.1	< 0.1	< 0.005

The data represent mean values of the results obtained in four to eight cultures.

lectin/0.2 ml for 24 h at 37 °C) has revealed (Table 3) that the tumor cells are most sensitive to PA-I, followed by Con A, while almost insensitive to PA-II [even in 48 h cultures, (Avichezer and Gilboa-Garber 1991)] and PHA.

The cytotoxic activity of the PA-I lectin was shown to be both dose and time-dependent and inhibitable by the addition of methyl-α-D-galactoside (0.12 M) to the culture medium (Avichezer and Gilboa-Garber 1991).

Remarks and Quality Controls. The use of methyl-α-D-galactoside instead of galactose, and fucose instead of mannose for cultures with PA-I and PA-II, respectively, is advised in order to prevent toxic effects of the sugars on the cell cultures. In any case, controls without the lectins but with the specific sugars used for the lectin-neutralization are indispensable.

40.3.4 Effects of Lectins on the In Vivo Tumorigenicity of the Cancer Cells

The higher sensitivity of transformed cells to lectins could be shown not only in vitro, but also in vivo – in suppression of transplanted tumor development. Such suppression may be mediated by several different mechanisms including: **Background and Objectives**

a) A direct cytotoxic effect of the lectin by inhibition of the tumor cell protein synthesis through the inactivation of ribosomal RNA – as clearly demonstrated with toxic galactose-binding lectins such as those of *Ricinus communis* (ricin), *Abrus precatorius* (abrin) and *Viscum album* (viscumin).
b) Stimulation and/or perturbation of gene expression in the tumor cells. Con A, for example, has been reported to induce a reversion of DNA hypomethylation (which is inversely correlated with gene expression) in tumor cells, (by enhancement of the cellular cytosine-5 methyl transferase activity).
c) Inhibition of the settling ("adhesion") and spreading of the tumor cells (which are frequently dependent on lectin–sugar interactions) by competing with the tumor cell endogenous surface lectins or with those derived from the host cell tissues and fluids.
d) Enhancement of the host immune defense mechanisms by activation of cytotoxic T lymphocytes, macrophages, natural killer cells and polymorphonuclear leukocytes for suppression of the tumors.

The effect of the *P. aeruginosa* lectins (as compared to the plant lectins PHA and Con A) on the in vivo tumorigenicity of 3LL and AKR cells has been evaluated at the following levels:

a) Incidence of primary tumors
b) Average size of primary tumors
c) Average size of metastatic tumors
d) Rate of the mice mortality

Materials   5% CO_2 humidified and thermostated incubator at 37 °C
1 ml sterile syringes
27G × 3/4″ sterile needles
Inbred AKR/Cu and C57Bl/6J mice (6–10 weeks old)
Lectins (Sect. 40.2)
Saline solution (0.85% NaCl)
D-galactose

Procedure   Tumor cell suspensions (containing 1×10^6 cells/ml RPMI medium-prepared as described above are "pretreated" by incubation with an equal volume of either saline or lectin solutions (around 75 µg/0.2 ml for AKR lymphoma and 20 µg/0.2 ml for 3LL cells) without or with 0.3 M galactose (for PA-I). Following incubation in a 5% CO_2 humidified atmosphere at 37 °C for 1 h, $2 \times 10^5/0.2$ ml tumor cells, exhibiting at least 80% viability, are inoculated s.c. in the back of the syngeneic mice. Each experimental group consists of at least five animals.

Tumor development is evaluated by recording the incidence and average size (of horizontal and vertical diameters in mm, estimated by palpation) of the local and metastatic tumors, at least two to three times per week. Mortality of the mice is recorded daily and usually its kinetics and extent correlate the tumor incidence. The mean survival time (MST – the sum of survival times of all the mice divided by the total number in each experimental group) is calculated and compared to that of the untreated mice.

Results

Tumorigenicity of Untreated and Lectin-Treated AKR Lymphoma Cells in AKR Mice. Subcutaneous (s.c.) inoculation of untreated and lectin (75 µg/0.2 ml)-pretreated (for 90 min) AKR lymphoma cells into AKR mice, has revealed that pretreatment of the cells by the bacterial PA-I and PA-II lectins is much less effective than by PHA and Con A lectins (Table 4). PHA and Con A are particularly more effective for the depression of the LM variant (Table 4, and Leibovici et al. 1987). Their effects are exhibited in significant delay in the primary tumor incidence; reduction in the average size of both the primary and metastatic tumors and prolongation of the life span of the mice.

Effects of the Lectins on 3LL Cells. Pretreatment of 3LL cells with PA-I, Con A and PHA (20 µg/0.2 ml) for 1–2 h before their s.c. inoculation into the mice, considerably reduced their tumorigenicity (PA-I > Con A > PHA) (Table 5 and Fig. 1). PA-II has not been found to be effective in this system. The lectin antitumoral effects could be abolished by the addition of the lectin-specific sugars to the incubation medium. The addition of D-galactose (0.3 M) to the PA-I-cell mixture prevented the effect (Gilboa-Garber et al. 1986). It is interesting to note that if the PA-I lectin is not preincubated

Table 4. Tumorigenicity of the LM and HM-AKR lymphoma cells pretreated with lectins (75 µg/0.2 ml) before their s.c. inoculation into AKR mice

AKR Tumor type	Lectin	Tumor incidence[a] (%)		Average tumor diameter[a] (mm)		Mortality[b] (%)
		Primary	Secondary	Primary	Secondary	
LM	None	100		22.8		100
	PA-I	60		6.9*		100
	PA-II	100		11.2*		100
	PHA	0		0*		60
	Con A	0		0*		60
HM	None	100	80	8.8	0.95	100
	PA-I	100	40	4.1**	0.40**	100
	PA-II	80	30	2.2*	0.05*	80
	PHA	100	10	3.0*	0.15*	100
	Con A	0	0	0*	0*	80

p values: * < 0.005; ** < 0.01
[a]Tumor incidence and diameter express the findings after 17 days (for LM) and 13 days (for HM).
[b]Mortality expresses the overall results after 50 days and therefore correlates better tumor incidence, since it also includes delayed tumor appearance.

Table 5. Tumorigenicity of the 3LL cells pretreated with lectins (20 µg/0.2 ml) before their s.c. inoculation into C57B1 mice

Lectin pretreatment	Tumor incidence[a] (%)	Average tumor diameter ± SD[a] (mm)	p value	Mortality[b] (%)
None	100	33.0 ± 2.2		100
PA-I	0	0	< 0.0005	0
PA-II	100	33.0 ± 11.6	Insignificant	100
PHA	80	23.2 ± 11.7	< 0.05	80
Con A	20	6.0 ± 12.0	< 0.0025	20

[a]Findings on the 19th day after inoculation.
[b]Overall results after 2 months.

with the 3LL cells, but injected i.p. (100–200 µg/mouse) concomitantly with their s.c. injection, and again 5 days later, it does not suppress tumorigenicity. On the other hand, s.c. injection of PA-I (90 µg/mouse) directly into the site of tumor development 3 days after the 3LL cell inoculation led to a significant reduction in the tumor mass and in the mice mortality (similar injections of the lectins 2 h after the cell inoculation did not inhibit the tumor development). Higher lectin doses were found to be more effective than lower ones, and when more than 3×10^5 3LL cells were inoculated, they overcame the lectin antitumoral effect.

Remarks
a) Both male and female mice (6–10 weeks old) can be used for the in vivo experiments, however, for highly reproducible results, the use of males is recommended.

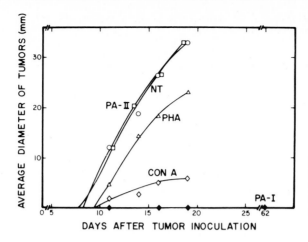

Fig. 1. Tumorigenicity of 3LL cells pretreated with the lectins (20 µg/0.2 ml) before their s.c. inoculation into C57Bl mice as compared to that of nontreated (*NT*) cells

b) The lectins are useful not only for the suppression of tumor cell growth, but also for decreasing the animal susceptibility to syngeneic transplanted tumors, resembling vaccines based on irradiated cells: intravenous (i.v.) inoculation of C57Bl mice with lectin-treated 3LL cells (5×10^3 cells/0.2 ml, pretreated by incubation at a concentration of $5 \times 10^4/0.2$ ml with 30 µg lectin, at $37\,^\circ$C for 60 min and then diluted with 1.8 ml RPMI medium), followed by s.c. inoculation of the mice, 2 months later, with a challenge of untreated 3LL cells, led to a delay in the s.c. 3LL tumor development and to prolongation of the animal life span (Table 6). The most pronounced life span delay was obtained with PHA and Con A. PA-I was less effective and PA-II exerted a very weak effect.

S.c. immunization of more than 30 mice with PA-I-treated 3LL cells (Sect. 40.3.4) and then 2 months later, a s.c. challenge with 2×10^5 untreated 3LL cells indicated no significant delay in the 3LL tumor growth.

Quality Controls
a) The viable cell number has to be determined (using the trypan blue exclusion test as described under Sect. 40.1) before exposure of the tumor cells to the lectins, since cell agglutination by the lectin may disturb accurate counting.

Table 6. Immunogenicity of lectin-pretreated 3LL cells inoculated i.v. into C57B1 mice 2 months before the s.c. untreated 3LL cell challenge

Lectin used	Tumor incidence[a] (%)	Average tumor diameter $\pm$ SD[a] (mm)	p value	MST $\pm$ SD[b] (days)	p value
None	100	52.5 ± 2.5		24.4 ± 3.8	
PA-I	100	42.7 ± 4.0	< 0.01	29.0 ± 2.7	< 0.05
PA-II	100	34.0 ± 16.0	> 0.1	26.2 ± 3.6	> 0.2
PHA	60	38.6 ± 4.9	< 0.0025	32.2 ± 1.2	< 0.005
Con A	60	26.5 ± 18.9	< 0.025	32.0 ± 2.4	< 0.005

[a]Tumor incidence and diameter on the 13th and 25th day, respectively.
[b]Overall results after 50 days.

b) For statistical evaluation of the results, the student's *t*-test is used, and *p* values higher than 0.05 are regarded as nonsignificant.

40.3.5 In Vitro Cell-Mediated Antitumoral Effect of PA-II Lectin on 3LL Cells

Lymphocytes exhibiting lectin-dependent cellular cytotoxicity (LDCC) against tumor cells can be generated in vitro by their culturing in the presence of mitogenic lectins (e.g., Con A, PHA). The *P. aeruginosa* PA-I and PA-II lectins have been shown to exhibit remarkable mitogenic activity on human peripheral T lymphocytes, similar to that obtained with Con A and PHA. PA-II lectin has also been found to stimulate murine splenocytes (maximal effect at a concentration of 15 µg/ml) and its stimulatory effect was shown to be inhibitable by L-fucose (excluding the possibility that the effect is due to nonspecific bacterial components). Since PA-II lectin did not exhibit a direct cytotoxic effect on 3LL tumor cells (Sect. 40.3.2 and 40.3.3), it was possible to test its ability to induce cell-mediated cytotoxic activity in murine splenocytes, in vitro.

Background and Objectives

5% CO_2 humidified thermostated incubator at 37 °C
24 "U"-shaped well plates (e.g., from Nunc, Denmark or Cel-Cult, England)
Fresh murine splenocyte suspension (Sect. 40.1.3)
L-fucose
PA-II lectin (Sect. 40.2)
DPBS (Sect. 40.1.1)
Methanol
Giemsa solution (8% in DPBS)

Materials

The 3LL tumor cells (3×10^4 cells in a final volume of 1 ml of Medium A supplemented with mercaptoethanol and glutamine as described in Section 40.1.1 and with 10% FCS) are cultured in the absence or presence of murine (C57Bl) splenocyte suspension (in the same culture medium) in the following numbers (and cell ratios):

Procedure

(a) 3×10^6 (1 : 100); (b) 1.5×10^6 (1 : 50); (c) 7.5×10^5 (1 : 25); and (d) 3.75×10^5 (1 : 12.5), with or without the PA-II lectin (15 µg/ml). The specificity of the lectin effect is further checked by its prevention in presence of 0.05 M L-fucose in the medium. The mixed cell cultures are grown in a 24-well plate and incubated at 37 °C for 72 h under a 5% CO_2 humidified atmosphere. At the end of the incubation period, the supernatants containing nonadherent cells are discarded, and the adherent cells are gently rinsed twice with DPBS, fixed with methanol (2 min), stained with Giemsa (8%, 40 min) and washed twice with DPBS.

Results

PA-II lectin which did not exhibit a direct toxic effect on 3LL tumor cells was found to exert, at a concentration of 15 µg/ml, a profound inhibition of the tumor cell growth in the presence of murine splenocytes (Fig. 2 and Avichezer and Gilboa-Garber, 1991).

This PA-II induced splenocyte-mediated cytolysis was completely inhibited by 0.05 M L-fucose in the culture medium. The same splenocytes, in the absence of PA-II lectin, did not inhibit the 3LL cell growth (Fig. 2 and Avichezer and Gilboa-Garber, 1991).

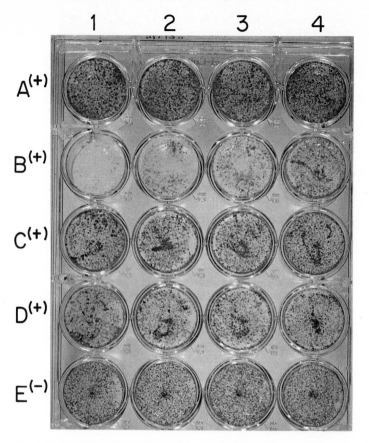

Fig. 2. Suppression of 3LL cell growth by PA-II lectin-dependent cell cytotoxicity. The 3LL cells were cultured in the absence (− , row E) and presence (+ , rows A–D) of splenocytes in the following ratios 1 : 100 (1); 1 : 50 (2); 1 : 25 (3) and 1 : 12.5 (4), without or with the lectin either alone or with L-fucose. Rows A (*1–4*) and E (*1*) : without lectin or L-fucose; B (*1–4*) and E (*2*): with lectin (15 µg/ml); C (*1–4*) and E (*3*): with L-fucose (0.05 M); D (*1–4*) and E (*4*): with both lectin and L-fucose

Remarks and Quality Control

a) The in vitro cell-mediated antitumoral effect of PA-II lectin on 3LL cells was found to be critically related to the specific hemagglutinating activity of the PA-II lectin sample used. The preparations to be used for these experiments have to exhibit a hemagglutinating titer around 1024, at the concentration of 15 µg protein/ml.

b) The cell ratio between the target 3LL cells and the effector splenocytes is also crucial: optimal at 1 : 100 and almost ineffective at 1 : 12.5 (Fig. 2).

Acknowledgments. We would like to acknowledge the assistance and collaboration of Prof. Judith Leibovici from the Department of Pathology, Tel-Aviv University, in the in vivo studies, and to thank Avrille Goldreich and Sharon Victor for the preparation of the manuscript. This work is a part of Dr. D. Avichezer's Ph.D. thesis.

References

Avichezer D, Gilboa-Garber N (1987) PA-II, the L-fucose and D-mannose binding lectin of *Pseudomonas aeruginosa* stimulates human peripheral lymphocytes and murine splenocytes. FEBS Lett 216: 62–66

Avichezer D, Gilboa-Garber N (1991) Antitumoral effects of *Pseudomonas aeruginosa* lectins on Lewis lung carcinoma cells cultured in vitro without and with murine splenocytes. Toxicon 29: 1035–1313

Avichezer D, Katcoff D, Garber NC, Gilboa-Garber N (1992) Analysis of the amino acid sequence of the *Pseudomonas aeruginosa* galactophilic PA-I lectin. J Biol Chem 267: 23023–23027

Cryz SJ Jr (1984) *Pseudomonas aeruginosa* infections. In: Germanier R (ed) Bacterial vaccines. Academic Press, New York, pp 317–351

Drysdale BE, Zacharchuk CM, Shin HS (1983) Mechanism of macrophage-mediated cytotoxicity: production of a soluble cytotoxic factor. J. Immunol 131: 2362–2367

Gilboa-Garber N (1982) *Pseudomonas aeruginosa* lectins. In: Ginsburg V (ed) Methods in enzymology, Vol. 83. Academic Press, New York, pp 378–385

Gilboa-Garber N (1988) *Pseudomonas aeruginosa* lectins as a model for lectin production, properties, applications and functions. Zbl Bakt Hyg A270: 3–15

Gilboa-Garber N, Garber N (1992) Microbial lectins. In: Allen HJ, Kisailus EC (eds) Glycoconjugates. Marcel Dekker Inc, New York, pp 541–591

Gilboa-Garber N, Avichezer D, Leibovici J (1986) Applications of *Pseudomonas aeruginosa* galactophilic lectin (PA-I) for cancer research. In: Bøg-Hansen TC, Van Driessche E (eds) Lectins: biology, biochemistry, clinical biochemistry, Vol. 5. Walter de Gruyter, Berlin, pp 329–338

Gross L (1957) Development and serial passages of highly potent strain of mouse leukemia virus (23080). Proc Soc Exp Biol Med 94: 767–771

Leibovici J, Avichezer D, Gilboa-Garber N (1987) Effect of lectins on tumorigenicity of AKR lymphoma cells of varying malignancy. Anticancer Res 6: 1411–1416

Levene C, Gilboa-Garber N, Garber N (1993) Lectins in blood banking. In: Doyle RJ, Slifkin M (eds) Applications of lectins in microbiology. Marcel Dekker Inc, New York (in press)

Sabath LD (ed) (1980) *Pseudomonas aeruginosa*, the organism, diseases it causes, and their treatment. Hans Huber Publishers, Bern

Sugiura K, Stock CC (1955) Studies in a tumor spectrum III. The effects of phosphoramides on the growth of a variety of mouse and rat tumors. Cancer Res 15: 38–51

41 Determination of Lectin-Dependent Alterations of Cellular Parameters by Immunophenotyping During Adjuvant Lectin Application

J. BEUTH, H.L. KO, H.-J. GABIUS, and G. PULVERER

The obvious ability of certain lectins to activate (non)specific defence mechanisms supports the assumption that lectin-carbohydrate interactions may induce clinically beneficial immunomodulation. Recent investigations proved that lectins, especially small, nontoxic doses of the galactoside-specific lectin from mistletoe (ML), have immunomodulatory potency (Gabius 1987; Beuth et al. 1991, 1992a, b; Gabius and Gabius 1991).

The immune system generally provides protection from infectious pathogens and malignancies via mechanisms of recognition and elimination. Although there are a variety of nonspecific defence mechanisms, only T and B lymphocytes are capable of recognizing foreign antigens by their molecular features and mounting an antigen-specific response. Recognition of distinct lymphocyte (sub)populations has been achieved both through the discovery of selective expression of specific cell surface antigens on lymphoid cells and by detection of different functional properties of cells bearing these antigens. Recent advances by fluorescence-activated cell sorting have led to an extensive classification of human lymphocytes (particularly T cells) into defined subpopulations. Furthermore, certain cell surface antigens appear to mediate functional properties or to indicate the state of differentiation of the respective cells, and monoclonal antibodies to these antigens can be used to identify and isolate cells bearing these molecules.

The most precise method of detection and separation involves identification and separation of cells with a single or a combination of fluorochrome-conjugated monoclonal antibodies using fluorescence-activated cell sorter (FACS). The major advantage of this technology (immunophenotyping) is that it permits cell separation not only on the basis of presence or absence of a particular antigen, but also in dependence on the amount of antigen.

41.1 Biological Response Modifier

Mitogenic and/or immunogenic plant lectins can be commercially obtained (e.g., phytohemagglutinin, PHA; concanavalin A, Con A; pokeweed mitogen, PWM) or can be isolated, as outlined by Rüdiger, Chapter 4, this volume. As an example, the procedure to determine alterations of cellular parameters by immunophenotyping after clinical administration of ML will be presented. Affinity chromatography on immobilized lactose as ligand, coupled to the resin after divinyl sulfone activation, was employed to purify the lectin from mistletoe extract, followed by analytical procedures, as described in detail elsewhere (Gabius 1990). Appropriate patients (e.g., cancer patients) may receive lectin injections subcutaneously or intravenously (1–2 ng per kg body weight twice a week, which proved to be optimal in preceding studies). Patients receiving ML therapy with their expressed consent showed no relevant side effects except for moderate flu-like symptoms (without fever and chills) shortly after ML administration.

41.2 Basic Immunophenotyping of Competent Cells

Commercially available SIMULTEST kits generally consist of vials containing murine monoclonal antibody reagent conjugated with fluorescein-isothiocyanate (FITC) and phycoerythrin (PE), and one vial of lyzing solution. The use of two fluorochromes permits simultaneous two-color analysis, because each fluorochrome emits light at a different wavelength, when excited at 488 nm. When the monoclonal antibody reagents are added to human/murine whole blood, the fluorochrome-labeled antibodies bind specifically to antigens on the surface of adequate leukocytes. Lyzing solution is used to lyze red blood cells (RBC) prior to flow cytometry analysis. The stained lymphocytes emit fluorescence when excited by the flow cytometer's laser beam. Spectral emission is simultaneously assessed at 530 nm for FITC (green) and 575 nm for PE (red-orange) stained cells. **Reagents**

Monoclonal antibody reagents for basic immunophenotyping assays (e.g., Reagents A-G; described below) can be purchased (e.g., from Becton-Dickinson, Heidelberg, FRG) and contain murine immunoglobulins, conjugated to either FITC or PE, in buffered saline with gelatin and sodium azide.

Reagent A: LeucoGATE (CD 45/CD 14) is used to define and evaluate the light scatter gate that distinguishes lymphocytes from granulocytes, monocytes, and unlyzed or nucleated RBC or debris.

Reagent B: this isotype control reagent is used to set markers around the nonlymphocyte population. It contains FITC-labeled IgG_1 anti-keyhole limpet hemocyanin (KLH) and PE-labeled IgG_{2a} anti-KLH murine monoclonal antibodies.

Reagent C: anti-Leu-4/anti-Leu 12 (CD 3/CD 19) is used to enumerate T and B lymphocytes. It contains FITC-labeled anti-Leu-4 for identification of T lymphocytes and PE-labeled anti-Leu-12 (B lymphocytes).

Reagent D: anti-Leu-3a/anti-Leu-2a (CD 4/CD 8) is used to characterize T lymphocyte subsets. It contains FITC-labeled anti-Leu-3a for the identification of helper/inducer T lymphocytes and PE-labeled anti-Leu-2a (suppressor/cytotoxic lymphocytes).

Reagent E: anti-Leu-4/HLA-DR (CD 3/HLA-DR) is used to enumerate T lymphocytes, DR-positive non-T lymphocytes (primarily B lymphocytes) and activated T cells. It contains FITC-labeled anti-Leu-4 for identification of T lymphocytes and PE-labeled anti-HLA-DR for identification of DR-positive non-T lymphocytes and activated T cells.

Reagent F: anti-Leu-4/anti-Leu-11c + Leu-19 (CD 3/CD 16 + CD 56) is used to identify T and natural killer (NK)-cells. It contains FITC-labeled anti-Leu-4 to identify T lymphocytes and PE-labeled anti-Leu-11c and anti-Leu-19 (NK cells).

Reagents G: buffered lyzing solution which contains diethylene glycol and formaldehyde. When stored at room temperature, the concentrate ($10 \times$) is stable until expiration date. For use, dilution of the concentration with room-temperature reagent-grade water (1:10) is obligatory. Prepared solution is stable for 1 month, if stored in glass containers at room temperature.

41.3 Blood Samples

Blood should be drawn aseptically by venipuncture into sterile K_3EDTA-containing tubes. A minimum of 1–2 ml of anticoagulated whole blood is required for FACS analysis and may be stored at room temperature for up to 6 h until ready for staining. A white blood cell (WBC) count and a differential count should be performed. Acceptable WBC concentration ranges are from 2×10^3 WBC/mm^3 to 100×10^3 WBC/mm^3. Samples with counts $> 10^3$ WBC/mm^3 must be diluted (with physiological saline containing 0.1% sodium azide). For samples with counts $< 2 \times$ WBC/mm^3, more than 1 ml blood may be required, and cells may be concentrated. Briefly, whole blood samples are stained with Reagent A through F (as described in the preceding paragraph). Diluted Reagent g (lyzing solution) is then used to lyze RBC following staining. If WBC count is appropriate, cell separation is not required before staining. Tubes should be protected from direct light, the procedure should be performed at room temperature using reagents at room temperature.

A detailed description of the staining and fixing of the cells will further characterize the technical procedure of basic immunophenotyping:

1. For each blood sample, appropriate tubes (12×75 mm; Becton-Dickinson) should be labeled (A-F) and provided with identification code.
2. 20 µl of Reagents A-F should be given into the respective tube (A-F).
3. 100 µl of well-mixed, anticoagulated whole blood should be added into the bottom of each tube. This sample should be vortexed at low speed for 3 s and incubated for 15–30 min at room temperature.
4. 2 ml of appropriately diluted lyzing solution (Reagent g) should be given into each tube, vortexed and incubated for 10–12 min at room temperature.
5. Tubes should be centrifuged ($300g$; 5 min) at room temperature.
6. Supernatant should be aspirated, leaving approximately 50 µl of residual fluid to avoid any stirring up of pellet.
7. 2 ml PBS (containing 0.1% sodium azide) should be added to each tube to resuspend the pellet. After vortexing at low speed (3 s), centrifugation ($200g$; 5 min) follows.
8. Supernatant should be aspirated leaving approximately 50 µl of fluid.
9. Cell pellet and residual fluid should be resuspended in 0.5 ml of 1% paraformaldehyde (Becton-Dickinson) and vortexed at low speed (3 s).
10. Tubes are ready to be analyzed with FACScan (Becton-Dickinson) and should be stored at 2–8 °C in the dark prior to analysis, which should be performed within 24 h of staining.

41.4 Flow Cytometry

The following system proved to provide reliable and reproducible results. Other systems may well exhibit different characteristics. Fluorescence-activated cell (FAC)Scan (Becton-Dickinson) is used, and is equipped for three-color fluorescence detection and two-parameter light scatter detection. The FACScan flow cytometer has to be prepared for sample analysis with CaliBRITE beads and AutoCOMP software. The stained sample tubes are then run on the flow cytometer and analyzed with the adequate software. The following items (1–4) will give a detailed description of how to use the Becton-Dickinson FACScan flow cytometer:

1. Use CaliBRITE beads to set photomultiplier tubes, adjust fluorescence compensation, and check detector sensitivity.
2. Use the LeucoGATE tube (A) to gate on lymphocytes with appropriate software (e.g., IMK Plus software for basic immunophenotyping). This software automatically sets a lymphocyte gate to eliminate most debris, monocytes, and granulocytes. If there is inadequate separation between cell populations, a message will appear. The software (IMK Plus) automatically counts $\geq 15\,000$ cells to obtain approximately 2000 lymphocytes in the light scatter.
3. Use CONTROL tube (B) to set fluorescence intensity markers. SIMULTEST software (e.g., IMK Plus) automatically sets fluorescence markers and notifies the operator with a message if there is $> 5\%$ nonspecific binding. Markers should be set around the negative population (the cluster of events which are low in both green and red-orange fluorescence).
4. Use the gates and markers established with tubes A and B to analyze tubes C-F. Thus, the FACScan system with appropriate software automatically calculates each reported lymphocyte subset as a percentage of total lymphocytes.

Flow Cytometry

41.5 Limitations of FACS Immunophenotyping

Although FACS immunophenotyping of competent cells may be regarded to be an adequate (therapy) control for certain indications (e.g., during immunomodulatory therapy), some limitations of the method and its application should be considered:

1. Stored and fixed cells should be assayed within 24 h.
2. Previously fixed cells and cells refrigerated prior to staining may give aberrant results.
3. Confounding variables (e.g., medication) may result in misleading data.
4. Laboratories must establish their own normal reference ranges for each parameter.
5. If the results are to be expressed in absolute counts, an independent differential count and WBC count must also be run on the same sample of blood.
6. FACS information must always be combined with other information and interpreted by an experienced diagnostician.
7. Variations in lymphocyte gate settings will change the relative amounts of subsets assayed.
8. Blood samples with $> 100 \times 10^3$ cells/mm^3 and $< 2 \times 10^3$ cells/mm^3 require special handling.
9. Cross-reactions with nonlymphocyte elements may (vary rarely) occur.
10. Samples of nucleated (immature) RBC show incomplete lysis and may increase the amount of debris.
11. Blood samples containing leukemic cells which have volume and scatter characteristics similar to mature (normal) cells will be counted as such.

Limitations

41.6 Clinical Application

Knowledge about the effector mechanisms involved in host defence against tumor growth and dissemination is crucial for the design of effective therapeutic approaches. Specific immune responses against tumor-associated antigens mediated by T lymphocytes (Fidler and Kripke 1980; Scollay et al. 1984) and nonspecific immune responses mediated by macrophages and NK-cells (Talmadge et al. 1980; Hanna 1983) appear to participate in natural or acquired resistance against neoplastic disease. Increases of such mechanisms by immunomodulators, e.g., plant extracts, may thus be beneficial in therapy. However, the precise biochemical definition of active substances and their effective dose is required to allow standardization and a rational guideline for critical evaluation is obligatory. These steps are essential to ensure reproducible efficiency of extracts despite encountered fluctuations of concentrations of diverse components.

As a step towards critically assessing its immunomodulatory capacity in patients, the effects of the administration of purified galactoside-specific lectin from mistletoe (ML) on lymphocyte subsets and expression of activation markers in a group of ten breast cancer patients have recently been monitored (Beuth et al. 1992a,b). As shown in Table 1, repeated subcutaneous injections of ML (twice a week; 1 ng per kg body weight; over a period of 4 weeks) resulted in increased counts of certain lymphocyte

Table 1. Quantitative analysis of peripheral blood lymphocyte subsets and expression of activation markers in ten breast cancer patients prior to (a) and after termination (b) of regular administrations of the immunomodulatory dose of the galactoside-specific lectin from mistletoe (ML-1; twice a week). (Beuth et al. 1992a)

Patients ($n = 10$)		No. of lymphocytes per µl of peripheral blood ($\times 10^2$)					
		Total lymphocytes	T cells	B cells	Helper T cells	Cytotoxic T cells	NK cells
1	a	5.1	3.8	0.5	2.4	1.5	0.7
	b	9.9	7.4	0.7	4.9	2.1	1.8
2	a	17.6	14.2	2.9	10.1	4.1	2.2
	b	28.2	18.6	3.0	13.8	4.8	3.5
3	a	18.2	13.2	3.0	8.7	4.5	2.0
	b	22.4	17.9	2.6	10.3	6.5	2.8
4	a	6.5	5.2	0.3	3.0	1.3	0.8
	b	6.8	6.4	0.4	4.2	2.1	1.1
5	a	7.0	5.2	0.4	2.9	1.1	1.0
	b	7.6	5.7	0.2	3.6	2.0	1.8
6	a	24.6	16.1	3.3	8.6	6.5	1.6
	b	32.6	18.5	2.7	9.8	7.1	2.7
7	a	5.6	3.9	0.5	2.4	1.5	0.7
	b	7.3	5.5	0.8	3.2	1.3	1.5
8	a	22.0	15.0	2.7	11.6	2.2	4.5
	b	26.6	19.3	3.3	14.4	2.0	6.4
9	a	6.0	3.8	0.8	3.4	1.5	0.6
	b	8.8	6.5	0.8	4.9	1.9	1.5
10	a	6.8	5.9	0.3	3.8	2.1	1.1
	b	8.6	7.8	0.6	5.4	3.5	2.2

NK = Natural killer.

subpopulations, which are believed to be involved in anticancer immunity. Enhancement of the overall percentage of lymphocytes in the blood counts was significant in several patients ($n = 2$). In detail, most of the patients treated with the immmunoactive ML dosage revealed enhanced counts of peripheral blood lymphocytes, pan T cells, helper/inducer T cells, and NK-cells, as determined in FACS staining experiments with monoclonal antibodies, as described. Counts of peripheral blood B lymphocytes as well as suppressor/cytotoxic lymphocytes revealed only slight alterations during ML treatment, which appeared to be within the normal biological range. Moreover, ML treatment obviously induced an enhanced expression of HLA-DR and interleukin (IL)-2 receptors, which are closely correlated with the activation of lymphatic cells, especially those of the T cell lineage.

Thus, immunophenotyping of competent cells and FACS analysis of blood samples proved to be an adequate method for the intended phenotyping after lectin application (Beuth et al. 1992a,b). Only thorough clinical trials will answer the pertinent question of whether or not any impact of such effects on the course of the disease can be expected. Further in vitro/in vivo studies on the molecular trigger mechanism(s) following lectin-carbohydrate interactions and the involved biosignaling pathways are in progress, the methodologies being described in this book.

References

Beuth J, Ko HL, Gabius HJ, Pulverer G (1991) Influence of treatment with the immunomodulatory effective dose of the β-galactoside-specific lectin from mistletoe on tumor colonization in BALB/c-mice for two experimental model systems. IN VIVO 5: 29–32

Beuth J, Ko HL, Gabius HJ, Burrichter H, Oette K, Pulverer G (1992a) Behavior of lymphocyte subsets and expression of activation markers in response to immunotherapy with galactoside-specific lectin from mistletoe in breast cancer patients. Clin Invest 70: 658–661

Beuth J, Ko HL, Tunggal L, Gabius HJ, Steuer M, Uhlenbruck G, Pulverer G (1992b) Das Lektin der Mistel in der adjuvanten Therapie beim Mammakarzinom — erste klinische Erfahrungen. Med Welt, in press

Fidler IJ, Kripke ML (1980) Tumor cell antigenicity, host immunity and cancer metastasis. Cancer Immunol Immunother 7: 201–205

Gabius HJ (1987) Endogenous lectins in tumors and the immune system. Cancer Invest 5: 39–46

Gabius HJ (1990) Influence of type of linkage and spacer on the interaction of β-galactoside-binding proteins with immobilized affinity ligands. Anal Biochem 189: 91–94

Gabius S, Gabius HJ (1991) Mistelpräparate als Immunstimulatoren in der Onkologie. Nieders Ärztebl 64: 16–24

Hanna N (1983) Natural killer cell-mediated inhibition of tumor metastasis in vivo. Surv Synth Path Res 2: 69–81

Scollay R, Chen WF, Shortman K (1984) The functional capabilities of cells leaving the thymus. J Immunol 132: 25–30

Talmadge JE, Meyers KM, Prieur DH, Starkey JR (1980) Role of NK-cells in tumor growth and metastasis in beige mice. Nature 284: 622–624

42 Lectin-Dependent Alteration in Availability of Nutrients in Serum and Erythrocytes

H.H. THALMANN and A.N. SAGREDOS

The binding of lectins to cell surfaces can cause transmissions of signals. They may increase the efficiency of processes of absorption at the cell membranes for different ingredients of the cell metabolism (vitamins, minerals, trace elements, fatty acids, amino acids) (Thalmann 1991). As a model, the effect of the galactoside-specific mistletoe lectin on the consumption of nutritive ingredients is exemplified, because it has potential value in tumor therapy (Gabius et al. 1991, 1992).

An increased absorption of these ingredients into the cells changes the cell milieu and leads – recognizable by an increased consumption of ingredients – to an increase of the metabolism in the cells.

The need of these cell ingredients is in the case of cancer patients especially high (Thalmann 1991, 1992). Deficiencies of cell ingredients can lead to a lassitude of the defensive system (for example against tumors, bacteria, virus, and fungus).

Physiological quantities of ingredients lead to a physiological cell milieu. This influences the physiological cell metabolism with impacts on the defence system.

The cell milieu is determined by qualitative and quantitative analysis of the profile of ingredients in the red blood corpuscles as well as in the serum (Sagredos and Leitner 1985).

The individual need of ingredients will be determined with the help of the cell milieu medicine (Thalmann 1992; Thalmann et al. 1992).

42.1 Experimental Procedure

Sampling One hundred ml of blood are taken from each patient (after abstinence from food, alcohol, and mineral water for 12 h), avoiding any hemolysis. Fifty ml blood is not treated, 50 ml blood is mixed with 0.25 ml of a lectin-containing mistletoe solution (35 ng lectin) (Thalmann 1991); the content of vitamins, minerals, trace elements, fatty acids, and amino acids is determined in the serum as well as in the erythrocytes according to the following procedures.

Vitamins The vitamins are examined in the full blood, serum, or erythrocytes according to the following methods: the vitamins B_1, B_2, B_6, niacin, pantothenic acid, and biotin are determined microbiologically; the vitamins B_{12}, folic acid and 25 OHD_3 radioimmunologically and the vitamins A and E according to the HPLC-method. Because of a too-fast disintegration, vitamin C is determined in the fresh urine. The microbiological test strains used and the precision of the method are documented in Table 1 (Sagredos and Leitner 1991)

Minerals and Trace Elements For determination of the minerals and trace elements, the methods of atomic absorption (AAS) and plasmaemission (DCP technique) were used. For the analysis of Na, K, Ca, Mg, P, Fe, Zn, and Cu the method of DC plasmaemission (DC-

Table 1. Methods and precision of the determination of vitamins

Vitamins	Principle of the method		Precision ± %
Vitamin B_1	Microbiological with *Lactobacillus viridescens*	ATCC 12706	4.2
Vitamin B_2	Microbiological with *Lactobacillus casei*	ATCC 7496	3.2
Vitamin B_6	Microbiological with *Saccharomyces carlsbergensis*	ATCC 9080	3.6
Nicotic acid	Microbiological with *Lactobacillus plantarum*	ATCC 8014	1.2
Biotin	Microbiological with *Lactobacillus plantarum*	ATCC 8014	3.9
Pantothenic acid	Microbiological with *Lactobacillus plantarum*	ATCC 8014	4.9
Folic acid	RIA	according to (7)	9.4
Vitamin B_{12}	RIA	according to (7)	4.3
Vitamin 250HD3	RIA	according to (7)	11
Vitamin A	HPLC	according to (5)	4.1
Vitamin E	HPLC	according to (5)	3.5

Table 2. Methods and precision of the determination of minerals and trace elements

Minerals trace elements	Principle of the method	Wave length nm	Precision ± %
Na	Plasma emission, DCP-technique	568.2	2.5–5.2
K	Plasma emission, DCP-technique	404.4	3.4–4.6
Ca	Plasma emission, DCP-technique	430.2	1.8–3.8
Mg	Plasma emission, DCP-technique	383.8	1.4–3.9
P	Plasma emission, DCP-technique	214.9	0.9–4.3
Fe	Plasma emission, DCP-technique	259.9	4.0–7.8
Zn	Plasma emission, DCP-technique	213.8	4.2–6.6
Cu	Plasma emission, DCP-technique	324.7	3.7–7.0
Se	Atom absorption	196.0	5–8

Plasmaemissions-spectralphotometer, model Spectraspan III ex Fisons) was preferred: samples of erythrocytes and serum are treated in a mixture of nitric acid and sulfuric acid at 130 °C and the intensity of element-specific emission is measured at a DC-plasma-emission-spectralphotometer (DCP-technique) after addition of cesiumchloride.

These elements are directly determined in urine according to the DCP technique after dilution in hydrochloric acid and addition of cesiumchloride.

The trace element selenium is determined at AAS (atomic absorption hydridesystem, model SP 9 ex Philips) after mineralization in a mixture of nitric acid, sulfuric acid, and perchloric acid (Sagredos and Leitner 1986). The precision of the methods is seen in Table 2 (Sagredos and Leitner 1991).

Fatty acids

Fatty acids in the serum and in the erythrocytes are determined by capillary gaschromatography after sapornification with potassium methanolate at 4 °C and esterification with BF3-methanolate. Heptadekanoic-acid-methylester serves as internal standard (Sagredos et al. 1991).

Amino acids

The content of the whole amino acids is measured via amino acid analysator after hydrolysis with 6 N hydrochloric acid. Norvaline serves as internal standard.

42.2 Results and Remarks

Presence of lectin apparently has an impact on the relation of serum/erythrocyte content for each group of substances as given in detail in Tables 1 and 2.

Interestingly, a decrease of the cholesterin content could be observed in the analysis with mistletoe lectin-healed versus untreated blood at an average of 13.3%. The triglycerids decreased by 7.25%.

Thus, several individual substances within this panel show lectin-dependent quantitative alterations in their content in serum and erythrocytes.

As tumor patients show an increased need of ingredients, an increased absorption will influence the cell milieu (Thalmann and Sagredos 1992). Only detailed analysis of the changes under therapy will ensure the balance needed to be kept by precisely chosen supplements. By this cell milieu medicine a physiological cell milieu can be attained (Thalmann et al. 1992). This is a basis for defense tumor diseases as prophylaxis and accompanying therapy.

Acknowledgments. We gratefully acknowledge the fundamental work of Dr. H.J. v. Leitner to find the median levels of the cell ingredients.

References

Gabius S, Kayser K, Gabius HJ (1991) Analytische, immunologische und tierexperimentelle Voraussetzungen für die klinische Prüfung der auf Lektingehalt standardisierten Misteltherapie. Dtsch Z Onkol 23: 113–119

Gabius HJ, Walzel H, Joshi SS, Kruip J, Kojima S, Gerker. V, Kratzin H, Gabius S (1992) The immunomodulatory β-galactoside-specific lectin from mistletoe: partial sequence analysis, cell and tissue binding and impact on intracellular biosignalling of monocytic leukemia cells. Anticancer Res 12: 669–676

Sagredos AN, Leitner v H-J (1985) Die Laboruntersuchungen von Mineralstoffen, Spurenelementen und Vitaminen in Körperflüssigkeiten und ihre Bedeutung für Diagnose und Therapie. Referat vor Arbeitsmedizinern, Timmendorfer Strand

Sagredos AN, Leitner HJ (1986) Zusammenhänge zwischen der Erythrozyten-und Serum-Fettsäurenzusammensetzung und dem Gehalt an Selen und Vitamin E von Hypertonikern und Normotonikern. Fette – Seifen – Anstrichmittel 88. Jahrgang, Dez. 1986

Sagredos AN, Leitner HJ (1991) On the normal median values of vitamins, minerals and trace elements in body fluids and their diagnostic and therapeutic significance. 4th Intern. Symp. Clin. Nutritition Heidelberg

Sagredos AN, Leitner HJ, Thalmann HH (1991) Zur Fettsäuren-Zusammensetzung in Serum und Erythrozyten bei gesunden Probanden und bei verschiedenen Krankheiten. 47th Annual Meeting of DGF in Braunschweig

Thalmann HH (1991) Veränderung der Zellmembran-Resorption durch Mistellektine als Basisreaktion einer Abwehrsteigerung. Vortrag: Medizinische Woche Baden-Baden, 31.10.1991

Thalmann HH (1992) Thalmanns Ernährungstherapie des Zellmilieus, Dokumentation d.bes. Therapierichtungen der natürlichen Heilweisen in Europa, BdV, III, 3c

Thalmann HH, Sagredos AN (1992) Individuelle Therapie mit Zellbausteinen (Mineralstoffe, Spurenelemente, Vitamine, Fettsäuren, Aminosäuren) bei onkologischen Erkrankungen, Vortrag 5. wissenschaftl. Kongreß der Gesellschaft für biolog. Krebsabwehr, Heidelberg, 16.6.1991.

Thalmann HH, Sagredos AN, Leitner v H-J (1992) Zellmilieu-Medizin, Dokumentation d.bes. Therapierichtungen der natürlichen Heilweisen in Europa, Bd. II

Lectins and Cell Adhesion

43 Bacterial Adhesion to Immobilized Carbohydrates

C. Yu and S. Roseman

A wide variety of bacterial genera express lectins (Mirelman 1986), and these proteins are apparently required for phenomena such as pathogenesis, biofouling, and biodegradation (Kuehn et al. 1992). We have recently reported that marine *Vibrio* utilize lectins as part of an adhesion/deadhesion system which functions as a nutrient sensing device (Yu et al. 1991; Castell et al. 1979). The regulation and biogenesis of bacterial lectins have attracted much attention, but are beyond of the scope of this report (Kuehn et al. 1992). The present chapter describes a method for determining the lectin activity in bacteria (Yu et al. 1991), as well as the kinetics of adhesion.

Lectin activity is measured by the ability of the cells to adhere to immobilized carbohydrates. The cells are labeled with radioactive isotopes and the number of attached bacteria is readily determined. Under appropriate conditions, cell proteins and DNA can be labeled to high specific radioactivities, thus increasing the counting efficiency, accuracy, and sensitivity of the assay. Various isotopic compounds can be used to label bacteria, but $[^{32}P]$-Pi was routinely used because it is inexpensive and gave high cell specific activities in growth media containing low concentrations of phosphate. After several generations of cell growth in $[^{32}P]$-Pi-containing medium, most of the incorporated $[^{32}P]$-Pi is found in the nucleic acid and phospholipid fractions of the cell.

43.1 Experimental Part

6-aminohexyl D-pyranosides of N-acetylglucosamine (GlcNAc-AH), glucose (Glc-AH), mannose (Man-AH), and galactose (Gal-AH) are obtained from Sigma Chemical Co. (St. Louis, Mo.). Affi-Gel 10 is obtained from Bio-Rad Co. (Richmond, CA). *Vibrio* are grown in 50% artificial seawater (g deionized water): NaCl, 11.8; Na_2SO_4, 2.0; $NaHCO_3$, 0.1; KCl, 0.33; KBr, 0.048; H_3BO_3, 0.013; $MgCl_2,6H_2O$, 5.3; $SrCl_2,6H_2O$, 0.02; $CaCl_2,2H_2O$, 0.74. When used for synthetic culture media, 50%-ASW is supplemented with 0.002% K_2HPO_4, 0.1% NH_4Cl, 0.5% appropriate carbon source and buffered with 50 mM HEPES to a final pH 7.5. **Materials**

The filter holder for the adhesion assay is constructed of a Lucite cylinder (2.2 cm internal diameter × 4.5 cm height). A 4 cm × 4 cm Nitex screen (35 μm pore size, Tetko Inc., Elmsford, N.Y.) is fixed to one end of the cylinder by a lucite ring as shown in Fig. 1.

1. Affi-Gel 10 beads are washed with ice-cold deionized water in a fritted glass funnel (Pyrex, 10–15 μm pore size). Gentle suction is applied to facilitate washing. Care must be taken not to dry the beads. **Preparation of Immobilized Ligands**
2. The activated beads (5 g wet wt.) are transferred to a vial containing 60 μmol of the 6-aminohexyl glycoside in 5 ml of 0.1 M $NaHCO_3$ at pH 8.5, and the slurry is shaken at room temperature for 2 h.

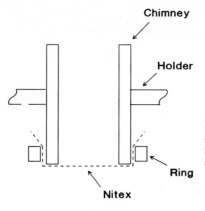

Fig. 1. Filter apparatus. Dimensions of the filter holder are described in *Materials*. Several of these holders can be mounted on a long piece of Plexiglas so that multiple samples can be processed at the same time

3. Unreacted 6-aminohexyl D-pyranosides in the reaction mixtures are determined with fluorescamine, using 6-aminohexanol as a standard (Castell et al. 1979), giving (by difference) the quantity of coupled glycoside.
4. Ethanolamine (0.1 M) is then added to react with the remaining hydroxysuccinimidyl groups on the activated beads, and the slurry maintained at 4 °C overnight.
5. The beads are then washed in a fritted glass-funnel with buffered 50%-ASW and stored at 4 °C in 0.2% sodium azide in 50% ASW.

Under these conditions, about 7 µmol of glycoside were covalently linked per g wet wt. of beads.

Preparation of Labeled Cells

1. *Vibrio furnissii* is inoculated into buffered 50%-ASW containing 0.5% Bacto-Peptone and 0.1% yeast extract. The culture is shaken overnight at 25 °C.
2. A 0.5 ml aliquot of the overnight culture is inoculated into 10 ml of labeling medium: buffered 50%-ASW containing 0.5% sodium lactate (pH 7.5) as carbon source, 0.1% NH_4Cl, and 20 µCi of $[^{32}P]$-Pi.
3. The culture is shaken at 25 °C and growth of the cells monitored by measuring absorbance at 540 nm.
4. When the A_{540} value is between 0.3–0.4, the cells are harvested by centrifuging at 3000g for 5 min at 4 °C. The resulting loosely packed cell pellet is washed three times with equal volumes of ice-cold buffered 50%-ASW. Washed cells are resuspended in ice-cold buffered 50%-ASW.
5. Specific activities of the cell suspensions are determined by the following equation:

$$\frac{\text{Total dpm per ml of cell suspension}}{(A_{540})\ (5 \times 10^8 \text{ cell per ml per unit } A_{540})} = \text{dpm per cell.}$$

6. After the specific activities of the cells are determined, appropriate quantities of lactate, NH_4Cl and $[^{32}P]$-Pi are added to the suspension to give a complete labeling medium. This mixture is stored on ice until used

A summary of the steps is schematically given in Fig. 2.

Adhesion Assay

1. Glycoside derivatized Affi-Gel 10 beads are thoroughly washed with 50% ASW in a fritted glass funnel to remove sodium azide, and the beads packed under gentle suction. Care must be taken not to dry the beads. Aliquots (10 mg each) of wet packed beads are dispensed into a series of 10 × 75 mm disposable test tubes.

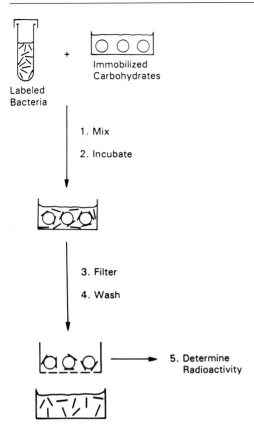

Labeled
Bacteria

Immobilized
Carbohydrates

1. Mix

2. Incubate

3. Filter

4. Wash

5. Determine
   Radioactivity

Fig. 2. Scheme for the cell-adhesion assay

2. To initiate the reaction, 100 µl aliquots of labeled cells are added to each tube, and the cell-bead mixture incubated at room temperature.
3. The screen and lower portion of the chimney are immersed in 50%-ASW. At the indicated time, 4 ml of 50%-ASW are added to each test tube to terminate the reaction, and the suspension immediately poured on the 35 µm Nitex screen.
4. The test tube and the Nitex screen are then washed twice with 4 ml of 50%-ASW. Excess liquid retained on the screen is removed by blotting the bottom on paper towels.
5. The screen and beads are transferred to a scintillation vial for determining the quantity of radioactivity.
6. The number of cells adhering to the beads is equal to the total radioactivity divided by the specific activity of the cells.

43.2 Results

A time course of adhesion of *V. furnissii* to carbohydrate derivatized Affi-Gel 10 is shown in Fig. 3. The cells adhere to GlcNAc-AH, Glc-AH, and Man-AH derivatized beads. There is no significant adhesion to Gal-AH or aminohexanol-derivatized beads.

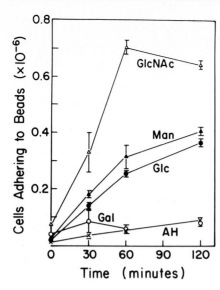

Fig. 3. Adhesion of *V. furnissii* to glycoside-derivatized beads. Adhesion was measured by the standard assay procedure in lactate-50%-ASW (+ ^{32}Pi) growth medium. *Error bars* represent the range of duplicate samples

43.3 Remarks

1. Affi-Gel beads consist of a cross-linked agarose matrix, so cells expressing a galactose-specific lectin will give a high background.
2. The components of the reaction medium must be determined empirically. For *V. furnissii*, the complete growth medium is required to express and maintain lectin activity.
3. Complex medium can also be used for labeling and in adhesion assays. Sugars used as a carbon source in the growth medium may have to be removed since they can act as competitive inhibitors against adhesion to immobilized carbohydrates.

References

Castell JV, Cervera M, Marco R (1979) A convenient micromethod for the assay for primary amines and proteins with fluorescamine: a reexamination of the conditions of reaction. Anal Biochem 99: 379–391

Kuehn MJ, Heuser J, Normark S, Hultgren SJ (1992) P pili in uropathogenic *E. coli* are composite fibres with distinct fibrillar adhesive tips. Nature 356: 252–255

Mirelman D (1986) Microbial lectins and agglutinins: Properties and biological activity. Wiley Series in Ecological and applied microbiology (Series Ed., Mitchell R), Wiley Interscience, New York.

Yu C, Lee A, Bassler BL, Roseman S (1991) Chitin utilization by marine bacteria: a physiological function for bacterial adhesion to immobilized carbohydrates. J Biol Chem 266: 24260–24267

44 Quantitative Micro-Adhesion Assay on Polystyrene Matrices

F.-G. HANISCH, F. DRESSEN, and G. UHLENBRUCK

The interaction of cell surface carbohydrates as carriers of biological information and their complementary carbohydrate binding proteins plays an important role in several cell-cell recognition processes. Receptor-mediated endocytosis of asialo-glycoproteins after recognition by hepatic binding protein, homing of cells like the hematopoetic stem cells or lymphocytes, egg-sperm binding, interactions among cells of the immune system (Brandley and Schnaar 1986), controlling of cellular proliferation (Lotan et al. 1985), the recruitment of neutrophils to sites of inflammation (Walz et al. 1990), and myelin formation (Zanetta et al. 1990) are examples of lectin-carbohydrate interactions in higher vertebrates. Moreover, glycoconjugate-lectin interactions have been postulated to participate in the establishment of metastatic lesions (Lotan and Raz 1988). To investigate the expression of lectin receptors on eukaryontic cells, several assays for demonstrating cell-binding to immobilized glycoconjugates have been developed. Cell adhesion to glycolipids immobilized on silica gel thin layer chromatography plates or on PVC plates was examined by Swank-Hill (Swank-Hill et al. 1987) and Blackburn (Blackburn and Schnaar 1983; Blackburn et al. 1986), respectively, while studies with glycoproteins were performed on glycosylated derivatives of polyacrylamide gels (Schnaar 1984) or on nitrocellulose as matrix (Lochner et al. 1990).

The establishment of quantitative cell adhesion assays on different matrices is affected by several critical parameters which necessitate a series of precautions and optimizations. Among these, specific cell adhesion is abolished by exposure to liquid-air interfaces, which renders the washing procedures more difficult. Moreover, there is a critical balance of parameters that influence the mechanical separation of unspecifically and specifically bound cells. Finally, using nonradiometric detection systems most assays suffer from lower sensitivities, background interferences with the blocking reagent in protein assays, or from the nonapplicability in a microplate formate.

44.1 Objectives

The method presented in this chapter allows for a fast (5–6 h) and sensitive determination of cell adhesion to glycoproteins, as well as to glycolipids immobilized to polystyrene 96-microwell plates. Moreover, cell binding is easily detected by colorimetric analysis in a microplate reader using viable cell staining with MTT. The method should be applicable to analyze quantitatively the lectin patterns of different cell types and to perform inhibition assays in small volumes and with low amounts of competing sugars for defining the binding specificities of cellular lectins.

44.2 Materials and Cells

Materials   Polystyrene 96-microwell plates were obtained from Nunc (Wiesbaden, FRG), adhesive plate sealers were from Flow Laboratories (Meckenheim, FRG). All glycolipids and all neoglycoproteins were purchased from Sigma (Munich, FRG). 3-(4,5-dimethylthiazol-2-yl)-2,5-diphenyltetrazolium bromide (MTT) was from Sigma (Munich). All other chemicals were of the highest available grade from standard commercial sources.

All media and media-supplements used in cell-culture were from Flow Laboratories (Meckenheim, FRG); PBS: 40 mM KH_2PO_4, 16 mM Na_2HPO_4, 135 mM NaCl, pH 7.2; TBS: 20 mM Tris, 150 mM NaCl, pH 7.2.

Cells and culture conditions   Chang liver cells were purchased from Flow Laboratories (Meckenheim, FRG). They were cultured in MEM, containing 10% heat inactivated fetal calf serum, 2 mM glutamine, 1% nonessential amino acids, 2.5 µg/ml amphotericin B, and 2% penicillin-streptomycin. Human breast carcinoma cells (MDA MB 231) were obtained from the American Type Culture Collection (Maryland, USA). They were grown in the same medium described above for the cultivation of Chang liver cells, but supplemented with 0.16 IE insulin/ml and 10 mM HEPES. Cloned Balb/c fibrosarcoma cells were established as described elsewhere (Hanisch et al. 1990).

Cells were grown in RPMI 1640 containing 10% fetal calf serum, 2 mM glutamine, 2.5 µg/mL amphotericin, and 2% penicillin-streptomycin. All cells were cultured as monolayers at 37 °C in a humid atmosphere enriched with 5% CO_2.

For the binding assay, cells were harvested with 0.03% EDTA in PBS, washed three times with serum-free RPMI-DM (Dutch-Modification) and resuspended in RPMI-DM supplemented with 2 mM glutamine. Before use they were allowed to recover for 30 min at 37 °C.

44.3 Procedure

Immobilization   *Glycoconjugate Immobilization.* Fifty µl of a glycoprotein solution in carbonate buffer pH 9.6 or 50 µl of a glycolipid solution in 50% ethanol/water were added to each polystyrene well and allowed to dry at 37 °C overnight. Wells were washed three times with 200 µl PBS and then incubated for 1 h at 37 °C with 200 µl PBS containing 5% BSA to block residual active surface. Wells were washed three times with RPMI-DM before use.

Cell Adhesion   *Cell Adhesion Assay.* To each well were added $7–8 \times 10^4$ cells unless specifically indicated. The volume of the cell suspension in each well was brought up to 420 µl with prewarmed RPMI-DM until a slightly positive meniscus could be seen. Plates were carefully sealed by rolling on an adhesive plate sealer without introducing air bubbles, and were incubated for 105 min at 37 °C. Subsequently, sealed plates were centrifuged in inverted position (well bottoms upward) for 10 min at 400g. After the centrifugation plates were righted, plate sealers were removed and wells were aspirated by a 8-channel vacuum washer, leaving 80 µl of medium in each well. This step has to be accomplished immediately after rightening the plate to prevent the sedimentation of nonadherent cells remaining in suspension.

Detection and Quantitation of Adhesive Cells. Cells were detected by measuring the   **Quantitation**
absorbance of blue formazan, generated by enzymatic reduction of the tetrazolium
salt MTT. One hundred µl of MTT (1 mg/ml in Dulbecco's Medium without phenol
red, supplemented with 20 mM HEPES) was added to each well and incubated for
90–120 min at 37 °C. Subsequently, the medium was replaced by 100 µl 1-propanol
and the plate was vigorously shaken to ensure complete solubilization of the blue
formazan. Optical density was measured in a Dynatech ELISA-Reader at 570 nm.

44.4 Results

44.4.1 Procedure

The conditions for cell binding were optimized for mouse fibrosarcoma cells,
L1LM12, with regard to the critical parameters: matrix, coating conditions, and
amount of immobilized glycoconjugates, blocking conditions, incubation conditions, and the discrimination between specifically and unspecifically bound cells.
Moreover, a detection system based on the reduction of the dye MTT was
established which allows the rapid and sensitive quantification of adhesive cells in a
microplate reader. The general applicability of the assay was proved by using two
further indicator cells: human breast carcinoma cells (MDA-MB 231), and Chang
liver cells.

Detection of Adhesive Cells. A simple and rapid procedure of cell detection was
described by Denizot (Denizot and Lang 1986) and Miller (Miller and McDevitt
1991). It depends on the reduction of the tetrazolium salt MTT (3-(4,5-dimethyl-
thiazol-2-yl)-2,5-diphenyltetrazolium bromide) to blue formazan by the mitochondrial enzyme succinate dehydrogenase in living cells (Slater et al. 1963). Activity
proved to be sufficient and quantitatively comparable in all cell types tested. Cell-
specific differences could be standardized by reference measurements in each binding
assay. The rate of adhesive cells was calculated as percentage of cells added. Increase
in blue formazan formation was linear over an incubation period of approximately
3 h. In standard assays, an incubation period of 90–120 min proved to be sufficient.

Coating Efficiencies and Ligand Densities. The amount of glycoproteins or glycolipids bound to each well was quantified by using a modified micro BCA-assay
(Tuszynski and Murphy 1990) and a resorcinol sulfuric acid micromethod for the
determination of neutral sugars (Monsigny et al. 1988), respectively.
    Contrasting with the adsorption of BSA to the plastic surface, which decrease if
concentrations exceeding 12 µg/well were added, the glycosylated BSA derivatives
showed no decrease in adsorption up to the highest concentration added (50 µg/
well). The amounts of glycolipids adsorbed to the wells were proportional to the
amount of glycolipid added over a broad concentration range. While neutral
glycolipids remained stably adsorbed to the wells during the washing procedure,
some leaking of charged gangliosides occurred, resulting in decreased surface
densities.

Blocking Conditions for Inactivation of the Plastic Surface. Free protein binding sites
were blocked with PBS containing 5% BSA to prevent unspecific cell-binding.
Addition of BSA to the incubation medium during cell binding was not necessary.
Unspecific cell binding under standard conditions was below 10% of cells added.

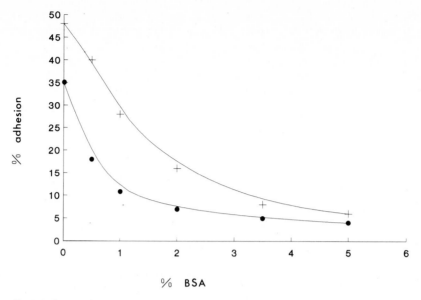

Fig. 1. Influence of BSA-concentration in the blocking solution on unspecific cell binding to polystyrene. Unspecifically bound cells were detached by applying a RCF of 400 g (L1LM12 fibrosarcoma cells), ●; or of 550 g (Chang liver cells), +

Below 2.5% BSA the plate binding of the different cell types increased considerably (Fig. 1).

Separation of Unspecifically and Specifically Bound Cells. Cell adhesion to immobilized glycoconjugates is very sensitive to shear forces and to exposure to air/liquid interfaces (Swank-Hill et al. 1987). Sufficient reproducibility was obtained by using centrifugal forces to separate unspecifically bound cells from the well bottom, as described by McClay (McClay et al. 1981). The discrimination between unspecific and specific binding is strongly dependent on the relative centrifugal force (RCF). Specific cell binding of mouse fibrosarcoma cells decreased at a RCF > 500 g, while a RCF < 250 g led to an increase of unspecific background and concomitant with this to a decrease of assay sensitivity. In standard assays a RCF of 400 g was used.

An incubation time of 105 min was found to be necessary to ensure sufficient cell binding (Fig. 2). A shortening of this incubation period by enhancement of cell sedimentation as described by McClay (McClay et al. 1981) led to nonreproducible and high background levels. When cells were centrifuged on plates (right side up) with a RCF of 8 g, a dislodgement force of 1200 g was not sufficient to lower the background to a level < 10% of total indicator cells. Preferably, after addition to the wells, cells are allowed to sediment by gravity.

A centrifugation time of 10 min was sufficient to dislodge unspecifically bound cells from the surface (data not shown). Further washing was not necessary.

Incubation Time. Maximum cell adhesion was measured after 90–120 min following addition of cell suspensions to the wells (Fig. 2). A prolongation of the incubation

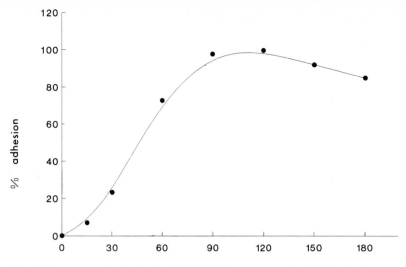

Fig. 2. Time course of adhesion by L1LM12 fibrosarcoma cells. The incubation was stopped at the indicated times by inverting the plates and removing nonadherent cells by centrifugation

time led to a decrease in cell binding; 105 min was chosen as standard incubation time.

Cell Concentration. A maximum of cell binding is reached at a cell concentration of $8–10 \times 10^4$ cells/well. Exceeding 10^5 cells/well the absolute number of bound cells decreased (data not shown); 8×10^4 cells/well were chosen in standard assays.

Incubation Medium. Sufficient binding occurred in cell culture mediums like DMEM or RPMI, which have to be supplemented with HEPES. Of significant influence was the addition of glutamine to the incubation medium which led to a two- to sixfold increase in the binding rate, dependent on the glycoconjugate coated.

44.4.2 Binding Patterns of the Tested Indicator Cells

Three types of nonrelated cell lines were used to prove the general applicability of the micro adhesion assay and to analyze the lectin patterns of the various cells: mouse fibrosarcoma cells L1LM12, (Fig. 3), Chang liver cells, and human breast carcinoma cells, MDA-MB 231. Characteristic differences in binding patterns could be shown, due to the specific lectin equipment of the three cell types.

44.4.3 Specificity of the Binding

In addition to the cell-characteristic binding patterns which already indicate the specificity of the adhesion, two independent lines of evidence corroborated the carbohydrate-lectin-mediated specificity of cell adhesion. A sharp decrease in cell adhesion was demonstrated after treatment of the glycoproteins with 1 mM sodium metaperiodate. Moreover, inhibition assays were carried out by incubation of the cells in the presence of potential inhibitors: free sugars and glycosides. Cell binding to β-Gal-HSA was inhibited by free D-galactose and by various β-galactosides.

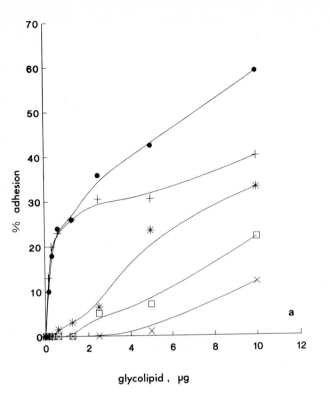

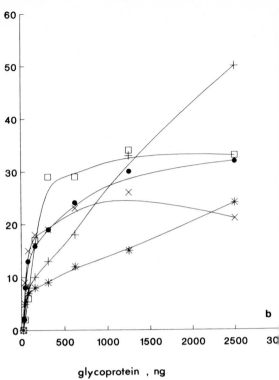

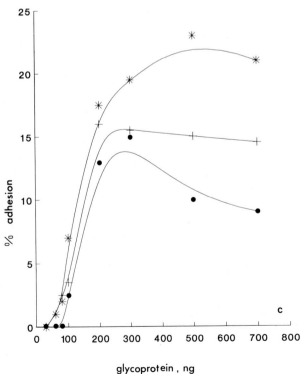

Fig. 3a–c. Cell adhesion of L1LM12 fibrosarcoma cells on glycolipids (**a**) and on glycoproteins (**b, c**). **a** ●, asialo GM1; +, Gb4-Cer; *, Gb3-Cer; □, β-Lac-Cer; ×, βGal-Cer **b** ●, β-Gal-BSA; +, β-GalNAc-BSA; *, β-GlcNAc-BSA; □, β-Glc-BSA; ×, α-Man-BSA **c** ●, α-Fuc-BSA; +, β-Lac-BSA; *, α-Gal-BSA

44.4.4 Troubleshooting

When using different indicator cells, some of the critical parameters have to be varied:

1. The conditions of centrifugation have to be optimized for each cell line. For example, at optimal RCF established for the separation of unspecifically bound mouse fibrosarcoma cells, Chang liver cells showed a strong background binding, and, accordingly, increased detachment forces had to be applied. However, care should be taken in raising the RCF above a critical value, which may vary from cell to cell.

2. Similarly, cell-dependent variations have to be considered during cell sedimentation to the well bottoms by gravity. The time period of settling and thus the incubation time may be influenced by the sizes and masses of the cells used, and have to be shortened or prolonged in individual cases.

References

Blackburn CC, Schnaar RL (1983) Carbohydrate-specific cell adhesion is mediated by immobilized glycolipids. J Biol Chem 258: 1180–1188

Blackburn CC, Swank-Hill P, Schnaar RL (1986) Gangliosides support neutral retina cell adhesion. J Biol Chem 261: 2873–2881

Brandley BK, Schnaar RL (1986) Cell-surface carbohydrates in cell recognition and response. J Leuk Biol 40: 97–111

Denizot F, Lang R (1986) Rapid colorimetric assay for cell growth and survival: modifications to the tetrazolium dye procedure giving improved sensitivity and reliability. J Immunol Meth 89: 271–277

Hanisch F-G, Sölter J, Jansen V, Lochner A, Peter-Katalinic J, Uhlenbruck G (1990) Glycosphingolipid expression on murine L1-fibrosarcoma cells: analysis of clonal in vivo and in vitro selected sublines with different lung colonization potential. Br J Cancer 61: 813–820

Lochner A, Hanisch F-G, Uhlenbruck G (1990) Carbohydrate-mediated cell adhesion to glycoproteins immobilized on nitrocellulose. Anal Biochem 190: 108–115

Lotan R, Raz A (1988) Lectins in cancer cells. Ann NY Acad Sci 551: 385–414

Lotan R, Lotan D, Raz A (1985) Inhibition of tumor cell colony formation in culture by a monoclonal antibody to endogenous lectins. Cancer Res 45: 4349–4353

McClay DR, Wessel GM, Marchase RB (1981) Intercellular recognition: quantitation of initial binding events. Proc Natl Acad Sci USA 78: 4975–4979

Miller RR, McDevitt CA (1991) A quantitative microwell assay for chondrocyte cell adhesion. Anal Biochem 192: 380–383

Monsigny M, Petit C, Roche AC (1988) Colorimetric determination of neutral sugars by a resorcinol sulfuric acid micromethod. Anal Biochem 75: 525–530

Schnaar RL (1984) Immobilized glycoconjugates for cell recognition studies. Anal Biochem 143: 1–13

Slater TF, Sawyer B, Strauli U (1963) Studies on succinate-tetrazolium reductase systems. Biochim Biophys Acta 77: 383–393

Swank-Hill P, Needham LK, Schnaar RL (1987) Carbohydrate-specific cell adhesion directly to glycosphingolipids separated on thin-layer chromatography plates. Anal Biochem 163: 27–35

Tuszynski GP, Murphy A (1990) Spectrometric quantitation of anchorage-dependent cell numbers using the bicinchoninic acid protein assay reagent. Anal Biochem 184: 189–191

Walz G, Aruffo A, Kolanus W, Bevilacqua M, Seed B (1990) Recognition by ELAM-1 of the sialyl-Le^x determinant on myeloid and tumor cells. Science 250: 1132–1135

Zanetta JP, Warter J-M, Kuchler S, Marschal P, Rumbach L, Lehmann S, Tranchant C, Reeber A, Vincendon G (1990) Antibodies to cerebellar soluble lectin CSL in multiple scleriosis. Lancet 335: 1482–1484

45 Lectin-Mediated Binding Activities of *Bradyrhizobium japonicum*

S.C. Ho and J.T. Loh

Bradyrhizobium japonicum binds to soybean root in a polar fashion (Halverson and Stacey 1986; Ho et al. 1990a). This binding is galactose-specific, suggesting a lectin-mediated binding phenomenon. Further study on the *B. japonicum* binding property leads to the discovery of four carbohydrate-binding activities: (a) heterotypic binding to soybean root; (b) heterotypic binding to cultured soybean SB-1 cells; (c) homotypic autoagglutination; and (d) adsorption to synthetic beads derivatized with lactose (Ho et al. 1990a). All of these binding activities can be inhibited by galactose or lactose, but not by galactose derivatives, such as N-acetyl-D-galactosamine. These observations suggest a common recognition mechanism involving a lectin residing on the bacterial surface. The purification of a carbohydrate-binding protein from *B. japonicum* further substantiates this notion (Ho et al. 1990b). This purified lectin, designated BJ38 ($M_r \sim 38\,000$), showed saccharide specificity that correlated well with that of *B. japonicum* binding activities. It is proposed that BJ38 may mediate the carbohydrate-binding activities of *B. japonicum*. Micrographs demonstrating four binding activities of *B. japonicum* and a model depicting the BJ38-mediated carbohydrate-specific binding of *B. japonicum* are shown in Figs 1 and 2, respectively. The methods summarized in this chapter are based in part on those reported by Ho et al. (1988, 1990a,b).

45.1 Bacterial Culture Conditions and the General Binding Characteristics of *B. japonicum*

Culture Conditions

The binding activities of *B. japonicum* are highly dependent on the growth conditions of the bacteria. They are very sensitive to the growth phase, medium composition, shaking conditions, temperature, and pH. The bacteria exhibit optimal binding activities at late exponential phase of growth (Ho et al. 1988). Yeast extract in the growth medium higher than 1% is inhibitory. When sodium chloride concentration is higher than 50 mM, or when the temperature is 37 °C or higher, the growth of the bacteria is hampered, which is followed by a loss in binding activities. High speed agitation (such as 200 rpm in a gyratory shaker normally used for culturing *E. coli*) will also greatly inhibit binding. The binding activities of *B. japonicum* are observed within a narrow pH range from 5 to 8. The following conditions are maintained throughout all binding studies of *B. japonicum*. The bacteria are initially cultured in 50 mM lactose agar plates with yeast extract-gluconate medium (YEG), which contains (per liter):

1.28 g K_2PO_4	0.77 mg $MnSO_4 \cdot H_2O$
0.2 g $Mg_2SO_4 \cdot 7H_2O$	0.15 mg $CuSO_4 \cdot 5H_2O$
7.35 mg $CaCl_2 \cdot 2H_2O$	2.49 mg $Na_2MoO_4 \cdot 2H_2O$
28 mg sequestrene	0.23 mg $CoCl_2 \cdot 6H_2O$
5.0 g gluconic acid	0.46 mg $Na_2B_4O_7 \cdot 10H_2O$
1.0 g yeast extract	0.38 mg Na_3VO_4
2.5 mg Na_2EDTA	0.1 mg $NaSeO_3$
4.39 mg $ZnSO_4 \cdot 7H_2O$	pH 6.0

These bacteria are transferred to 50 ml YEG in a 125 ml Erlenmeyer flask and cultured to the late exponential phase of growth ($A_{620} = 1.7–2.0$) with gyratory shaking at 120 rpm at 30 °C.

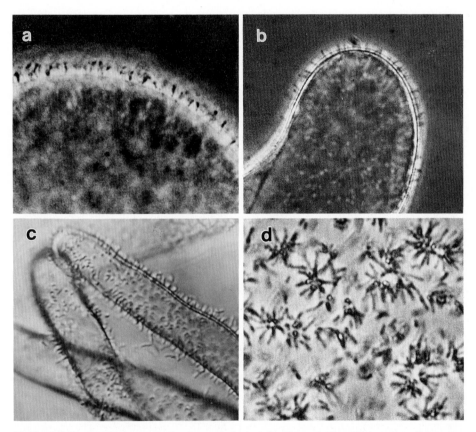

Fig. 1a–d. Micrographs demonstrating four carbohydrate binding activities of *B. japonicum*. *B. japonicum* cells bind to **a** a Lac-Sepharose bead, **b** a SB-1 cell, **c** soybean root hairs, and **d** other *B. japonicum* cells

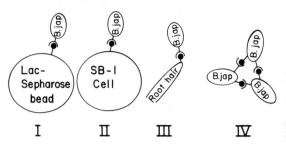

Fig. 2. Models (I-IV) depicting the bacterial lectin BJ38 mediating the carbohydrate-binding activities of *B. japonicum.* —⊂ represents BJ38, and —•, glycoconjugates, involved in the binding.

45.2 Quantitation of Binding of *B. japonicum*

Quantitation The binding activities can be quantitated by radimmunoassay using the antibody generated against the heat-killed *B. japonicum* (anti-Brj antibody, Ho et al. 1988). The antigen can be prepared by washing a mid-exponential phase culture of *B. japonicum* and resuspending the cells in phosphate-buffered saline (PBS; 10 mm sodium phosphate 0.137 M NaCl, 3 mM KCl, 1 mM $MgCl_2$, 1 mM $CaCl_2$ at pH 7.4). The bacteria is heated for 10 min in a sand bath at 97 °C and then stored at -20 °C in 0.2 ml aliquots (1 mg of protein per ml). One aliquot of this sample is mixed with 0.8 ml distilled water and 1 ml of complete Freund adjuvant. The emulsified antigen is injected subcutaneously to a New Zealand White rabbit. Booster injections can be given at biweekly intervals for 2 months with the same antigen fraction in incomplete Freund adjuvant. One week after the last injection, blood is collected. The immunoglobulin fraction can be isolated by protein A-Sepharose affinity chromatography. This immunoglobulin fraction is designated as anti-Brj, which contains antibodies mainly reactive to the lipopolysaccharide of *B. japonicum.*

For radioimmunoassay (Ho et al. 1988), anti-Brj can be radioiodinated by chloramine-T method (Langone 1980). After radioiodination, the free ^{125}I is removed by passage through an AG 1×8 ion-exchange column (Bio-Rad Laboratories, Richmond, CA). The titer of the antibody can be determined by the amount of antibody required to saturate the surface antigens of certain number of bacteria (e.g. 1×10^7 bacteria). For a typical assay, the samples are incubated in a final volume of 1 ml with a concentration of anti-Brj 60 µg/ml and 1×10^5 cpm in PBS containing 0.1% bovine serum albumin (PBS-BSA). After incubation at room temperature for 3 h with shaking, the cells are washed three times with 2 ml PBS-BSA. For bacteria bound to the soybean roots, SB-1 cells, or Lac-Sepharose beads, the samples can be washed the same way as for the removal of unbound bacteria, as stated below. For free bacteria, to establish, for example, the standard curve for the quantitation, the bacterial samples are washed three times by 2 ml PBS-BSA by filtration on Millipore membranes (0.22 µm pore size, Millipore Corp., Bedford, MS). The membranes are presoaked in 5% BSA to prevent nonspecific adsorption. Filtration is performed with mild vacuum by a water aspirator, such that the bacteria are retained on the membrane. After washing, the amount of radioactivity associated with each sample is then determined by gamma counter. For the standard curve, radioactivity associated with each bacterial sample is plotted against the number of bacteria used for the assay. The standard curve is linear from 8×10^5 to 1×10^7 bacteria (Ho et al. 1988). The number of bacteria bound to soybean roots, SB-1 cells, or Lac-Sepharose beads can be determined from the radioactivity using the standard curve.

45.3 Binding of *B. japonicum* to Cultured Soybean SB-1 Cells

The SB-1 cells, originally derived from soybean roots [*Glycine max* (L.) Merr. CV. Mandarin] are cultured in 50 ml of 1B5C medium in 125 ml Erlenmeyer flasks on a gyratory shaker (160 rpm) in the dark at 26 °C. To test the binding of *B. japonicum* to SB-1 cells, SB-1 cells (0.5 ml of the cells suspension, 20% vol/vol, about 10^6 cells) of a 2-day-old culture are added to a 35 mm culture dish containing 1 ml of 1B5C medium and then 0.5 ml of *B. japonicum* of appropriate concentration is inoculated into the dish. The mixtures are incubated for 16 h at 26 °C. After incubation, cells and are transferred to polystyrene tubes (12 × 75 mm) and washed three times with 2 ml of PBS-BSA by centrifugation (640 *g* for 1 min) and resuspension to remove unbound bacteria. The relative amounts of bacteria bound are compared by radioimmunoassay with anti-Brj antibody, or by direct observation of the bound bacteria under a microscope.

Binding

45.4 Binding of *B. japonicum* to Soybean roots

Soybean seeds are surface sterilized and germinated at 26 °C in the dark by wrapping them inside the sterile paper towels moistened with water. After 4 days, segments of the roots, about 1 cm long, are excised aseptically from the root hair zone. These segments are divided into groups of ten in 35 mm culture dishes and rinsed once with 2 ml of 1B5C medium. The samples are treated with 2 ml of bacterial suspension. After incubation for 16 h at 26 °C, the root segments are washed six times with 2 ml PBS-BSA for 10 min each to remove unbound bacteria. The samples are then assayed for bound bacteria by radioimmunoassay with anti-Brj. On the other hand, the bacterial binding can be directly observed under a phase contrast microscope.

Binding

45.5 Assay of *B. japonicum* Autoagglutination

The ability of *B. japonicum* cells to interact with one another can be evaluated by culturing the bacteria on solid agar medium (YEG) at 30 °C for 5 days. The bacteria can be picked up by a sterile toothpick and smeared on a microscope slide with a drop of water. The samples can be observed under the microscope. To test whether the autoagglutination is carbohydrate-specific, the bacteria are cultured in agar medium in the presence of various saccharides for 5 days. Autoagglutination is obvious when cells are joined together in star-like rosette structures. When individual cells are separated from each other, this would indicate inhibition of autoagglutination.

Autogglutination

45.6 Binding of BJ38 to Synthetic Beads

Various kinds of disaccharides can be coupled to aminoethyl polyacrylamide beads (aminoethyl Bio-Gel P100; Bio-Rad Laboratories, Richmond, CA) using the sodium

Binding

cyanoborohydride method (Baues and Gray 1977). The reaction mixture, containing lactose (100 mg), NaBH$_3$CN (50 mg) and 7 ml Bio-Gel P100 in 0.2 M potassium phosphate (pH 9.0), is agitated at room temperature for 4 days. The beads are washed by 0.2 M sodium acetate and excess amino groups are blocked by N-acetylation by addition of 4 ml acetic anhydride per 20 ml gel suspension in 0.2 M sodium acetate. After 30 min, the beads are sequentially washed with water, 0.1 N NaOH, water, and PBS. Similarly, saccharides can be coupled to Sepharose 4B beads by activating the beads with epichlorohydrin (Matsumoto et al. 1979). The resulting epoxide can be aminated with ammonia to generate terminal amino groups for coupling with the disaccharide of interest in the presence of cyanoborohydride as stated above. The lactose-derivatized Sepharose 4B (Lac-Sepharose) can also be used for affinity purification of BJ38 (see below).

The binding of *B. japonicum* to saccharide-derivatized beads are assayed by the same way as that for the SB-1 cells. The beads are first suspended in water (50% vol/vol). Aliquots (200 µl) are transferred to 35 mm culture dishes, followed by the addition of 0.8 ml of bacterial suspension. The samples are incubated for 16 h at 26 °C. After incubation, the samples are transferred to polystyrene tubes (12 × 75 mm) and washed to remove unbound bacteria with 2 ml of PBS-BSA by centrifugation (460 g for 1 min) and resuspension. The amount of bacteria bound to the beads is then determined by radioimmunoassay with anti-Brj antibody.

45.7 Isolation of the Bacterial Lectin BJ38 from *B. japonicum*

Isolation   Similar to the binding activities of the bacteria, the recovery of BJ38 is highly dependent on the growth conditions of the bacteria, in particular the growth phase, speed of agitation, and composition of the medium. The following stringent growth conditions are optimized for BJ38 isolation (Ho et al. 1990b).

B. japonicum is cultured on an agar plate in YEG, supplemented with 50 mM lactose. The bacteria are transferred to 50 ml YEG in a 125 ml Erlenmeyer flask and cultured for one day. This culture is then inoculated into 2 l of YEG and cultured for two days on a gyratory shaker at 120 rpm at 30 °C. Aliquots of this culture (300 ml) are then inoculated into six flasks, each containing 1.5 l of YEG. The bacteria are further cultured for about 30 h until the absorbance at 620 nm is achieved to a value of 1.7–2.0. The bacterial culture is harvested by centrifugation at 8000 rpm for 15 min, and is frozen overnight. The frozen bacteria are resuspended in 50 ml PBS, passed through the French press once at 1600 psi, and then centrifuged at 15 000 rpm for 30 min with a Sorvall SS-34 rotor. The supernatant is collected and applied to a Lac-Sepharose column (1.5 × 7 cm) equilibrated with PBS. The column is then washed with 400 ml of PBS. The lectin bound to the column is then eluted with 50 ml of 0.1 M lactose in PBS (Lac-PBS).

45.8 Radioiodination of BJ38

Radioiodination   Radioiodination of lectins by chloramine T method can serve two purpose. First, it offers high sensitivity for detection of low quantities of protein. As a matter of fact, BJ38 was first identified by radioiodination of the lactose-eluted fractions after affinity purification. Second, because of the resulting high radiospecific activity for protein labeling, it can be used as a probe for following the lectin binding activities. Radioiodination of proteins by chloramine T results in labeling the tyrosine residues. Some of the lectins depend on a tyrosine residue in sugar binding activity. Therefore, radioiodination with chloramine T method may result in inactivation of some lectins. The most common procedure for radioiodination of lectins is carried on in the presence of hapten sugar, so as to protect the sugar binding site. After radiolabeling, the lectin can be subjected to an affinity column to recover the active lectin. The following procedure has been used for BJ38 labeling.

A sample of BJ38 (100 µl) in Lac-PBS, directly obtained from the lactose elution after affinity purification, is added to 1 mCi Na^{125}I. Chloramine T (25 µl, 1 mg/ml in Lac-PBS) is added to start the reaction. After 1 min, 50 µl of sodium metabisulfite (1 mg/ml in Lac-PBS) is added to stop the reaction. The sample is then applied to a Sephadex G-25 column (2.5×11 cm) to remove the lactose and free ^{125}I. The radioactive lectin eluted at the void volume is then reisolated by a second Lac-Sepharose column. This second affinity purification is to ensure that the labeled lectin recovered still maintains its sugar-binding activity and to eliminate other contaminants. Figure 3 shows a typical profile for the affinity chromatography of the iodinated sample. The first peak mainly contains other contaminants during the first affinity purification. The second peak contains the lactose specific eluted BJ38. For BJ38 labeling, a radiospecific activity of 100 mCi/mg protein can be obtained. A final concentration of 0.1% BSA can be added as carrier protein and the sample is dialyzed against PBS to remove the lactose. After that, ^{125}I-BJ38 can be used for various binding studies. When this radioactive sample is reapplied to the Lac-Sepharose column to test for its sugar binding activity, more than 95% of the radioactivity can be readsorbed onto the affinity column and be specifically eluted

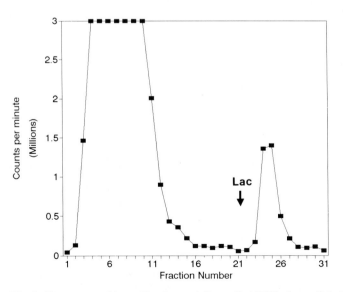

Fig. 3. Chromatographic profile of a partially purified BJ38 after radioiodination on a Lac-Sepharose column. The *first peak* contains radioactivity did not bind to the column. The *second peak* contains ^{125}I-BJ38 specifically eluted by Lac-PBS

by Lac-PBS. Furthermore, sodium dodecyl sulfate polyacrylamide gel electrophoresis can be used to determine the molecular weight and the purity of the sample after autoradiography.

45.9 Binding of BJ38 to Soybean Root, SB-1 Cells, and Lac-Sepharose Beads

Binding   Soybean roots (ten segments excised from the roots hair zone), SB-1 cells (1×10^6 cells), or Lac-Sepharose beads (200 µl of the bead suspension in PBS (1:1 vol/vol)), are placed in 35-mm culture dishes. Various reagents to be tested are added, followed by ^{125}I-BJ38 (0.1 ml with 1.5×10^4 cpm) in a final volume of 2 ml. The samples are incubated with gentle shaking (50 rpm) at room temperature for 4 h. The samples are then washed three times with 2 ml of PBS. The amount of radioactivity associated with the samples is then determined.

45.10 Binding of BJ38 to *B. japonicum*

Binding   *B. japonicum* suspension (0.2 ml) is added to each culture tube and incubated with various reagents and ^{125}I-BJ38 (1.5×10^4 cpm) in a final volume of 2 ml at room temperature for 4 h with gentle shaking. Individual samples are filtered through Millipore membrane filters that have been presoaked in PBS containing 5% BSA to prevent nonspecific binding. Unbound radioactivity is removed by washing three times with 2 ml of PBS-BSA. The radioactivity retained on the membrane is determined.

Acknowledgment. This work was supported by grant GM45200 from the National Institutes of Health and by the Michigan Agricultural Experimental Station.

References

Baues RJ, Gray GR (1977) Lectin purification on affinity columns containing reductively aminated disaccharides. J Biol Chem 252: 57–60

Halverson LJ, Stacey G (1986) Signal exchange in plant-microbe interactions. Microbiol Rev 50: 193–225

Ho SC, Ye W, Schindler M, Wang JL (1988) Quantitative assay for binding of *Bradyrhizobium japonicum* to cultured soybean cells. J Bacteriol 170: 3882–3890

Ho, SC, Wang JL, Schindler M (1990a) Carbohyrate binding activities of *Bradyrhizobium japonicum*. I. Saccharide-specific inhibition of homotypic and heterotypic adhesion. J Cell Biol 111: 1631–1638

Ho SC, Schindler M, Wang JL (1990b) Carbohydrate binding activities of *B. japonicum*. II. Isolation and characterization of galactose-specific lectin. J Cell Biol 111: 1639–1643

Langone JJ (1980) ^{125}I-labeled protein A: reactivity with IgG and use as a tracer in radiommunoassay. Methods Enzymol 70: 356–361

Matsumoto I, Mizuno Y, Seno N (1979) Activation of Sepharose with Epichlorohydrin and subsequent immobilization of ligand for affinity adsorbent. J Biochem 85: 1091–1098.

46 Identifying a Role for Carbohydrate Recognition in Regulating Neuronal Architecture

B. Zipser and R.N. Cole

A multi-faceted approach must be considered in developing an experimental strategy to determine the role of carbohydrate-binding proteins (or lectins) and oligosaccharides in the regulation of neuronal architecture. Numerous studies provide indirect evidence that interactions between cell surface oligosaccharides and carbohydrate-binding proteins regulate various aspects of neuronal development (reviewed in Jessell et al. 1990). Unfortunately, no simple single experiment will demonstrate that this type of carbohydrate interaction is involved in the formation of the nervous system. Only a combination of approaches, as used in nonneuronal systems, provides strong direct evidence for the recognition of carbohydrates by carbohydrate-binding proteins during cell-cell interactions (Ofek and Sharon 1988; Shur 1989; Yednock and Rosen 1989). This multi-dimensional strategy involves masking the suspected oligosaccharide with antibodies or lectins which bind to specific carbohydrate antigens; altering the oligosaccharide structure with enzymes that remove or add saccharides; and blocking the binding of the suspected carbohydrate-binding protein with competing saccharides. Obtaining at least three independent lines of evidence is necessary to demonstrate a role for a carbohydrate recognition in regulating neuronal architecture.

Care must be taken in choosing a neuronal system to examine carbohydrate recognition between its oligosaccharides and carbohydrate-binding proteins. Often, cultures of dissociated neurons are used to identify the underlying molecular mechanisms of neuronal pathfinding and synapse formation. Although important hints on how neurons read a molecular terrain are learned from studies involving isolated neurons, in reality, we can only know how neurons read molecular signals by studying their interactions in an intact brain. This is especially true when these molecular signals involve carbohydrate-binding proteins or carbohydrates because their expression and distribution change with the partial neuronal deafferentation found in brain slices or the full isolation of neurons used in dissociated cultures (Dontenwill et al. 1985; Jessell et al. 1990; Naegele and Katz 1990; Scott et al. 1990).

46.1 Experimental Strategy for Investigating Carbohydrate Recognition

Building a strong case for carbohydrate recognition in the patterning of the nervous system requires several independent lines of evidence from manipulations of endogenous lectins and oligosaccharides. Here, we discuss an experimental strategy involving two perturbations of carbohydrate chains and one perturbation of carbohydrate-binding proteins.

46.1.1 Antibody Perturbation of Carbohydrate Epitopes

Antibodies which recognize a specific carbohydrate epitope are commonly used as experimental reagents. It is preferable to use Fab fragments instead of whole IgGs. Whole IgG antibodies, because of their two antigen binding sites, can cross-link membrane surface molecules in a process called patching. Thus, a perturbation effect elicited by whole IgGs may be due to a general distortion of the membrane topology. Cleaving the IgG at the hinge region produces fragments (Fab fragments) that have only one binding site and cannot cross-link molecules. Therefore, a Fab fragment should only inhibit the function of the molecule to which it binds. One caveat is that the Fab fragment may not bind to the molecule's actual functional domain. The binding of the Fab fragment may cause a steric obstruction or a conformational change in the nearby functional domain. Hence, a perturbation elicited by a Fab fragment which binds to a carbohydrate epitope is necessary but not sufficient to demonstrate that the carbohydrate epitope is involved in generating the neuronal architecture.

A developing nervous system can be extremely sensitive to the presence of Fab fragments. At low concentrations (in the nanomolar range), Fab fragments are powerful tools for identifying critical choices in the selection of pathways and connectivities. However, at high concentrations (in the micromolar range), they cause developmental arrest. Presumably, low concentrations of Fab fragments only interrupt specific adhesion/recognition events of neurons. In contrast, high concentrations of Fab fragments are toxic to neurons and, therefore, elicit effects that obscure or mislead interpretations of the physiological functions of lectins and oligosaccharides.

To control for the specificity of a perturbation effect elicited by one type of Fab fragment, other types of Fab fragments should be applied to the developing nervous system. Fab fragments used as a control should recognize a different surface epitope and not elicit the same perturbation effect.

46.1.2 Enzymatic Perturbation of Carbohydrate Epitopes

Enzymes that alter the carbohydrate structures on the membranes of neural cells are used to independently examine the role of a carbohydrate epitope in a carbohydrate recognition event. An obvious enzyme to use first is a general glycosidase because little structural information on the endogenous oligosaccharide structure is necessary. For example, the involvement of asparagine-linked (N-linked) oligosaccharides in a carbohydrate recognition event can be assayed by applying N-glycanase [peptide-N^4-(N-acetyl-β-glucosaminyl) asparagine amidase]. This glycosidase removes N-linked carbohydrate chains from most glycoproteins. With additional information on the structure of the carbohydrate epitope of interest, more specific glycosidases or glycosyltransferases (Ichikawa et al. 1992) can be used to experimentally manipulate the carbohydrate environment in which neurons select pathways and connectivities.

As with Fab fragments, the enzyme dosage can be critical. An enzyme such as N-glycanase used at high dosages is expected to affect the performance of many different cell types because of the ubiquity of N-linked carbohydrate chains on cell surfaces. The general removal of carbohydrate chains may affect additional supporting tissues that in turn alter the neuronal architecture. Thus, to control for the specificity of a perturbation effect elicited by a glycosidase, its impact on cell types, besides those targeted, needs to be examined. For example, in the leech central nervous system (CNS), both sensory neurons and glial cells express cell-specific N-linked carbohydrate chains on their cell surface glycoproteins. Applied at low dosages at the correct developmental time, however, N-glycanase only affects the development of neuronal architecture, with no obvious effect on the glial cell architecture (Zipser and Cole 1991). This suggests that a small reduction in the surface concentrations of N-linked carbohydrate structures at this developmental stage has a critical impact only on neuronal differentiation and not on glial cell differentiation.

46.1.3 Sugar Perturbation of Carbohydrate-Binding Proteins (or Lectins)

Carbohydrate recognition events can be perturbed by blocking carbohydrate-binding proteins with saccharides that compete with their endogenous carbohydrate ligands. Carbohydrates synthetically linked to proteins, termed neoglycoproteins, are particularly useful reagents for these sugar blocking studies.

Monosaccharides conjugated to proteins elicit perturbation effects at micromolar concentrations. Not conjugated to a protein, the same monosaccharides may not elicit the perturbation effect, even at a 1000-fold higher concentration. For example, 50 mM α-methyl-mannoside does not affect the architecture of the leech CNS whereas 26 µM of same saccharide conjugated to 1 µM of BSA strongly perturbs this neuronal architecture (Zipser and Cole 1991). There are at least two possible reasons for the effectiveness of neoglycoproteins. First, the availability of multiple monosaccharides on the neoglycoprotein may promote the binding of these saccharides to the typically multivalent lectins. As a result, simple monosaccharides are successful in competing with the endogenous oligosaccharides that may possess complex structures. Secondly, much of the monosaccharide applied to the experimental preparation may be metabolized and not take part in blocking the carbohydrate-binding protein. The use of neoglycoproteins overcomes these problems. More physiologically relevant concentrations of saccharides (µM), which are not available to metabolic processes, can be presented to endogenous carbohydrate-binding proteins.

It is important, however, to know the saccharide's linkage in the neoglycoprotein. The binding of many carbohydrate-binding proteins is sensitive to various substitutions at different carbons in a saccharide. Therefore, a monosaccharide that blocks the binding of a carbohydrate-binding protein in vitro may not effectively block the same carbohydrate-binding protein if it is linked to a protein through a structurally important carbon in the saccharide.

The appropriate control for the specificity of the action of a particular saccharide is the demonstration that other saccharides do not elicit the same perturbation effect. For example, to control for the specificity of a mannose effect, the action of other sugars such as galactose needs to be tested. Neoglycoproteins containing mannose or galactose are commercially available (see Sec. 46.2.2. and following)

46.2 An Application of this Experimental Strategy

Application of the above experimental strategy to define the role of carbohydrate recognition will be discussed using the leech nervous system as an example. The embryonic leech nervous system can be cultured intact. This intact embryonic nervous system is fully permeable to macromolecules such as antibodies, enzymes, or glycoproteins making it possible to explore the molecular mechanisms underlying neuronal pathfinding and synapse formation.

46.2.1 Description of an Embryonic Nervous System Illustrating this Experimental Strategy

Description A functional class of leech neurons, the sensory afferents, can be distinguished by a mannose-containing epitope that is N-linked to a 130 kDa protein and is reactive with the monoclonal antibody, Lan3-2. Sensory afferents project as a tightly fasciculated bundle through peripheral nerves, but upon arriving in the CNS, they defasciculate (or unbundle) into the synaptic neuropile (Fig. 1A). This defasciculation allows the previously bundled sensory afferents to project as single axons through the synaptic neuropile as they search for their postsynaptic targets. We investigated whether the axonal defasciculation is mediated by the mannose-containing Lan3-2 epitope on the sensory afferents.

We found that axonal defasciculation is inhibited (Fig. 1B): (a) by masking the Lan3-2 mannose-containing epitope with Lan3-2 Fab fragments; (b) by cleaving the N-linked carbohydrate moieties from surface proteins with N-glycanase; and (c) by competing for a putative mannose-binding protein with mannose that was covalently linked to BSA (mannose-BSA). These three perturbation experiments, detailed below, lead us to the following hypothesis: The defasciculation of sensory afferent neurons results from recognition/adhesion between their mannose-containing Lan3-2 epitope and a putative mannose-binding protein in the synaptic neuropile.

46.2.2 Equipment, Reagents, and Solutions

Equipment Culture dishes: 35×10 mm, Falcon Petri dishes (Fisher Scientific) filled with 2 ml Sylgard (Dow Chemical Corp., Midland, Ml) are prepared for embryonic culturing by sterilizing under UV lights for at least 1 h.
Laminar flow hood for sterile culture
Microscope with bright field or fluorescence optics and oil immersion lens (25–100 ×).
A video camera and computer with imaging analysis capabilities to quantifying perturbation effects (see Sec. 46.3.2.6).

Reagents for perturbation Antibodies: monoclonal antibodies (mAb) were generated against the leech nervous system as described in Zipser and McKay (1981). Fab fragments are prepared from whole monoclonal IgGs according to Good et al. (1980)
Enzymes: N-glycanase (Genzyme Corp., Boston, MA)
Neoglycoproteins: mannose-BSA (26 mol of p-aminophenyl α-D-mannopyranoside/mol albumin (Sigma A-4664), galactose-BSA [26 mol 2-amido-2-deoxy-D-galactose/mol albumin (Sigma-5908)]

Solutions PBS: 0.9% NaCl in a 50 mM sodium phosphate buffer, pH 7.4
Defined culture medium: 1% ITS + TM (Collaborative Research Inc., Bedford, MA), 10 nM CR epidermal growth factor (Collaborative Research Inc., Bedford,

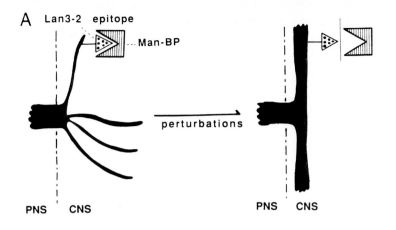

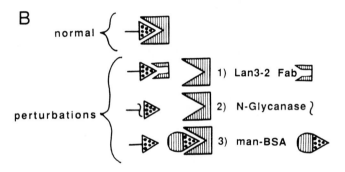

Fig. 1A,B. Hypothesis of carbohydrate recognition mediating the defasciculation of sensory afferents in the synaptic neuropile. **A** During development, sensory afferent axons grow as a fasciculated bundle through peripheral nerves (*PNS*) into the central ganglia (*CNS*), where they defasciculate, projecting as single axons through the synaptic neuropile. Even though the mannose-containing *Lan3-2 epitope* is evenly distributed on the sensory afferent surface, the defasciculation event is restricted to the synaptic neuropile. Our experimental evidence is consistent with a mannose-binding protein (*Man-BP*) being present in the synaptic neuropile. Recognition/adhesion between the Lan3-2 epitope of sensory afferents and the Man-BP in the synaptic neuropile results in the region-specific defasciculation. Perturbations of this mannose-specific recognition inhibits the defasciculation, and, instead, sensory afferents continue to grow through the CNS as a bifurcated fasciculated tract, appearing as a T. **B** The normal recognition between the Lan3-2 epitope and the Man-BP is inhibited in the presence of *1* Lan3-2 Fab fragments: *2* N-glycanase, a glycosidase; and *3* mannose-BSA (man-BSA), a neoglycoprotein. The Lan3-2 Fab fragments bind to and mask the mannose-containing Lan3-2 epitope. N-glycanase cleaves N-linked carbohydrate domains removing the Lan3-2 epitope. And, man-BSA competes with the Lan3-2 epitope for the mannose-binding protein. (Zipser and Cole 1991)

MA), $100 \mu M$ vitamin C, and 0.15% glucose in Leibovitz-15 (Sigma Chemical Comp., St. Louis, MO) medium. The potassium concentration of this defined medium is adjusted to 15 mM

Fixative: 4% paraformaldehyde in 0.1 M phosphate buffer, pH 7.4

Immunocytochemistry staining solutions: antibodies, avidin-HRP and avidin-rhodamine are diluted into PBS containing 3% BSA and 2% Triton X-100. Primary antibodies are monoclonal antibodies generated against the leech nervous system. Secondary antibodies are biotinylated F(ab′) 2 against mouse-IgG

(DAKO Corp., Santa Barbara, CA) and are visualized with avidin-HRP (Vector Laboratories, Inc. Burlingame, CA) or avidin-rhodamine (Vector Laboratories, Inc. Burlingame, CA)

Anti-bleach medium: mounting medium for fluorescence is prepared according to Johnson and C. Nogueira Araujo (1981)

DAB solution: 0.5% 3,3'-diaminobenzidine (Sigma Chemical Co., St Louis, MO) in 50 mM Tris buffer (pH 7.4) diluted to 0.02% with PBS, and filtered through a Whatman #1 filter

Other solutions: 0.3% hydrogen peroxide; 70, 95, and 100% ethanol; 100% methanol; xylene; and Permount (Fisher Scientific)

46.2.3 Procedures

46.2.3.1 Preparing Embryos For Perturbation Experiments

Preparing Embryos The germinal plates from 15 sibling leech embryos from one 10.5-day-old cocoon (incubated at 20 °C) are divided into control and experimental groups. Each group contains at least three embryos. One group is immediately fixed in paraformaldehyde and immunocytochemically stained (see Sect. 46.2.3.3) to determine the stage of sensory neuron differentiation at the onset of the experiment. Embryos belonging to the other groups are rinsed several times in plain sterile Leibovitz-15 medium and pinned by their epithelial envelopes to the separate Sylgard-coated culture dish. One group is cultured in defined culture medium to determine the extent of normal development that takes place during the culturing period. The remaining groups of embryos are cultured in the presence of different experimental reagents added to the defined culture medium. All groups are cultured for up to 1.5 days at 20 °C under sterile conditions.

46.2.3.2 Perturbing Neuronal Growth

Antibodies *Antibody Perturbations.* Embryos are grown in the presence of Lan3-2 Fab fragments which mask the mannose-containing Lan3-2 epitopes of sensory afferents (Fig. 1B). Fab fragments are applied at concentrations of 6–12 nm. At this concentration range axons do not defasciculate but retain their normal growth rate. In contrast, 150 nM of Fab fragments arrest axonal growth. To control for the specificity of the Lan3-2 Fab fragment effect, a control group of embryos is grown in the presence of 6-12 nM of Fab fragments which bind to a muscle surface antigen. The muscle-specific Fab fragments do not affect the development of sensory neurons.

Enzymes *Enzyme Perturbations.* Embryos are cultured in the presence of 4 U of N-glycanase. N-Glycanase cleaves the asparagine-linked carbohydrate chains containing the Lan3-2 epitope (Fig. 1B). To determine whether N-glycanase specifically alters sensory afferent projection or whether other cell types also are affected, the development of glial cells in N-glycanase-treated embryos are examined. N-glycanase at this concentration does not affect the normal development of glial processes.

Sugar *Sugar Perturbations.* Embryos are cultured in the presence of 1–4 µM of mannose-BSA. Mannose-BSA should compete with the Lan3-2 epitope for a mannose-binding protein (Fig. 1B). To control for the specificity of the mannose effect, a control group of embryos is cultured in the presence of $1-4\,\mu M$ galactose-BSA. Sensory neurons project normally in the presence of this concentration of galactose-BSA.

46.2.3.3 Analyzing Perturbations of Neuronal Growth

Antibody Staining of Embryos. Embryos are fixed for 0.5 h at room temperature in paraformaldehyde, immersed in methanol for 30 min to destroy endogenous peroxidases, permeabilized with xylene for 2 min, washed twice with 100% methanol (2 min each), and returned into PBS. Embryos are incubated with primary antibody (mAb Lan3-2 specific for sensory afferents or mAb Laz6-56 specific for glial cells) overnight on a rotatory shaker at room temperature. The following day embryos are rinsed briefly three times in 2% Triton X-100 in PBS and placed in a 1:100 dilution of biotinylated F(ab') 2 against mouse IgG for 2 h. Embryos are again rinsed briefly in 2% Triton X-100 in PBS and incubated in a 1:150 dilution of avidin-HRP or a 1:1000 dilution of avidin-rhodamine for 1 h. Embryos are then rinsed briefly in PBS to remove Triton X-100. Avidin-HRP treated embryos are incubated in a DAB solution containing 0.015% hydrogen peroxide. The reaction is terminated when nonspecific background staining begins to appear. The stained embryos are rinsed briefly with PBS, dehydrated in sequential 2 min steps through 70, 95 and 100% ethanol, cleared in xylene for 5 min, and embedded in Permount. Avidin-rhodamine treated embryos are mounted directly from PBS into anti-bleach mounting medium, which reduces bleaching of fluorescence. The coverslip is sealed over the embryo with nail polish and the slide is stored at $-20\,°C$ in the dark.

Antibody Staining

Quantification of Changes in Neuronal Architecture. Qualitatively, the difference between sensory afferent projections in the presence versus the absence of a perturbing reagent will be obvious as viewed through a microscope. Quantification, however, will be desired for publication. We suggest two possible methods.

Quantification

Method 1. The alterations in neuronal architecture can be quantified by measuring the area of the projections of HRP-stained sensory afferents under a Nikon $100 \times$ oil immersion objective. The video images of sensory afferent projections can be captured through a DAJE-MTI series 68 video camera (Michigan City, IN) using Jandel analysis software, PCVisionplusR Frame Grabber version (Imaging Technology, Inc. Jandel Scientific)

Method 2. Alternatively, the changes in the volume of sensory afferent projections can be obtained by analyzing fluorescently labeled sensory afferents using confocal microscopy such as the Odyssee system (Noran instruments, Middleton, WI) equipped with Image-1 software (Universal Imaging Corp., W. Chester, PA)

46.3 One Last Consideration

It may be possible to perturb the interactions of a carbohydrate epitope only during a particular developmental stage. For example, in the leech embryo, axonal defasciculations are inhibited during early development when axons express the Lan3-2 carbohydrate epitope on their peripheral membrane protein (gp130). Later in development, these axons begin to express the Lan3-2 carbohydrate epitope on their integral membrane proteins (gp 103 and 95). At this stage, axonal defasciculations are no longer readily inhibited with the experimental strategy described here (Zipser and Cole 1991).

References

Dontenwill M, Roussel G, Zanetta JP (1985) Immunohistochemical localization of a lectin-like molecule, R1, during the postnatal development of the rat cerebellum. Dev Brain Res. 17: 245–252

Good AH, Wofsy L, Kimura J, Henry C (1980) Purification of immunoglobulins and their fragments. In: Mishell BB and Shiigi SM (ed) Selected methods in cellular immunology. W. H. Freeman and Company, New York, pp. 284–286

Ichikawa Y, Look GC, Wong C-H (1992) Enzyme-catalyzed oligosaccharide synthesis. Anal Biochem 202: 215–238

Jessell TM, Hynes MA, Dodd J (1990) Carbohydrates and carbohydrate-binding proteins in the nervous system. Ann Rev Neurosci 13: 227–255

Johnson GD, C. Nogueira Araujo GM (1981) A simple method of reducing the fading of immunofluorescence during microscopy. J Immunol Meth 43: 349–350

Naegele JR, Katz LC (1990) Cell surface molecules containing N-acetylgalactosamine are associated with basket cells and neuroglialform cells in cat visual cortex. J Neurosci 10: 540–557

Ofek I, Sharon N (1988) Lectinophagocytosis: a molecular mechanism of recognition between cell surface sugars and lectins in the phagocytosis of bacteria. Infection and Immunity 56: 539–547

Scott LJC, Balsamo J, Sanes JR, Lilien J (1990) Synaptic localization and neural regulation of an N-acetylgalactosaminyl transferase in skeletal muscle. J Neurosci 10: 346–350

Shur B (1989) Expression and function of cell surface galactosyltransferase. Biochim Biophys Acta 988: 389–409

Yednock TA, Rosen SD (1989) Lymphocyte homing. Adv Immunol 44: 313–378

Zipser B, Cole RN (1991) A mannose-specific recognition mediates the defasciculation of axons in the leech CNS. J Neurosci 11: 3471–3480

Zipser B, McKay R (1981) Monoclonal antibodies distinguish identifiable neurons in the leech. Nature 289: 549–554

47 Identifying Cell-Cell Adhesion-Inhibitory Antibodies to Cell Surface Proteins Using Divalent Primary and Monovalent Secondary Antibodies – a Method Developed Using Cellular Slime Molds

W.R. Springer

Cellular slime mold has been used as a model for many developmental processes including cell-cell and species-specific adhesion. The organism is the subject of a number of books and countless research articles. The reader is referred to Loomis (1982), Raper (1984) and Spudich (1987) for three reviews covering all aspects of the cellular slime molds from their taxonomy to the isolation of specific proteins.

The study of cell-cell adhesion in the cellular slime mold, *Dictyostelium discoideum*, was begun by Gerisch (1980) in the mid-1960s. Through his and his colleagues efforts, and those of many investigators in a number of independent laboratories, a great deal is known about adhesion in this particular species as well as in other *Dictyostelium* and *Polysphondylium* species. A critical part of the experimental findings hinges on evidence from various types of cell-cell adhesion assays.

A brief description of the life cycle of the cellular slime molds explains why they are excellent organisms for the study of cell adhesion. When the individual foraging amoeba have depleted their bacterial food supply, they begin to form large multicellular aggregates which eventually differentiate into structures made up of dormant spores perched on top of a stalk of evacuated cells. It is during the aggregation stage that different kinds of adhesions develop, as measured by assays like the one described here. It is presumed that each type of adhesion is mediated by specific surface molecules, many of which have been found to be glycoconjugates. Antibodies which bind to these molecules should interfere with the adhesion. Thus one can suggest a role in adhesion for a molecule recognized by adhesion-inhibiting antibodies.

The advantage of in vitro assays like this one over observations of in situ aggregation is that aggregation can also involve directed migration along surfaces or via chemotactic gradients, both of which are eliminated when cells are shaken in liquid suspension. Such in vitro assays also allow the easy addition of exogenous materials to be tested as inhibitors of the adhesion measured. Similar methods have, of course, been developed for other homotypic cell adhesions and are used regularly.

The method described here was optimized for cellular slime molds by Jim McDonough while we were collaborators in the laboratory of Sam Barondes at the University of California, San Diego (McDonough et al. 1980). We have used the assay with little change for more than 10 years, and have found it to be a rapid and reliable means of quantifying adhesion under various conditions of interest. The assay and variations have been used to measure the effect on adhesion of monovalent Fab fragments on antibodies raised against surface proteins. Fab fragments are used because, being monovalent, they are not able to cause agglutination of cells by cross-linking two cells together, as can happen with intact antibody. Because of a reduction in avidity often observed with monovalent Fab fragments, an inhibitory effect may require large amounts of material. We found that we could use much smaller amounts of intact, divalent antibodies to inhibit the measured adhesion if we prevented these antibodies from agglutinating the cells by assaying in the presence of univalent second antibody directed against the particular species of primary antibody (Springer and Barondes 1980). The increased sensitivity and the elimination of the need to prepare Fab fragments by this modification allowed the rapid screening of large numbers of small amounts of antibody preparations such as arise during the production of monoclonal antibodies.

47.1 The Adhesion Assay

Cells   Culturing and preparing cells for the assay. Since the focus of this chapter is the adhesion assay itself, and the reader may not necessarily be planning to use cellular slime molds, I will give an abbreviated description of the culture methods and will refer the reader to complete and detailed methods found in chapters in Spudich (1987) and Raper (1984).

   Although there are some strains of cellular slime molds which can grow in defined medium, the majority require co-culture with a bacterial food source. Cells are grown by collecting spores from a stock plate of 1 to 2 weeks of age, using a flamed loop and mixing with a stationary phase culture of bacteria (*K. pneumoniae*) grown in a nutrient-rich broth. An aliquot of this mixture is spread on nutrient agar and incubated at room temperature until a slight clearing of the bacterial lawn is observed. Cells at the stage at which bacteria are still present are considered vegetative and show little or no adhesion under the appropriate assay conditions. These cells are forced to differentiate by removing the bacteria through differential centrifugation (three washes in cold water at $250g$) and placing 0.5 ml at 2×10^8/ml onto a filter pad consisting of an AABP 047 00 black nitrocellulose filter on top of an AP10 047 00 cellulose pad (Millipore Corp. Bedford Mass.) soaked with 1.5 ml PDF (see below). A black filter is used to make it easier to observe aggregation, which is quite evident without magnification. When various species are allowed to aggregate on filter pads directly from the vegetative stage, a variable preaggregation phase causes the different strains to form aggregates in as little as 5 h to as much as 20 h. Since this makes it difficult to obtain cultures at a specific stage of development at a specific time of day, we have found it convenient to starve cells overnight in suspension first, then place them on filter pads. Overnight suspension cells are arrested at a stage of development which allows rapid aggregation in 1 to 2 h after being plated onto filter pads for all species. Washed cells are resuspended to 10^7 /ml in PDF and rotated at 200 rpm at room temperature for 16 to 20 h. These cells are then centrifuged, resuspended and placed on filter pads as above.

10X PDF

Buffers   Dissolve 54.4 g of KH_2PO_4 in 1 l of water (solution A). Dissolve 34.9 g of K_2HPO_4 in 500 ml water (solution B). Gradually add solution A to about 400 ml of solution B until the pH is 6.5. Measure out 1 l of the pH 6.5 buffer and add 14.9 g KCl, 6.1 g $MgSO_4 \cdot 7H_2O$, and 5 g streptomycin sulfate. Label as 10X PDF and store refrigerated. PDF is prepared by diluting 1 part of 10X PDF with nine parts water

10X SPS

Dissolve 22.4 g $Na_2HPO_4 \cdot 7H_2O$ in 500 ml water (solution A). Dissolve 45.4 g KH_2PO_4 in 2 l (solution B). Gradually add solution A to solution B until the pH is 6. Label as 10X SPS and store refrigerated. SPS is made by diluting one part 10X SPS with nine parts water.

Assay buffer

Dissolve the contents of a 5 ml vial (Cappel, Organon Teknika, Durham, NC) of the appropriate Fab fragments of goat IgG made against the species of the primary antibody being testing (in the case of monoclonal antibodies this would be mouse) in 100 ml of SPS containing 2 mg/ml bovine serum albumin. Store frozen

Low speed centrifuge capable of 250 g **Equipment**
Liquid dispensers, various sizes **and Supplies**
Rotary shaker
Vortex mixer
Electronic particle counter
15 dram plastic snap cap vials
12×75 mm and 50 ml conical centrifuge tubes

Table 1 provides an example of the data collected from an assay to test eight **Procedure** concentrations of an inhibitory antibody, and suggestions for tube and vial labeling and timing. All incubations are to be performed on ice. The assay itself is done at room temperature. Prior to harvesting the cells, 12×75 mm centrifuge tubes are labeled (Table 1), 0.25 ml of serial twofold dilutions of the appropriate primary antibody are added, and stored on ice. The corresponding vials are labeled (Table 1), filled with 0.5 ml assay buffer and placed on the rotary shaker.

Cells at the appropriate stage of aggregation on filter pads are removed by slipping the filter only into a 50 ml conical centrifuge tube, adding 2 ml of ice-cold SPS, and vortexing. The filter is removed and the cell suspension counted. Counting is done by adding 50 µl of cells to 250 µl of cold SPS (this is the same dilution as will be used to incubate the cells with antibody), vortexing, and counting 50 µl of this suspension in 10 ml of cold saline (0.9% NaCl) in the electronic particle counter. The counter is adjusted so that the maximum number of single cells are counted. Because the cells slowly shrink in the saline, counts should be performed within 2 min of adding the saline. The cells should remain cold and vortexed just prior to counting, to assure the maximum number of single cells. The concentration of cells in the original 50 ml centrifuge tube is adjusted to give a count of 20 000 to 50 000 for a 0.5 ml sample through the counter. This works out to somewhere between 5 and 12 $\times 10^7$/ml. After the concentration is properly adjusted, 50 µl of these cells are added to each of the previously prepared antibody dilutions, mixed well and incubated on ice for 30 min. During the incubation, duplicate 50 µl samples from each incubation tube are counted (three counts each) in 0.5 ml SPS and 10 ml ice-cold saline to obtain the 0 time values of Table 1. If there is not sufficient time to complete these within the 30 min, they can also be obtained at the end of the assay. The cells are quite stable in ice-cold SPS for many hours.

The assay itself consists of removing 50 µl aliquots of cells from an incubation tube and adding them to a 15 dram plastic vial containing 0.5 ml of assay buffer at room

Table 1. Completed data sheet for a typical experiment

Experiment No. 1273   Date 5/7/92

Title: Inhibition of *D. purpureum* adhesion by monoclonal antibody d-41.

Vial	0 Min Count			Ab[a] (µg/ml)	10 Min Count			
1A[b]	35 556,	35 891,	36 002	0	(0)[c] 9168,	8790,	9523	[10][c]
B	36 120,	35 764,	35 213		(2) 8965,	9270,	8999	[12]
2A	34 567,	34 201,	33 973	500	(4) 37 899,	37 765,	38 352	[14]
B	32 533,	31 984,	31 847		(6) 37 656,	37 904,	38 276	[16]
3A	35 710,	35 298,	35 901	250	(8) 36 257,	36 737,	36 123	[18]
B	36 128,	36 013,	35 678		(10) 37 450,	37 100,	36 568	[20]
4A	36 237,	35 678,	35 567	125	(12) 35 567,	35 156,	36 023	[22]
B	34 789,	34 235,	35 235		(14) 33 134,	32 341,	33 341	[24]
5A	36 676,	37 356,	36 680	63	(16) 32 789,	33 125,	33 002	[26]
B	36 245,	36 670,	36 000		(18) 31 892,	32 012,	31 345	[28]
6A	34 967,	34 567,	35 124	32	(20) 21 368,	22 109,	22 034	[30]
B	36 267,	35 134,	36 007		(22) 23 367,	22 890,	21 683	[32]
7A	35 145,	35 013,	36 045	16	(24) 10 387,	10 328,	11 006	[34]
B	36 080,	36 663,	35 255		(26) 9977,	10 002,	9767	[36]
8A	34 900,	35 560,	36 520	8	(28) 8369,	8458,	9045	[38]
B	35 159,	35 093,	35 471		(30) 9534,	9457,	9023	[40]

[a] Concentration of the antibody in the incubation tube. [b] This number corresponds to the incubation tube from which the sample was taken. 1A and 1B would be the labeled vials for the duplicate determinations. [c] The numbers in parentheses and brackets are the time in min at which the sample was added to the vial (parentheses) and the time at which the vial was counted (brackets).

temperature while the vial is still being shaken on the rotary shaker. This needs some practice, but is easily accomplished if one uses a Rainin EDP or similar micropipetor. An aliquot is removed from the appropriate tube and added to a vial every 2 min (See Table 1). After shaking for 10 min, the vial is removed from the shaker, 10 ml of ice-cold saline is added, and the vial counted as for the 0 time points. Every 2 min, another vial is removed until the assay is complete (Table 1).

Variables and controls

Precision is critical during the dilution steps. The concentration of cells in the antibody incubation and the volume added to the vial determine the number of cells allowed to adhere and counted at the 0 time point. Small variations in cell number will be compensated for in the reduction of the data (see below), but large variations may change the actual amount of adhesion due to changes in the number of collisions which can occur between cells. When cells are much above 50 000 per 0.5 ml, two cells often enter the counting aperture at the same time, resulting in a lower cell count.

To check for nonspecific inhibition one should include as a negative control, at least when developing the assay, an antibody which binds to an irrelevant surface molecule, or if this is not available, an antibody from the same species and class as the primary antibody. When a new lot of Fab fragments are used in the assay buffer, the ability of the preparation itself to inhibit adhesion should be tested by comparing adhesion in assay buffer with or without the Fab. We have found some preparations to have effects which can be reduced or eliminated by dialyzing the preparation against SPS. Other conditions which affect results are salt concentrations greater than 50 mM, pH greater than 8 and metabolic poisons such as azide in the incubations, all of which result in little or no adhesion.

47.2 Results

Table 1 shows what could be the typical results of an experiment in which eight dilutions of an antibody were tested for inhibition. For each antibody concentration four sets of data are recorded, the number of cells seen by the counter at time zero, triplicate counts of two aliquots, and the number seen at 10 min, triplicate counts of two aliquots. The order is useful to see early on whether things are working properly. The first data obtained are a measure of the strength of the overall adhesion (in the presence of no antibody), while the second set (highest antibody concentration) suggests that concentrations are within the inhibitory range. The results of the 0 min counts are fairly consistent and depend, as discussed above, upon the precision of the pipeting. However, in one case, tube 2 (500 µg/ml antibody) the counts are significantly lower than the rest. This often occurs at high antibody concentration because the divalent antibody is capable of strongly agglutinating the cells resulting in some aggregates even at time 0 after vortexing. One can generally tell that this is what is happening, and not just a smaller initial cell number, by looking at the 10 min counts. In this case the counts from tube 2 are significantly higher than the average 0 min counts for all of the samples. This occurs because, even though the cells are kept cold and vortexed just prior to counting at 0 min, untreated cells have a small residual adhesiveness which can only be blocked by very high levels of primary antibody in the presence of secondary Fab.

Further interpretation of the results requires reduction of the data to a useful form as shown in Table 2. In this table, the counts of the cells at 0 and 10 min have been averaged, rounded to the nearest hundreds, and last two digits dropped to simplify calculations. Percent adhesion is then determined by the formula $(100) (S_0 - S_{10})/S_0$, where S_0 and S_{10} are the number of single cells (converted as just indicated) at time 0

Table 2. Reduction of the data from Table 1

	Experiment No. 1273	Date 5/7/92

Title: Inhibition of *D. purpureum* adhesion by monoclonal antibody d-41.

Ab (µg/ml)	Percent Adhesion		Percent Inhibition	
0	$(100) (358–91)/358 =$	74.6	$(100) (74.6–74.6)/74.6 = 0$	
500	$(100) (332–380)/332 =$	–14.5	$(100) (74.6–(–15.4))/74.6 =$	119
250	$(100) (358–367)/358 =$	–2.5	$(100) (74.6–(–2.5) 0/74.6 =$	103
125	$(100) (353–343)/353 =$	2.8	$(100) (74.6–2.8)/74.6 =$	96.3
63	$(100) (366–324)/366 =$	11.5	$(100) (74.6–11.5)/74.6 =$	84.6
32	$(100) (353–222)/353 =$	37.1	$(100) (74.6–37.1)/74.6 =$	50.3
16	$(100) (357–102)/357 =$	71.4	$(100) (74.6–71.4)/74.6 =$	4.3
8	$(100) (355–90)/355 =$	74.7	$(100) (74.6–74.7)/74.6 =$	–0.1

and 10 min for each antibody concentration respectively. Finally percent inhibition is similarly determined using the formula $(100)(A_0 - A_{Ab})(A_0)$ where A_0 and A_{Ab} are the percent adhesion in the presence of no antibody (or control) and the given concentration respectively (Table 2).

47.3 Interpreting the Data

Figure 1 plots the data of Table 2. If the antibody can inhibit adhesion, the greater the concentration the less the adhesion, and, therefore, the greater the inhibition observed. Because adhesion is a property of the development and health of the cells, the maximum percent adhesion can vary from day to day. It is, therefore, generally easier to combine data from various days using the percent inhibition formula which normalizes the results. One should, however, be concerned when the maximum adhesion varies more than say 20% of normal.

The plots of the data from Table 2 do not take into account the anomalies caused by high antibody concentrations as discussed above and which result in adhesions of less than 0 and inhibitions greater than 100%. If the 0 time counts for the rest of the samples are close to each other (indicating little variability in pipeting), it is possible to recalculate the percent adhesion and inhibition using the counts at 10 min in 500 µg/ml antibody as S_0 for *all* tubes, essentially defining 100% inhibition as that found using the highest concentration of antibody. This gives the second set of graphs in Fig. 1. In practice, since one is generally more interested in the concentrations that cause partial inhibition, it is acceptable to consider any adhesion below 0 as 0 and resulting in 100% inhibition. One can see in Fig. 1 one that slopes and 50% values for both curves are very similar. For comparison purposes, an inhibitory titer

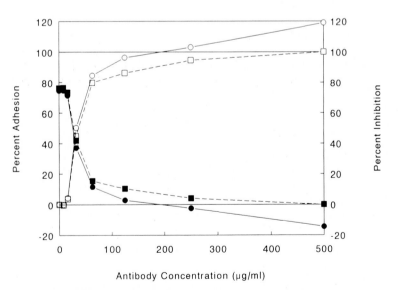

Fig 1. Graphical display of the assay data. The data from Table 2 (*circles*) or the data from Table 1 recalculated using S_{10} from the 500 µg/ml antibody determination as S_0 for all other determinations (*squares*) is plotted as percent adhesion (*closed symbols*) and percent inhibition (*open symbols*)

can be defined as the concentration resulting in 50% inhibition or if there are insufficient data, the highest concentration to give less than 80% inhibition.

47.4 Modifications

The assay conditions of shear stress as determined by rpm as well as, in the case of *Dictyostelium discoideum* and some other species, the addition of EDTA distinguishes more than one type of adhesion which may be inhibited by different specific antibodies. Therefore when working with a new species or cell types, attention to parametrics and assay conditions may be rewarded. The assay is also capable of determining the inhibitory activity of other materials such as proteins or carbohydrates in place of the antibodies.

47.5 Recommendations

This method was developed to screen large numbers of antibodies for inhibitory activity such as might be found during the preparation of monoclonal antibodies from hybridoma cultures. Culture medium can be used directly but must be diluted at least 5 fold to reduce the ionic strength to levels tolerated by the cells. When an antibody has been determined to inhibit adhesion as intact divalent immunoglobulin by this assay, it is best to determine if it is also capable of inhibiting as univalent Fab by incubating such a Fab preparation in place of the primary immunoglobulin. In this case the assay buffer need not contain univalent secondary antibody.

47.6 List of Suppliers

1. Secondary antibodies
   Fab fragments from goat antimouse IgG (# 55 477) and antirabbit IgG (# 55 640) can be obtained from Cappel Research Products, Organon Teknika, 100 Akzo Avenue, Durham, NC 27 704. The Fab are sold in 5 ml aliquots varying in protein concentration from lot to lot, but are always diluted as given in the procedures.
2. Electronic particle counter
   We use a Coulter Electronics (Hialeah, FL) Model ZBI counter.
3. Slime mold and bacterial cultures
   Cultures can be obtained from laboratories working with them or by purchasing them from the American Type Culture Collection, 12 301 Parklawn Drive, Rockville MD 20 852.
4. The 15 dram vials used in the assay can be purchased from most laboratory suppliers and are usually listed as "snap cap vials".

References

Gerisch G (1980) Univalent antibody fragments as tools for the analysis of cell interactions in *Dictyostelium*. In: Moscona AA, Monroy A (eds) Current topics in developmental biology. Academic Press New York pp 243–269

Loomis WF (ed) (1982) The development of *Dictyostelium discoideum*. Academic Press, New York

McDonough JP, Springer WR, Barondes SH (1980) Species-specific cell cohesion in cellular slime molds. Exp Cell Res 125: 1–14

Raper KB (1984) The dictyostelids. Princeton University Press Princeton

Springer WR, Barondes SH (1980) Cell adhesion molecules: detection with univalent second antibody. J Cell Biol 87: 703–707

Spudich JA (ed) (1987) *Dictyostelium discoideum*: molecular approaches to cell biology. Academic Press Orlando (Methods in cell biology, vol 28)

48 A Modified Capillary Method for Measuring Bacterial Chemotaxis

C. YU and S. ROSEMAN

The chemotactic systems in bacteria are responsible for the sensing of chemicals in the surrounding medium. They enable the bacterium to swim towards or away from a chemoeffector. The systems in *Escherichia coli* and *Salmonella typhymurium* have been intensively investigated. Chemoreceptor proteins are present in the periplasmic space and can be soluble or integral membrane components. The latter act directly as signal transducers, while the soluble protein receptor/ligand complexes interact with membrane-bound signal transducing proteins. In both cases, the intracellular signal is transmitted by the Che proteins to the flagellar motor. These signals alter the direction of rotation of the flagella, resulting in migration of the cells in a desired direction (Macnab 1987).

Several assays have been used for measuring the chemotactic response of bacteria to potential effectors. Commonly used methods are swarm plates and capillary assay. The swarm plate assay is a qualitative test for metabolizable chemoattractants (Adler 1976). Bacteria are inoculated in the center of a soft agar plate containing the compound to be tested. During cell growth, the chemoattractant around the colony is depleted, generating a concentration gradient. Cells that show positive responses to the attractant migrate outward, forming a "swarm ring" (a dense ring of cells at the perimeter of the growth zone).

In the capillary assay, a capillary tube filled with the potential chemoattractant is immersed into a cell suspension in liquid medium. In this case, a concentration gradient is formed at the mouth of the capillary tube. Cells showing a positive response will migrate toward the tip of the capillary and eventually swim into the tube, and the cell number in the capillary is then measured. This assay, although more tedious to perform, gives a quantitative determination of the chemotactic response. It can be used for testing nonmetabolizable attractants, since the concentration gradient is pregenerated, and when suitably modified can also be used to quantitate the response to repellants. The assay can also be used to test for the expression of a specific chemotactic response with cells grown under different conditions. In the following sections, we will describe in detail a simple, modified capillary assay for measuring chemotactic responses to different chemoattractants by the marine bacterium, *Vibrio furnissii* (Adler 1973; Bassler et al. 1991). The method has also been used with *E. coli*.

48.1 Procedure

1. *V. furnissii* is inoculated into buffered 50% artificial sea water (ASW) (Yu et al. 1991) containing 0.5% Bacto-Peptone and 0.1% yeast extract. The culture is shaken at 25 °C overnight.
2. The overnight culture (0.5 ml) is inoculated into 10 ml of labeling medium: buffered 50%-ASW containing 0.5% sodium lactate (pH 7.5) as carbon source and 50 μCi of [^{32}P]-Pi.

Preparation of Labeled Cells

3. The culture is shaken at 25 °C and the growth monitored by measuring absorbance at 540 nm.
4. Cells are harvested in early exponential growth phase, when the A_{540} is between 0.3–0.4.
5. Cells are collected by centrifuging at 3000g for 5 min at room temperature. The resulting loosely packed cell pellet is washed three times with equal volumes of buffered 50%-ASW. Washed cells are resuspended in buffered 50%-ASW and maintained at room temperature. The cell suspension should be used as quickly as possible.
6. The specific activity of the cells is determined from the following equation:

$$\frac{\text{Total dpm per ml of cell suspension}}{(A_{540})\,(5 \times 10^8 \text{ cell per ml per } A_{540})} = \text{dpm per cell}$$

Similar procedures can be used to label other species of bacteria; the growth medium is, of course, altered to meet the growth requirements of the cells. The growth medium should always contain a limiting chemical quantity of phosphate for efficient incorporation of [^{32}P]-Pi. Other labeled compounds, such as radiolabeled amino acids mixture, can be used, but are usually not as cost effective as [^{32}P]-Pi.

In *V. furnissii* and *E. coli*, little to no [^{32}P] is lost from the cells during storage or during the assay.

Capillary Assay

1. Solutions of chemoattractants to be tested are prepared in buffered 50%-ASW at final concentrations ranging from 500 mM to 1 µM.
2. Micropipets (5 µl, Fisher #21-164-2A) are filled with chemoattractants by dipping the tips into the solutions, which are drawn into the pipets by capillary action. The top of the pipet is then sealed in a flame, and the filled pipets are stored in the chemoattractant solution.
3. Aliquots of cell suspensions (150 µl), are dispensed into small test tubes (6 × 50 mm).
4. To initiate the assay, a micropipet filled with chemoattractant and wiped on the outside is placed vertically into the test tube with the open end of the pipet immersed in the cell suspension.
5. The pipet is held in position by a small piece of adhesive "clay" (Fun-Tak, Beecham Products, Dayton, OH) at the mouth of the test tube (Fig. 1).

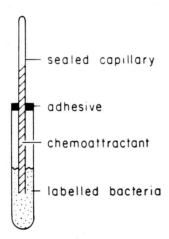

sealed capillary

adhesive

chemoattractant

labelled bacteria

Fig. 1. Modified capillary assay for measuring chemotaxis. A detailed description of the procedure is given in the experimental section

6. After incubation at 28–30 °C for 1 h, each micropipet is removed from the cell suspension, the outside of the pipet rinsed with H_2O, and wiped dry.
7. The contents of the pipet are then quantitatively expelled into vials containing liquid scintillation fluid, and radioactivity in the vial determined by counting in a Liquid Scintillation Spectrophotometer.
8. The number of bacterial cells in the pipet is determined from the specific activity.

The number of bacteria in the capillary tube can also be determined by one of several viable cell count methods if non-radiolabeled cells are used.

48.2 Results

Figure 2 gives the response of *V. furnissii* to various amino acids. The number of bacteria that entered capillary tubes containing only the buffer was also determined. The latter values represent the number of bacteria that migrated into the micropipet by chance during the incubation period, and is the background of the assay.

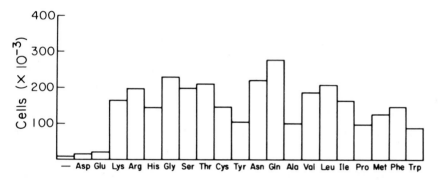

Fig. 2. Chemotaxis to amino acids. A single culture of *Vibrio furnissii* was grown in lactate-labeling medium, harvested, washed, and tested for chemotaxis to all 20 amino acids at 10 mM in the modified capillary assay. (−) indicates the number of bacteria that migrated into a capillary tube lacking amino acids

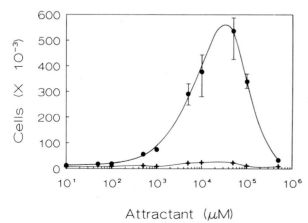

Fig. 3. Chemotactic response of *V. furnissii* to varying concentrations of N-acetylglucosamine. *V. furnissi* was grown in lactate-labeling medium in the presence (●) or absence (+) of 0.6 mM N-acetylglucosamine as inducer for the receptor. The assay was performed as described in the text

Figure 3, shows the tactic response of *V. furnissii* to N-acetylglucosamine. Lactate-grown cells exhibited low levels of taxis towards GlcNAc, whereas induction of the cells by the sugar elicited a 20-fold increase in response. The induced cells show a typical bell-shaped curve with increasing concentrations of the attractant in the capillary (Adler 1973). The threshold and optimal concentrations for taxis to GlcNAc under these experimental conditions were 10 µM and 30 mM respectively.

48.3 Remarks

The centrifugation conditions for collecting labeled cells must be determined empirically for each strain of bacterium. Minimal gravitational force and time of centrifugation should be used, so that the cell pellet can be resuspended with gentle pipeting. Excessive shearing of cells during resuspension will result in loss of flagellae and a reduced chemotactic response.

It is extremely important to keep the mixtures free of any vibration during the incubation period. During the assay, the bacteria respond to a steep gradient at the mouth of the micro-pipet, and disturbance of the gradient produces negative results.

Experiments are conducted in triplicate. Frequently, there is excellent agreement between replicate samples, but the error bars in Fig. 3 show the range of variation that can occur. In addition, there is quantitative variation from one cell preparation to another (same strain), which may reflect variations in growth conditions and washing steps. Thus, it is important to test any given variable, such as the effect of induction, in a single experiment.

References

Adler J (1973) A method of measuring chemotaxis and use of the method to determine optimum conditions for chemotaxis by *Escherichia coli*. J Gen Microbiol 74: 77–91

Adler J (1976) The sensing of chemicals by bacteria. Sci Am 234 (4): 40–47

Bassler BL, Gibbons PJ, Yu C, Roseman S (1991) Chitin utilization by marine bacteria: chemotaxis to chitin oligosaccharides by *Vibrio furnissii*. J Biol Chem 266: 24268–24275

Macnab RM (1987) Motility and chemotaxis. In: Nerdhardt FC, Ingraham JL, Low KB, Magasanik B, Schaechter M, Umbarger HE (eds) *Escherichia coli* and *Salmonella typhimurium*: cellular and molecular biology. American Society for Microbiology, Washington, D.C., Vol. 1, pp. 732–759

Yu C, Lee AM, Bassler BL, Roseman S (1991) Chitin utilization by marine bacteria: a physiological function for bacterial adhesion to immobilized carbohydrates. J Biol Chem 266: 24260–24267

Molecular Genetics of Lectins

J.B. MURPHY and M.E. ETZLER

The use of recombinant DNA technology is proving to be valuable for studies of lectin structure, biosynthesis, and function. At present, the complete primary structures of a number of plant and animal lectins have been deduced from the sequences of their corresponding mRNAs, and several laboratories are proceeding with site-directed mutagenesis studies in an effort to determine the relationship of lectin structure to function (for recent reviews see Etzler 1992; Lotan 1992). A variety of studies with transgenic organisms has been initiated to determine the factors regulating the biosynthesis of lectins and their intracellular targeting, and it is anticipated that future molecular genetic studies may greatly aid in elucidating the elusive function(s) of these molecules.

All of the above approaches are dependent on the ability to obtain lectin complementary or genomic DNA clones. In this chapter we provide a guide for choosing the type of DNA to clone, as well as procedures for preparing and screening the appropriate libraries. Since space limitations do not enable us to cover all approaches that can be used, we have selected those procedures that should be most useful for a novice in this area. We assume that the reader has a working knowledge of basic molecular biology techniques. Several comprehensive manuals are available to provide further assistance with these procedures (Maniatis et al. 1982; Ausubel et al. 1987–89).

49.1 General Considerations for Library Selection

Several factors must be considered in choosing between cloning lectin cDNA or genomic DNA, including the intended use of the clone, availability of starting material, and types of probes available for screening. Genomic clones are essential for studying lectin gene structure and regulation, but the possibility of introns may pose difficulties if sequences of these clones are used to deduce lectin primary structure. It is becoming increasingly apparent that plants as well as animals may contain multiple lectin genes that are differentially expressed, both spatially and temporally. A genomic library gives one the potential for isolating clones for all of these genes, whereas a cDNA library will yield clones of only those genes transcribed in the tissue that was the source of transcripts used for preparation of the cDNA.

Since cDNA libraries reflect the relative abundance of various transcripts in the source tissue, they are particularly useful in those cases in which lectin mRNA is prevalent. The frequency of cDNA for that lectin in the library is likely to be much higher than in a genomic library. Indeed, the first plant lectin cDNAs were isolated from cDNA libraries prepared from mRNA obtained from legume seeds in midmaturation, where the lectin mRNA is superabundant (for review see Etzler 1992).

The availability of probes is also a major consideration in choosing a library. If DNA clones of related lectins are available, these probes can be used to screen either cDNA or genomic libraries. If these nucleic acid probes are not available, one can construct oligonucleotides for use as probes of cDNA libraries or use antibodies specific for the lectin to screen expression cDNA libraries.

49.2 Preparation of cDNA for Library Construction

49.2.1 Preliminary Considerations

Intact mRNA is essential for the preparation of a useful cDNA library. It is therefore important to insure that all solutions and containers used in the RNA isolation are free of RNAse activity. This condition can be achieved by treating all reagents and containers used in the preparation with diethylpyrocarbonate, followed by the autoclaving of solutions or baking of glassware.

There are many different methods for preparing RNA that are described in detail in the literature (see Maniatis et al. 1982; Ausubel et al. 1987–89). The isolation of plant RNA, however, presents a particular problem because of the presence of starch and polyphenols that can interfere with the isolation procedure. We have found the most useful method for isolating plant RNA to be the method of Taylor and Powell (1982), which involves the use of hexadecyltrimethylammonium bromide (CTAB).

Poly(A) + RNA may be isolated from total RNA by chromatography on oligo dT cellulose (Aviv and Leder 1972). About 5 μg of mRNA is sufficient for the preparation of cDNA described below. If possible, this mRNA should be checked by Northern analysis using a labeled nucleotide probe for the lectin to assure the presence of lectin mRNA. The resulting band should be sharp and show no smearing, which would be indicative of RNA degradation. If the lectin transcript is present in very small amounts, it may be difficult to detect by Northern analysis. In such cases, the quality of the mRNA can be checked by using a probe for a transcript of an abundant protein, such as actin.

49.2.2 Preparation of cDNA

Materials   mRNA (at least 5 μg)

10 × first strand buffer: 500 mM Tris-HCl, pH 8.3, containing 300 mM KCl, 100 mM MgCl$_2$ and 4 mM dithiothreitol

10 × second strand buffer: 200 mM Tris-HCl, pH 7.4, containing 50 mM MgCl$_2$, 1 M KCl, 100 mM (NH$_4$)$_2$SO$_4$, and 1.5 mM β-NAD

Buffer A: 1 mM Tris, 0.01 mM EDTA, pH 7.6

Buffered phenol (prepared as described in Maniatis et al. 1982)

Chloroform-isoamylalcohol 24 : 1

dNTP: mixture containing 10 mM of each of the 4 deoxynucleotide triphosphates

α-^{32}P-dCTP: 3000 Ci/mmol

Oligo(dT)$_{12-18}$ or random hexanucleotide primers

Sephadex G-50 spun column (commercially available or prepared as described in Maniatis et al. 1982)

RNAse inhibitor

AMV reverse transcriptase

RNase H

E. coli DNA polymerase

T4 DNA polymerase

1. Heat 5 µg of mRNA at 65 °C for 10 min and cool on ice.
2. Mix in following order:
   10 µl 10 × first strand buffer
   5 µl dNTP
   H_2O to give a final volume of 100 µl
   1 µg of oligo(dT) or random hexanucleotide primers
   50 units of RNAse inhibitor
   RNA
   10 µCi α-^{32}P-dCTP
3. Add 20 units of AMV reverse transcriptase per µg of RNA and incubate at 42 °C for 90 min.
4. Remove 1 µl aliquot and determine percentage of incorporation of counts into TCA-precipitable material. This value can be used to calculate the amount of first strand DNA. The expected yield is 15–30%, i.e., approximately 1 µg of DNA per 5 µg of RNA.
5. Extract remainder of above sample with buffered phenol and then with chloroform–isoamylalcohol 24 : 1
6. Pass aqueous layer over a G50 spun column which has been washed with buffer A. Collect eluate and evaporate under vacuum to about 20 µl.
7. Add: 10 µl of 10 × second strand buffer
   10–20 µCi α-^{32}P-dCTP
   1 µl dNTP
   Adjust volume to 100 µl with H_2O
8. Add 1 unit of RNase H and 20 units of *E. coli* DNA polymerase. Incubate at 16 °C for 1 h and then at room temperature for 1 h.
9. Add 10 units of T4 DNA polymerase and dNTP to a final concentration of 50 µM. Incubate for 15 min at room temperature followed by 15 min at 37 °C.
10. Heat at 68 °C for 10 min to inactivate the enzymes. Analyze aliquots as above for incorporation of counts into TCA-precipitable material. The yield should be approximately 70–90% of the first strand yield.
11. Extract the sample with buffered phenol and chloroform-isoamyl alcohol and pass through a G50 spun column as described in steps 5 and 6 above.

Note. The cDNA obtained above may be checked for size and blunt ends by autoligating a small aliquot with T4 DNA ligase and comparing its mobility with the unligated DNA on an agarose gel. The DNA should be in the size range of 300–10 000 bp, with the majority being >1000 bp. The auto-ligated sample should have a significantly larger molecular weight than the unligated sample, indicating the ability of the cDNA to auto-ligate through its blunt ends.

49.2.3 Attachment of Adaptors

The cDNA prepared above must now be linked to adaptors to allow its insertion into an appropriate cloning vector. The choice of adaptors is dependent upon the restriction sites available for insertion of the DNA into the vector. Since blunt-ended *Not*I/*Eco*R1 adaptors are commonly used for insertion of DNA into lambda vectors, we describe the attachment of these adaptors to the cDNA.

Materials   cDNA (up to 1 µg)

Blunt-ended *Not*I/*Eco*R1 adaptors with a phosphorylated blunt end and a non-phosphorylated overhang (commercially available)

100 mM ATP, pH 7.0

10 × ligase buffer: 660 mM Tris-HCl, pH 7.5, containing 50 mM MgCl$_2$, 10 mM dithioerythritol and 10 mM ATP

Sepharose-400 spun column (commercially available or prepared as described in Maniatis et al. 1982)

T4 DNA ligase, specific activity of at least 5 units/µl

Polynucleotide kinase

Procedure   1. To the cDNA prepared above add:
   10 µl of 10 × ligase buffer
   0.05 OD$_{260}$ units of blunt-ended *Not*I/*Eco*R1 adaptors
   H$_2$O to give final volume of 100 µl after addition of ligase in step 2
2. Add 2–3 units of T4 DNA ligase.
3. Incubate 12–16 h at 12 °C and then heat at 68 °C for 10 min to inactivate the enzyme. Cool to room temperature.
4. Add 10 µl 100 mM ATP and nine units of polynucleotide kinase.
5. Incubate for 30 min at 37 °C. (This step replaces the phosphate groups on the *Eco* R1 end of the adaptors)
6. Remove unreacted adaptors and small (< 400 bp) DNA using a S-400 spun column. Collect eluate and precipitate with ethanol.

At this point the cDNA is ready to be ligated into an *Eco* R1 site in the linker region of an appropriate vector.

49.3 Preparation of Genomic DNA Fragments for Library Construction

49.3.1 Preliminary Considerations

Methods for isolation of DNA are described in detail in the literature (Maniatis et al. 1982; Ausubel et al. 1987–89). To avoid contamination with mitochondrial and chloroplast DNA, it is advisable to first isolate nuclei before proceeding with the DNA isolation (Jofuku and Goldberg 1988). The genomic DNA must then be fragmented and fractionated to obtain segments ranging in length from 15 to 22 kb. To insure an adequate representation of genes in the DNA library, it is essential that the fragments be generated by limited digestion with a restriction enzyme such as *Sau* 3AI that recognizes tetranucleotide sites found frequently and randomly throughout the genome.

49.3.2 Limited Digestion of DNA with *Sau* 3AI (Jofuku and Goldberg 1988)

Reagents   5 × reaction buffer: 30 mM Tris-HCl, pH 7.5, containing 250 mM NaCl, 30 mM MgCl$_2$ and 500 µg/ml bovine serum albumin

Sau 3AI

Enzyme storage buffer: 10 mM Tris-HCl, pH 7.5, containing 50 mM KCl, 0.1 mM
   Na_2EDTA, 1 mM dithiothreitol, 500 μg/ml bovine serum albumin and 50% (w/v)
   glycerol
0.5 M Na_2EDTA
0.5% Agarose
DNA markers in 10–25 kb size range
25 × TEA: 1 M Tris, 0.025 M Na_2EDTA, 0.5 M Na acetate, pH 8.2

To establish the time required for obtaining appropriate sized fragments, set up a   **Procedure**
test reaction as follows:

1. Dilute *Sau* 3AI to 1 unit/μl of enzyme storage buffer.
2. Add five units of diluted *Sau* 3AI to 5 μg of DNA in a total volume of 50 μl of 1 ×
   reaction buffer.
3. Incubate at 37 °C for various time points ranging from 90 s to 1 h.
4. At each of the above time points, remove 10 μl of the reaction mixture and add to
   1 μl of 0.5 M Na_2EDTA to inactivate the enzyme. Store on ice until samples from
   all time points have been collected.
5. Mix each of the above samples with 2 μl of gel loading dye and load onto a 0.5%
   agarose gel. Carry out electrophoresis in 1 × TEA buffer for 16 h at 25 V.
6. Choose a set of conditions from the above trial that provides a maximum number
   of DNA fragments in the size range of 12–22 kb.

After establishing the reaction conditions, scale up the reaction ten fold to
generate DNA fragments for use in the next step. After addition of the 0.5 M
Na_2EDTA, the fragments can be stored at 4 °C until proceeding with the fractionation procedure below.

49.3.3 Fractionation by Sucrose Gradient Centrifugation
(Jofuku and Goldberg 1988)

2.5 × salts: 2.5 M NaCl, 12.5 mM Na_2EDTA, 25 mM Tris-HCl, pH 7.5   **Materials**
68% Sucrose, filtered and autoclaved
10% Sucrose in 1 × salts
40% Sucrose in 1 × salts
3 M Potassium acetate, pH 6
TNE: 10 mM Tris-HCl, pH 7.5, containing 5 mM NaCl and 0.1 mM Na_2EDTA
Isopropanol

1. Prepare linear 10–40% sucrose gradients in 17 ml polyallomer tubes.   **Procedure**
2. Load the DNA fragments directly onto the gradients in less than 1 ml total
   volume.
3. Centrifuge at 27 000 rpm for 24 h at 4 °C.
4. Collect approximately 200 μl (ten drop) fractions by drop fractionation.
5. Analyze 10 μl aliquots of the fractions by electrophoresis on 0.5% agarose as
   described above.
6. Pool those fractions that contain *Sau* 3AI fragments between 12–22 kb in length.
7. Dialyze against TNE buffer.
8. Add 1/10 volume of 3 M potassium acetate and an equal volume of isopropanol.
9. Precipitate DNA overnight at − 20 °C.

10. Centrifuge at 20 000 rpm for 45 min at 0 °C.
11. Wash the DNA with cold 70% ethanol.
12. Centrifuge at 20 000 rpm for 45 min at 0 °C.
13. Dry the pellet in vacuo.
14. Resuspend the DNA in 50–100 µl TNE and store at 4 °C.

49.4 cDNA and Genomic Library Preparation

49.4.1 Preliminary Considerations

A variety of vectors is now commercially available for cDNA and genomic library construction. Most of these vectors are based on the bacteriophage lambda due to its high cloning efficiency. Two standard vectors commonly used for cDNA library construction are λgt10 for libraries to be screened with nucleic acid probes and the λgt11 expression vector for libraries that are to be screened with antibodies (Young and Davis 1983). Another insertion vector, λZAP, constructed by Short (1988), allows the construction of a more versatile cDNA library that can be screened either by nucleic acid probes or by antibodies. This vector can accept up to 10 kb of DNA at six unique cloning sites that are located in the N-terminal region of the *lacZ* gene. It is also engineered to enable in vivo excision of the cloned DNA and its rescue as plasmid subclones thereby eliminating the need for subcloning.

Genomic libraries require replacement vectors that can accommodate large inserts ranging from 9 to 23 kb. The λEMBL vectors (Frischauf et al. 1983) meet this requirement and are commonly used for genomic cloning.

Kits containing lambda vectors and packaging reagents are commercially available from such companies as Amersham (Arlington Heights, IL), Boehringer Mannheim Corp. (Indianapolis, IN), Pharmacia Biotechnology International AB (Uppsala, Sweden) and Stratagene (La Jolla, CA). Since each of these kits comes with a set of specific instructions, we provide only a general description that is applicable for the use of any of the above lambda vectors.

Materials

cDNA or genomic DNA prepared as above
Lambda vector of choice
Packaging mix for lambda vector of choice
Host bacteria for lambda vector of choice
SM buffer: 50 mM Tris-HCl, pH 7.5, containing 100 mM NaCl, 8 mM $MgSO_4$ and 0.01% gelatin
LB medium: 10 g Bactotryptone, 5 g yeast extract, 5 g NaCl in 1 l H_2O, adjusted to pH 7.5 with 1 N NaOH.
Agar plates (150 mm diameter) prepared with LB medium: 20 g of Bactoagar per liter of LB medium
   Note: Depending upon the host strain used, these plates may be overlaid with a top layer of 0.8% agarose plus or minus antibiotics, maltose and $MgSO_4$. See manufacturer's directions for exact composition of medium
IPTG: 100 mM Isopropyl β-D-thiogalactopyranoside in H_2O (store at − 20 °C)
X-Gal: 5-Bromo-4-chloro-3-indolyl-β-D-galactopyranoside, 250 mg/ml in dimethyl-formamide (store at − 20 °C)

1. Ligate cDNA or genomic fragments into appropriate vector arms according to manufacturer's directions.

2. Package an aliquot of the above ligation mix according to manufacturer's directions and serially dilute the resulting phage suspension.

3. Infect appropriate host cells with the above dilutions of the phage suspension according to the manufacturer's instructions.

4. Add top agarose containing IPTG, X-Gal and the appropriate antibiotics, $MgSO_4$ and maltose necessary for the host cells.

5. Plate above cells on LB agar plates and incubate at 37 °C overnight.

6. Determine the efficiency of packaging with the various dilutions of phage by counting plaques. Also determine the ratio of white to blue plaques, which is a measure of the quality of the library. A good library should contain > 90% white plaques; these white plaques signify that the phage contain inserts.

7. Package the remainder of the ligation mixture and mix the infected cells with 0.8 % agarose containing the appropriate antibiotics, $MgSO_4$ or maltose required for the host cells. Plate this mixture on 150 mm diameter plates containing LB agar using a dilution determined by the test packaging above to give 5×10^4 pfu per plate.

8. The plates are now ready to be screened as described below.

Procedure

49.5 Screening Libraries with DNA Probes

49.5.1 Preliminary Considerations

Three types of DNA probes that can be used for library screening are: (1) DNA fragments from a related lectin gene; (2) synthetic oligonucleotide probes constructed on the basis of lectin amino acid sequence information; or (3) DNA probes constructed by polymerase chain reaction (PCR) using DNA primers designed on the basis of nucleic acid or amino acid sequence information. We recommend the use of the first type of probe whenever possible; this choice minimizes problems of degeneracy and artifacts that can be encountered with the other two types of probes and also enables the preparation of labeled probe with high specific activity. If synthetic oligonucleotide probes are used, it is important to choose a portion of amino acid sequence for use in probe construction that enables the synthesis of a probe with as little degeneracy as possible. It is also advisable to make the probe at least 20 nucleotides long and as rich in G/C as possible, particularly at the 5' and 3' ends.

DNA fragments to be used as probes can be labeled by random primer labelling according to the procedure of Feinberg and Vogelstein (1984). We have found that specific activities of 10^8–10^9 cpm/μg DNA can be obtained if ^{32}P-dATP with a specific activity of 800 Ci/mmol is used in this procedure.

Since synthetic oligonucleotides constructed by conventional oligonucleotide synthesizers have no 5' phosphate groups, they may be labeled directly using γ-^{32}P-ATP and polynucleotide kinase according to the procedure of Maxam and Gilbert (1980). Specific activities achieved by the 5' end-labeling procedure are lower than those achieved by the random-primer labeling method used for DNA fragments.

49.5.2 Making Replicate Filters

Materials Nitrocellulose filters, sterile, 137 mm diameter
Whatman 3 MM paper
0.5 M NaOH containing 1.5 M NaCl
1 M Tris-HCl, pH 7.0, containing 1.5 M NaCl
20 × SSC: 3 M NaCl in 0.3 M sodium citrate
Labeled DNA probe

Procedure 1. Incubate plates with top agarose containing phage and host cells for 6–12 h at 37 °C until plaques are large but not confluent.
2. Store plates at 4 °C for 1 h to prevent agarose from being removed during filter lift below.
3. Place sterile nitrocellulose filters on plates for 10 min and label carefully with four asymmetric dots to establish orientation of filter with plate.
4. Carefully remove filters and transfer to Whatman 3 MM paper soaked with 0.5 M NaOH, 1.5 M NaCl for 2 min.
5. Neutralize filters by transferring to Whatman 3 MM paper soaked with 1 M Tris-HCl, 1.5 M NaCl, pH 7.0 for 2 min.
6. Wash filters in 3 × SSC.
7. Dry filters at room temperature and then bake in a vacuum oven at 80 °C for 90 min.
8. Hybridize filters with DNA as described below.

49.5.3 Screening Filters Using a DNA Fragment

Materials Whatman 3 MM paper
Deionized formamide: add 20 g AG501-X8D resin (Bio-Rad, Richmond, CA) to 500 ml of formamide and stir in hood for 1 h. Filter through Whatman number 1 filter paper and store at − 20 °C
50 × Denhardt's: 1% Ficoll, 1% bovine serum albumin, 1% polyvinylpyrrolidone, 0.02% NaN_3 in H_2O. Store at 4 °C.
Denatured herring sperm DNA: 10 mg/ml in H_2O. Sonicate to dissolve and boil for 10 min. Store at − 20 °C
Poly(A) : 10 mg polyadenylic acid/ml H_2O. Store at − 20 °C
Pre-hybridization buffer: 5 × SSC, 0.01% NaPPi, 50 mM Na phosphate (pH 7.2), 5 × Denhardt's, 0.1% SDS, 200 µg/ml herring sperm DNA, 50 µg/ml poly (A) and 50% deionized formamide. Dissolve the NaPPi in SSC, $NaPO_4$, Denhardt's and H_2O. Add the formamide, followed by the SDS and poly (A). Boil the herring sperm DNA for 5 min, cool on ice and add
Hybridization buffer: make similar to pre-hybridization buffer above, but mix poly(A), herring sperm DNA and labeled DNA fragment (approximately 1 × 10^6 cpm per ml of hybridization solution) and boil for 2 min before adding to solution.
SM buffer: see above

Procedure 1. Wet the filters in 5 × SSC for 5 min and then add to freshly prepared pre-hybridization buffer heated to 42 °C. Use 6–7 ml solution per filter. Incubate with shaking at 42 °C for > 4 h

2. Transfer the filters to hybridization buffer pre-heated to 42 °C and hybridize overnight. The volume of buffer should be approximately 1 ml/filter
3. Wash the filter four to five times in 500 ml of 2 × SSC, 0.05% SDS, 0.01% NaPPi for 15 min at room temperature until the wash is close to background cpm
4. Wash the filters for 1 h at 55 °C (or higher temperature dependent on the probe size and its G/C content) in above wash buffer containing 0.1 × SCC
5. Blot filters on Whatman 3 MM paper. Do not dry completely. Wrap the filters in Saran Wrap and place on X-ray film. Mark film corresponding to marks placed on filters under Procedure, 49.5.2, step 3. These marks will enable the identification of positive clones below. Expose X-ray film at − 80 °C overnight
6. Using marks placed on film above, line up film with original marked plate from Procedure, 49.5.2, step 3. Select positive clones on agar plates corresponding to the exposed spots on film. Using a sterile Pasteur pipet remove an agar plug corresponding to each clone and place each into a tube containing SM buffer
7. Make serial dilutions of each of the above phage, and plate and screen as above until a pure plaque is obtained. This phage is a lectin DNA clone and can be stored at 4 °C in SM containing a few drops of chloroform

49.5.4 Screening Filters Using an Oligonucleotide Probe

Whatman 3 MM paper
Yeast tRNA: 11 mg/ml H$_2$O. Store at − 20 °C   **Materials**
Pre-hybridization buffer: 6 × SSC, 0.05% NaPPi, 1 × Denhardt's, 0.5% SDS and 100 µg/ml herring sperm DNA. Boil the herring sperm DNA 5 min and cool on ice before adding
Hybridization buffer: 6 × SSC, 0.05% NaPPi, 1 × Denhardt's and 20 µg/ml yeast tRNA. Add labelled oligonucleotide probe so solution will contain 1 × 10^6 cpm per ml.
SM buffer: see above.

Note: With oligonucleotide probes, the temperature of hybridization may depend   **Procedure**
on the probe length and its G/C content. See Ausubel et al. 1987–1989 for determining the appropriate temperature for the probe to be used

1. Wet the filters in 5 × SSC and transfer to pre-hybridization buffer heated to 37 °C. Incubate with gentle shaking at 37 °C for 2 h.
2. Add filters to hybridization buffer and incubate at 37 °C (or other appropriate temperature) overnight.
3. Wash filters in 500 ml of 0.05% NaPPi, 6 × SSC four to five times until the wash buffer is close to background cpm.
4. Wash filters at 50 °C (or the appropriate temperature for the size of the probe and its G/C content).
5. Blot filters on Whatman 3 MM paper. Do not dry completely. Wrap the filters in Saran Wrap and place on X-ray film. Mark film corresponding to marks placed on filters under Procedure, 49.5.2, step 3. These marks will enable the identification of positive clones below. Expose X-ray film at − 80 °C overnight.
6. Using marks placed on film above, line up film with original marked plate from Procedure, 49.5.2, step 3. Select positive clones on agar plates corresponding to the exposed spots on film. Using a sterile Pasteur pipet, remove an agar plug corresponding to each clone and place each into a tube containing SM buffer.

7. Make serial dilutions of each of the above phage, and plate and screen as above until a pure plaque is obtained. This phage is a lectin DNA clone and can be stored at 4 °C in SM containing a few drops of chloroform.

49.6 Screening Expression Libraries with Antibodies

Materials

Nitrocellulose filters 137 mm in diameter

$10 \times$ TBS: 0.5 M Tris-HCl, pH 7.4, containing 1 M NaCl

Blotto: a solution containing 50 g dry milk, 167 µl 30% antifoam A and 1 ml of 1 mg/ml merthiolate in 1 l of $1 \times$ TBS

Primary antibody solution: appropriate dilution in Blotto of antiserum or antibodies specific for the gene product of interest. Note. An $(NH_4)_2SO_4$ precipitated fraction of the host cell extract should also be added to this solution to adsorb any antibodies that may be directed to host cell proteins. This fraction is prepared by lysing 1 l of cells grown overnight with chloroform and precipitating with 60% $(NH_4)_2SO_4$. Dissolve the precipitate in 20 ml of $0.5 \times$ TBS and add to antibody solution up to 2–10% by volume

Secondary antibody solution: appropriate dilution of goat anti-rabbit IgG conjugated to alkaline phosphatase (commercially available) in Blotto

IPTG: 10 mM isopropyl β-D-thiogalactoside in H_2O (store at -20 °C)

BCIP: 50 mg/ml 5-bromo-4-chloro-3-indolyl phosphate in H_2O (store at -20 °C)

NBT: 50 mg/ml nitroblue tetrazolium in 70% dimethylformamide (store at 4 °C)

Color reagent: Just before use add 33 µl BCIP and 66 µl NBT to 10 ml 100 mM Tris-HCl, pH 9.5, containing 100 mM NaCl and 5 mM $MgCl_2$

SM buffer: see above

Procedure

1. Incubate plates containing infected host cells at 37 °C or 43 °C for 3–4 h to establish plaque formation prior to induction of the β-galactosidase fusion protein.
2. Soak nitrocellulose filters in sterile 10 mM IPTG and dry.
3. Place dried filters of top of plates containing phage and incubate 3 h at 37 °C to induce expression of the β-galactosidase fusion protein.
4. Carefully mark filters and plates with four asymmetric dots so they can be aligned at a subsequent step.
5. Remove filters and wash three times for 5 min each wash in $1 \times$ TBS.
6. Wash filters 15 min in Blotto to block nonspecific binding of antibodies to filters.
7. Incubate filters overnight at room temperature in a solution containing an appropriate dilution of the primary antibodies.
8. Wash filters three times in $1 \times$ TBS for 5 min.
9. Wash filters 15 min in Blotto.
10. Incubate filters for 2 h at room temperature in secondary antibody solution.
11. Wash filters three times in $1 \times$ TBS for 5 min.
12. Develop filters in color reagent.
13. Using marks placed on filters in step 4, select positive clones on agar plates corresponding to the colored spots on filter. Using a sterile Pasteur pipet, remove an agar plug corresponding to each clone and place each into a tube containing SM buffer.

14. Make serial dilutions of each of the above phage and plate and screen as above until a pure plaque is obtained. This phage is a lectin DNA clone and can be stored at 4 °C in SM containing a few drops of chloroform.

Acknowledgement. The authors thank Dr. John Harada for his helpful suggestions in the preparation of this chapter.

References

Ausubel FM, Brent R, Kingston RE, Moore DD, Seidman JG, Smith JA, Struhl K (1987–89) Current protocols in molecular biology. Vols 1, 2, John Wiley & Sons, New York

Aviv H, Leder P (1972) Purification of biologically active globin messenger RNA by chromatography on oligothymidylic acid-cellulose. Proc Natl Acad Sci USA 69: 1408–1412

Etzler ME (1992) Plant lectins: molecular biology, synthesis and function. In: Allen HJ, Kisailus EC (eds) Glycoconjugates: composition, structure and function, Marcel Dekker, Inc., New York, pp 531–539

Feinberg AP, Vogelstein B (1984) A technique for radiolabelling DNA restriction endonuclease fragments to high specific activity. Anal Biochem 137: 266–267

Frischauf A-M, Lehrach H, Poustka A, Murray N (1983) Lambda replacement vectors carrying polylinker sequences. J Mol Biol 170: 827–842

Jofuku KD, Goldberg RB (1988) Analysis of plant gene structure. In: Shaw CH (ed) Plant molecular biology: a practical approach, IRL Press, Oxford, pp 37–66

Lotan R (1992) β-Galactoside-binding vertebrate lectins: synthesis, molecular biology, function. In: Allen HJ, Kisailus EC (eds) Glycoconjugates: composition, structure and function, Marcel Dekker, Inc., New York, pp 635–671

Maniatis T, Fritsch EF, Sambrook J (1982) Molecular cloning. A laboratory manual, Cold Spring Harbor Laboratory, New York

Maxam AM, Gilbert W (1980) Sequencing end-labelled DNA with base-specific chemical cleavages. Meth Enzymol 65: 499–560

Short JM (1988) λ ZAP: a bacteriophage λ expression vector with in vivo excision properties. Nucl Acids Res 16: 7583–7600

Taylor B, Powell A (1982) Isolation of plant DNA and RNA. Focus 4: 4–6

Young RA, Davis RW (1983) Efficient isolation of genes by using antibody probes. Proc Natl Acad Sci USA 80: 1194–1198

50 Cell-Free Synthesis of Lectins

E. Van Damme and W. Peumans

In the early days of molecular biology, mRNA purification and in vitro translation techniques played an important role in the study of gene expression and its regulation. Moreover, these techniques made an important contribution to the development of modern recombinant DNA technology which nowadays is a most important tool in many areas of biological and biomedical research. At present, the purification of intact mRNA and the characterization of its translation products is still of primordial importance not only for a study of the regulation of gene expression but also for the isolation of the genes themselves. For instance, molecular cloning of an eukaryotic gene usually starts from a cDNA library constructed from an mRNA preparation. To make sure that this mRNA preparation contains the relevant messenger, its translation product has to be identified among the poly-peptides synthesized in a cell-free protein synthesizing system under direction of the mRNA preparation. Since there has been a steadily increasing interest during the last few years in both the isolation of plant lectin genes and the study of their expression, the preparation of high quality mRNA from lectin-containing plant materials and its subsequent translation in a cell-free protein synthesizing system has become an important step in the molecular approach to phytohemagglutinins. Although it is difficult to describe a simple and efficient method for the extraction of RNA, given the diversity of materials to start with, the aim of this contribution is to outline a general procedure which provided that some alterations are made, is in principle applicable to all plant tissues. Once a total RNA preparation has been obtained standard procedures can be followed for the isolation of mRNA and its translation in a cell-free system. Similarly, standard procedures can be used for the detection of the in vitro synthesized lectin polypeptides unless the carbohydrate binding activity and specificity of the lectin in question can be exploited in a straightforward affinity chromatography purification step.

50.1 RNA Isolation

50.1.1 Materials and Equipment

Recommendations When working with RNA, special care should be taken to avoid ribonuclease contamination of glassware and solutions, since minute amounts of this enzyme will be sufficient to destroy mRNA activity. Hence, extreme cleanliness should be observed in the working area to limit the risk of RNase contamination. The problem of endogenous ribonuclease activity in the biological material also has to be limited and controlled as will be discussed below in the footnotes to Protocol 1.

Glassware. RNase contamination on glassware and plasticware can be destroyed by baking in an oven (150 °C, 4 h) or autoclaving (121 °C, 20 min). Alternatively, equipment which cannot be heat-treated can be submerged in a solution of diethylpyrocarbonate (0.5%) overnight and subsequently rinsed extensively with sterile water to remove any residual diethylpyrocarbonate.

Solutions. All solutions should be prepared using the purest reagents available. If RNA preparations are to be performed on a regular basis, it is advisable to reserve one set of chemicals and solutions exclusively for the RNA work. RNase in solutions can be destroyed by treatment with diethylpyrocarbonate (0.2%, final concentration) for 12 h. Before use the diethylpyrocarbonate has to be removed from the solution by heating (100 °C for 15 min) or autoclaving since diethylpyrocarbonate can react with nucleic acids. Alternatively, solutions can be sterilized by passage through a membrane filter (0.22 μm pore size) or by autoclaving. This method is very suitable for Tris buffers which cannot be treated with diethylpyrocarbonate as the reagent reacts with primary amines.

50.1.2 Extraction of Total RNA

Several methods for the preparation of total plant RNA have been reported during the last few years. The procedure using phenol as protein denaturing agent in particular has proven to be very successful for plant material from different sources, and will be discussed below in detail. Alternatively, a protocol based on the use of high salt solutions, e.g., lithium chloride or guanidinium chloride can be applied (Draper and Scott 1988). Whatever procedure is used it may be necessary to adjust the protocol depending on the plant tissue (and its protein and polysaccharide content) to ensure maximal efficiency in RNA extraction.

Reagents and Equipment

Homogenization buffer: 100 mM Tris pH 9.0, 5 mM EDTA pH 8.0, 100 mM NaCl, 1% β-mercaptoethanol (to be added just before use)
TE buffer: 10 mM Tris pH 8.0, 1 mM EDTA pH 8.0
Phenol containing 0.1% 8-hydroxyquinoline equilibrated with 0.2 M Tris pH 8.8
Chloroform: isoamylalcohol (24/1)
Ethanol
10% SDS
4 M lithium chloride
3 M sodium acetate pH 5.2
5 M sodium chloride
Mortar and pestle
Refrigerated centrifuge
Cooled tubes

Protocol 1. *Extraction of Total RNA*

Protocol

1. Freeze the plant material as soon as possible after excision in liquid nitrogen. Store at − 80 °C until use[a].

---

[a]Since most plant tissues contain cells with thick cell walls and, in addition, are usually reinforced by very tough structures like, e.g., fibers, they are difficult to break up. Therefore it is recommended to freeze the tissue in liquid nitrogen or solid carbon dioxide before grinding. If possible, use only young plant materials for RNA extraction as they contain more RNA and less RNases.

2. Grind the plant material using a precooled (− 80 °C) mortar and pestle until a fine powder is obtained. Suspend the powder in 20 ml/g fresh weight tissue cold homogenization buffer[b].

3. Centrifuge the nuclei and cellular debris in a precooled centrifuge (1 min, 3000 rpm).

4. Add SDS to a final concentration of 0.5% and stir[c].

5. Add per 10 ml of extract obtained in step 4, 10 ml of phenol preheated to 60 °C. Stir for 15 min.

6. Add 10 ml of chloroform: isoamylalcohol (24/1)[d]. Stir for 15 min.

7. Separate the phases by centrifugation for 20 min at 4000 rpm (0 °C).

8. Remove the aqueous phase, avoiding any precipitated protein at the interface. Repeat the phenol/chloroform extractions until no interface is obtained[e].

9. Precipitate all nucleic acids from the aqueous phase by addition of 0.1 volume of 3 M sodium acetate and 2.5 volumes of cold ethanol. Leave at − 80 °C for at least 30 min or overnight at − 20 °C.

10. Centrifuge the nucleic acids in precooled glass tubes at 4000 rpm for 20 min (0 °C).

11. Wash the pellet subsequently using 70% ethanol, 100% ethanol, and 100% ethanol: ether (1/1).

12. Dry the pellet and dissolve it in a small volume of TE buffer[f].

13. Add an equal volume of 4 M lithium chloride[g]. Allow the solution to stand at − 20 °C overnight.

14. Spin down the RNA for 20 min at 4000 rpm (0 °C).

15. Subsequently wash the RNA pellet with 70% ethanol, 100% ethanol, and 100% ethanol: ether (1/1).

16. Dry the pellet and dissolve the RNA in water or TE buffer. Keep the RNA on ice at all times. Store at − 80 °C.

50.1.3 Preparation of Poly(A)-Rich RNA

Eukaryotic messenger RNA (mRNA) can be separated from ribosomal RNA (rRNA) and transfer RNA (tRNA) in a total RNA preparation by affinity chromatography on oligodT-cellulose (Siflow et al. 1979). Under high salt conditions and neutral pH the poly(A) tail on the mRNA will hybridize with the oligodT-cellulose whereas rRNA and tRNA will run through the column. The poly(A)-rich RNA can be eluted using low salt buffers.

---

[b]Homogenization is performed at low temperature and high pH to reduce RNase activity. The volume of homogenization buffer can be adapted depending on the water content of the tissue. If large amounts of plant tissue are to be extracted, a precooled Waring blender can also be used.

[c]If SDS is added to the homogenization buffer in step 2, all nuclei will be disrupted and their DNA content will be released into the solution. This will often result in a very viscous supernatant which is difficult to handle.

[d]Isoamylalcohol is added to the chloroform to minimize foaming.

[e]The interface can be extracted a second time with the homogenization buffer containing SDS to recover any RNA trapped in the protein fraction. Usually three to four phenol/chloroform extractions have to be performed to get rid of all proteins.

[f]The pellet can be dried under vacuum or by heating in a water bath.

[g]Only the RNA will precipitate in 2 M lithium chloride, the DNA will remain in the solution.

OligodT-cellulose
Running buffer: 10 mM Tris pH 7.6, 1 mM EDTA, 0.5 M sodium chloride
Loading buffer: 10 mM Tris pH 7.6, 1 mM EDTA, 0.1% SDS
Elution buffer: 10 mM Tris pH 7.6, 1 mM EDTA
5 M Sodium chloride
1 M Magnesium chloride
3 M Sodium acetate pH 5.2
0.1 N Sodium hydroxide
Ethanol

Materials
and Solutions

Protocol 2: *Preparation of Poly(A)-Rich RNA*

1. Suspend the oligodT-cellulose in running buffer, transfer it to a small column and allow the cellulose to settle down. Wash the cellulose extensively with running buffer[a].

Protocol

2. Dissolve the total RNA (Protocol 1, step 16) in a small volume of loading buffer[b].
3. Heat the sample at 70 °C for 5 min, add 0.1 volume of 5 M sodium chloride, and cool quickly to room temperature.
4. Load the RNA on the oligodT-cellulose column[c].
5. Extensively wash the column with running buffer to wash out any residual rRNA[d].
6. Elute the poly(A) RNA using 1 ml aliquots of elution buffer. Collect each fraction in a sterile test tube which is kept on ice. Determine the concentration of each fraction by reading its A_{260}[e].
7. Pool the fractions containing poly(A) RNA and precipitate the RNA by addition of 0.1 volume of 3 M sodium acetate, 2 volumes of cold ethanol and 0.001 volume of 1 M magnesium chloride. Allow the RNA to precipitate at − 20 °C overnight.
8. Collect the poly(A) RNA by centrifugation (4000 rpm, 20 min, 0 °C), rinse the pellet as described in protocol 1, dry it and dissolve it in sterile water. Store at − 80 °C[f].
9. The oligodT-cellulose column can be stored at 4 °C under 0.1 N sodium hydroxide.

50.2 In Vitro Translation

Although commercial wheat germs have been used for more than 20 years as a readily available and inexpensive source of efficient cell-free protein synthesizing

---

[a]Small particles should be removed from the oligodT-cellulose with a Pasteur pipet.
[b]The addition of SDS in the loading buffer is a useful precaution against ribonuclease activity. Do not cool the RNA sample below room temperature as the SDS will precipitate.
[c]Do not allow the column to run too fast as the poly(A) RNA will not bind. If desirable the RNA sample can be passed over the column a second time.
[d]The column is washed until the A_{260} of the effluent is zero.
[e]The yield and purity of an RNA preparation can be assessed by measuring the absorbance of a diluted sample. One A_{260} unit is equivalent to 37 µg/ml RNA, Clean RNA should have a ratio $A_{260} : A_{280}$ of 2.0 and $A_{235} : A_{260}$ less than 0.5.
[f]If desirable, a second affinity chromatography on oligodT-cellulose can be performed. To disrupt any remaining mRNA–rRNA complexes, a treatment with DMSO is recommended: The RNA pellet is dissolved in a one volume of 10 mM Tris pH 7.6. Add nine volumes of DMSO and one volume of loading buffer, and heat the sample at 70 °C for 2 min. Add 90 volumes of running buffer and pass the sample over the column as in steps 2–8.

systems, attempts to prepare extracts from raw wheat germs often result in totally inactive preparations (Roberts and Paterson 1973; Marcu and Dudock 1974; Carlier and Peumans 1976). Before discussing the possible reasons for the lack of activity of the extracts, it should be emphasized that wheat germs are not a well-defined biological material but rather refer to a particular fraction of wheat obtained during the milling process. Although this fraction contains predominantly germs (which correspond to the embryos of the kernels and appear as light-yellow platelets) it is always contaminated with white flour particles (which are derived from the endosperm of the kernels and contain mainly starch) and small pieces of bran (which appear as brown platelets and originate from both the seed coat and the aleurone layer). Since the milling process itself and the properties (such as water content, protein/starch content) of the wheat to be processed determine the composition of the raw wheat germ fraction commercial wheat germs are a highly variable biological material. Besides that, wheat germ preparations rapidly lose their potential activity upon storage. The reason for their instability is that commercial germs are squeezed between rollers during the milling process which leads to an extensive damage of the cells of the embryo tissues and hence makes them extremely susceptible to oxidative reactions. When stored at room temperature most of the activity of the germs is lost within a few hours. Even at $-20\,°C$ there is a rapid loss of the potential activity of the wheat germs; since the oxidative reactions cannot be inhibited by lowering the storage temperature (Carlier and Peumans 1976). It is imperative, therefore, to collect the wheat germs and prepare the extracts immediately after processing of the wheat in the mill. However, even when fresh germs are used, the resulting extracts are usually either poorly active or totally inactive. The most likely explanation for the lack of activity follows from detailed studies of the effect of nonembryo materials on the activity of cell-free extracts prepared from isolated wheat embryos. It has been shown, in fact, that especially endosperm fragments (which are abundantly present in commercial wheat germ preparations) are deleterious for the activity of the embryo extracts (Peumans et al. 1980). Although it was suggested that endosperm RNases and proteases might be responsible for the observed negative effect, it is evident now that particularly the ribosome-inhibiting proteins present in this tissue make the extracts inactive (Stirpe and Barbieri 1986).

To avoid all possible problems which are inherent to the use of commercial wheat germs, it is advisable to use isolated wheat embryos, since they are a most reliable source of highly active cell-free protein synthesizing extracts. Moreover, since usually only limited amounts of material are required, the isolation of the embryos is much less time-consuming than the testing of several batches of commercial germs.

50.2.1 Preparation of the Cell-Free System

Materials and Solutions

Wheat
RNase free glassware
Sephadex G25 (medium)
Centrifuge
Mortar and pestle
Whatmann 3 MM filter paper
DB buffer: 20 mM Hepes-KOH pH 7.8, 120 mM K-acetate, 2 mM Mg-acetate, 6 mM
β-mercaptoethanol

Solutions necessary for cell free translation system, see Table 1
Trichloroacetic acid (TCA)

Wheat grains used for the isolation of embryos should be of good quality in terms of viability and germination vigor. In case there is any doubt about the quality of the wheat, a simple germination test can give a fairly good indication whether a particular batch is suitable or not. To do so, 100 kernels are put in a petri dish on moistened filter paper and allowed to germinate at 25 °C in the dark. Normally, over 90% of the grains develop roots and shoots in less than 3 days. If not, the viability of the wheat is too low and the embryos are not suited for the preparation of a cell-free protein synthesizing system, since a major lesion associated with ribosome-bound initiation factors occurs in their protein synthesizing machinery.

The tenuously attached embryos can be removed from the wheat kernels by a short (about 3–5 s) blending in a coffee mill. Since a prolonged blending results in a more drastic disintegration of the grains the blending time should be kept as short as possible. To isolate the embryos from the mixture, the size fraction between 0.7 and 1.5 mm is sieved out and the bran is removed by gentle blowing. Thereafter, the embryos (which appear as light yellow comma-shaped particles about 1 mm in diameter and 2–3 mm in length) can be selected out either manually or by flotation on a mixture of cyclohexane/carbontetrachloride (1/2; v/v). On the latter mixture, the embryos flotate while the endosperm fragments settle at the bottom of the vessel (Johnston and Stern 1957). The embryos can be collected by decanting the mixture of organic solvents on a sieve and should be (air)-dried completely. When not used immediately, the embryos should be stored at −20 °C under which conditions they can be kept for years without any loss of activity. Depending on the wheat variety and the efficiency of the removal of the embryos from the kernels the total yield of embryos varies between 1 and 4 g per kg of wheat.

The preparation of cell-free extracts from wheat embryos is easy and only requires currently used laboratory equipment (Protocol 3). Moreover, since wheat embryo extracts are fairly stable, all operations can be done at room temperature.

Protocol 3: *Preparation of Wheat Embryo Extracts*

1. Homogenize wheat embryos (100 mg) in 0.5 ml of DB using a mortar and pestle.
2. Centrifuge the homogenate for 2 min at 10 000g. Remove the supernatant (about 0.3 ml) with a Pasteur pipet, carefully avoiding the upper fatty layer.
3. Layer the supernatant on top of a column of Sephadex G25 (medium) built up in a Pasteur pipet (7.0 × 0.5 cm) and equilibrated with DB. Elute the column under gravitational force and collect the peak of the excluded fraction (about 0.4 ml) in a tube which is kept on ice. Since the excluded fraction moves as an opaque, light brown band whereas the fraction containing the low molecular weight compounds displays an intense yellow color (because of the presence of flavonoids), the separation on the Sephadex G25 column can be followed visually.
4. Keep the excluded fraction on ice for 10 min and centrifuge for 4 min at 10 000g. Take off the supernatant, freeze in small in liquid nitrogen and store at −80 °C until use. For optimal activity in the cell-free translation mixture (cf. Table 1) the A_{260} of the final extract should be between 40 and 100.

50.2.2 Incubation Conditions

The incubation conditions described below have been optimized for the translation of polyA RNA from different plant sources. Although some authors reported that the optimal conditions may vary as a function of the mRNA, and hence have to be determined for each mRNA preparation, it is our experience that all plant mRNAs (including viral RNAs) are efficiently translated under exactly the same conditions. Only those mRNA preparations which are contaminated with other RNAs, DNA, polysaccharides, or proteins may require slightly different conditions. However, since the translation efficiency of the wheat embryo cell-free system is strongly inhibited by any of these contaminants, only a very low overall translation activity will be obtained, irrespective of the conditions used.

50.2.2.1 Preparation of the Incubation Mixture

Preperation   When done in a final volume of 500 µl, the amino acid incorporation mixture contains all components as indicated in Table 1.

Since a wheat embryo cell-free system prepared as described above contains some translatable endogenous mRNAs the background activity of the system can be fairly high. Although it is possible to remove or inactivate most of the endogenous messengers in the wheat embryo extract (see also below, Sect. 50.2.2.3), a strong reduction of the endogenous template activity can be achieved by a simple pre-incubation of the cell-free translation system. Therefore, the incubation mixture is prepared as described but prior to adding the exogenous mRNA and the labeled amino acid, it is incubated for 20–30 min at 25 °C. Once the RNA and the labeled amino acid are added the mixture is incubated for 90 min at 25 °C.

50.2.2.2 Measurement of the Total Amino Acid Incorporation

Measurement   Before analyzing the in vitro synthesized products, it is necessary to estimate the total amount of incorporated radioactive amino acid to make sure that both the extract and the added RNA are active. The extent of amino acid incorporation is

Table 1. Composition of the in vitro translation system

Stock solution	Added volume	Final concentration
80 mM creatine phosphate	10.0 µl	8 mM
250 µg/ml creatine kinase	2.5 µl	25 µg/ml
0.2 mM GTP	5.0 µl	20 µM
10 mM ATP	10.0 µl	1 mM
240 mM Hepes-KOH pH 8.0	25.0 µl	12 mM
200 mM dithiothreitol	5.0 µl	2 mM
800-24 mM KOAc-MgOAc	25.0 µl	40–1.2 mM
40 mM spermidine	5.0 µl	0.4 mM
19 amino acids (1 mM each)	50.0 µl	0.1 mM
Cell-free extract	200.0 µl	
Water	132.5 µl	
1 mg/ml RNA	15 µl	30 µg/ml
5 mCi/ml labeled amino acid	5 µl	50 µCi/ml

easily assessed by determining the TCA-insoluble radioactivity of small aliquots using the filter paper disk method.

Protocol 4. *Determination of Total Amino Acid Incorporation in Wheat Embryo Extracts by the Filter Paper Disk Method*

1. Pipet three aliquots of 5 μl each on small pieces (0.5 × 0.5 cm) of filterpaper (Whatman 3 MM) (numbered with a pencil) and dry on a hot plate (60 °C).   **Protocol**
2. Place the discs (all together) in a beaker of ice-cold 10% TCA containing 1 mM unlabeled methionine (in case radioactively labeled methionine was added to the cell-free translation mixture) and wash them with gentle swirling for 10 min.
3. Transfer the discs into a beaker of hot TCA (90 °C) and incubate for 15 min to hydrolyze charged tRNA.
4. Wash the disks successively in absolute ethanol, ethanol/diethylether (50/50; v/v) and diethylether, each for about 1 min.
5. Dry the disks on a hot plate.
6. Determine the radioactivity in a liquid scintillator using a toluene-based scintillation medium.

50.2.2.3 Reduction of the Endogenous Template Activity of the Wheat Embryo Cell-Free System

One of the main advantages of the wheat embryo cell-free system over other eukaryotic systems is its low endogenous activity. If the background activity is too high, several methods can be used to reduce the endogenous template activity of the wheat embryo cell-free system. Most of the endogenous mRNPs (95%) can be removed by adding 15 mM magnesium or calcium in the homogenization medium. In the presence of high concentrations of these divalent cations, the endogenous mRNPs form large aggregates which sediment during subsequent centrifugation of the homogenate (Peumans et al. 1978). Combined with a preincubation of the cell-free system prior to the addition of mRNA and labeled amino acid (cf. Sect. 50.2.2.1), this method results in a virtually complete abolishment of the endogenous activity. Alternatively, the endogenous messengers can be destroyed by treatment of the wheat embryo extracts with the Ca-dependent *Staphylococcus* nuclease, as has been described for the rabbit reticulocyte lysate (Pelham and Jackson 1976).

Protocol 5. Treatment of Wheat Embryo Extracts with Ca-Dependent *Staphylococcus aureus* Nuclease

1. Add 1 M CaCl$_2$ to the wheat embryo extract (cf. step 4, Protocol 3) to give a final   **Protocol**
   concentration of 1 mM.
2. Add *Staphyloccocus aureus* nuclease to 20 units/ml and incubate for 15 min at 20 °C.
3. Add 0.1 M EGTA (pH 7.5) to give a final concentration of 2 mM.
4. Freeze the nuclease-treated extract in liquid nitrogen and store at − 80 °C.

50.2.3 Product Identification

Once the RNA has been translated successfully, the translation product(s) of interest has to be indentified. If specific antibodies are available, they can be used to identify and quantify the translation products. Although in principle several immunological techniques can be used, immunoprecipitation is usually the most suitable method. In case lectins have to be identified amongst the translation products, affinity chromatography on immobilized sugars or glycoproteins can be an alternative technique, at least if the in vitro synthesized lectins display their specific sugar binding activity.

Immuno-precipitation   Since only minute quantities of proteins are synthesized in the cell-free system, direct precipitation of the relevant polypeptides with primary antibodies is virtually impossible. Therefore, immunoprecipitation is usually done in two steps. In the first step, primary antibodies are added to the incubation mixture after cell-free protein synthesis and allowed to react with the antigen. Subsequently, the antigen/primary antibody complexes are precipitated in a second step by (1) the addition of secondary antibodies raised against the primary antibodies or (2) by coprecipitation with carrier antigen and excess antibody. Alternatively, the complexes can be absorbed on Protein A-Sepharose. The precipitates formed during the second step are washed thoroughly to remove aspecific contaminations in the immunoprecipitates. Finally, the radioactivity in the precipitate is quantified (by liquid scintillation counting) and analyzed by SDS-PAGE (Laemmli 1970) and fluorography (Bonner and Laskey 1974). A generally applicable immunoprecipitation procedure is depicted in Protocol 6.

Protocol 6. *Immunoprecipitation of the in Vitro Synthesized Translation Products*

Protocol   Since wheat embryo extracts contain an N-acetylglucosamine-binding lectin which binds to several rabbit serum proteins (including the immunoglobulins) it is imperative to include 0.1 M N-acetylglucosamine in all solutions used throughout the immunoprecipitation procedure.

1. Dilute the reaction mixture immediately upon completion of the cell-free translation with five volumes of washing buffer containing 0.15 M NaCl, 20 mM Tris-HCl (pH 7.6), 1% Triton X100 and 0.1 M N-acetylglucosamine.
2. Add an appropriate volume of primary antiserum and incubate the mixture for 2–3 h at 25 °C, allowing the formation of antigen-antibody complexes.
3. Once the formation of antigen-primary antibody complexes is completed add: (1) Either an appropriate volume of secondary antibodies or carrier antigen and excess (primary) antibodies to immunoprecipitate the labeled antigen-antibody complexes or (2) Protein A-Sepharose to absorb the labeled antigen-antibody complexes. It is important to note that the amounts of carrier antigen, primary and secondary antibodies and Protein A-Sepharose must be determined empirically.
4. Incubate the mixtures under constant shaking, preferably in round-bottomed plastic tubes for 1 h at 25 °C.
5. Transfer the mixtures to microcentrifuge tubes, cool on ice for 1 h and spin at 12 000g for 2 min. Take off and discard the supernatant.
6. Wash the precipitate at least three times by resuspending it in washing buffer and recentrifugation.
7. Resuspend the pellet in washing buffer without detergent and centrifuge. Remove the supernatant and dissolve the final precipitate in an appropriate volume of sample buffer suitable for subsequent SDS-PAGE.

Once the immunoprecipitate has been obtained, the radioactivity can be determined by scintillation counting. In addition, the purity of the final product can be checked by SDS-PAGE and fluorography using standard procedures.

Affinity Chromatography

Lectin polypeptides synthesized in a wheat embryo cell-free system are usually properly folded and exhibit their normal carbohydrate-binding activity (Peumans et al. 1982; Van Damme et al. 1991). Therefore, affinity chromatography on the appropriate immobilized sugar or glycoprotein can be used as a highly specific purification technique. A detailed purification scheme for the in vitro synthesized mannose-binding lectin from snowdrop (*Galanthus nivalis*) is described in Protocol 7.

Protocol 7. *Affinity Chromatography of In Vitro Synthesized Snowdrop Lectin*

Protocol

1. After completion of the cell-free translation dilute the reaction mixture with five volumes of 0.5 M ammonium sulphate containing 20 mM Na-acetate (pH 6.5).
2. Load the mixture on a small column (0.2 ml bed volume) of mannose-Sepharose 4B built up in a Pasteur pipet.
3. Wash the column with 2 ml of 0.5 M ammonium sulphate containing 20 mM Na-acetate (pH 6.5).
4. Elute the lectin with 1 ml of 0.2 M mannose in 20 mM Na-acetate (pH 6.5).
5. Dialyze the eluted lectin against distilled water (for about 1 h) and freeze dry immediately.
6. Dissolve the freeze-dried lectin in a sample buffer suitable for SDS-PAGE.

The radioactivity in the affinity purified lectin can be determined by scintillation counting. In addition, its purity can be checked by SDS-PAGE and fluorography using standard procedures.

References

Bonner WM, Laskey RA (1974) A film detection method for ^{3}H-labelled proteins and nucleic acids in polyacrylamide gels. Eur J Biochem 46: 83–86

Carlier AR, Peumans WJ (1976) The rye embryo system as an alternative to the wheat system for protein synthesis in vitro. Biochim Biophys Acta 447: 436–444

Draper J, Scott R (1988) The isolation of plant nucleic acids. In: Draper J, Scott R, Armitage P, Walden R (eds) Plant genetic transformation and gene expression, Blackwell Scientific Publications, Oxford, p199–236

Johnston FB, Stern H (1957) Mass isolation of viable wheat embryos. Nature 179: 170–171

Laemmli UK (1970) Cleavage of structural proteins during the assembly of the head of bacteriophage T4. Nature (London) 227: 680–685

Marcu K, Dudock B (1974) Characterization of a highly efficient protein synthesizing system derived from commercial wheat germ. Nucl Acid Res 1: 1385–1397

Pelham AR, Jackson RJ (1976) An efficient mRNA-dependent translation system from reticulocyte lysates. Eur J Biochem 67: 247–256

Peumans WJ, Carlier AR, Caers LI (1978) Sedimentation properties of preformed messenger particles in dry rye embryo extracts. Planta 140: 171–176

Peumans WJ, Carlier AR, Caers LI (1980) Botanical aspects of cell-free protein synthesizing systems from cereal embryos. Planta 147: 307–311

Peumans WJ, Stinissen HM, Tierens M, Carlier AR (1982) In vitro synthesis of lectins in cell-free extracts from dry wheat and rye embryos. Plant Cell Rep. 1: 212–214

Roberts BE, Paterson BM (1973) Efficient translation of tobacco mosaic virus RNA and rabbit globin 9 S RNA in a cell-free system from commercial wheat germ. Proc Natl Acad Sci USA 70: 2330–2334

Siflow CD, Hammett JR, Key JL (1979) Sequence complexity of polyadenylated ribonucleic acid from soybean suspension culture cells. Biochemistry 18: 2725–2731

Stripe F, Barbieri L (1986) Ribosome-inactivating proteins up to date. FEBS Lett. 195: 1–8

Van Damme EJM, Kaku H, Perini F, Goldstein IJ, Peeters B, Yagi F, Decock B, Peumans WJ (1991) Biosynthesis, primary structure and molecular cloning of snowdrop (*Galanthus nivalis* L.) lectin. Eur J Biochem 202: 23–30

51 cDNA Cloning and Expression of Plant Lectins from the Legume Family

K. Yamamoto, Y. Konami, K. Kusui-Kawashima, T. Osawa, and T. Irimura

The primary purpose of this chapter is to outline our strategy for the cDNA cloning of leguminous lectins and for the expression of cDNA in *Escherichia coli*. In general, cDNA cloning of specific polypeptides is achieved through either information on polypeptide sequences, specific antibodies against polypeptides, cDNA-transfection-mediated changes of cellular functions, or sequence homology to related polypeptides. We obtained partial polypeptide sequences of *Bauhinia purpurea* lectin, generated oligonucleotide probes based on these sequences, and screened a cDNA library prepared from poly(A) + mRNA extracted from germinating seeds using these oligonucleotide probes. We used fundamental techniques widely used for cDNA cloning of a variety of polypeptides. These techniques are described in manuals and textbooks, such as Maniatis et al. (1989), and we do not include the details.

51.1 Flow Chart

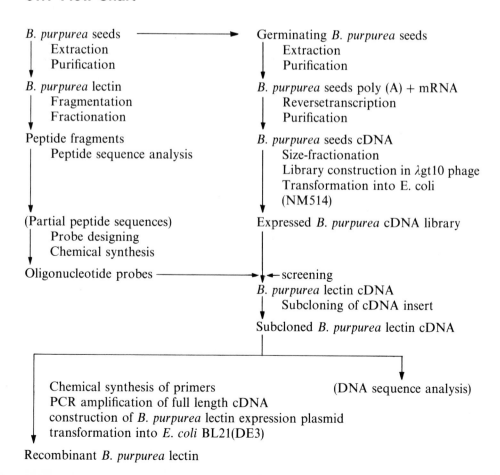

51.2 Materials

Materials
B. purpurea seeds are purchased from F. W. Schumacher (Sandwich, MA). Restriction enzymes and sequencing-grade proteases are purchased from Boehringer-Mannheim or Takara Shuzo (Kyoto). Custom-made oligonucleotides probes are supplied from Milligen (Tokyo) and further purified on a reversed phase HPLC equipped with C_{18} column (μBondasphere). mRNA purification kit from Pharmacia and cDNA library construction kit with λgt10 phage from Amersham are used. Subcloning plasmid pT7T3 18U, expression plasmid pET-8c, E. coli strains NM514, XL-1, and BL21(DE3) are also used.

HPLC having a capacity of gradient elution equipped with C_{18} reversed phase column (μBondasphere) is used to fractionate peptide fragments. Polypeptide sequence analysis is performed with a gas-phase protein sequencer. Standard apparatus for the molecular biology experiments, including a temperature regulator suitable for the polymerase-chain-reaction, is necessary.

51.3 Experimental Procedure

Peptide Sequence Analysis
Accurate information on partial polypeptide sequence is essential, although complete sequence is not necessary. One may refer our original articles on the peptide sequencing of plant lectins (Konami et al. 1990, 1990a,b). The isolation and partial polypeptide sequence analysis of the B. purpurea lectin has been described (Kusui et al. 1991). We have determined the N-terminal sequence of the B. purpurea lectin as TSSTLTGFTFPNFWSNTQEXGTEIIF. The N-termini of 9 Asp-N-fragments of the B. purpurea lectin are DNGRXYQLSH, DVPHITA, DTWPNTEWS, DPSKNQVVAVEF, DYGGXLGLFXYKTAT, DGFAFLAP, DLPKILPER-VRIGFSGGT, DGI, and DATSKII. Trypsin fragments of the B. purpurea lectin have N-terminal sequences IGFSGGTGFXETQYILSWSF, SIDVPHITAD, LTYD, LAPVDSSVK, HYQL, and YPHIGINVXSTVSVA. FPNFW and WPNTEW are chosen to prepare oligonucleotide probes for the screening of B. purpurea lectin cDNA because of their minimum degeneracy among all the sequence data we obtained.

Poly(A)+mRNA Isolation
B. purpurea seeds are moisturized and maintained in a warm ($\sim 37\,°C$) and moist place in the dark. Total RNA is isolated from germinated B. purpurea seeds (~ 30 g) by the phenol-sodium dodecyl sulfate method (Watanabe and Price 1982). Although guanidium/cesium chloride extraction is widely used for the isolation of total cellular RNA from mammalian cells, the phenol-sodium dodecyl sulfate method appears to be suitable to the RNA preparations from plant cells, especially from plant seeds. Otherwise, contaminating polysaccharides and lipids make the further subfractionation of RNA difficult. Poly (A) + mRNA is prepared by oligo-dT-cellulose column chromatography.

cDNA Preparation
Poly (A) + mRNA (5.0 μg) is used to construct a double-stranded cDNA by the method of Gubler and Hoffman (1983) using cDNA Synthesis System Plus (RPN 1256) from Amersham.

A cDNA library is prepared from the double-stranded cDNA (1.0 μg) into λgt10. All cloning steps are carried out according to the cDNA cloning system (Amersham), except that cDNA inserts are size-fractionated (greater than 500 base pairs) by electrophoresis in 1.0% agarose gel containing ethidium bromide prior to ligation. The prepared cDNA library is used to transform *E. coli* NM514.

Generation of cDNA Library

E. coli transformants in the library are propagated on L-broth plates. The amplified library is transferred to nylon membrane filters (Pall, Glen Cove, NY) according to Benton and Davis (1977). The replicated filters are hybridized with two synthetic DNA mixed probes (5′CCA [A/G] AA [A/G] TTNGG [A/G] AA3′ and 5′CCA [C/T] TCNGT [A/G] TTNGGCCAA3′), which represent all possible DNA sequences predicted from the amino acid sequence. The sequence of the first probe has been deduced from the N-terminal sequence and that of the second probe has been deduced from the sequence of a Asp-N fragment. Theoretically, N-terminal sequence is desirable in order to obtain full length cDNA. Hybridization and washings are carried out as described by Maniatis et al. (1989) at $(Tm - 4)\,°C$. Tm is calculated according to an equation $Tm = 4(G + C) + 2(A + T)$, where $G, C, A,$ and T indicate the number of corresponding nucleotides in the oligonucleotide probes. Following hybridization, the filters are exposed to Kodak XAR-5 film at $-80\,°C$. A cDNA library constructed from *B. purpurea* mRNA (4.0×10^5 pfu) is screened by hybridization with these synthetic mixed probes. In our previous experiment, two clones (BP1 and BP2) which gave strong hybridization signals with the two probes were detected from the library (2.0×10^4 pfu).

cDNA Cloning

Southern Blot Hybridization and Subcloning of cDNA Inserts. Clones positively hybridized with the above two probes are isolated and phages collected according to the standard procedures (Maniatis et al. 1989). After cDNA inserts are separated by EcoRI digestion, Southern blotting analysis (1975) is carried out using these two DNA probes. The cDNA inserts are analyzed in 1% agarose gel containing ethidium bromide. After electrophoresis, the cDNA inserts are transferred to Hybond-N and hybridization is carried out by the method of Southern. In our previous experiment, BP1 and BP2 were apparently identical and contained inserts of an approximate size of 1200 base pairs.

Hybridization and Subcloning

The cDNA inserts are ligated with the vector, pT7T3 18U (Pharmacia LKB Biotechnology, Uppsala, Sweden) using T4 DNA ligase, and then transformed into competent *E. coli* XL-1 cells. Recombinant plasmids are isolated by the alkaline lysis method (Birnboim and Doly 1979). Single-strand plasmid DNAs are prepared according to the method of Vieira and Messing (1987). Several deletion mutants suitable for sequencing are generated by using a deletion kit (Takara Shuzo, Kyoto). Then the DNAs are sequenced by the dideoxy chain termination method (Sanger et al. 1977). In our previous experiment, the DNA sequence analysis revealed that the cDNA insert comprised 1152 nucleotides of mRNA with a 30-nucleotide 3′ polyadenylic acid tail. The cDNA included an 870-nucleotide coding region with a single initiation codon at positions 17–19 (Fig. 1).

Construction of the Expression Plasmid in E. coli and Production of Recombinant B. purpurea Lectin. Various methods are applicable to construct prokaryotic expression vectors. In our previous experiment, a fragment of *B. purpurea* lectin cDNA, extending from nucleotide 101–889 of the sequence of clone BP1, and flanked

Expression Plasmid

```
           10        20        30        40        50        60        70        80        90
ATCATAATTCACAACCATGCTTCTCTACAACTCAAAATCCTATGTTCTTCAACTTATCTTCATAACTTTGTTGCTCACCCAACTTAACAA
                 M  L  L  Y  N  S  K  S  Y  V  L  Q  L  I  F  I  T  L  L  L  T  Q  L  N  K

          100       110       120       130       140       150       160       170       180
GGTGAAGTCAACAAGCTCAACCTTAACAGGCTTCACCTTCCCCAATTTCTGGTCAAATACCCAAGAAAATGGTACAGAAATAATCTTCCT
  V  K  S  T  S  S  T  L  T  G  F  T  F  P  N  F  W  S  N  T  Q  E  N  G  T  E  I  I  F  L
             1

          190       200       210       220       230       240       250       260       270
AGGCAATGCCACTTATACTCCTGGGGCTCTACGCCTTACCAGGATTGGTGAAGATGGCATCCCTCTGAAGAGCAACGCAGGCCAAGCTTC
  G  N  A  T  Y  T  P  G  A  L  R  L  T  R  I  G  E  D  G  I  P  L  K  S  N  A  G  Q  A  S

          260       290       300       310       320       330       340       350       360
ATATTCTCGCCCTGTGTTCCTTTGGGAGAGCACAGGCCATGTAGCAAGCTTTTACACTTCCTTCTCCTTTATTGTGAGAAGCATTGATGT
  Y  S  R  P  V  F  L  W  D  S  T  G  H  V  A  S  F  Y  T  S  F  S  F  I  V  R  S  I  D  V

          370       380       390       400       410       420       430       440       450
TCCACATATTACATCTGATGGCTTTGCCTTCTTTCTTGCACCCGTGGATTCTAGCGTCAAAGATTATGGAGGATGCCTGGGACTTTTCAG
  P  H  I  T  A  D  G  F  A  F  F  L  A  P  V  D  S  S  V  K  D  Y  G  G  C  L  G  L  F  R

          460       470       480       490       500       510       520       530       540
ATATAAAACTGCTACTGACCCATCAAAGAATCAAGTTGTTGCTGTTGAATTTGACACTTGGCCAAATACCGAATGGAGTGACCTACGTTA
  Y  K  T  A  T  D  P  S  K  N  Q  V  V  A  V  E  F  D  T  W  P  N  T  E  W  S  D  L  R  Y

          550       560       570       580       590       600       610       620       630
TCCACATATTGGAATAAATGTTAACTCCACTGTCTCCGTCGCAACTACGAGATGGGACAACGATGATGCCTATGTAACAAAATCGACAGC
  P  H  I  G  I  N  V  N  S  T  V  S  V  A  T  T  R  W  D  N  D  D  A  Y  V  T  K  S  T  A

          640       650       660       670       680       690       700       710       720
CCACATAACCTATGATGCCACATCCAAAATAATAACTGTTCTTTTAACTTATGATAATGGTAGACATTATCAACTATCTCATGTGGTGGA
  H  I  T  Y  D  A  T  S  K  I  I  T  V  L  L  T  Y  D  N  G  R  H  Y  Q  L  S  H  V  V  D

          730       740       750       760       770       780       790       800       810
TTTGCCAAAGATTCTTCCAGAACGGGTTAGAATTGGCTTCTCCGGGGGCACTGGATTTAATGAAACACAATATATTCTCTCTTGGAGCTT
  L  P  K  I  L  P  E  R  V  R  I  G  F  S  G  G  T  G  F  N  E  T  Q  Y  I  L  S  W  S  F

          820       830       840       850       860       870       880       890       900
CACTTCAACGTTGAATAGCACCAAAATCAGTGCCTTGACTCAGAAGTTAAGGTCCTCGGCCTCTTATTCCAGTATGTAAACTCTTATCTA
  T  S  T  L  N  S  T  K  I  S  A  L  T  Q  K  L  R  S  S  A  S  Y  S  S  M  *

          910       920       930       940       950       960       970       980       990
AATAAGGTACAACCAAGCCAGCCAATGTGGTTGGATATGTCCTGTGTCGTATGGACTATGTTGTGTTCTAGTATGTGATCTCTTTTCTAA

         1000      1010      1020      1030      1040      1050      1060      1070      108
ATAAGGATTCTATAATCCATGGTCGTGTTACGATTACTTCAGTGGTGGCCATCTAGTCAACTATTGTAGTTGGATAGGTGAAAATACTGT

         1090      1100      1110      1120      1130      1140      1150
TGTATTTTATGGATGCACTGTGTTGAAGTTTATAATGCTATTAAAAAAAAAAAAAAAAAAAAAAAAAAAAAAAAA
```

Fig. 1. The nucleotide sequence and the deduced amino acid sequence of the *B. purpurea* lectin cDNA clone. The nucleotides are listed in the 5′ to 3′ direction. The termination site is indicated by an *asterisk*. The signal sequence is indicated by the *double solid line* and the amino acid sequences obtained from protein sequence analyses by the *thin solid lines*

by artificial sites introduced for NcoI and BamHI digestion, was amplified by means of the polymerase chain reaction (Perkin-Elmer Cetus Gene Amp Kit) using the primers 5'ACCATGGCCACAAGCTCAACCTTA3' (containing the amino terminal sequence) and 5'GATCCTGTTACATACTGGAATAAGAG3' (antisense sequence containing the carboxyl terminal sequence and stop codon). After digestion with NcoI and BamHI, the DNA fragment generated by polymerase chain reaction was inserted between the NcoI and BamHI sites of pET-8c plasmid. This expression vector contains T7 phage promoter and ampicillin resistant gene.

The constructed plasmid is introduced into *E. coli* strain BL21 (DE3) cells grown on LB plates containing 50 μg/ml ampicillin. BL21 (DE3) cells containing pET-8c/BPA plasmid are transferred to a plastic tube with L-broth and grown to the mid log phase at 37 °C. Two hours after the addition of isopropyl-D-thiogalactoside at a concentration of 1 mM, the *E. coli* cells are collected by centrifugation. In our previous experiments the recombinant *B. purpurea* lectin expressed in *E. coli* forms insoluble aggregates. We have tried to use several solubilization methods without success. However, binding of ^{125}I-labeled Gal-BSA to insoluble aggregates of the recombinant *B. purpurea* lectin was easy to confirm. The binding was blocked by the presence of 100-fold excess of unlabeled Gal-BSA.

References

Benton WD, Davis RW (1977) Screening λgt recombinant clones by hybridization to single plaques in situ Science 196: 180–182.

Birnboim HC, Doly J (1979) A rapid alkaline extraction procedure for screening recombinant plasmid DNA. Nucleic Acids Res 7: 1513–1523.

Gubler U, Hoffman BJ (1983) A simple and very efficient method for generating cDNA libraries. Gene 25: 263–269

Konami Y, Yamamoto K, Osawa T (1990) The primary structure of the *Lotus tetragonolobus* seed lectin. FEBS Lett 268: 281–286

Konami Y, Yamamoto K, Osawa T (1991a) The primary structures of two types of the *Ulex europeus* seed lectin. J Biochem 109: 650–658.

Konami Y, Yamamoto K, Osawa T (1991b) Purification and characterization of two types of *Cytisus sessilifolius* anti-H(O) lectins by affinity chromatography. Biol Chem Hoppe-Seyler 372: 103–111

Konami Y, Yamamoto K, Osawa T (1991c) Purification and characterization of new type lactose-binding *Ulex europeus* lectin by affinity chromatography. Biol Chem Hoppe-Seyler 372: 95–102

Konami Y, Yamamoto K, Toyoshima S, Osawa T (1991d) The primary structure of *Labrunum alpinum* seed lectin. FEBS Lett 286: 33–38

Konami Y, Yamamoto K, Osawa T, Irimura T (1992a) Correlation between carbohydrate-binding specificity and amino acid sequence of carbohydrate-binding regions of *Cytisus*-type anti-H(O) lectins. FEBS Lett 304: 129–135

Konami Y, Yamamoto K, Osawa T, Irimura T (1992b) The primary structure of the *Cytisus scoparius* seed lectin and its carbohydrate-binding peptide. J Biochem 112: 366–375

Kusui K, Yamamoto K, Konami Y, Osawa T (1991) cDNA cloning and in vitro expression of *Bauhinia purpurea* lectin. J Biochem 109: 899–903

Maniatis T, Fritsh EF, Sambrook J (1989) Construction and analysis of cDNA libraries. Molecular cloning (2nd ed.) Cold Spring Harbor Laboratory, Cold Spring Harbor, NY

Sanger F, Nicklen S, Coulson AR (1977) DNA sequencing with chain-terminating inhibitors. Proc Natl Acad Sci USA 74: 5463–5467

Southern EM (1975) Detection of specific sequences among DNA fragments separated by gel electrophoresis. J Mol Biol 98: 503–517

Vieira J, Messing J (1987) Production of single-stranded plasmid DNA. Methods Enzymol 153: 3–11

Watanabe A, Price CA (1982) Translation of mRNAs for subunits of chloroplast coupling factor I in spinach. Proc Natl Acad Sci USA 79: 6304–6308

52 Production of Intact Recombinant Lectins in *Escherichia coli*

J. HIRABAYASHI, Y. SAKAKURA, and K. KASAI

Production of recombinant proteins in bacterial cells may be the most promising approach when biological constraints limit the quantity of target proteins from natural sources. They can also be expressed in eukaryote cells, if appropriate conditions are fulfilled. However, the prokaryote systems represented by an *Escherichia coli* system have many practical advantages: (1) facility of cell culture (less time-consuming and more economical), (2) a well-established expression mechanism (easily controllable and reproducible), and (3) high productivity (even a gram-order preparation is possible). On the other hand, eukaryote expression systems are useful when some post-translational modifications (e.g., N-terminal acylation, glycosylation, phosphorylation) are important. Bacterial systems are devoid of such modification machineries. However, the great facility of the *E. coli* system should be the first choice for an attempt to produce recombinant protein.

To express recombinant proteins, one may choose their expression either in their fusion forms or intact forms. The former approach is more convenient, if target genes (complementary DNA) have been cloned in pUC or Bluescript vectors, in which they are expressed as a fusion form with an α-peptide of *E. coli* β-galactosidase inducible by isopropylthio-β-galactoside (IPTG). In general, fusion proteins are stable in the host cells. However, the added unnecessary peptides are in many cases undesirable for further studies. Therefore, it is preferable to prepare intact proteins.

Expression of intact recombinant proteins is significantly affected by various transcriptional and translocational factors; i.e., copy number of a target gene, codon usage, selection of a bacterial promoter, and distance between the Shine-Dalgarno (SD) sequence and an initiation codon ATG (summarized by Maniatis et al. 1989). Among these, the distance between SD sequence and the initiation codon is known to be most critical. Therefore, it should be set properly (within a distance of five to ten bases) for the maximal binding of the transcribed mRNA to ribosome.

In this chapter, we illustrate using two examples of the production of intact recombinant lectins in *E. coli*; human 14-kDa (Hirabayashi et al. 1989a) and 30-kDa β-galactoside-binding lectins (Oda et al. 1991). They are characterized as metal-independent-type lectins (Kasai 1990), and suggested to be involved in various onco-developmental events (Barondes 1984; Raz and Lotan 1987). Another advantage of the *E. coli* system in the described cases is that lactose is not required for lectin extraction, which is necessary to solubilize lectins from natural sources (Hirabayashi and Kasai 1984; Hirabayashi et al. 1987). Although the derived recombinant lectins have not been acetylated like their natural forms, they proved to have maintained their original biological activity (hemagglutination), antigenicity and protein stability (Hirabayashi et al. 1989b).

52.1 Outlines of Procedure

The strategy for production of recombinant intact lectins consists of the first DNA-level experiment and the subsequent protein level experiment; i.e., construction of expression plasmids and purification of the expressed recombinant proteins from *E. coli* cells. Outlines for each process are summarized below.

52.1.1 Construction of Expression Plasmid

1. Introduction of a 5′-cloning site
   This is a key step in the whole procedure. There are two approaches to accomplish it:

   a) Procedure using a synthetic linker, described by an example of human 14 kDa lectin in III-A (Fig. 1).
   b) Oligonucleotide-directed mutagenesis, described by an example of human 30 kDa lectin in III-B (Fig. 2).

   It depends on each gene structure as to which procedure should be chosen. If there is a suitable unique restriction site downstream of the initiation codon within 50 bp, the first approach would be better, while a mutagenesis approach is applicable to any gene construct.
      Another convenient way would be a polymerase chain reaction (PCR)-aided mutagenesis to introduce the 5′-cloning site, though confirmation of the whole nucleotide sequence of the amplified segment is necessary.
2. Cloning of a target lectin gene into a bacterial expression vector.

52.1.2 Expression and Purification of Recombinant Lectin

1. Transformation of *E. coli* with the constructed expression plasmid.
2. Culture of the transformed *E. coli* cells and induction of lectin expression.
   It is important to use *lacI*-positive *E. coli* strains or *lacI*-carrying expression vector to fully repress expression of a target gene until an inducer IPTG is added.
3. Extraction of recombinant lectin from *E. coli* cells and subsequent purification by affinity chromatography.

52.2 Detailed Procedures

52.2.1 Procedure Using a Synthetic Linker

Production of recombinant human 14 kDa lectin (Hirabayashi et al. 1989b; see Fig. 1 for the scheme)

1. Cut at a unique *Hinf*I site, 27 nucleotides downstream of the initiation codon and at a unique *Nco*I site, 12-base downstream of the termination codon of human 14 kDa lectin cDNA (Hirabayashi et al. 1989a) which is subcloned in pUC18 (upper left in Fig. 1) after partial *Eco*RI digestion of the original λgt11 clone. Purify the excised 0.39 kbp fragment by 5% polyacrylamide gel electrophoresis (PAGE), as described by Maniatis et al. (1989).   **Synthetic Linker**
2. Prepare a 0.09 kbp fragment from another pUC18 subclone (upper right in Fig. 1) coding a C-terminal part of the lectin by cutting at the same *Nco*I site as in the above subclone and *Pst*I site that originates from the multicloning site of pUC18.
3. Synthesize two complementary oligonucleotides (35 mer sense and 34 mer antisense strands, Fig. 1). Anneal them by boiling in 50 mM Tris-HCl, pH 8.0, 0.5 M

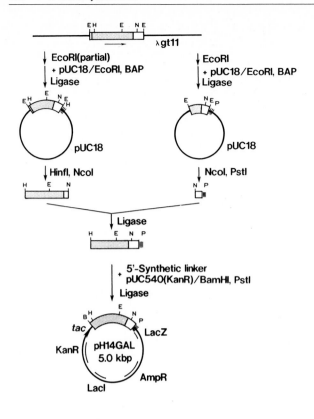

Synthetic linker

5' GATC ATG GCT TGT GGT CTG GTC GCC AGC AAC CTG 3'
3' G TAC CGA ACA CCA GAC CAG CGG TCG TTG GAC TTA 5'

Fig. 1. Scheme for construction of expression plasmid for human 14 kDa β-galactoside-binding lectin by the synthetic linker method. (Hirabayashi et al. 1989a)

NaCl for 5 min after phosphorylation with T4 polynucleotide kinase and ATP, and then reanneal each other in the same buffer (70 °C, for 10 min, then gradually cooling to room temperature). The resultant synthetic linker contains a 5'-*Bam*HI site and a 3'-*Hin*fI site. The initiation codon ATG is located immediately downstream of the *Bam*HI site, ten base apart from the SD sequence.

4. Ligate the 0.39 kbp fragment of (1) with the 0.9 kbp fragment of (2) and the synthetic linker of (3) using T4 DNA ligase[1]. Purify the resultant 0.51 kbp fragment by polyacrylamide gel electrophoresis.

5. Ligate the 0.51 kbp fragment with a *Bam*HI, *Pst*I-cut expression vector pUC540 (Kan), which is a derivative of the previously described expression vector pUC540 with added kanamycin resistance (Soma et al. 1987).

For this purpose, however, any expression vector will do if it is constructed for bacterial expression.

---

[1] The use of polyethyleneglycol is known to facilitate the ligation reaction (Hayashi et al. 1986). A kit for this purpose is also commercially available, e.g., from Takara Shyuzo Co. (Kyoto, Japan).

52.2.2 Oligonucleotide-Directed Mutagenesis

Production of human 30-kDa lectin (Sakakura 1991; see Fig. 2 for the scheme)

1. Clone a target lectin gene (1.14 kbp *Eco*RI fragment, originally obtained as a **Mutagenesis** λ gt10 clone, Oda et al. 1991) into a replicative form of M13 DNA. Confirm the direction of the inserted fragment by restriction enzyme analysis. Choose a clone encoding a (−) strand of the lectin sequence for this study.
2. Prepare a single-stranded (−) DNA to be used as a template for mutagenesis. Be sure to confirm purity of the obtained single-stranded DNA by agarose gel electrophoresis (Maniatis et al. 1989).

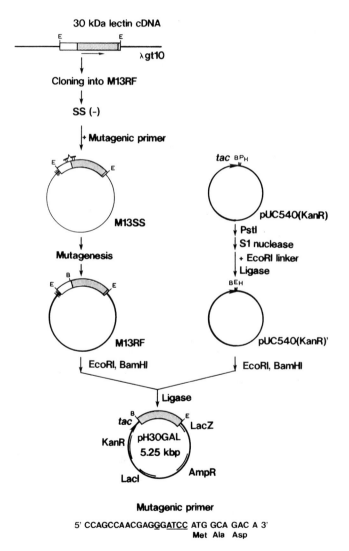

Fig. 2. Scheme for construction of expression plasmid for human 30 kDa β-galactoside-binding lectin by the oligonucleotide-directed mutagenesis. (Sakakura 1991)

3. According to the mutagenesis procedure described in the following chapter, introduce a unique 5'-*Bam*HI cloning site immediately upstream of the initiation codon of the human 30 kDa lectin gene using a mutagenic oligonucleotide (Fig. 2).

4. Choose mutagenized phage clones by dot-blot hybridization (described in the following chapter) using a ^{32}P-labeled mutagenic oligonucleotide as a probe (Maniatis et al. 1989).

5. Prepare a replicative form of M13 DNA from one of the positive clones. The resultant gene should have a newly generated unique *Bam*HI site immediately upstream of the initiation codon (Fig. 2). Double-digest the DNA with *Bam*HI and *Eco*RI, whereby two insert fragments (0.34 kbp and 0.80 kbp) are excised from the vector.

6. Ligate the 0.8 kbp *Bam*HI/*Eco*RI fragment with *Bam*HI/*Eco*RI-cut pUC540(Kan).

Since the pUC540(Kan) vector does not originally have an *Eco*RI site, a unique *Eco*RI site was replaced for a previous *Pst*I site, treating at this *Pst*I site, treating with S1 nuclease and ligating with a commercially available *Eco*RI linker (Fig. 2). In this case, dephosphorylation of the vector and purification of the 0.8 kbp gene fragment are not necessary

52.2.3 Expression and Purification of Recombinant Intact Lectin

Recombinant Lectin

1. Transform an appropriate *E. coli* strain[2] with the ligation products of A-6) or B-5) according to the procedure described by Hanahan (1983). Spread the transformed *E. coli* cells over agar plates containing ampicillin (50 µg/ml) and kanamycin (30 µg/ml).

2. Select *E. coli* clones which have a properly ligated expression plasmid by restriction enzyme analysis following small-scale (e.g., 10 ml) plasmid preparation, or colony-immunoblot analysis described below. Store *E. coli* cells as glycerol culture (− 80 °C).

3. Proliferate the *E. coli* clone in 2 × TY medium containing ampicillin and kanamycin at 37 °C to a stationary phase[3] in an appropriate jar fermenter (Fig. 3).

4. Induce expression by adding IPTG to give a final concentration of 0.1 mM. Allow further incubation for 2–4 h at 37 °C[4].

5. Collect *E. coli* cells by centrifugation (4 °C, 6 000 rpm, 20 min), and suspend the cells with at least 3 volumes of ice-cold extraction buffer (20 mM Na-phosphate, pH 7.2, 150 mM NaCl, 2 mM EDTA, 4 mM β-mercaptoethanol, 1 mM phenylmethylsulfonyl fluoride; designated PMSF-MEPBS).

---

[2] Use recombination-deficient strains (*recA*⁻) such as SCS-1, HB101, and JM109. Otherwise, a functional lectin gene may be lost during cell proliferation probably by a spontaneous recombination.

[3] It is recommended to first carry out a pilot experiment (e.g., 10 ml culture) to confirm that the selected clone really expresses the recombinant lectin with an expected molecular size by Western-blotting analysis using specific antibody.

For a large-scale culture (1–20 l), proliferate the cells in a step-wise manner; e.g., from 10 ml to 200 ml cultures, and finally to a 5 l culture. Otherwise, nonresistant mutant or wild-type clones will grow to be a predominant species in the absence of antibiotics.

[4] When an expression level is very low (< 0.1 mg/l culture), try to lower the temperature during the induction to 25–35 °C, which may improve the efficiency.

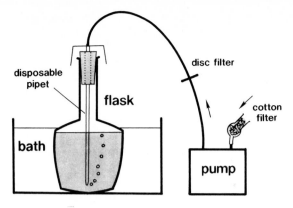

Fig. 3. A manual lab-jar fermenter for a 2–5 l culture of *E. coli*. Air pump (30 l/min; e.g., Minivac Dry PG-30, Yamato Sci. Co., Tokyo, Japan); 0.22 μm filter (Millipore, Milex-GS); water bath (37 °C)

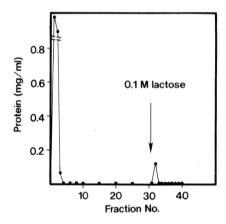

Fig. 4. Purification of recombinant intact human 30 kDa lectin by affinity chromatography. *E. coli* extract (approximately 1 g protein from a 4 l culture) containing the recombinant lectin was applied to a column of asialofetuin agarose (De Waard et al. 1976) equilibrated with MEPBS. *Arrow* indicates the start of the 0.1 M lactose elution

6. Disrupt the cells on ice by sonication using a Branson Sonifier 250. Use a microtip for a 1–5 l culture, or a standard tip for a 5–20 l culture[5].
7. Centrifuge the above material (15 000 rpm, 4 °C, 30 min), and collect the clear supernatant solution.
8. If necessary, repeat extraction procedures (6) and (7).
9. Collect the above extract and measure the protein concentration by the method of Bradford (1976). Apply the extract in a cold room (4–6 °C) to a column of asialofetuin-agarose (200 mg acid-treated fetuin coupled to 20 ml of BrCN-activated agarose gel, De Waard et al. 1976) which had been previously equilibrated with MEPBS.
10. Wash the column with MEPBS (> 500 ml) until no essential protein is eluted.
11. Elute the adsorbed recombinant lectin with MEPBS containing 0.1 M lactose (Fig. 4A for purification of recombinant human 30 kDa lectin).

---

[5] To prevent proteolysis, maintain the lyzate temperature below 10 °C during sonication, especially when using a microtip. After sonication, confirm microscopically that most of the cells are disrupted.

Lactose can be removed by either dialysis against $0.1 \times$ PBS containing 1 mM β-mercaptoethanol or gel filtration on a Bio-Gel P-6 (Bio-Rad, 50–100 mesh, 1.8×50 cm column) using the same buffer[6]. Thus, the obtained recombinant intact lectins proved to be pure by PAGE in the presence of sodium dodecyl sulfate (Fig. 4B), and show comparable sugar-binding activity to those of natural type lectins.

52.2.4 Colony-Immunoblot Analysis

The described procedure is based on immunological detection of an expressed recombinant lectin protein, and is convenient for rapid selection of positive *E. coli* clones.

Colony-Immunoblot

1. From a transformation plate, randomly pick up several dozen of colonies and transfer them using a sterile tooth-pick on a nitrocellulose membrane (grid-type, 82 mm diameter, Amersham) covered on a $2 \times$ TY agar plate containing ampicillin (50 µg/ml) and kanamycin (30 µg/ml). The membrane should have previously been immensed in sterile 0.1 M IPTG and air-dried on a sterile filter.
2. Incubate the plate at 42 °C for 2–4 h or at 37 °C for 6–16 h to express a recombinant gene[7].
3. All procedures hereafter are performed at room temperature.
   Peel off the membrane from the plate. Remove the gel if attached to the membrane. Lyze the *E. coli* cells on a filter permeated with 0.5 M NaOH, 1.5 M NaCl for 5 min similar to colony hybridization (Maniatis et al. 1989). Neutralize the cells by placing the membrane on another filter permeated with 1 M Tris-HCl, pH 7.5, 1.5 M NaCl for 5 min.
4. Briefly rinse the membrane with 50 mM Tris-HCl, pH 7.5, 150 mM NaCl, 1% (W/V) Tween-20 (designated TBS-Tween), then block a nonspecific binding site on the membrane by treating with TBS-Tween containing 1% (w/v) skim milk (100 ml) with shaking for 20 min. Do not dry the membrane until step 9 (color development).
5. Rinse the membrane TBS-Tween, then react with 10 ml of anti-human 30-kDa lectin antibody[8] (1000 × diluted serum with TBS-Tween) in a sealed bag for 16 h with gentle shaking.
   The above steps, (1)–(5), can be performed within a day.
6. Wash the membrane with TBS-Tween (2–3 min, three times).
7. React the membrane with 10 ml of horseradish peroxidase-conjugated (goat) anti-rabbit IgG antibody (1000 × dilution in TBS-Tween) for 2 h with gentle shaking.
8. Extensively wash the membrane with TBS (twice for 2 min, then twice for 5 min; Tween is not necessary).

[6] Dialysis of human 30 kDa lectin should be performed using a buffer with higher ionic strength such as $0.2 \times$ PBS, because this protein has a tendency to stick to the dialysis tube.
[7] At the same time, streak with the colonies another agar plate for storage, and incubate at 37 °C for 16 h.
[8] When necessary, the antibody solution should be passed through an *E. coli* lyzate column to remove anti-host antibodies which may increase a background signal.

9. Develop the color of the positive clones which express immunologically reactive proteins with H_2O_2 and an appropriate chromogenic substrate such as 3,3'-diaminobenzidine[9].

References

Barondes SH (1984) Soluble lectins: a new class of extracellular proteins. Science 223: 1259–1264

Bradford MM (1976) A rapid and sensitive method for the quantitation of microgram quantities of protein utilizing the principle of protein-dye binding. Anal Biochem 72: 248–254

De Waard A, Hickman S, Kornfeld S (1976) Isolation and properties of β-galactoside-binding lectins of heart and lung. J Biol Chem 251: 7581–7587

Hanahan D (1983) Studies on transformation of *Escherichia coli* with plasmids. J Mol Biol 166: 557–580

Hayashi K, Nakazawa M, Hiraoka N, Obayashi A (1986) Regulation of inter- and intramolecular ligation with T4 DNA ligase in the presence of polyethylene glycol. Nucl Acid Res 14: 7617–7631

Hirabayashi J, Kasai K (1984) Human placenta β-galactoside-binding lectin: Purification and some properties. Biochem Biophys Res Commun 122: 938–944

Hirabayashi J, Kawasaki H, Suzuki K, Kasai K (1987) Complete amino acid sequence of 14 kDa β-galactoside-binding lectin of chick embryo. J Biochem (Tokyo) 101: 775–787

Hirabayashi J, Ayaki H, Soma G, Kasai K (1989a) Cloning and nucleotide sequence of a full-length cDNA for human 14 kDa β-galactoside-binding lectin. Biochim Biophys Acta 1008: 85–91

Hirabayashi J, Ayaki H, Soma G, Kasai K (1989b) Production and purification of a recombinant human 14 kDa β-galactoside-binding lectin. FEBS Lett 250: 161–165

Kasai K (1990) Biochemical properties of vertebrate 14K beta-galactoside-binding lectins. In: Franz H (ed) Advances in lectin research, vol 3. VEB Verlag Volk und Gesundheit, Berlin, pp 10–35

Maniatis T, Fritsch EF, Samblook J (1989) In: Molecular cloning: a laboratory manual (2nd ed), Cold Spring Harbor Laboratory, Cold Spring Harbor, New York

Oda Y, Leffler H, Sakakura Y, Kasai K, Barondes SH (1991) Human breast carcinoma cDNA encoding a galactoside-binding lectin homologous to mouse Mac-2 antigen. Gene 99: 279–283

Raz A, Lotan R (1987) Endogenous galactose-binding lectins: a new class of functional tumor cell surface molecules related to metastasis. Cancer Metastasis Rev 6: 433–452

Sakakura Y (1991) Biochemical and molecular biological studies of vertebrate β-galactoside-binding lectins. *Thesis* (Teikyo Univ. Kanagawa, Japan)

Soma G, Kitahara N, Tsuji Y, Kato M, Oshima H, Gatanago T, Inagawa H, Noguchi K, Tanabe Y, Mizuuo D (1987) Improvement of cytotoxicity of tumor necrosis factor (TNF) by increase in basicity of its N-terminal region. Biochem Biophys Res Commun 148: 629–635

---

[9] For this purpose, highly sensitive, nontoxic chromogens for peroxidase are commercially available as a kit; e.g., Wako POD immunostain set (Wako Pure Chemicals, Co., Tokyo, Japan) and Konica immunostaining kit IS50B (Konica, Co., Tokyo, Japan).

53 Construction of Mutant Lectin Genes by Oligonucleotide-Directed Mutagenesis and Their Expression in *Escherichia coli*

J. Hirabayashi and K. Kasai

Recent progress in molecular biology has made it possible to produce invaluable recombinant proteins rapidly once their corresponding genes have been cloned. There are also many extended fields for this technology. To understand the structural and functional basis of a target protein, mutagenesis approaches are extremely useful. Although various procedures have been developed, oligonucleotide-directed mutagenesis may be the most convenient and powerful approach, where a single amino acid can be replaced genetically so that a clearer result can be obtained than has ever been attained by conventional chemical modification experiments.

Upon mutagenesis, however, target amino acid residues should be substituted to residues as homologous as possible to maintain the original peptide conformation. If a mutagenesis had no significant effect on structure and function of a target protein, one may conclude that the target residue is not important. On the other hand, there can be two interpretations if a mutagenesis led to a great reduction in biological activity of a target protein. One possibility is that the target residue is really critical for the function, and the other is that even such a conservative substitution caused a local but significant conformation change leading to a decrease in activity. Upon any mutagenesis experiment, one should keep in mind the latter possibility.

It would be necessary to purify inactive mutant proteins to examine the above possibility. Since conventional affinity adsorbents are no longer useful for such inactivated proteins, this is an important problem. For this purpose, however, new affinity techniques (generally called affinity-tag procedures) have recently been developed, which utilize additional functional tags (affinity tags), such as protein A (for immuno-affinity chromatography), oligohistidine (for metal-chelate chromatography), and single arginine (for anhydrotrypsin-agarose chromatography) (Hirabayashi and Kasai 1992). These tags can be removed with specific proteases after purification. Various expression vectors for this purpose are also commercially available.

53.1 Outlines of Procedure

Site-directed mutagenesis is performed using a synthetic oligonucleotide (16–30 mer) that has a few mismatched bases and a template single-stranded (M13 viron) DNA. In the described strategy summarized in Scheme 1 (see also Fig. 1), a full-length target lectin gene (cDNA) is assumed to have already been cloned. The scheme consists of three parts; (1) cloning a target gene into the M13 vector and preparation of single-stranded DNA (steps 1 and 2 in Scheme 1), (2) oligonucleotide-directed mutagenesis (steps 3–8), and (3) expression of the derived mutant gene in *Escherichia coli* (steps 9–11).

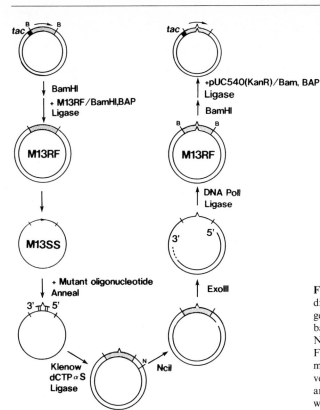

Fig. 1. Scheme for oligonucleotide-directed mutagenesis. The mutagenesis procedure is essentially based on the method described by Nakamaye and Eckstein (1986). For expression of the derived mutant genes, an expression vector pUC540 (Kan) (Hirabayashi and Kasai 1991) is utilized here, which carries a *tac* promoter

Scheme 1. Strategy for mutagenesis study

Steps
1. Cloning a target gene into a replicative form (RF) of M13 DNA
2. Preparation of single-stranded (SS) viron DNA
3. Annealing of a mutant oligonucleotide primer
4. Extension by Klenow with dCTP α S and ligation
5. Nicking with *Nci* I
6. Deletion with exonuclease III (ExoIII)
7. Repolymerization with DNA polymerase I and ligation
8. Transformation of *E. coli*
9. Preparation of a mutant replicative form of DNA
10. Cloning into an expression vector
11. Expression and evaluation of mutant lectins

There are various procedures for oligonucleotide-directed mutagenesis (Kramer and Frits 1987; Kunkel et al. 1987). The procedure described here is originally according to Nakamaye and Eckstein (1986). A key to this method is the use of a phosphorothioate derivative (dCTP α S) in the (−) strand synthesis followed by nicking of only the (+) strand (Fig. 1). This chemistry is based on the observation that a phosphorothioate group is resistant towards cleavage by a group of restriction enzymes such as *Pvu*II and *Nci*I. Nakamayae and Eckstein recommended the use of *Nci*I combined with bacterial exonuclease III, which yields high mutagenesis efficiencies (70–90%). A kit for this purpose is also available from Amersham.

53.2 Detailed Procedures

53.2.1 Cloning into M13 DNA and Preparation of Phage Stock

Cloning
Excise a target lectin gene from an original expression plasmid with an appropriate restriction enzyme. Ligate it with a replicative form of the M13-series DNA which has been digested with the same restriction enzyme and dephosphorylated with bacterial alkaline phosphatase. Transform with this construct an *E. coli* strain having an F-pilli episome which is necessary for the phage infection (e.g., JM109 and TG1; Messing 1983). To facilitate selection of the recombinant clones, perform plaque formation on agar plates with a top agarose gel containing isopropylthio-β-galactoside (IPTG) and 5-bromo-4-chloro-3-indoryl-β-galactoside (X-Gal), as described by Messing (1983).

Pick up several colorless recombinant plaques on the plates with sterile toothpicks and put them in 1.5 ml of 2 × TY. To each tube add a 1/100 volume of a full-growth culture of the same strain *E. coli* used above for the transformation. Proliferate infected cells by shaking the tubes at 37 °C for 3–5 h[1]. Prepare phage stocks according to Scheme 2.

The obtained phage stock solutions can be stored at 4 °C for at least 3 months without a significant loss of titer.

53.2.2 Preparation of Single-Stranded (SS) Viron DNA

Preparation
Single-stranded (SS) M13 viron DNA is used as a template for mutagenesis. The described procedure (Scheme 3), originally established by Messing (1983), is composed of two processes; i) concentration of M13 phage by polyethyleneglycol (PEG) precipitation, and ii) removal of the viral coat protein with phenol. Here, the procedure (Scheme 3) is moderately scaled up for multiple mutagenesis experiments.

Scheme 2. Preparation of phage stock solution

Cell suspension (1.5 ml)
↓ cfg (12 000 rpm, rt, 5 min)
1.4 ml supernatant to another tube
↓ cfg (as above)
1.0 ml supernatant to another tube
↓ + 0.2 ml 20% (w/v) PEG-6000, 2.5 M NaCl
↓ Voltex-mix, stand (rt, 15 min)
↓ cfg (as above)
Phage precipitate: remove the bulk of PEG solution
↓ cfg (12 000 rpm, rt, 2 min)
Phage precipitate: remove totally PEG solution
↓ + 0.1 ml TE buffer[a]
"Phage stock solution"

[a]10 mM Tris-HCl, pH 7.5, 1 mM EDTA.

[1]M13 phage does not lyze *E. coli* cells through infection like λ phages, but infected cells grow more slowly than uninfected cells (Messing 1983).

Scheme 3. Preparation of single-stranded M13 DNA

$2 \times$ TY (50 ml) in a 300 ml flask
   ↓ + 0.5 ml *E. coli* cells (full-growth)
   ↓ culture, 37 °C till OD_{550} = 0.3 (approx. 80 min)
   ↓ + 50 µl phage stock solution
   ↓ culture, 37 °C, 4 h[a]
Divide into two 40 ml tubes
   ↓ cfg (10 000 rpm, 4 °C, 30 min)
Sup (20 ml × 2) to another pair of 40 ml tubes
   ↓ + 20% (w/v) PEG-6000, 2.5 M NaCl (4 ml/tube)
   ↓ Voltex-mix, stand (4 °C, 1 h)
   ↓ cfg (10 000 rpm, 4 °C, 20 min)
Phage precipitate: remove the bulk of PEG solution
   ↓ cfg (10 000 rpm, 4 °C, 5 min)[b]
Phage precipitate: remove totally PEG solution
   ↓ + TE buffer (0.125 ml/tube), Voltex-mix
   ↓ combine (0.25 ml) and transfer to a 1.5 ml tube
   ↓ cfg (12 000 rpm, rt, 5 min)
Sup: remove cell debris, if any
   ↓ + 20% (w/v) PEG-6000, 2.5 M NaCl (0.1 ml)
   ↓ Voltex-mix, stand (rt, 15 min)
   ↓ cfg (12 000 rpm, 4 °C, 5 min)
Phage precipitate: remove the bulk of PEG solution
   ↓ cfg (12 000 rpm, 4 °C, 2 min)
Phage precipitate: remove totally PEG solution
   ↓ + TE buffer (0.25 ml)
   ↓ cfg (12 000 rpm, rt, 2 min)
Sup: remove cell debris, if any
   ↓ + TE-bufferized phenol (0.1 ml)
   ↓ Voltex-mix, stand (rt, 15 min), Voltex-mix
   ↓ cfg (12 000 rpm, rt, 3 min)
Aqueous phase
   ↓ extract with ether (0.25 ml), three times
   ↓ extract with chloroform (0.25 ml), twice
Aqueous phase (0.25 ml)
   ↓ + 25 µl 3 M Na-acetate, pH 6
        600 µl EtOH
   ↓ − 80 °C, 1 h
   ↓ cfg (12 000 rpm, 4 °C, 10 min)
ppt (single-stranded DNA)
   ↓ rinse with 80% EtOH
   ↓ 50 µl of TE

---

[a] Do not shake vigorously. It is important to maintain a rather moderate shaking (around 120 rpm) during the proliferation of infected cells. A too vigorous shaking will damage the cells, resulting in considerable contamination of cellular RNA and DNA which cannot be removed by the subsequent procedure.
[b] Centrifuge directly and remove PEG solution completely.

Determine the yield by measuring A_{260} (a 1% single-stranded DNA solution gives A_{260} of 27), and confirm the purity by standard agarose-gel electrophoresis. Approximately 100 µg single-stranded DNAs, which are free from contaminating RNA and chromosomal DNA, are obtained.

53.2.3 Site-Directed Mutagenesis

Site directed Mutagenesis

1. 5′-phosphorylation of a mutagenic primer. Synthesize a mutagenic oligonucleotide (16–40 mer). Quantify the yield by measuring absorbance at 260 nm after purification (e.g., polyacrylamide gel electrophoresis, gel filtration[2], reversed-phase chromatography, ion-exchange chromatography).
   Phosphorylate the purified oligonucleotide according to Scheme 4.
2. Annealing. Anneal the phosphorylated mutageneic primer with a template DNA according to Scheme 5.
3. Elongation and ligation. To the above solution, add chemicals and enzymes to perform the elongation and ligation reactions according to Scheme 6.
   Remove 1 µl of the reaction product to analyze using standard agarose gel electrophoresis (Maniatis et al. 1989). Also, confirm the generation of a replicative form IV DNA (covalently closed circular form). If no corresponding DNA band is observed, the following procedures should not be performed.
4. Removal of single-stranded DNA. Add to the above product 170 µl of water and 30 µl of 5 M NaCl. To remove SS DNA, filter the solution (250 µl) using a 0.2 µm nitrocellulose filter module (Centrex, product No. 02300, Schleicher & Schuell) by centrifugation (1500 rpm, room temperature, 10 min). To wash the filter, add 100 µl of 0.5 M NaCl and centrifuge again. To the filtrate (350 µl), add 700 µl of ethanol to precipitate DNA ($-80\,°C$, 1 h). Dissolve the precipitated DNA in a solution of 20 mM NaCl, 10 mM Tris-HCl, pH 8.0, 1 mM EDTA.

Scheme 4. Phosphorylation a mutagenic primer

Oligonucleotide 10 pmol/µl	5 µl (50 pmol)
10 × kination buffer[a]	3 µl
water	21 µl
T4 polynucleotide kinase (10 units/µl)	1 µl
Total	30 µl

↓ 37 °C, 15 min

↓ 70 °C, 10 min (to inactivate the enzyme)

[a]10 × kination buffer:

0.5 M Tris-HCl, pH 7.5
0.1 M $MgCl_2$
0.1 M β-mercaptoethanol
10 mM ATP

[2]For this, we use a high-performance gel filtration column, Asahipak GS-220 (exclusion size determined for pullulan, 4,000 Da; column size, 7.6 × 250 mm; Asahi Chemicals, Co., Tokyo, Japan) with a volatile solvent (20 mM triethylammonium acetate, pH 7, 20% acetonitrile). To determine the yield, the integrated peak areas are compared with that of a reference oligonucleotide of which the exact concentration has been given.

Scheme 5. Annealing of a mutagenic primer and a template DNA

Single-stranded template M13 DNA (1 µg/µl)	5 µl (2 pmol)
Phosphorylated mutagenic primer (1.6 pmol/µl)	2.5 µl (4 pmol)
5 × annealing buffer[a]	4 µl
Water	8.5 µl
Total	20 µl

    ↓ 70 °C, 3 min
    ↓ 37 °C, 30 min
    ↓ 0 °C

[a]5 × annealing buffer:

0.6 M Tris-HCl, pH 8.0
0.6 M NaCl.

Scheme 6. Elongation and ligation of (−) strand

Sample of (2)	20 µl
0.1 M MgCl$_2$	5 µl
Nucleotide mixture 1[a]	25 µl
Klenow	6 units
T4 DNA ligase	6 units
Total	approx. 50 µl

    ↓ 16 °C, overnight

[a]Nucleotide mixture 1 (+ dCTP α S):

0.5 mM dCTP α S (Amersham)
0.5 mM dATP
0.5 mM dGTP
0.5 mM dTTP
2 mM ATP.

Scheme 7. Nicking of the (+) strand by *Nci*I digestion

solution of (4)	10 µl
Nci I buffer[a]	65 µl
Nci I.	6 units
Total	approx. 76 µl

    ↓ 37 °C, 90 min

[a]*Nci*I buffer:

12 mM Tris-HCl, pH 8.0
32 mM NaCl
 7 mM MgCl$_2$
12 mM dithiothreitol.

5. Nicking of the (+) strand. Transfer 10 μl of the solution of (4) to another reaction tube. Nick the (+) strand according to Scheme 7.
   Remove 10 μl to monitor the reaction by agarose gel electrophoresis. Confirm the appearance of nicked (open circular) DNA in place of the replicative form type IV DNA.
6. Deletion of 3′ region of the (+) strand. To the remaining solution of (5), add chemicals to delete the (+) strand according to Scheme 8.
   Briefly centrifuge the reaction tube. Remove 15 μl to analyze using agarose gel electrophoresis. Confirm the presence of gapped DNA which migrate significantly faster than the nicked DNA of (5).
7. Repolymerization and ligation. To the above solution (75 μl), add chemicals to repolymerize the (+) strand according to Scheme 9.
   Remove 15 μl for the analysis using agarose gel electrophoresis to confirm the appearance of a replicative form IV.

Scheme 8. Deletion of the (+) strand by exonuclease III digestion

Solution (5)	66 μl
0.5 M NaCl	12 μl
Exonuclease III buffer[a]	10 μl
Exonuclease III	50 units
(Freshly diluted with exonuclease III buffer)	
Total	90 μl

   ↓ 37 °C 30 min (exactly)[b]
   ↓ 70 °C 15 min (inactivate the enzyme).

[a]exonuclease III buffer:

90 mM NaCl
60 mM Tris-HCl, pH 8.0
 6 mM $MgCl_2$
10 mM dithiothreitol.
[b]Based on the assumption of about 100 nucleotides digested per minute with the enzyme.

Scheme 9. Repolymerization and ligation of the (+) strand

Sample solution of (6)	75 μl
Nucleotide mixture 2[a]	13 μl
100 mM $MgCl_2$	5 μl
DNA Polymerase I	3 units
T4 DNA ligase	2 units
Total	approx. 95 μl

   ↓ 16 °C, 3 h

[a]Nucleotide mixture 2:

0.5 mM dCTP
0.5 mM dATP
0.5 mM dGTP
0.5 mM dTTP
2 mM ATP.

8. Transfection of *E. coli*. Prepare competent cells of an appropriate *E. coli* strain having F pilli such as JM107 and SMH-50 according to Hanahan (1983). TG1 has been recommended by Amersham because of its high transfection efficiency with phosphorothioate-containing double-stranded DNA. Use 2 or 10 µl of the repolymerization product of (7) for the transfection. After heat shock (42 °C, 45 s), chill the tube containing the transfected cells. To this tube, add 200 µl of log-phase TG1 cells in 2 × TY medium. Transfer the cell suspension to a 15 ml glass tube containing 3 ml of prewarmed (50 °C), 0.7% top agarose solution. Spread the cells over a prewarmed (37 °C) 2 × TY agar plate. After the top gel is hardened, incubate the plate at 37 °C for 16 h to allow plaque formation.

53.2.4 Selection of Mutant M13 Phage Clones

To confirm mutagenesis, dot-blot hybridization described below is a convenient method, whereby up to 100 phage clones can be assayed briefly and systematically using a ^{32}P-labeled mutagenic oligonucleotide probe. However, more strict confirmation by direct nucleotide sequencing (e.g., by dideoxy chain termination method; Maniatis et al. 1989) is strongly recommended. **Selection**

Prepare phage stock solutions for 10–50 candidate clones by the procedure III-A. Blot 1 µl of each solution and a control solution derived from a nonmutant phage clone on a nylon membrane (e.g., Amersham, Hybond N). Air-dry the blot at room temperature, and then bake it at 80 °C for 2 h.

Prehybridize the blot in 6 × SSC, 10 × Denhardt's solution (Maniatis et al. 1989). 0.2% sodium dodecyl sulfate at 67 °C for 1 h. Remove the prehybridization solution, and then hybridize the blot with a 5-^{32}P-labeled mutagenic primer as a probe at room temperature for 1 h. For this, label 15 pmol of the oligonucleotide with T4 polynucleotide kinase and [γ-^{32}P] ATP (3000 Ci/mmol). Filter the labeled solution (e.g., through Millipore Mylex GV filter unit) prior to hybridization, which greatly reduces a background signal.

After hybridization, wash the blot membrane with a low stringency; three times at room temperature with an excess of 6 × SSC (5 min for each washing). Wrap the wet membrane with Saran Wrap, and carry out the first autoradiography by exposing an X-ray film for 1–16 h. Do not dry the membrane until the subsequent high stringency washing.

After the first autoradiography, wash the membrane again with higher stringency; several times with 6 × SSC at $T_m - 5$ °C and monitoring the radioactivity. Perform the second autoradiogaraphy under the same conditions.

All clones give signals after the low-stringency washing, but only mutant clones should give stable signals after the high-stringency washing (see Fig. 2). If the used negative control still gives a stable signal in the second autoradiogram and there is no significant difference between the first and second autoradiograms, further wash

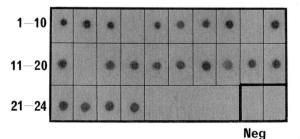

Neg

Fig. 2. An example of dot blot hybridization. Twenty-four phage clones were blotted on a nylon membrane (Amersham Hybond N), and hybridized with ^{32}P-labeled oligonucleotide used for the mutagenesis. As a negative control, a nonmutant clone was also analyzed, which gave no hybridization signal after a stringency washing (6 × SSC, 55 °C in this experiment)

the membrane with a slightly higher stringency[3]. Choose clones giving stable signals even after the high-stringency washing.

53.2.5 Preparation of a Mutant Replicative Form of M13 DNA

Preparation   Prepare a replicative form of mutant M13 DNA. Culture *E. coli* cells used for transformation in $2 \times$ TY medium (50 ml) with 5 μl of each phage stock solution of positive clones obtained above. After a 16 h culture, harvest the infected cells and prepare a replicative form of recombinant M13 DNA according to a conventional alkali-SDS method for plasmid preparation (Maniatis et al. 1989). Yields are 20–50 μg. Confirm purity of the obtained DNA and the presence of an insert fragment by standard agarose gel electrophoresis after restriction enzyme digestion.

53.2.6 Cloning a Mutant Lectin Gene into an Expression Vector

Cloning   Excise an insert DNA fragment from the recombinant M13 DNA obtained above by an appropriate restriction enzyme. After heat-inactivation of the enzyme, extract DNA with phenol/chloroform and precipitate with EtOH. Ligate the insert fragment with an original expression vector[4], which has been cut with the same restriction enzyme and dephosphorylated[5]. Transform an appropriate strain of *E. coli* (e.g., HB101, JM109) with the above ligation construct. Choose clones which have a properly ligated plasmid by colony immuno-blot selection (described in the preceding chapter, Hirabayashi et al., Chap. 52, this Vol.).

53.2.7 Expression of a Mutant Lectin Protein in *E. coli* and Its Evaluation

Expression   Express the mutant lectin protein in *E. coli* as described in the preceding chapter.

In Table 1 mutagenesis experiments of human 14 kDa β-galactoside-binding lectin, as an example, are summarized. Among the mutants, all Cys-Ser mutants (C2S, C16S, C42S, C60S, C88S, and C130S), in which either of six cysteine residues of the human lectin is substituted with serine, proved to be fully active in terms of sugar-binding activity. A mutant, in which a conserved tryptophan at a position of 68 is substituted with tyrosine (W68Y), shows a slightly reduced affinity to lactose, but is still adsorbed firmly on asialofetuin-agarose. These results unambiguously show that none of these oxidative residues plays a critical role in sugar binding. In contrast, replacement of some conservative hydrophilic residues (i.e., Asn^{46}, Glu^{71} and Arg^{73}) to homologous amino acids resulted in almost complete loss of activity, suggesting relative importance of these hydrophilic residues in the carbohydrate recognition of the β-galactoside-binding lectin.

---

[3]Be sure to read a thermometer directly soaked in the washing solution. When T_ms for both mutant and nonmutant sequences are relatively high and closely similar (e.g., 72 and 70 °C, respectively), stringency control may be difficult. Design a mutagenic primer so as to facilitate the stringency control.

[4]It is not necessary to purify the insert fragment if the used expression vector contains relevant antibiotic resistant gene(s) for selection (e.g., ampicillin or tetracycline resistance genes).

[5]Dephosphorylation is not necessary if a forced cloning is performed using heterogeneous cloning sites.

Table 1. Summary of mutagenesis of human 14 kDa β-galactoside-binding lectin

Mutant	Primer used for mutagenesis[a] (Amino acid sequence)	Temp[b] (°C)	Mutagenesis efficiency (%)	Host for expression	I_{50} of lactose[c] (mM)	Yield[d] (mg/l culture)
Wild type				Y1090	0.54	2.00
C2S	CC ATG GCT **TCT** GGT CTG G Met Ala **Ser** Gly Leu	58	80	HB101	0.32	0.93
C16S	CT GGA GAG **TCC** CTT CGA GTG Gly Glu **Ser** Leu Arg Val	64	88	TG1	0.68	0.42
C42S	C AAC CTG **TCC** CTG CAC TTC Asn Leu **Ser** Leu His Phe	60	75	TG1	0.50	1.92
C60S	CC ATC GTG **TCC** AAC AGC AAG Ile Val **Ser** Asn Ser Lys	66	95	Y1090	0.50	0.32
C88S	CA GAG GTG **TCC** ATC ACC TTC Glu Val **Ser** Ile Thr Phe	62	33	TG1	0.34	0.27
C130S	C AAG ATC AAA **TCT** GTG GCC TTT G Lys Ile Lys **Ser** Val Ala Phe	68	96	TG1	0.72	0.15
N46D	CTG CAC TTC **GAC** CCT CGC Leu HIs Phe **Asp** Pro Arg	60	92	TG1	ND[e]	—
W68Y	C GGG GCC **TAC** GGG ACC G Gly Ala **Tyr** Gly Thr	64	8	HB101	1.1	2.15[f]
E71Q	G GGG ACC **CAG** CAG CGG G Gly Thr **Gln** Gln Arg	62	79	TG1	ND[e]	—
R73H	CC GAG CAG **CAC** GAG GCT G Glu Gln **His** Glu Ala	66	ND[g]	Y1090	ND[e]	—

[a]Mutated nucleotides and amino acids are shown in bold type.
[b]Melting temperature calculated by an empirical formula; $T_m = 2(A + T) + 4(G + C)$.
[c]Affinity to lactose in terms of lactose concentration required for 50% inhibition in the lectin-asialofetuin-binding assay (Hirabayashi and Kasai 1991).
[d]Representative yield from several experiments.
[e]Asialofetuin-binding activity not detectable.
[f]Improved yield by changing the expression host from TG1 (Hirabayashi and Kasai 1991) to HB101.
[g]Mutagenesis confirmed by direct nucleotide sequencing.

References

Hanahan D (1983) Studies on transformation of *Escherichia coli* with plasmids. J Mol Biol 166: 557–580

Hirabayashi J, Kasai K (1991) Effect of amino acid substitution by site-directed mutagenesis on the carbohydrate recognition and stability of human 14 kDa β-galactoside-binding lectin. J Biol Chem 266: 23648–23653

Hirabayashi J, Kasai K (1992) Arginine-tail method: a new affinity tag procedure utilizing anhydrotrypsin agarose. J Chromatogr 597: 181–187

Kramer W, Frits HJ (1987) Oligonucleotide-directed construction of mutations via gapped duplex DNA. Methods Enzymol 154: 350–367

Kunkel TA, Roberts JD, Zakour RA (1987) Rapid and efficient site-specific mutagenesis without phenotypic selection. Methods Enzymol 154: 367–382

Maniatis T, Fritsch EF, Samblook J (1989) In: Molecular cloning: a laboratory manual (2nd ed). Cold Spring Harbor Laboratory, Cold Spring Harbor, New York

Messing J (1983) New M13 vectors for cloning. Method Enzymol 101: 20–78

Nakamaye KL, Eckstein F (1986) Inhibition of restriction endonuclease Nci I cleavage by phosphorothioate groups and its application to oligonucleotide-directed mutagenesis. Nucl Acid Res 14: 8765–8785

54 Mammalian Lectin as Transforming Growth Factor

K. Yamaoka, S. Ohno, H. Kawasaki, and K. Suzuki

Experiments on cultured cells have revealed a variety of growth factors and growth inhibitors, some of which, depending on cell type and culture conditions, act both as a growth factor and a growth inhibitor (Tucker et al. 1984;Roberts et al. 1985). The in vivo functions of growth regulatory factors are assumed to stimulate fetal and placental growth during development, to regulate growth and differentiation of continuously regenerating tissues, and to stimulate tissue repair processes (Buckley et al. 1985; Goustin et al. 1985; Masui et al. 1986). However, the molecular mechanisms which regulate the activity of growth regulators remain almost totally unsolved.

Transforming growth factor (TGF) was originally defined as a growth factor that confers a malignant phenotype on untransformed cells and induces them to form progressively growing colonies in soft agar (DeLarco and Todaro 1978). Among several proteins purified by means of the TGF activity, TGF_α was found to be an analog of epidermal growth factor (EGF), while TGF_β turned out to be an unrelated protein (Anzano et al. 1983). TGF_β is a multifunctional growth factor which can stimulate proliferation in some cells (Roberts et al. 1981), while it is a potent inhibitor of proliferation in many cells (Tucker et al. 1984; Roberts et al. 1985). In addition, TGF_β regulates many other cellular phenomena that are not directly related to cell proliferation (Ignotz and Massague 1986; Roberts et al. 1986). We have previously purified a protein from an avian sarcoma virus transformed rat NRK cells, 77N1 (Yamaoka et al. 1984), which shows stimulation of DNA synthesis and the promotion of anchorage independent growth of BALB3T3 A31 cells (Hirai et al. 1983). Since many properties of the TGF are distinguishable from those of TGF_α (Ozanne et al. 1980; Marquardt and Todaro 1982; Twardzik et al. 1983) and TGF_β (Roberts et al. 1981; Anzano et al. 1983;Assoian et al. 1983), we termed it $TGF_\gamma2$ (Yamaoka and Hirai 1986). $TGF_\gamma2$ is a heat- and -acid-labile protein with a molecular mass of 14 kDa. It does not compete with EGF for receptor binding and does not require EGF for colony formation. To clarify the molecular nature of $TGF_\gamma2$, we have determined the amino acid sequence of $TGF_\gamma2$. The sequence of $TGF_\gamma2$ revealed an unexpected identity to a rat lung β-galactoside binding protein (GBP) (Clerch et al. 1988), a lectin ubiquitously distributed in vertebrates (Waard et al. 1976; Powell 1979; Beyer et al. 1980; Ohara and Yamagata 1986; Ohyama et al. 1986). Over-expression of GBP in BALB3T3 A31 cells by introduction of GBP cDNA paralleled with the increase in the amount of a 14 kDa protein recognized by anti-$TGF_\gamma2$ antibody and showed several transformed phenotype (Yamaoka et al. 1991). These results indicate that GBP is directly involved in the regulation of cell proliferation through its TGF activity.

54.1 Partial Amino Acid Sequence of TGF$_\gamma$2 and Its Identity to Rat Lung GBP

Partial amino acid sequences of TGF$_\gamma$2 were determined using peptides derived from trypsin digestion (T1 and T2) or cyanogen bromide cleavage (C3) of the purified protein. A computer search of the sequence database revealed that all three peptide sequences are completely contained in the protein sequence of rat lung GBP (Clerch et al. 1988) (Fig. 1). The sequence identity of TGF$_\gamma$2 and GBP clearly indicates that they are identical or closely related proteins.

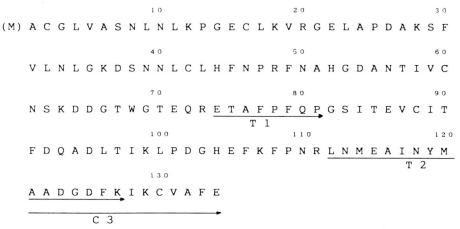

```
                    1 0                    2 0                        3 0
(M)  A  C  G  L  V  A  S  N  L  N  L  K  P  G  E  C  L  K  V  R  G  E  L  A  P  D  A  K  S  F

                    4 0                    5 0                        6 0
     V  L  N  L  G  K  D  S  N  N  L  C  L  H  F  N  P  R  F  N  A  H  G  D  A  N  T  I  V  C

                    7 0                    8 0                        9 0
     N  S  K  D  D  G  T  W  G  T  E  Q  R  E  T  A  F  P  F  Q  P  G  S  I  T  E  V  C  I  T
                                           T 1

                    1 0 0                  1 1 0                      1 2 0
     F  D  Q  A  D  L  T  I  K  L  P  D  G  H  E  F  K  F  P  N  R  L  N  M  E  A  I  N  Y  M
                                                                     T 2

                    1 3 0
     A  A  D  G  D  F  K  I  K  C  V  A  F  E

                              C 3
```

Fig. 1. Partial amino acid sequence of peptide fragments of TGF$_\gamma$2 and homology with rat lung GBP. Partial amino acid sequences determined by direct sequencing of TGF$_\gamma$2 peptides obtained by trypsin digest (*T1*, *T2*) and cyanogen bromide cleavage (*C3*). Peptides were purified by HPLC using a reverse phase TSK gel ODS-120T column with an acetonitrile gradient in 0.1% TFA. The residue (single amino acid code) number is taken from that reported by Clerch. (Clerch et al. 1988)

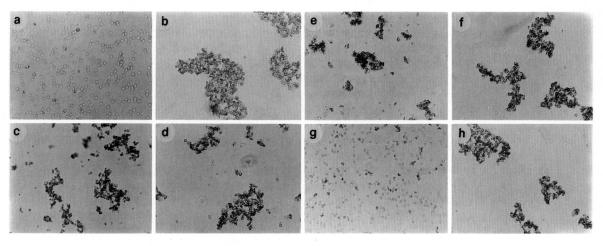

Fig. 2a–h. Hemagglutination activity of TGF$_\gamma$2. Rabbit erythrocytes were incubated with **a** phosphate buffered saline; **b** TGF$_\gamma$2; **c** TGF$_\gamma$2 + α-D(+)-melibiose; **d** TGF$_\gamma$2 + D(+) galactose; **e** TGF$_\gamma$2 + N-acetyl-D-galactosamine; **f** TGF$_\gamma$2 + methyl-α-D-galactopyranoside; **g** TGF$_\gamma$2 + thiodigalactoside; **h** TGF$_\gamma$2 + p-nitrophenyl-β-D-galactopyranoside. TGF$_\gamma$2 was used at the dose of 100 nM and saccharides were used at a dose of 0.25 mM. Agglutination activities were determined 1 h later and photographs were taken microscopically

54.2 TGF$_\gamma$2 Shows the GBP Activity

In order to further analyze the relationship of these two activities, we first examined purified TGF$_\gamma$2 for GBP activity by means of a hemagglutination assay. As shown in Fig. 2, purified TGF$_\gamma$2 agglutinates trypsin-treated rabbit erythrocytes. Thiodigalactoside was the most potent inhibitor of the hemagglutination activity of TGF$_\gamma$2 at the dose of 0.25 mM. Other saccharides were much less potent, and they have no effects even at concentrations higher than 50 mM. These results indicate that TGF$_\gamma$2 has a high affinity for galactose in the glycoside form, and thus the GBP activity.

54.3 Overexpression of a GBP cDNA in A31 Cells Leads to the Expression of the Transformed Phenotype

54.3.1 Preparation of GBP cDNA Expression Plasmid

Rat GBP cDNA was isolated by a PCR cloning method (Saiki et al. 1988). cDNA was synthesized using poly (A) $^+$RNA from 77N1 cells as a template and oligo (dT) as a primer, followed by amplification by a polymerase chain reaction (PCR) using oligonucleotides PR1 and PR2 (PR1, 5'-TCTCTAGAAATCTCTTCGCTTCAA TCATGGC-3' and PR2,5'-TCTCTAGATTCACTCAAAGGCCACACACTTA-3') as primers. PR1 and PR2 contain the sequences -18 to $+5$ and that complementary to $+387$ to $+409$ of rat lung GBP cDNA, respectively, in addition to the XbaI recognition sequence in their 5'-portions. The amplified fragment cloned into the XbaI site of pUC18 was then sequenced and confirmed to encode a full- length GBP cDNA. The GBP cDNA was inserted into the EcoRI site of the expression vector pSRD (Ohno et al. 1988) which contains the SV40 early promoter/enhancer and HTLV-IR sequence to construct GBP cDNA expression on plasmid pSRD-GBP.

54.3.2 Transfection

A31 cells were carried out by the standard calcium phosphate coprecipitation procedure (Wigler et al. 1979).

Solutions Solutions (sterilized by 0.22 µm filter)

1. $2 \times$ HEPES-buffered saline ($2 \times$ HBS); 280 mM NaCl/10 mM KCl/1.5 mM Na$_2$HPO$_4 \cdot$2H$_2$O/12 mM dextrose/50 mM HEPES (100 ml solution; NaCl 1.63 g/HEPES 1.19 g/KCl 0.074/Na$_2$HPO$_4 \cdot$2H$_2$O 0.027 g/dextrose 0.21 g) Adjust the pH 7.05 with 0.5 N NaOH
2. 2 M CaCl$_2$
3. $0.1 \times$ TE (pH8.0); 1 mM Tris-HCl (pH8.0) /0.1 mM EDTA (pH 8.0)
4. DNA; 40 µg/ml in $0.1 \times$ TE (pH 8.0)

Procedure **On day 1:** Exponentially growing A31 cells were harvested by trypsinization and reseeded at a cell density of 5×10^5 cells/plate (60-mm plastic plate) in 5 ml of 10% fetal bovine serum containing Dulbecco's Modified Eagle Medium (FBS-DMEM). The culture was incubated for 24 h at 37 °C in a humidified incubator in an atmosphere of 5% CO_2.

On day 2: Preparation of the calcium phosphate-DNA coprecipitation. Place 219 μl of DNA in $0.1 \times TE$ (containing 15 μg of GBP cDNA expression plasmid pSRD-GBP and 1 μg of pSV2neo which carries a neomycin resistance gene) and 250 μl of $2 \times HBS$ in plastic tube. Slowly add 31 μl of 2 M $CaCl_2$ with gentle mixing. Standed the mixture for 30 min at room temperature.

In this procedure precipitation occurred. The precipitate (calcium phosphate-DNA suspension) was added to the cell culture. The cultere plate was gently swinged to mix the precipitate and medium well. Incubate the transfected cells for 24 h at 37 °C in a humidified incubator in an atmosphere of 5% CO_2.

On day 3: The medium and precipitate mixture was removed by aspiration. Cells were washed once with phosphate-buffered saline, and 5 ml of 10% FBS-DMEM were added and incubated for more 24 h under the same conditions.

On day 4: The cells were trypsinized and reseeded in G418 (400 μg/ml) containing 10% FBS- DMEM and incubated in the same conditions for 2 weeks. This medium was changed every 3–4 days during incubation to remove dead cells. During this incubation time, the G418-resistant cells grow in colonies.

After 2 weeks. Several G418-resistant clones were isolated. Transfection of cells with pSV2neo alone resulted in the isolation of a G418-resistant clone cl.20.

54.3.3 Detection of GBP

Preparation of Cell Extract. Six independent G418-resistant cell clones (clones 1–6) were isolated. Each cell clone was grown in 100 mm plastic plate with 10 ml of DMEM supplemented with 10% fetal bovine serum and antibiotics. Each cell was scraped from the substratum with a rubber policeman, collected by centrifugation at 450g for 10 min after washing with 0.9% NaCl solution, three times. Cells were weighed and suspended 10% weight/volume in 10 mM Tris-HCl pH 7.4 containing 1 mM phenylmethylsulfonyl fluoride. These cell suspensions were sonicated by Sonifier (Bronson) for 45 s and the homogenates were centrifuged at 100 000g for 1 h. The supernatant was stored as the cell extract. **Cell Extract**

Immunoblot Analyses. To ascertain the expression of introduced GBP cDNA in these transfected cells, we analyzed the cell extracts by immunoblotting using an antiserum raised against $TGF_\gamma 2$ (Yamaoka and Hirai 1986). Figure 3 shows that A31 cell clones transfected with GBP cDNA produce increased amounts of a protein reactive with the antiserum against $TGF_\gamma 2$. The antiserum recognized a 14 kDa protein band in the extract of $TGF_\gamma 2$-producing rat 77N1 cells, but very little in A31 cells. The antiserum also reacted with a 14 kDa protein in the cell extracts of clones 1, 3, and 6 of A31 cells while, clones 2, 4, and 5 gave only faint bands. **Immunoblot**

Assay for DNA Synthesis. $TGF_\gamma 2$ was originally characterized by the ability to induce reversible stimulation of DNA synthesis of A31 cells (Hirai et al. 1983). Then, we analyzed the mitogenic activities of the cell extracts from these clones in A31 cells. A31 cells were trypsinized, counted, and plated at 3×10^4 cells/well in 100 μl DMEM supplemented with 0.5% fetal bovine serum and antibiotics in a millititer plate **DNA Synthesis**

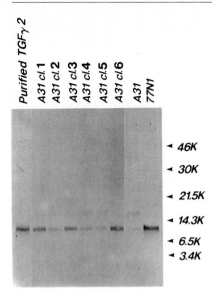

Fig. 3. Immunoblot detection of overexpressed GBP in GBP cDNA transfected cell clones. Cell extracts from each clone were electrophoresed on 15% polyacrylamide gels under reducing conditions and analyzed by immunoblot analyses with anti-TGF$_\gamma$2 antibody

(Millititer STGV; 96-well filtration plate; Millipore), and cultured at 37 °C in a humidified atmosphere of 5% CO_2 in air. After 24 h, DNA synthesis was stimulated by the addition of the 1–10 µl of test samples to each well. After 18 h, the cell were pulse-labeled with 10 µl of 5 kBq. [^{125}I] -Iododeoxyuridine (New England Nuclear) per well for 4 h. After 4 h, the culture plate was filtrated using millititer vacuum holder, to remove culture medium and radio active materials and washed with 200 µl of saline five times. When 200 µl of 5% trichloroacetic acid (TCA) was added to each well and stood at 4 °C for 15 min, the precipitation of DNA was made. The culture plate was filtrated and washed with 200 µl of 5% TCA five times. Then, each well was cut by puncher (Millipore), and the incorporation into acid-insoluble material was determined by gamma counter. As shown in Table 1, the cell extracts from clones 1, 3, and 6 showed a dose-dependent stimulation of DNA synthesis. As with TGF$_\gamma$2 (Yamaoka et al. 1984), this activity was abolished by heating at 100 °C for 3 min or

Table 1. DNA synthesis and hemagglutination activities

Cell extract	[^{125}I] Iododeoxyuridine incorporation ($\times 10^3$ cpm) Dose (µg)				Hemagglutination activity
	4	12	20	40	
A31	0.6	0.6	0.3	0.3	—
Clone 1	0.7	2.5	4.0	7.4	+
2	0.5	0.1	0.2	0.1	—
3	1.4	1.3	3.6	6.9	+
4	0.5	0.1	0.1	0.6	—
5	0.4	0.2	0.1	0.1	—
6	2.3	9.1	19.4	39.1	+
77Nl	4.8	18.0	35.3	47.7	+

by treatment with 1 M acetic acid or 5 mM dithiothreitol. The cell extracts of clones 1, 3, and 6 also showed hemagglutination activity specific for GBP that was inhibited by β-galactoside.

Assay for Colony Formation. The transfected A31 cell clones (clones 1–6) were examined for growth properties. We examined these clones for their ability to form colonies in soft agar, because the acquisition of anchorage-independent growth correlates with in vivo tumorgenicity. One milliliter of 0.7% Noble agar (Difco) in DMEM with 5% fetal bovine serum was poured as base layer agar in a 35 mm plastic plate. When the base layer was solidified at room temperature, the top layer was made by pouring 1 ml of 0.35% agar in DMEM with 5% fetal bovine serum containing $0.5–2.0 \times 10^3$ transfected or A31 cells onto the base layer, and again solidified at room temperature. The cultures were incubated at 37 °C in a humidified atmosphere of 5% CO_2 in air for 1 week and viable colonies were stained with 1 ml of 0.05% iodonitrotetrazolium per plate overnight at 37 °C (Rosenthal et al. 1986). Colony efficiency was calculated by the number of colonies with a diameter greater than 0.05 mm, determined under a microscope for each sample. The clones 1, 3, and 6, which were morphologically transformed in the monolayer culture, grew well, and showed anchorage-independent growth in soft agar, with colony efficiencies of 18.5%, 7.5%, and 49.8%, respectively (Table 2). Clones 2, 4, and 5 did not show these activities. Thus the expression of the 14 kDa protein correlates with morphological transformation and anchorage-independent growth in soft agar.

Colony Formation

54.4 TGF$_\gamma$ 2/GBP cDNA-Expressing A31 Cells Are Tumorigenic in Nude Mice

In order to assess the tumorigenicity of these cDNA expressing cells in vivo, A31 clone 6, that expressed the highest level of 14 kDa TGF$_\gamma$ 2/GBP protein, was injected subcutaneously into nude mice. As shown in Table 3, mice injected with clone 6 cells developed 100% tumor formation after 2 weeks. On the other hand, mice injected with parental A31 cells failed to form tumors. These results indicate that GBP-expressing A31 cells are transformed.

Table 2. Colony formation of GBP cDNA transfected cell clones in soft agar

Clone number	Number of colonies with		
	500 cells	1000 cells	2000 cells
A31	0	0	0
1	137	274	397
2	0	0	0
3	26	29	150
4	0	0	0
5	0	0	0
6	171	344	956
20	0	0	0

Table 3. Tumor formation in nude mice by cDNA expressing A31 cell clone 6

Cell	Cells injected per mouse	Mouse with tumor	Tumor size (cm)
A31 Clone 6	5×10^6	3/3	1.2×2.0
A31	5×10^6	0/2	

Table 4. The effect of saccharides on DNA synthesis activity

Saccharide	concentration (mM)				
	0.5	1	2	10	50
α-D(+) Melibiose	98.0	108.9	90.7	nd.	nd.
D(+) Galactose	105.0	110.5	94.5	nd.	nd.
N-Acetyl-D-galactosamine	111.3	99.6	94.2	nd.	nd.
Methyl-α-D-galactopyranoside	91.8	102.0	96.6	nd.	nd.
Thiodigalactoside	99.2	109.8	110.0	168.0	165.5
Lactose	97.6	99.8	98.1	100.6	109.3
p-Nitrophenyl-β-D-galactopyranoside	150.4	92.2	96.1	nd.	nd.

25 ng of $TGF_\gamma 2$ with each dose of saccharides was determined DNA synthesis activity on parental A31 cells. Results are expressed as the percentage of activity after treatments. nd. not determined.

54.5 TGF Activity Is Not Mediated Through GBP Activity

Since β-galactoside inhibits the hemagglutination activity of $TGF_\gamma 2$, we determined the effects of saccharides on the growth factor activity using purified $TGF_\gamma 2$. Table 4 shows that the activity of purified $TGF_\gamma 2$ to stimulate DNA synthesis of A31 cells was not affected by the addition of any saccharides, including thiodigalactoside, while the hemagglutination activity was completely inhibited by this saccharide at a dose of 0.25 mM. These results indicate that the TGF activity is not mediated through the GBP activity and that TGF and GBP activities of $TGF_\gamma 2$ and GBP are ascribable to distinct molecular determinants.

References

Anzano MA, Roberts AB, Smith JM, Sporn MB, DeLarco JE (1983) Sarcoma growth factor from conditioned medium of virally transformed cells is composed of both type α and type β transforming growth factors. Proc Natl Acad Sci USA 80: 6264–6268

Assoian RK, Komoriya K, Meyere CA, Miller D and Sporn MB (1983) Transforming growth factor in human platelets; identification of a major storage site; purification and characterization. J Biol Chem 258: 7155–7160

Beyer EC, Zweig SE and Barondes SH (1980) Two lactose binding lectins from chicken tissues. J Biol Chem 255: 4236–4239

Buckley AB, Davidson JM, Kamerath CD, Wolt TB and Woodward SC (1985) Sustained release of epidermal growth accelerates wound repair. Proc Natl Acad Sci USA 82: 7340–7344

Clerch LB, Whitney P, Hass M, Brew K, Miller T, Werner R and Massaro D (1988) Sequence of a full-length cDNA for rat lung β-galactoside- binding protein: primary and secondary structure of the lectin. Biochem 27: 692–699

DeLarco JE and Todaro GJ (1978) Growth factors from murine sarcoma virus-transformed cells. Proc Natl Acad Sci USA 75, 4001–4005

Goustin AS, Betsholtz C, Pfeifer-Ohlsson S, Persson H, Rydnert J, Bywater M, Holmgren G, Herdin C-H, Westermark B and Ohlsson R (1985) Coexpression of the sis and myc proto-oncogenes in developing human placenta suggests autocrine control of trophoblast growth. Cell 41: 301–312

Hirai R, Yamaoka K and Mitsui H (1983) Isolation and partial purification of a new class of transforming growth factors from an avian sarcoma virus-transformed rat cell line. Cancer Res 43: 5742–5746

Ignotz RA and Massague J (1986) Transforming growth factor-β stimulates the expression of fibronectin and collagen and their incorporation into the extracellular matrix. J Biol Chem 261: 4337–4341

Marquardt H and Todaro GJ (1982) Human transforming growth factor: Production by a melanoma cell line, purification and characterization. J Biol Chem 257, 5220–5225

Masui T, Wakefield LM, Lechner JE, LaVeck MA, Sporn MB and Harris CC (1986) Type β transforming growth factor is the primary differentiation-inducing serum factor for normal human bronchial epithelial cells. Proc. Natl Acad Sci USA 83: 2438–2442

Ohara T and Yamagata T (1986) Isolation and characterization of galactose-binding proteins from new born mice. Biochim Biophys Acta 884: 344–354

Ohno S, Akita Y, Konno Y, Imajoh S and Suzuki K (1988) A nobel phorbol ester receptor/protein kinase, nPKC, distantly related to the protein kinase C family. Cell 53: 731–741

Ohyama Y, Hirabayashi J, Oda Y, Ohno S, Kawasaki H, Suzuki K and Kasai K (1986) Nucleotide sequence of chick 14K β-galactoside-binding lectin mRNA. Biochem Biophys Res Commun 134: 51–56

Ozanne B, Fulton RJ and Kaplan PL (1980) Kirsten murine sarcoma virus transformed cell lines and a spontaneously transformed rat cell line produce transforming growth factor. J Cell Physiol 195: 163–180

Powell JT (1979) Purification and properties of lung lectin. Biochem J 187: 123–129

Roberts AB, Anzano MA, Lamb LC, Smith JM and Sporn MB (1981) New class of transforming growth factors potentiated by epidermal growth factor: Isolation from non-neoplastic tissues. Proc Natl Acad Sci USA 78: 5339–5342

Roberts AB, Anzano M, Wakefield LM, Roche NS, Stern OF and Sporn MB (1985) Type β transforming growth factor: A bifunctional regulator of cellular growth. Proc Natl Acad Sci USA 82: 119–123

Roberts AB, Sporn MB, Assoian RK, Smith JM, Roche NS, Wakefield LM, Heine UI, Liotta LA, Falanga V, Kehrl JH and Fauci AS (1986) Transforming growth factor type β: rapid induction of fibrosis and angiogenesis in vivo and stimulation of collagen formation in vitro. Proc Natl Acad Sci USA 83: 4167–4171

Rosenthal A, Lindquist PB, Bringman TS, Goeddel DV and Derynck R (1986) Expression in rat fibroblasts of a human transforming growth factor α cDNA results in transformation. Cell 46: 301–309

Saiki RK, Gelfand DH, Stoffel S, Scharf SJ, Higuchi R, Horn GT, Mullis KB and Erlich HA (1988) Primer-directed enzymatic amplification of DNA with a thermostable DNA polymerase. Science 239: 487–491

Tucker RF, Shipley GD, Moses HL and Holley RW (1984) Growth inhibitor from BSC-1 cells closely related to platelet type transforming growth factor. Science 226: 705–707

Twardzik DR, Todaro GJ, Reynolds Jr, FH and Stephenson JR (1983) Transforming growth factors (TGFs) produced by cells transformed by different isolates of feline sarcoma virus. Virology 124: 201–217

Waard A, Hickman S and Kornfeld S (1976) Isolation and properties of β-galactoside binding lectins of calf heart and lung. J Biol Chem 251: 7581–7587

Wigler M, Sweet R, Sim GK, Wold B, Pellicer A, Lacy E, Maniatis T, Silverstone S and Axel R (1979) Transformation of mammalian cells with genes from procaryotes and eucaryotes. Cell 16: 777–785

Yamaoka K and Hirai R (1986) Characterization of intra- and extra-cellular transforming growth factor from an avian sarcoma virus-transformed rat cells. Biochem Intern 13: 863–872

Yamaoka K, Hirai R, Tsugita A and Mitsui H (1984) The purification of an acid- and heat-labile transforming growth factor from an avian sarcoma virus transformed rat cell line. J Cell Physiol 119: 307–314

Yamaoka K, Ohno S, Kawasaki H and Suzuki K (1991) Overexpression of a β-galactoside binding protein causes transformation of BALB3T3 fibroblast cells. Biochem Biophys Res Commun 179: 272–279

55 Identification and Expression of a Protozoan Invasion Factor: Penetrin, the Heparin-Binding Protein from *Trypanosoma cruzi*

E. Ortega-Barria and M.E.A. Pereira

Adhesion of microbial pathogens to the host cell surface is an early and critical step in the process of infection. Cell surface carbohydrates and complementary microbial carbohydrate-binding proteins have been implicated in the attachment of bacteria, viruses, and parasites to target tissues. Of particular interest are a novel group of heparin-binding proteins that are involved in the recognition, adhesion, and invasion of eukaryotic cells by obligate intracellular organisms, such as *Trypanosoma cruzi* and Herpes simplex virus (WuDunn and Spear 1989; Ortega-Barria and Pereira 1991). In the same context, recently have been suggested that a heparan sulfate-like proteoglycan present on the surface of *C. trachomatis* is required for attachment to host cells (Zhang and Stephens 1992). For *T. cruzi*, host cell adhesion and invasion is preceded by migration of the parasite from the site of insect bite, through the extracellular matrix to the target host cell (Ortega-Barria and Pereira 1992). Thus, it is likely that *T. cruzi* express this and possibly other adhesion molecules to facilitate the formation of receptor-mediated contacts with extracellular matrix components and subsequently with cell surface receptors.

We have approached the identification of *T. cruzi* factors relevant in host cell adhesion and entry, by studying the ability of intact parasites to recognize and adhere to inert plastic surfaces coated with chemically defined carbohydrate ligands. Using this model, we demonstrated the specific adhesion of *T. cruzi* trypomastigotes to immobilized heparin and heparan sulfate proteoglycan and its inhibition by addition of the appropriate soluble proteoglycans. Furthermore, heparin and heparan sulfate proteoglycan were able to block binding of trypomastigotes to fixed host cells in vitro and inhibited infection of cultured mammalian cells, suggesting that the initial recognition of the host cell is at least in part via a surface heparin-like receptor. Subsequently, heparin covalently linked to an insoluble support was used for purification by affinity chromatography of a trypomastigote surface membrane protein. Finally, antibodies raised against the purified heparin-binding protein (HBP) were used to screen a *T. cruzi* genomic DNA library in λ Zap, and clone the gene from trypomastigotes that conferred upon a noninvasive strain of *E. coli* the ability to enter cultured mammalian cells (Ortega-Barria and Pereira 1991).

In this chapter we describe the purification of the heparin-binding protein from *T. cruzi*, and its use in studying specific cell recognition and adhesion events. In addition, we outline a procedure developed to identify and characterize different gene products responsible for the entry of microorganisms into nonphagocytic cells by analyzing the uptake of live recombinant bacteria into cultured mammalian cell lines (Isberg and Falkow, 1985; Ortega-Barria and Pereira 1991).

55.1 Schematic Outline of Procedures

(Fig. 1a, b).

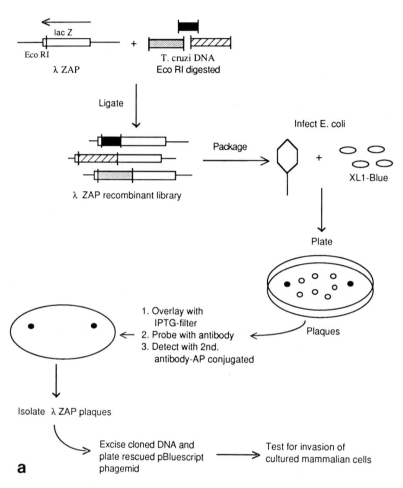

Fig. 1. a Screening of a *T. cruzi* trypomastigote genomic DNA library in *λ* ZAP. DNA was isolated from culture-form trypomastigotes and partially digested with Eco RI. The fragments were inserted into the single Eco RI site of *λ* ZAP. Recombinant molecules were packed in vitro and planted on *E. coli* XL1-Blue. Plaques were screened using a 1:100 dilution of mice anti-HBP, followed by a second antibody conjugated to alkaline phosphatase. Inserted *T. cruzi* DNA was automatically excised, and the rescued pBluescript phagemid was used to infect XL1-Blue bacteria. Finally, isolated recombinant *E. coli* were tested for its ability to invade a monolayer of cultured mammalian cells. **b** Isolation of recombinant *E. coli* strains that are able to enter nonphagocytic eukaryotic cells. XL1-Blue was grown overnight, and added onto a confluent monolayer of Vero cells for 3 h at 37 °C. Then, bacteria unable to adhere or invade cultured cells were removed by washing, and gentamicin was added to kill extracellular bacteria. After 2 h at 37 °C, the monolayer was washed, internalized recombinant XL1-Blue was released with 1% Triton X-100, and plated for individual colonies on agar plates (Fig. 1b see page 502)

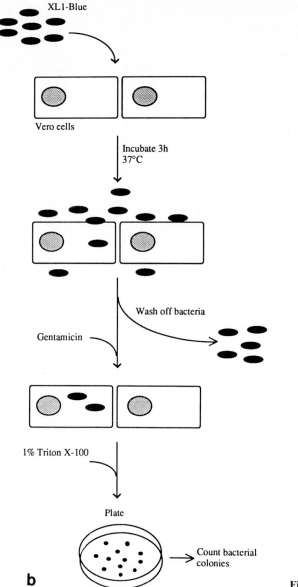

b

Fig. 1. (continued)

55.2 Equipment, Reagents and Solutions

Equipment
  – Ultracentrifuge (Beckman L5-50B with 70.1 Ti fix angle rotor)
  – Clinical centrifuge (Beckman TJ-6) with adaptor for conical tubes of capacities from 15 to 50 ml and carriers for microtiter plates
  – Sonicator (Branson 450 with microtip probe)
  – Equipment for column chromatography

- Equipment for SDS-PAGE
- Scintillation counter (Beckman LS 3801)
- Gamma counter (Beckman 5500)
- Inverted microscope (Nikon)
- Flat-bottom 96-well bacteriological grade plates (Falcon, flexible polyvinyl chloride)
- 24-well tissue culture plate (Bellco)
- Hemacytometer
- Spectrophotometer (Beckman 35)
- Reciprocal shaker (Lab-Line Instruments, Inc.)

Adhesion Assay Reagents
- 1% (w/v) heat-treated BSA in PBS, pH 7.2 (85°C, 10 min or 56 °C, 60 min)
- 1% (w/v) BSA in RPMI 1640, pH 7.2
- EGTA/HBSS- buffer [20 mM ethylene-glycol-bis (β-aminoethyl ether) N,N'-tetraacetic acid; Ca^{2+} and Mg^{2+} free Hanks' balanced salt solution, pH 7.7]
- Tissue culture reagents (GIBCO Laboratories)
- Vero cells (ATCC CCL 81)
- CHO K1 cells (ATCC CCL 61)
- Nu-serum (Collaborative Research Inc.)
- Dialysed Nu-serum
- Glutaraldehyde grade I (Sigma)

Reagents
and Solutions

Iodination and Purification Reagents
- Phosphate buffered saline (20 mM sodium phosphate; 150 mM sodium chloride, pH 7.2)
- Cyanogen bromide (Sigma)
- Sepharose CL-4B-200 (Sigma)
- Radioisotopes (ICN Radiochemicals)
  [^{35}S] methionine (1113 mCi/mmol)
  ^{125}Iodine (892 mCi/mmol)
- Triton X-100 (Sigma)
- CHAPS (Sigma)
- NaCl buffer (NaCl from 0.250 to 2.0 M; 20 mM sodium phosphate, pH 7.2)
- Antiprotease cocktail (10 mM pepstatin; 10 mM leupeptin; 10 mM iodocetamide, 10 µg/ml soybean trypsin inhibitor; 2 mM PMSF)
- Heparin and heparan sulfate proteoglycan (Sigma)
- Iodobeads (N-chlorobenzenosulfomide sodium salt, Pierce Chemical Co.)
- *T. cruzi* trypomastigotes, Silvio X-10/4 clone maintained by infection of cultured Vero cells
- Dialysis buffer (0.2 M ammonium bicarbonate)

Bacterial Invasion Assay Reagents
- *E. coli* XL-1 Blue transformed with pBluescript without insert
- HBP-6 (*E. coli* XL-1 Blue transformed with pBluescript containing the penetrin gene)
- 7F (*E. coli* XL-1 Blue transformed with pBluescript containing the neuraminidase gene)
- Gentamicin (Schering)
- Vero cells (ATCC CCL 81)

- CHO K1 cells (ATCC CCL 61)
- 1% (w/v) BSA in RPMI 1640, pH 7.2

Antibody Screening of λ ZAP DNA Genomic Library
- *T. cruzi* trypomastigotes genomic DNA library in λ Zap (Stratagene)
- Nitrocellulose (Schleicher and Schuell, 0.45 µm pore, 150 mm)
- Blocking solution (5% powdered skin milk in PBS, pH 7.2)
- IPTG: 10 mM isopropyl-β-thiogalactopyranoside in water (United States Biochemical)
- AP buffer (165 µg/ml BCIP, 330 µg/ml NBT, 100 mM NaCl, 5 mM MgCl, 100 mM Tris-HCl, pH 9.5)
- Nitroblue tetrazolium (NBT), 5-bromo-4-chloro-3-indolyl-phosphate (BCIP) (Research Organic)
- Rabbit antimouse lgG alkaline phosphatase-conjugated (Promega)
- LB Medium (1L):

Bacto-tryptona	10 *g*
Bacto-yeast extract	5 *g*
NaCl	10 *g*
pH 7.5	

- SM buffer (1L):

NaCl	5.8 *g*
$MgSO_4$	2.0 *g*
Tris-HCl, pH 7.5	150 ml

- YT medium (1L):

Bacto tryptone	8 *g*
NaCl	5 *g*
Yeast extract	5 *g*
pH 7.2–7.4	

- Top agar (0.7% agarose in LB broth)
- LB plate (1.5% Difco agar in LB broth; 80 ml/150 mm plate)

55.3 Iodination of *T. cruzi* Cell Surface Membrane Proteins

We labeled the external surface proteins of *T. cruzi* by using the solid-state iodination reagent N-chloro-benzenesulfonamide (Iodobeads), that add cationic iodine (1^+) to tyrosine residues in proteins, without loss of *T. cruzi* HBP antigencity or biological activity. The ^{125}I serves as a simple label for proteins that are represented in small amounts on the plasma membrane of the cell, facilitating its detection by gamma counting. Subsequently, we fragmented labeled-parasites by sonication and separated cell membranes by differential centrifugation and detergent solubilization as a prelude for the purification of outer membrane proteins by affinity chromatography on heparin-Sepharose. Alternatively, another solid-phase agent, lodogen, may be used as a surface label for cells.

Protocol 1. Collect tissue culture trypomastigotes from supernatants of *T. cruzi* infected Vero cell monolayers, centrifuge at 2500 rpm for 15 min at 4 °C in a clinical centrifuge, and carefully remove the medium by aspiration.

2. Wash parasites three times by centrifugation as above, by using cold PBS pH 7.2 to eliminate any residual culture medium protein and resuspend in the same buffer at 2.5×10^8 cells/ml.

Note. It is important that cell viability is high both before and after iodination, since internal structures of dead cells will be labeled quickly.

3. Transfer parasites to a snap-cap polystyrene tube, and four Iodobeads (1.2 µmol), 500 µCi of carrier-free ^{125}I, and incubate 15 min on ice with intermittent agitation.
4. Using a Pasteur pipet, transfer parasites to a 15 ml conical tube containing cold PBS, pH 7.2 and wash three times as above.
5. Use an aliquot to determine the amount of incorporated radioactivity per cell (Johnstone and Thorpe 1987).

Warning. Iodination require special handling precautions, and should be performed in a designated area in a vented enclosure with activated charcoal filtration installed inside a fume hood behind lead shielding.

55.4 Isolation of *T. cruzi* Membrane Proteins

1. Lyse trypomastigotes by sonication on ice with four pulses of 20 s each at an output of 5 and 50% of the cycle.   **Protocol**

Note. Check cell integrity with phase microscopy after each pulse to ascertain that a homogeneous preparation has been obtained. Avoid foaming and overheating of the lysate, it is harmful for proteins.

2. Remove large particles (nuclei and unbroken cells) by centrifugation at $175\,g$ for 5 min at 4 °C.
3. Centrifuge supernatant at $100\,000g$ 60 min at 4 °C.
4. Resuspend the pellet in 2.5 ml of 1% Triton X-100 in antiprotease cocktail for 2 h on ice.

Note. Tubes may be stirred, but excessive agitation should be avoided since foam formation is associated with the denaturation of proteins.

5. Detergent-insoluble material is removed by centrifugation at $100\,000g$ 60 min at 4 °C.
6. The pellet is re-extracted with 2.5 ml of 1% Triton X-100 as in (4) and ultracentrifuged as described in (5).
7. Supernatants containing solubilized membrane proteins are pooled, mixed with an equal amount of antiprotease cocktail, and used for affinity chromatography.

Note. CHAPS can be removed by dialysis and may be replaced by Triton X-100 when a detergent-free preparation is required for assays involving live cells.

55.5 Purification of *T. cruzi* Heparin-Binding Protein

1. Heparin is coupled to cyanogen bromide-activated Sepharose 4B-CL as described (March et al. 1974).   **Protocol**

Warning. Cyanogen bromide is volatile and extremely toxic if inhaled or adsorbed through the skin. Sepharose activation should be performed in a designated area inside a fume hood, all equipment used should be soaked in 1 M NaOH overnight, and then washed off.

2. ^{125}I-labeled soluble surface proteins are mixed with an equal volume of heparin-Sepharose and incubated overnight at 4 °C with gentle shaking.
3. Centrifuge at 175 g 10 min at 4 °C. Resuspend in the original volume of 0.5% Triton X-100 in PBS, pH 7.2.
4. Pour the slurry into a 0.8 × 4 cm disposable column and wash extensively with 0.05% Triton X-100 in PBS, pH 7.2.
5. Bound proteins are eluted by step-wise addition of 25 ml of 0.25, 0.5, 1.0 and 2.0 M NaCl.
6. Collect 2-ml fractions, monitoring eluted radioactivity in a gamma counter.
7. Pool radioactive peaks, dialyze extensively against 0.2 M ammonium bicarbonate at 4 °C, and lyophilize.

Note. HBP may be stored lyophilized or in 0.1% BSA, RPMI 1640 for up to 2 weeks at 4 °C without loss of its biological activity.

8. Run a 10% SDS-PAGE, transfer to nitrocellulose paper, and expose to X-OMAT AR film.

General Comments
- Determine the specific activity of the purified HBP by TCA precipitation and protein concentration (Johnstone and Thorpe 1987).
- The hemagglutinating activity of HBP may be assessed by preparing serial double dilutions of the affinity purified protein, adding an equal volume of 4% red blood cells, and incubating 1 h at room temperature (Burger 1974).
- Stain SDS-PAGE fractionated proteins, cut out the appropriate size band. The gel pieces are homogenized, mixed with an equal volume of Freuds' complete adjuvant and used for mice immunization (Cavallesco and Pereira 1988).

55.6 Cell Attachment Assay

Most of the methods that study cell-cell or cell-matrix interaction require that a suspension of dissociated cells be made prior to assessing the adhesiveness of cells. Attachment of single cells to a substrate may be evaluated by labeling the cells with an appropriate isotope, counting the radioactivity of a sample with a known number of cells and use this value to measure the number of cells that bind to a test surface (Walther and Öhman, 1974; Curtis and Lackie 1991). The test substrate may be prepared by passive adsorption of soluble proteins onto plastic surfaces. Matrix glycoproteins and collagen generally adsorb in a biologically active conformation useful for studying the interactions of cells with surfaces of potentially adhesive molecules. On the contrary, small polypeptides and oligopeptides, that have a significant rate of desorption, should be linked covalently to the substrate surface (Curtis and Lackie 1991).

Preparation of Cells

1. Remove culture medium and wash cell monolayers two times with 10 ml of methionine-free RPMI 1640.
2. Add 10 ml of fresh methionine-free RPMI 1640 containing 2% dialyzed Nu-serum and incubate 30 min at 37 °C.
3. Add 50 μCi/ml [^{35}S]methionine and incubate 1 h at 37 °C.
4. Wash cell monolayers three times with 10 ml HBSS$^-$.
5. Add 10 ml of EGTA/HBSS$^-$ buffer and incubate 20 min at 37 °C.

Cell Preparation

Note. Check cell layers every 5 min under phase contrast microscopy until a single cell suspension (> 90%) is obtained. If longer periods of incubation are required, cells may be incubated at 4 °C and 1% BSA should be added to the dissociation buffer.

6. Collect single cell suspensions, disperse gently by pipeting and wash three times by centrifugation at 1000 rpm 10 min at 4 °C with 1% BSA in RPMI 1640, pH 7.2, and adjusted to 1×10^6 cells/ml in the same medium.
7. Keep cells on ice until needed for the adhesion assay.

General Comments

– Dissociate subconfluent cells preferentially with EGTA or EDTA in HBSS$^-$ buffer. The affinity of the chelators for divalent cations increase at pH 7.7.
– If trypsin solution is used (0.05% trypsin, 0.53 mM EDTA in HBSS$^-$), dispersed cells may be washed in a medium which contain serum to halt proteolytic activity. However, peptide inhibitors such as leupeptin (5 μl of a 1 mg/ml solution per ml of trypsin) are preferred, because the effect of serum components in cell adhesion. Subsequently, a period of recovery (i.e., 2 h at 37 °C on microbiological grade plastic dishes) is necessary before adhesion may be tested.
– Avoid mechanical methods of dissociation (i.e., rubber policeman), that leads to extensive cell damage and lysis.
– Check single cell suspension for viability by the exclusion of trypan blue.
– Determine the amount of incorporated radioactivity per cell by TCA precipitation (Johnstone and Thorpe 1987).
– Avoid the use of serum in the assay medium, since it may affect cell adhesion (Rottmann 1974).

Preparation of Substrata and Cell Attachment

1. Dilute proteins in PBS, pH 7.2 at 50 μg/ml.
2. Add 100 μl to the first well of a 96-well microtiter plate and make double serial dilutions.
3. Incubate overnight at 4 °C.
4. Recover unbound protein, wash two times with PBS, pH 7.2 and add 100 μl of 1% heated BSA in PBS, pH 7.2.
5. Incubate 2 h at room temperature.
6. Wash two times with PBS, pH 7.2, add 100 μl 1% BSA in RPMI 1640, and incubate at 37 °C until needed.
7. Remove medium, immediately add labeled single cell suspension, and incubate 1 h at 37 °C.
8. Unbound cells are removed by aspiration and washing three times with warm PBS, pH 7.2.

Cell Attachment

Note. Add warm (37 °C) PBS carefully against the side walls of the microplate wells and remove by gentle aspiration via a multichanel pipet, held against the wall.

9. Add 100 µl of 2.0 N NaOH, transfer to 5 ml of aqueous counting scintillant fluid, and measure in a Beckman liquid scintillation counter.

Note. Alternatively, individual wells may be cut out and added into scintillation vials to determine the amount of associated radioactivity.

General Comments
– Heat-treated BSA prevent nonspecific adhesion of cells to plastic substrata. Centrifuge heated BSA at 35 000 *g* for 30 min and use the supernatant.
– Maintain temperature at 37 °C throughout the experiment, since cell adhesion is highly sensitive to temperature changes.
– After washing, bound cells may be fixed by addition of cold 4% paraformaldehyde in PBS, pH 7.2, stained with Diff-Quik, and counted under light microscopy.
– As positive control for cell attachment utilize fibronectin or laminin at 10 µg/ml for coating. As negative control BSA should be used at equivalent concentrations to that of the test protein.
– Calculate the amount of plastic-immobilized protein by measuring recovered radioactivity (unbound protein) in a gamma counter. If the immobilized protein is an enzyme (i.e., *T. cruzi* transsialidase), calculate the amount of bound protein by measuring residual enzyme activity in the coated wells, and based on the specific enzyme activity calculate the amount of protein bound to the substrate.

55.7 Antibody Screening of λ Zap Genomic DNA Library

Eukaryotic and prokaryotic genes may be isolated by screening *E. coli* expression libraries with antibody probes using the bacteriophage vector λ ZAP (Sambrook et al. 1989; Snyder et al. 1989). With this objective, a genomic DNA library is first constructed in λ ZAP and plated on bacterial cells to form phage plaques. IPTG-induced antigens are transferred onto nitrocellulose filters and then probed with antibodies to detect the desired recombinant (Snyder et al. 1989). Successful immunoscreening depends on the quality of the antibody. Thus, polyclonal antibodies are preferred because they recognize different determinants on the same protein. However, monoclonal antibodies or mixture of monoclonal antibodies against different epitopes on the same molecule may be used successfully.

Grow Plating Bacteria
1. Streak out *E. coli* XL1-Blue for single colonies on LB plates and incubate overnight at 37 °C.
2. Starting with a single colony, grow XL1-Blue bacteria in LB/tetracycline media overnight at 37 °C in a sterile glass flask with vigorous shaking on an orbital rocker to an $OD_{600} = 0.5$.
3. Centrifuge at 2500 rpm 10 min at 4 °C and *gently* resuspend to its original volume with 10 mM $MgSO_4$.

Infect Bacteria with Library
4. Mix the packed λ ZAP library suspension containing ∼50 000 recombinant bacteriophage with 600 µl of XL1-Blue in polystyrene snap-cap tubes.
5. Incubate 15–30 min at 37 °C with gently intermittent shaking.

6. Add 6.5 ml of melted LB top agar (48 °C) to each aliquot of infected bacteria, mix by inverting two times, and spread evenly onto 150-mm LB plates of bottom agar.
7. After top agarose harden, incubate plates overnight at 42 °C.

Note. With other host bacteria (i.e., BB4) that supports vigorous growth of λ ZAP, this step may take only 6 h.

8. Carefully overlay plates with a dry IPTG-treated nitrocellulose filter.
9. Incubate plates 2–4 h at 42 °C.
10. Mark the position of filter on the plate with a syringe needle containing ink.
11. Carefully remove filters from plates and block 1 h at room temperature with 5% powdered skin milk, PBS, pH 7.2.

Note. Store agar plates at 4 °C until needed for plaque isolation. Do not allow the membrane to dry out during any of the subsequent steps.

12. Rinse filters 1–2 times in PBS, pH 7.2. Add first antibody (mouse anti-HBP 1:100 in 1% powdered skin milk, PBS, pH 7.2) and incubate 2 h at room temperature. **Detection of Recombinants**
13. Wash filters in 15–20 ml of PBS, pH 7.2 three times for 15 min each.
14. Add second antibody (rabbit antimouse lgG alkaline phosphatase-conjugated 1:7500 dilution) and incubate 1 h at room temperature.
15. Wash filters in 15–20 ml of PBS, pH 7.2 three times for 15 min each.
16. Add alkaline phosphatase color development substrate (NBT/BCIP). Positive plaques appear as circles with clear centers.

Note. The phosphatase color development reaction can continue for hours or even overnight at 4 °C.

17. Remove positive plaques using the large end of a Pasteur pipet, place each agar plug in 1 ml SM buffer and add 20 μl CHCl₃ in a sterile microfuge tube. **Isolation of Recombinants**
18. Vortex the tube to release λ ZAP particles into the SM buffer, and store at 4 °C (stable for up to a year at 4 °C).
19. Repeat the same screening procedure at least three times, until all plaques on the plate are positive (Mierendorf et al. 1987).

General Comments

– Plates work best after 2 days of preparation.
– Plaques are grown until 1 mm in diameter to allow clean transfer to nitrocellulose filters.
– Screen ~1 × 10⁶ plaques using 20 aliquots of ~50 000 particles each.
– Twelve hours of growth are usually required before λ ZAP plaques are visible on a XL1-Blue bacterial lawn.
– The plaque density should be in the range 2.5–5 × 10⁴ plaques per 150 mm to maximize the efficiency of screening and minimize overcrowding.
– Before use, rinse nitrocellulose filters in a solution of 10 mM IPTG dissolved in distilled water, and allow to air dry on Whatmann 3 MM filter paper.
– If the top agar stick to the nitrocellulose membrane rather than to the bottom agar, chill the plate at 4 °C for 1 h, and then remove the filter.
– The antibody solution may be used several times with little reduction in titers and is stored with 0.02% sodium azide at 4 °C after each use.

Troubleshooting

1. Signal weak or absent. Primary antibody may be of low affinity or of low titer, increase the concentration of the antibody and extend incubation time at 4 °C (Mierendorf et al. 1987).
2. Background too high. Color development reaction was too long. Improper blocking, try alternative blocking agents (i.e., 3% BSA; 0.05% Tween 20/0.5% BSA; 1% gelatin). Anti-IgG conjugated antibody concentration too high.

55.8 Automatic Excision of λ ZAP Vector

After isolating λ ZAP plaques, the plasmid sequences contained in λ ZAP can be automatically excised from the phage vector by simultaneous infection of *E. coli* with a helper bacteriophage and converted to a plasmid vector, phagemid pBluescript (Stratagene). Secreted phagemid is separated from the *E. coli* strain used for excision of the cloned DNA by heating at 70 °C, and used to infect a new *E. coli* host.

Protocol
1. In a 50 ml conical tube mix:
   – 200 µl XL1-Blue (OD_{600} = 1.0)
   – 200 µl isolated λ ZAP phage (containing $> 1 \times 10^5$ particles)
   – 10 µl R408 helper phage (7×10^{10} pfu/ml)

Note. As negative control use XL1-Blue and R408 helper phage alone, without the isolated λ ZAP phage.

2. Incubate 15 min at 37 °C.
3. Add 5 ml 2X YT media in a snap-cap polypropylene tube and incubate 6 h at 37 °C with intermittent shaking.
4. Heat tube at 70 °C for 20 min and centrifuge at 2500 rpm 15 min at 4 °C.
5. Save the supernatant containing the pBluescript phagemid and store at 4 °C (stable up to 2 months).
6. To plate the rescued phagemid, mix in a microfuge tube:
   – 200 µl XL1-Blue host (OD_{600} = 1.0)
   – 200 µl pBluescript phagemid from step (5)
7. Incubate 15 min at 37 °C with intermittent shaking.
8. Centrifuge 2 min in a microfuge and resuspend in 100 µl LB/ampicillin media.
9. Plate 100 µl on LB/ampicillin agar plates and incubate at 37 °C overnight.
10. Colonies appearing on the plate contain the pBluescript plasmid with the cloned DNA insert.
11. Analyze HBP gene insert from recombinant *E. coli* using miniprep plasmid DNA and treatment with Eco RI (Sambrook et al. 1989).
    a. Verify gene identity by:
    – Assaying for gene activity
    – Analyzing HBP gene product by SDS-PAGE and immunoblot with anti-HBP antibody.
    – DNA sequencing and compare deduced amino acid sequence with the known protein sequence.

General Comments

- Plaques isolated from XL1-Blue lawn may have lower titer, requiring amplification before proceeding onto excision (Stratagene).
- Recombinant bacteria may be stored at $-70\,°C$ in 30% glycerol in LB/ampicillin media.

Troubleshooting
1. Unsuccessful rescue
    a) Increase the incubation time in step 3.
    b) Make a high titer stock (amplify) of the phage for the rescue assay (Stratagene).

55.9 Bacterial Invasion Assay

Infection of tissue culture monolayers provide a simple in vitro system that mimics microbial invasion in vivo. Cultured mammalian cells may be used as a model to study recombinant *E. coli* invasiveness. The method is useful to isolated genes involved in cell adhesion and invasion from microorganism in which the appropriate function can be expressed and localized in the surface of recombinant bacteria (Isberg and Falkow 1985; Falkow et al. 1987; Ortega-Barria and Pereira 1991).

1. Plate Vero cells at 2.0×10^5 cells/well 18 h before invasion experiment (24 well/plate, Bellco TC 24).   **Protocol**
2. Streak recombinant clones on LB/ampicillin plates and grow overnight at $37\,°C$, next day start culture from a single colony in 50 ml LB media.
3. Growth recombinant *E. coli* clones in 25 ml LB/amp medium overnight at $37\,°C$ with shaking at ~150 rpm, on a rotatory shaker.
4. Centrifuge bacteria at 2500 rpm 15 min at $4\,°C$, wash 1x with 1% BSA RPMI 1640, pH 7.2, and resuspend in 2 ml of the same media.
5. Determine the number of bacteria at OD_{600} and adjust to 2.0×10^7/ml.
6. Just before invasion assay, wash cell monolayers 2x with 1% BSA in RPMI 1640, and incubate at $37\,°C$ until needed.
7. Add 2.0×10^7 bacteria (1 ml) to each well, and incubate the plate 5 min at room temperature on a rotatory shaker.
8. Centrifuge the plate 10 min at 1000 rpm at $4\,°C$ and incubated 2 h at $37\,°C$ in a CO_2 incubator.
9. Wash the cell monolayer $3 \times$ with sterile warm PBS, pH 7.2 and incubated 2 h at $37\,°C$ with 2 ml of 0.2% BSA-RPMI 1640 containing $75\,\mu g$/ml of gentamicin.
10. Wash $4 \times$ with sterile warm PBS, lyze cells by addition of 0.2 ml sterile distilled H_2O, followed by 0.2 ml 1% Triton X-100, and pipeted up and down to disrupt cells.
11. Plate $50\,\mu l$ of the cell lysate on agar/ampicillin plates, incubated overnight at $37\,°C$, and count number of colonies.

Note. If necessary, the intracellular localization of bacteria may be confirmed by electron microscopy. Epithelial cell lines (CHO K1) may also be used in experiments of bacterial invasion.

References

Burger MM (1974) Assays for agglutination with lectins. Methods Enzymol 32: 615–621

Cavallesco R, Pereira MEA (1988) Antibody to *Trypanosoma cruzi* neuraminidase enhances infection in vitro and identifies a subpopulation of trypomastigotes. J Immunol 140: 617–625

Curtis ASG, Lackie JM (eds) (1991) In: Measuring cell adhesion. John Wiley and Sons. Chichester, England

Falkow S, Small P, Isberg R, Hayes SF, Corwin D (1987) A molecular strategy for the study of bacterial invasion. Rev Infect Dis 9: S450–S455

Isberg R, Falkow S (1985) A single genetic locus encoded by *Yersinia pseudotuberculosis* permits invasion of cultured animal cells by *Escherichia coli* K-12. Nature 317: 262–264

Johnstone A, Thorpe R (eds) (1987) In: Immunochemistry in practice. Blackwell Scientific Publications. Oxford

March SC, Parikh I, Cuatrecasas P (1974) A simplified method for cyanogen bromide activation of agarose for affinity chromatography. Anal Biochem 60: 149–152

Mierendorf RC, Percy C, Young R (1987) Gene isolation screening λgt11 libraries with antibodies. Methods Enzymol 152: 458–469

Ortega-Barria E, Pereira MEA (1991) A novel *T. cruzi* Heparin-binding protein promotes fibroblast adhesion and penetration of engineered bacteria and trypanosomes into mammalian cells. Cell 67: 411–421

Ortega-Barria E, Pereira MEA (1992) Entry of *Trypanosoma cruzi* into eukaryotic cells. Infectious Agent and Disease 1: 136–145

Rottmann W (1974) Coulter electronic particle counter assay. In Measurement of cell–cell interactions. Roseman S. Rottmann W, Walther B, Öhman R, Umbreit J. Methods Enzymol 32: 597–611

Sambrook J, Fritsch EF, Maniatis T (1989) Molecular cloning. A laboratory manual (Cold Spring Harbor, New York: Cold Spring Harbor Laboratory)

Snyder M, Elledge S, Sweetser D, Young R, Davis RW (1989) λ gt11: Isolation with antibody probes and other applications. In: Recombinant DNA methodology Academic Press, Inc. 309–330

Stratagene Predigested Lambda Zap/Eco RI Instruction manual. Stratagene, San Diego, CA

Walther B. Öhman R (1974) Cell layer assay. In: Measurement of cell-cell interactions. Roseman S, Rottmann W, Walther B, Öhman R, Umbreit J. Methods Enzymol 32: 597–611

WuDunn D, Spear PG (1989) Initial interaction of herpes simplex virus with cell is binding to heparan sulfate. J Virol 63: 52–58

Zhang JP, Stephens RS (1992) Mechanism of *C. trachomatis* attachment to eukaryotic host cells. Cell 69: 861–869

Subject Index